FOR STUDENTS

- Career Opportunities
- Career Fitness Program
- Becoming an Electronics Technician
- Free On-Line Study Guides (companion web sites)

FOR FACULTY*

- Supplements
- On-Line Product Catalog
- Electronics Technology Journal
- Prentice Hall Book Advisor

The Prentice Hall Book Advisor is intended to help you, the instructor, find the right textbooks for your students.

* Please contact your Prentice Hall representative for passcode

OPERATIONAL AMPLIFIERS WITH LINEAR INTEGRATED CIRCUITS

Fourth Edition

William D. Stanley
Old Dominion University

Prentice
Hall

Upper Saddle River, New Jersey | Columbus, Ohio

Library of Congress Cataloging-in-Publication Data

Stanley, William D.
   Operational amplifiers with linear integrated circuits/William D. Stanley.—4th ed.
     p. cm.
   Includes bibliographical references and index.
   ISBN 0-13-032013-7 (alk. paper)
    1. Operational amplifiers. 2. Linear integrated circuits. I. Title.

TKL 7871.58.06 S74 2002
621.381′535—dc21

00-068802

Editor in Chief: Stephen Helba
Product Manager: Scott J. Sambucci
Associate Editor: Kate Linsner
Production Editor: Rex Davidson
Design Coordinator: Robin G. Chukes
Cover Designer: Galen Ludwick
Cover Art: Visual Edge 2000
Production Manager: Pat Tonneman
Production Coordinator: Techbooks

This book was set in Times Roman by TechBooks and was printed and bound by R.R. Donnelley & Sons Company. The cover was printed by Phoenix Color Corp.

Electronics Workbench™ and MultiSim™ are trademarks of Electronics Workbench.

Prentice-Hall International (UK) Limited, *London*
Prentice-Hall of Australia Pty. Limited, *Sydney*
Prentice-Hall Canada, Inc., *Toronto*
Prentice-Hall Hispanoamericana, S.A., *Mexico*
Prentice-Hall of India Private Limited, *New Delhi*
Prentice-Hall of Japan, Inc., *Tokyo*
Prentice-Hall Singapore Pte. Ltd.
Editora Prentice-Hall do Brasil, Ltda., *Rio de Janeiro*

10 9 8 7 6 5 4 3 2 1
ISBN 0-13-032013-7

Dedicated to the late
Joseph S. Reeves,
who introduced me to the
exciting world of electronics
when I was a teenager

PREFACE

As in the first, second, and third editions, the primary objectives of this book are

- To establish the general methods for analyzing, modeling, and predicting the performance of operational amplifiers and related linear integrated circuits
- To develop the reader's facility in designing realistic circuits to perform specified operations
- To acquaint the reader with many of the common circuit configurations and enable the student to select available devices to use with these circuits

All or portions of this book are usable for at least the following groups:

- Upper-division engineering technology students (junior or senior level)
- Lower-division engineering technology students (following the basic course sequence in discrete electronic devices and circuits)
- Applied engineering students
- Practicing design engineers, technologists, and technicians

Engineering or engineering technology students who have had a course in circuit analysis employing frequency response analysis methods and a course in calculus should be able to cover the entire book. Calculus is used where appropriate, but virtually all results and design approaches can be understood *without calculus*. The large number of example problems and exercises make the book also valuable for self-study by practicing technical personnel.

The primary emphasis throughout the book is on developing the reader's facility for analyzing and designing the various circuit functions, rather than on simply presenting a rote collection of existing circuits or showing numerous wiring diagrams for specialized integrated circuit modules. This establishes a foundation for understanding new developments as they arise. Since new devices constantly appear on the market and existing ones quickly become obsolete, we study only a few devices in detail. Those selected for this purpose have withstood the test of time and are widely used. Because the best way to adapt to new technology is to have a firm grasp of the basic principles, this book has been organized toward that goal.

v

Chapter 1 provides some general models of linear amplifier circuits, definitions, and parameters. Students often miss these general concepts from the detailed material covered in basic electronic courses. This common deficiency is a case of the classical pattern of "not seeing the forest for the trees," and the material provided here will help solve that problem.

Chapter 2 begins the analysis and design of operational amplifier circuits using ideal model assumptions. While the reader will not be able to see all the limitations of the circuits at this point, actual workable designs can be produced almost immediately from the information in this chapter, including amplifiers of various types, current sources, summing circuits, and various other applications.

Chapters 3 and 4 are devoted to the practical limitations of realistic operational amplifiers and the associated effects on operating performance. Chapter 3 emphasizes dc effects, including the effects of finite gain, finite input impedance, nonzero output impedance, input offset voltage, and input bias currents. Chapter 4 emphasizes ac effects, that is, those effects that are frequency dependent. Primary topics include closed-loop bandwidth relationships, effects of slew rate, and common-mode rejection. Chapter 5 considers additional linear applications, including frequency-dependent circuits such as integrators, differentiators, and phase shift networks. Instrumentation amplifiers and operation with a single power supply are also included.

Comparator circuits form the basis for Chapter 6, including both open-loop comparators and comparators with feedback (Schmitt triggers). Along with the use of general purpose op-amps, a representative dedicated IC comparator (LM311) will be considered.

Timers and oscillators are the topics of Chapter 7, which covers both analysis and design of an op-amp multivibrator and an op-amp square-wave/triangular wave generator. The Barkhausen criterion is introduced, and the Wien bridge and phase shift oscillators are included. The IC circuits considered are the 555 timer and the 8038 function generator chip.

Chapter 8 is devoted to active filters, with emphasis on the widely used Butterworth function for low-pass filters and the standard resonance characteristic for band-pass filters. Finite-gain low-pass and high-pass filter design data are included, and we also discuss the infinite-gain band-pass circuit design. The last portion of the chapter deals with the state-variable filter, for which low-pass, band-pass, high-pass, and band-rejection filters can be realized. After completing this chapter, the student should be able to design and implement a variety of practical active filters.

Rectifier, diode, and power supply circuits form the basis for Chapter 9. Precision rectifier, holding, and limiting circuits are covered in the first part of the chapter, and the latter part of the chapter is devoted to power supply circuits. The development of the voltage regulator concept is included with this material. We also consider the use of a representative IC regulator (LM117/317).

Chapter 10 considers the timely topic of data conversion, which we may think of as the bridge between the analog and digital worlds. We study both digital-to-analog and analog-to-digital conversion and some of the most common circuits. The concepts of voltage-to-frequency and frequency-to-voltage conversion are also discussed. Various miscellaneous topics that do not fit the objectives of earlier chapters appear in Chapter 11, including multipliers, logarithmic amplifiers, and phase-locked loops, along with application of an IC phase-locked loop (NE/SE565).

Changes in the Fourth Edition

The general organization of the text has remained virtually the same since the second edition. Some major changes were made then, but it appears that the information organization and flow have been quite satisfactory since that point.

The most significant change for the fourth edition has been the replacement of PSPICE example problems at the end of most chapters with Multisim (Electronics Workbench) example problems. Multisim has made great strides in recent years and is now being widely used in engineering and engineering technology programs to model a variety of linear and nonlinear circuits. It is very user-friendly, and students can acquire proficiency in its operation very quickly. Moreover, users develop the circuit through the creation of a schematic diagram. Virtual instruments are available and their functions closely parallel that of real instruments. Thus, Multisim is very close to being a true "virtual laboratory."

The Multisim examples are provided in the last sections of the chapters and generally utilize circuits that were analyzed or designed in the chapter. However, these sections may be omitted without loss of continuity for instructors who do not wish to cover this subject.

Appendix E provides an abbreviated treatment of the basic concepts of schematic creation using Multisim for readers not having this background. Multisim example problems throughout the text provide further information about the use of various functions and models.

Problems

Most chapters contain end-of-chapter problems organized in three categories: Drill, Derivation, and Design. Drill problems make up the largest category. These are designed to test the reader's understanding of chapter principles. Problems are organized loosely in groups of two based on similar principles. Based on this grouping, answers to selected odd-numbered problems appear at the back of the book. Derivation problems require the reader to develop some of the results given in the text or to extend the results to new situations. Design problems require the reader to determine element values for a known circuit or a new circuit that will meet some prescribed specifications. Some of the design problems have a unique solution; others have more than one possible solution. Answers accompanied by an asterisk (*) are *typical solutions;* other valid solutions may be possible.

Laboratory Exercises   Most chapters also provide additional laboratory exercises, with a total of 52 exercises in Chapters 2 through 8. These exercises will provide sufficient work for the core of a supporting laboratory. It is assumed that the student who uses the laboratory exercises will already be familiar with basic test equipment and instrumentation such as voltmeters, ammeters, and the oscilloscope; consequently, statements such as "measure the voltage . . ." often appear without regard to the instrument type or procedure.

Safety   Although most of the laboratory exercises use relatively low voltages, there are always risks associated with the use of electrical equipment. Consequently, you should always follow the safety procedures learned in basic laboratory courses. You should have this instruction before you perform the exercises in this book.

Equipment

Minimal equipment is required to perform the exercises, and equipment is of the type normally found in most basic instructional laboratories. Further, the exercises have been designed to provide as much flexibility as possible. To provide full coverage, the following units should be available:

Circuit Trainer or "Breadboard" A variety of solderless units are available, some of which also provide suppliers and a function generator. Obviously, hookup wire is also required.

Oscilloscope Any modern dual-trace general-propose oscilloscope should be adequate. In fact, a single-trace oscilloscope can be used, but phase shift measurements are somewhat more involved. For brevity, the oscilloscope is referred to in some exercises as a *CRO* (cathode-ray *o*scilloscope).

Multimeter A general-purpose multimeter capable of measuring both dc and ac voltages and currents as well as passive resistance is appropriate. Either a digital or an analog instrument can be used. References to the multimeter in laboratory exercises are often made on the basis of the function involved; for example, "use a voltmeter to measure...."

Function Generator The desired unit should provide the choice of a sine wave, a square-wave, or a triangular wave over a frequency range from several hertz to at least 100 kHz. The output amplitude should be adjustable from zero to a level of at least several volts. A variable dc offset, which may be used in conjunction with the output signal or as a separate adjustable dc voltage, is a desirable feature.

Dual Power Supply The unit should provide either two fixed voltages of $+15$ V and -15 V or adjustable voltages that can be set at these levels.

Fixed 5-V DC Voltage This standard digital TTL level is available in many circuit trainers and is convenient for several of the exercises.

Oscilloscope The oscilloscope is probably the most important single instrument in measurements associated with linear circuits. In many exercises, you will be asked to sketch the forms of various waveforms and to show their time relationships. In virtually all cases, you should employ dc coupling for this purpose. This is important because the base line of the trace represents a zero-voltage level, and it will be necessary to measure voltages above and below this level. The ac coupling input position should be used only when it is desirable to amplify a small time-varying component without the dc level forcing the signal to shift off the screen.

To show the time relationships, it is necessary that you establish a time reference on the trace. Often, the input voltage is the logical choice for this reference. For example, if the input signal is a sine wave, the beginning of a cycle is an ideal point at which to establish the reference.

Suppliers *Resistors* The actual number of different values required is modest, but for maximum flexibility, an assortment of resistors with 5% tolerances in $\frac{1}{4}$- or $\frac{1}{2}$-W sizes covering the range from about 100 Ω to 5.6 MΩ is recommended. The most popular size is 10 kΩ, and you will need a minimum of five of that size.

Capacitors   An assortment of nonpolarized low-tolerance capacitors such as polystyrene or mylar in the range from about 0.001 μF to about 0.1 μF is recommended for critical applications such as active filters, oscillators, phase shift circuits, and so forth. In addition, larger-value polarized electrolytic (e.g., tantalum) and smaller-value disc ceramic capacitors may be used in noncritical by-pass applications.

Potentiometers   A few potentiometers are desirable (and sometimes necessary) when a resistance must be adjusted to an exact value. Although the ranges are flexible, some suggested values are 0–1 kΩ, 0–10 kΩ, and 0–100 kΩ.

Electronic Components   You will need these electronic components:

5	741	Operational amplifiers
1	311	Comparator
1	555	Timer
1	317	Regulator
2		General purpose diodes

CONTENTS

6   COMPARATORS   267

7   OSCILLATORS AND WAVEFORM GENERATORS   305

1

GENERAL AMPLIFIER CONCEPTS

OBJECTIVES

After completing this chapter, the reader should be able to:

- Discuss and compare the areas of *digital electronics* and *linear* (or *analog*) electronics.
- Discuss the operation of an ideal amplifier and define its *voltage gain*.
- Define the two ideal independent source models and the four controlled (or dependent) source models, and sketch the schematic forms.
- Draw the Thevenin and Norton circuit models for realistic independent sources.
- Draw the circuit model for a complete amplifier and define *input impedance*, *output impedance*, and *voltage gain*.
- Draw a complete circuit model for the combination of an amplifier with source and load, showing loading effects at the input and output terminals, and determine the loaded gain.
- Draw the complete circuit model for the cascade arrangement of several amplifier stages along with source and load, showing end loading effects and stage interaction effects, and determine the loaded gain.
- Define *power gain* and *decibel power gain*, and determine either quantity if the other is known.
- Define *decibel voltage gain* in terms of linear voltage gain, and determine either quantity if the other is known.
- Determine the decibel gain associated with a cascade of several stages.
- Define *impedance*, *resistance*, *reactance*, *admittance*, *conductance*, and *susceptance*.
- Show the steady-state ac phasor forms for each of the circuit components.
- Define *steady-state transfer function*, *amplitude response*, and *phase response*.

1

- State the mathematical form of the transfer function of the one-pole low-pass model, define its major parameters, and sketch the forms of the amplitude and phase response functions.
- Sketch the Bode plot break-point approximations of amplitude and phase functions of the one-pole low-pass model.

1–1 INTRODUCTION

The primary objective of this introductory chapter is to present some of the most general properties of amplifier circuits as well as their circuit models. Most readers are probably familiar with some of these properties from previous studies of electronic circuits and devices. As students seldom see the "big picture" when studying individual properties of various amplifier circuits, however, the intent here is to look at the amplifier as a complete system and focus on input, output, and gain characteristics. Such properties apply to virtually all types of amplifiers whether they are implemented with bipolar junction transistors, field effect transistors, or operational amplifiers.

We will examine individual amplifier models and discuss the effects of input and output loading. We will also consider the effects of interaction between stages in cascade arrangements.

In covering the concept of decibel gain, we will define both decibel power gain and decibel voltage gain and explore computations utilizing these definitions. We will also develop decibel gain relationships for cascade arrangements.

After reviewing the fundamental concepts of steady-state ac circuit analysis, we will introduce the concept of the steady-state transfer function as a means for modeling a circuit's frequency response. We will also define the *amplitude* and *phase* response functions.

We will also develop and discuss in some detail the one-pole, low-pass, roll-off frequency response model. This form is quite important because many amplifier circuits (including operational amplifiers) are dominated by this type of response over a wide frequency range.

1–2 LINEAR VERSUS DIGITAL ELECTRONICS

There are many ways to classify the various divisions of electronics, most of which are ambiguous because of the complexity of the field and the overlap between the different application areas. One particular classification scheme, however, deserves some attention here because of its relevance to the focus of this book. At a relatively broad level, electronics can be separated into the divisions of (1) **digital electronics** and (2) **linear electronics.** There is a temptation to call the second category *analog electronics*, but in accordance with widespread usage, the term *linear electronics* will be used.

Digital electronics is concerned with all phases of electronics in which signals are represented in terms of a finite number of digits, the most common of which is the binary number system. Digital electronics also includes all arithmetic computations on such numbers, as well as associated logic operations. The distinguishing feature of digital electronics is the representation of all possible variables by a finite number of digits. Obvious examples in which digital electronics plays the major role are computers and calculators.

Linear electronics is concerned with all phases of electronics in which signals are represented by continuous or *analog* variables. Linear electronics also includes all signal-processing functions (for example, amplification) associated with such signals.

Actually, the term *linear* is a misnomer, since many of the circuits classified as such are nonlinear in nature. On a slightly humorous note, a better term might be *nondigital electronics* to indicate that a large percentage of electronic applications other than digital are often classified under the category of linear electronics. However, the classification term *linear electronics* has become so imbedded within the electronics industry that its usage will no doubt continue. The reader should realize, however, that many nonlinear circuits are classified in this category.

This book is devoted to the consideration of linear integrated electronic circuits and devices. This major segment of the linear electronics field has widespread application to many specialty fields.

There are a number of electronic circuits that involve a combination of digital and linear electronics, and some of the most important of these are considered in this book. Foremost among such circuits are analog-to-digital and digital-to-analog converters. Such devices are used in the interfacing areas between analog and digital circuits, and they utilize both linear and digital circuit principles.

One of the most important applications in the field of linear electronics is the process of **amplification.** This operation was one of the very earliest applications of electronic devices, and it still remains an essential operation in virtually all phases of the industry. Consequently, many of the linear applications in this book either directly or indirectly involve amplification.

Ideal Amplifier

An ideal linear amplifier is characterized by the fact that the output signal is directly proportional to the input signal, but the level will be changed in the process. Amplifiers in system form are often represented by a block diagram such as shown in Figure 1–1. The input voltage signal is denoted as $v_i(t)$, and the output voltage signal is denoted as $v_o(t)$. The quantity t represents time, and the functional forms $v_i(t)$ and $v_o(t)$ represent the fact that both voltages are functions of time; that is, they vary in some fashion as time passes. When it is not necessary to emphasize this functional notation, the parentheses and t will be omitted, in which case v_i and v_o may be used to represent the quantities involved. The functional notation was introduced at this point so that the reader can recognize it throughout the book when it occurs.

FIGURE 1–1

Block diagram representation of a linear amplifier.

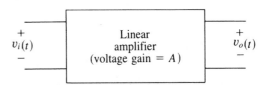

$+$
$v_i(t)$
$-$

Linear
amplifier
(voltage gain $= A$)

$+$
$v_o(t)$
$-$

In this case, the quantity A represents the voltage gain. For an ideal linear amplifier, it can be defined as

$$A = \frac{v_o(t)}{v_i(t)} \tag{1-1}$$

If the amplifier is not perfectly linear, the basic definition of (1–1) is no longer correct. For the moment, we will avoid that situation and assume ideal linear amplification. If the gain and input voltage are known, the output signal is

$$v_o(t) = Av_i(t) \tag{1-2}$$

The input and output ideal signal relationships are illustrated by the waveforms shown in Figure 1–2. To simplify this illustration, a signal consisting of pulselike segments is assumed in (a). The corresponding form of the output of an ideal linear amplifier is shown in (b). All points on the output waveform are A times the corresponding points on the input waveform. For example, if $A = 5$ and the input voltage is 2 V, the output voltage is $5 \times 2 = 10$ V.

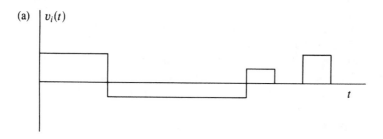

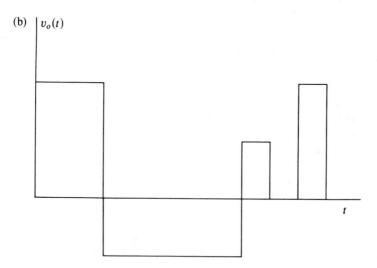

FIGURE 1–2
Input and output waveform examples for an ideal linear amplifier.

It should be stressed that virtually all amplifiers require a dc power input in order to provide amplification. It is customary to show only signal levels on many block diagrams for signal-processing analysis, and any dc power supplies (or *bias* supplies, as they are often called) are understood to be present. Such a diagram could cause someone to assume incorrectly that the amplifier is creating energy. Most active amplifier devices permit a small signal input to control a larger signal output, but the extra power is furnished by the dc power supply.

EXAMPLE 1–1

The voltage gain of an essentially linear amplifier is determined from oscilloscopic measurements. The waveforms shown in Figure 1–3 are obtained based on a sinusoidal input. Determine the voltage gain A.

Solution
The definition of (1–1) is applied to determine the voltage gain of an amplifier. For a sine wave, peak values, peak-to-peak values, or root-mean-square (rms) values may be used, provided that *both* input and output waveforms are expressed in the same units. For oscilloscopic measurements, peak-to-peak values are usually the easiest to apply and will be used here.

Let V_{ipp} represent the peak-to-peak value of the input voltage and let V_{opp} represent the peak-to-peak value of the output voltage. Based on the scaling factors of the two channels, these voltages are V_{ipp} = 0.05 V/division $\times$ 4 divisions = 0.2 V and V_{opp} = 2 V/division $\times$ 3 divisions = 6 V. The voltage gain A is

$$A = \frac{V_{opp}}{V_{ipp}} = \frac{6}{0.2} = 30 \tag{1–3}$$

Input
0.05 V/division

Output
2 V/division

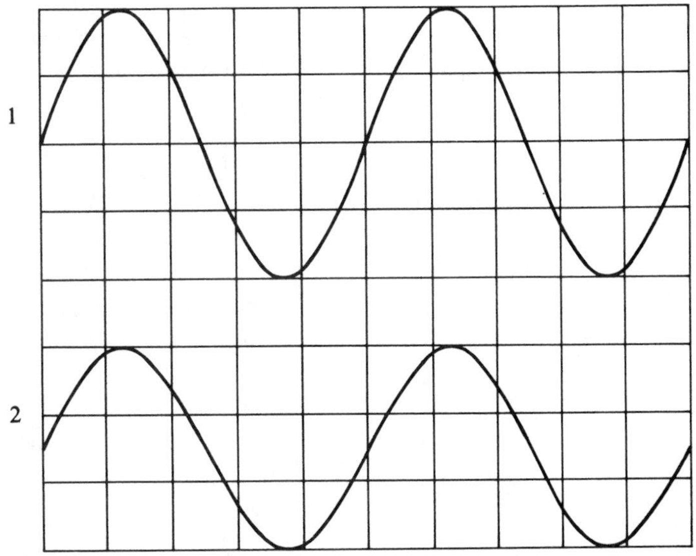

FIGURE 1–3
Waveforms of Example 1–1.

1-3 SOURCE MODELS

All sources of electrical energy can be represented in terms of either voltage sources or current sources. Figure 1–4 shows the forms of the two ideal source models; the ideal voltage source model is shown in (a), and the ideal current source model is shown in (b).

The ideal voltage source is assumed to generate a voltage v_s, which is independent of the current flowing through the source. The $(+)$ and $(-)$ terminals show the respective positive and negative reference terminals for the voltage. The ideal current source is assumed to generate a current i_s, which is independent of the voltage across the source. The arrow indicates the assumed positive reference direction for the current. Both v_s and i_s may, in general, vary with time; however, the key property of these ideal sources is that their values are independent of any external loads connected to the devices.

Practical sources may be modeled by a combination of an ideal source and one or more passive circuit components. In a wide variety of situations at relatively low frequencies, the passive component is resistance. The two forms of circuit models applicable to this situation are shown in Figure 1–5.

The Thevenin model of Figure 1–5(a) represents the practical source by an ideal voltage source v_s in series with a resistance R_s. The Norton model of Figure 1–5(b) represents the practical source by an ideal current source i_s in parallel with a resistance R_s.

Either of the preceding models may be used to represent a practical source provided that the only significant passive circuit parameter is resistance. It is easy to convert from one form to the other. The current i_s in the Norton model is related to the voltage v_s in the Thevenin model by

$$i_s = \frac{v_s}{R_s} \tag{1-4}$$

The current i_s is the current that would flow through a short circuit connected across the output.* Conversely, the voltage v_s is related to the current i_s by

$$v_s = R_s i_s \tag{1-5}$$

The voltage v_s is the voltage that would be measured across the terminals under open-circuit conditions.

FIGURE 1–4
Ideal voltage and current source models.

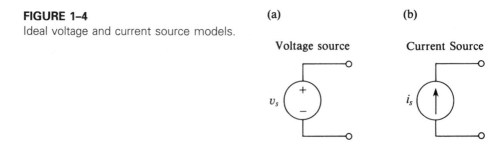

(a)

Voltage source

(b)

Current Source

v_s + −

i_s

---

*This process is primarily a "paper exercise" and should not be attempted in the laboratory for most realistic sources.

FIGURE 1–5
Thevenin and Norton models for more practical
sources.

(a) Thevenin model

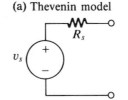

(b) Norton model

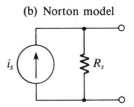

Although either model in Figure 1–5 can be used to represent any source, the Thevenin model is usually more convenient when the internal resistance is very *small* compared to any external load resistance to be connected. Conversely, the Norton model is usually more convenient when the internal resistance is very *large* compared to any external load resistance to be connected.

When the internal resistance is so small that it may be neglected, the actual source can be approximated by an ideal voltage source model. Conversely, when the internal resistance is so large that it may be neglected, the actual source can be approximated by an ideal current source model. Thus, *an ideal voltage source is said to have zero internal resistance, and an ideal current source is said to have infinite internal resistance.*

All of the preceding source models are considered **independent** sources, since the value of v_s or i_s does not depend on some other circuit variable. (We will consider *dependent* or *controlled* sources in the next section.)

EXAMPLE 1–2

The open-circuit voltage of a certain transducer output is 80 mV, and the equivalent output resistance is 2 kΩ. Determine two circuit models for the source.

Solution
From the information as given, the source may be modeled directly by its Thevenin form, which is an 80-mV voltage source in series with 2 kΩ, as shown in Figure 1–6(a). To determine the corresponding current source i_s, the voltage and current are first expressed in their basic units, and we have

$$i_s = \frac{v_s}{R_s} = \frac{80 \times 10^{-3}\,\text{V}}{2 \times 10^3\,\Omega} = 40 \times 10^{-6}\,\text{A} = 40\,\mu\text{A} \tag{1–6}$$

The Norton form is shown in Figure 1–6(b).

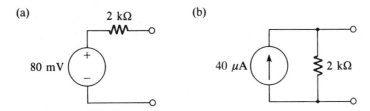

FIGURE 1–6
Source models of Example 1–2.

1-4 CONTROLLED-SOURCE MODELS

Linear active devices are used to amplify and control signals in many different ways. For example, the physical movements of a stylus on a phonograph record generate a small signal at the output of the cartridge, which represents the recorded music. After sufficient amplification by linear devices, the signal controls the output of a power amplifier stage, which provides enough power to the speaker to convert the electrical energy to acoustical energy. Throughout this system all stages should ideally have linear relationships; that is, the output of each stage should be a constant times the input.

Consider a linear active device with one set of input signal terminals and one set of output signal terminals. Depending on the electronic device and how it operates, either voltage or current at the input may be the controlling variable. Further, the output controlled may be either voltage or current. Thus, there are four possible combinations of input-output control, and all of these occur in actual systems:

1. Voltage-controlled voltage source
2. Voltage-controlled current source
3. Current-controlled voltage source
4. Current-controlled current source

We will investigate each of these four conditions in this section and present certain models to represent their behavior. The models given will be considered in the most idealized forms. At this point we will not consider how the control functions are actually implemented. There are various ways to achieve these operations using bipolar junction transistors, field effect transistors, operational amplifiers, vacuum tubes, and so on. We will be dealing strictly with the ideal input-to-output control operations in their simplest mathematical forms. Thus, the relatively simple looking models shown in the figures could represent the effect of complex circuits containing many individual active and passive circuit components.

Although the same symbol (circle) used to represent independent sources may also be used for dependent sources, there is a growing trend toward a different symbol for dependent sources. The symbols in Figure 1–7 are appearing in many parts of the literature; we will use them here because they are easy to recognize.

Voltage-Controlled Voltage Source (VCVS)

The most common combination is a voltage-controlled voltage source (VCVS), whose idealized model is shown in Figure 1–7(a). An independent signal voltage v_i is assumed to be applied across the input terminals. The action of the control circuit is to create a signal voltage v_o at the output given by

$$v_o = Av_i \qquad \textbf{(1-5)}$$

The voltage source on the right is an example of a **_dependent_** or **_controlled_** source because it is dependent on a variable in a different part of the circuit. The output voltage is thus a constant times the input voltage, a basic requirement for a linear amplifier. The quantity A

is the *voltage gain,* and it is dimensionless. This model could be used to represent the ideal linear amplifier discussed in Section 1–2.

Voltage-Controlled Current Source (VCIS)

The idealized model of a VCIS is shown in Figure 1–7(b). The input-controlling variable is the signal voltage shown on the left. However, the controlled variable in this case is the output current, whose operation is represented by the dependent current source on the right. The value of this current at the output is

$$i_o = g_m v_i \qquad (1\text{–}8)$$

The constant g_m is the *transconductance* of the device, and it has the units of siemens (S).

Current-Controlled Voltage Source (ICVS)

The idealized model of an ICVS is shown in Figure 1–7(c). Unlike the previous two models, in this case the input signal current i_i is the controlling variable. The controlled variable is the output voltage, which is represented by the dependent voltage source on the right. The value of this voltage at the output is

$$v_o = R_m i_i \qquad (1\text{–}9)$$

The constant R_m is the *transresistance* of the device, and it has the units of ohms (Ω).

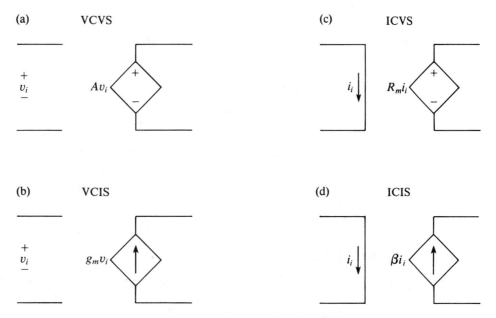

FIGURE 1–7
Four possible models of ideal controlled (or dependent) sources in electronic circuits.

Current-Controlled Current Source (ICIS)

The last of the four possible combinations is the ICIS, and its idealized model is shown in Figure 1–7(d). The controlling variable is the input current shown on the left. The controlled variable is the output current, which is represented by the dependent current source shown on the right. The value of the output current is

$$i_o = \beta i_i \qquad (1\text{–}10)$$

The constant β is the *current gain,* and it is dimensionless.

Various electronic devices approximate the behavior of the different source models just discussed. For example, bipolar junction transistors in their ideal form may be represented by the ICIS model. Conversely, field effect transistors may be represented very closely by the VCIS model.

In many applications, it is desirable to design circuitry to perform one of the operations previously discussed. For example, consider a device in which the output signal voltage is of interest. It may be desired to transfer this signal to a load that responds primarily to current. In this case, a voltage-controlled current source having a suitable value of g_m would be required. In later chapters, we will learn how to design circuits using modern linear integrated circuits that will perform all the basic processing operations just discussed.

1–5 COMPLETE AMPLIFIER SIGNAL MODEL

The four ideal controlled-source models defined in Section 1–4 are used in representing various linear signal amplification functions. Consequently, a number of possible complete circuit models arise in practice. However, at this point we will focus on one particular complete amplifier model because of its widespread usage and because some of its parameters are similar to those used in other circuits as well. As previously discussed, only signal quantities will be shown in the diagrams.

A general block diagram of a linear amplifier with one set of input terminals and one set of output terminals is shown in Figure 1–8(a). A corresponding signal model that can be used to represent a wide variety of complete amplifier circuits is shown in Figure 1–8(b). We will assume that all passive parameters are resistive in this simplified model. The effects of reactive elements (capacitance and inductance) will be considered in Section 1–8.

It is customary in discussing complete amplifier models to use the general term *impedance* for the input and output effects, even when resistance is the only parameter assumed (which is the most common situation). We will use the symbol R with appropriate subscripts, however, to delineate that the impedance is assumed to be resistive.

Input Impedance

The input impedance is the effective impedance across the two input terminals as "seen" by a signal source. For the resistive case of Figure 1–8(b), the input impedance is simply R_{in}. For this resistive case, we can state that

$$R_{\text{in}} = \frac{v_i}{i_i} \qquad (1\text{–}11)$$

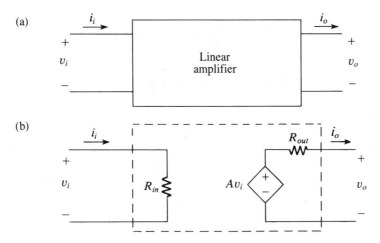

FIGURE 1–8
Block diagram of a linear amplifier and a common form of a signal model utilizing a VCVS.

For the more general case with reactive components, transform or phasor voltage and current (rather than instantaneous quantities) must be used in the definition, and a complex input impedance $\overline{Z}_{in}$ is required in the model.

The input impedance is important in determining the fraction of any available signal voltage that actually appears across the amplifier terminals when the source has an internal impedance. In general, the voltage actually appearing across the amplifier terminals will be lower than the available source voltage due to the interaction between the source and input impedances.

Output Impedance

The output impedance is the impedance portion of the Thevenin or Norton equivalent circuit as viewed at the output terminals. In Figure 1–8(b), a Thevenin form is used for the output portion of the amplifier circuit, and the output impedance is seen to be a resistive value R_{out} in this case. In the more general case, a complex output impedance $\overline{Z}_{out}$ is required in the model.

The output impedance is important in determining the change in output signal as any external load connected to the output terminal changes. Thus, if two amplifier stages are connected, the output impedance of the first stage interacts with the input impedance of the second stage.

Voltage Gain

In the model of Figure 1–8(b), a voltage-controlled voltage source represents the effective voltage amplification of the circuit. With no load connected across the output, the output voltage is A times the input voltage. Thus, the **open-circuit voltage gain** is readily determined from the circuit diagram to be A. Under loaded conditions, the voltage gain will be reduced, as will be demonstrated later.

With the given polarity of the controlled source in Figure 1–8(b) and with A positive, the voltage gain is noninverting; that is, the output voltage has the same sign as the input voltage (in phase). However, if *either* the polarity of the controlled source is reversed *or* if A is negative (but not both), the gain is *inverting*; that is, the output voltage is inverted in sign with respect to the input voltage (out of phase).

EXAMPLE 1–3

A certain amplifier is characterized by the following properties: the input impedance = 1500 Ω (resistive); the output impedance = 250 Ω (resistive); and the open-circuit voltage gain = 50. Draw an equivalent circuit model for the amplifier.

Solution
The equivalent circuit will be of the form in Figure 1–8(b) and is shown in Figure 1–9.

FIGURE 1–9
Amplifier model of Example 1–3.

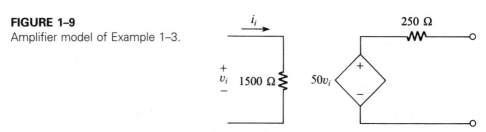

1–6 CASCADE OF AMPLIFIER STAGES

A model representing a large number of possible linear amplifier circuits was given in the last section. The parameters required to represent an amplifier by this model are the open-circuit voltage gain, the input impedance, and the output impedance. We will now investigate the loading and interaction effects that occur when an amplifier is cascaded with other amplifier circuits, a load, or a source with internal resistance.

Input and Output Loading Effects

First, consider the process of connecting the input terminals of an amplifier to a source with nonzero internal impedance and simultaneously connecting it to a finite load impedance. The connections involved are shown in Figure 1–10(a). An equivalent circuit employing the model of the last section coupled with the external parameters is shown in Figure 1–10(b). To simplify notation, we use R_i and R_o for input and output impedances, respectively.

The voltage v_i at the amplifier terminals can be expressed in terms of the open-circuit source voltage v_s and the voltage division between the source internal impedance R_s and the amplifier input impedance R_i. We have

$$v_i = \frac{R_i}{R_i + R_s} v_s \qquad (1\text{–}12)$$

(a)

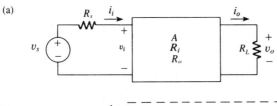

(b)

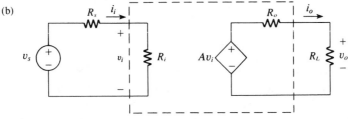

FIGURE 1–10
(a) Connection of an amplifier to a signal source and load, and (b) the linear signal model.

The output voltage v_o can be expressed in terms of the VCVS Av_i and the voltage division between the amplifier output impedance R_o and the external load resistance R_L. This voltage is

$$v_o = \frac{R_L}{R_L + R_o} Av_i \tag{1–13}$$

Substitution of v_i from (1–12) in (1–13) yields

$$v_o = \left(\frac{R_L}{R_L + R_o}\right)\left(\frac{R_i}{R_i + R_s}\right)Av_s \tag{1–14}$$

Let A_{so} represent the net voltage gain from the source open-circuit voltage to the output load voltage. From (1–14), this quantity is

$$A_{so} = \frac{v_o}{v_s} = \left(\frac{R_L}{R_L + R_o}\right)\left(\frac{R_i}{R_i + R_s}\right)A \tag{1–15}$$

By rearranging the terms in (1–15) according to a left-to-right circuit order, we find that the following pattern emerges:

$$A_{so} = \left(\frac{R_i}{R_i + R_s}\right) \times A \times \left(\frac{R_L}{R_L + R_o}\right) \tag{1–16a}$$

$$\begin{array}{ccc} \text{input} & \text{open-circuit} & \text{output} \\ A_{so} = \text{loading} \times & \text{voltage} & \times \text{loading} \\ \text{factor} & \text{gain} & \text{factor} \end{array} \tag{1–16b}$$

The input loading factor is simply the voltage divider relationship reflecting the fraction of the source voltage appearing across the amplifier input terminals. The output loading factor is the voltage divider relationship reflecting the fraction of the open-circuit output voltage appearing across the load resistance. The open-circuit voltage gain in the ideal gain case would be achieved with no loading at either input or output.

The reader should not try to memorize an equation such as (1–16a). Rather, the concept delineated by (1–16b) should be carefully noted. With some practice, an expression of the form of (1–16a) can be written down in one step if the general process of (1–16b) is understood.

Cascade Loading Effects

Next we will extend the discussion to include more than one amplifier stage in the cascade connection. Specifically, we will consider two amplifier stages, but the trend that emerges should allow the reader to extend the concept to an arbitrary number of stages.

Consider the circuit shown in Figure 1–11(a), consisting of two amplifier circuits, a source of value v_s with internal impedance R_s, and a load resistance R_L. The input and output impedances of the first stage are denoted as R_{i1} and R_{o1}, respectively, and the corresponding quantities for the output stage are denoted as R_{i2} and R_{o2}. Finally, the open-circuit gain for the first stage is A_1, and the corresponding value for the second stage is A_2.

The expanded equivalent circuit is shown in Figure 1–11(b). The voltage v_{i1} at the input to the first stage is

$$v_{i1} = \frac{R_{i1}}{R_{i1} + R_s} v_s \qquad (1\text{--}17)$$

(a)

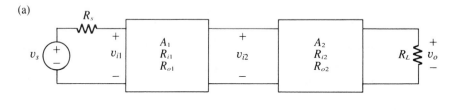

(b)

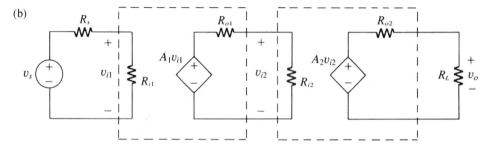

FIGURE 1–11
(a) Cascade connection of two amplifier stages to a signal source and load, and (b) the linear signal model.

The voltage v_{i2} at the input to the second stage is

$$v_{i2} = \frac{R_{i2}}{R_{i2} + R_{o1}} \times A_1 v_{i1} \tag{1-18}$$

Finally, the output voltage v_o across the load is

$$v_o = \frac{R_L}{R_L + R_{o2}} \times A_2 v_{i2} \tag{1-19}$$

A relationship between v_o and v_s can be obtained by first substituting v_{i1} from (1–17) in (1–18) and then substituting that expression for v_{i2} in (1–19). The voltage gain A_{so} between source and output can then be determined by dividing both sides by v_s. The result is

$$A_{so} = \frac{v_o}{v_s} = \left(\frac{R_L}{R_L + R_{o2}}\right) \left(\frac{R_{i2}}{R_{i2} + R_{o1}}\right) \left(\frac{R_{i1}}{R_{i1} + R_s}\right) A_1 A_2 \tag{1-20}$$

The terms in (1–20) can be rearranged in left-to-right circuit order as follows:

$$A_{so} = \left(\frac{R_{i1}}{R_{i1} + R_s}\right) \times A_1 \times \left(\frac{R_{i2}}{R_{i2} + R_{o1}}\right) \times A_2 \times \left(\frac{R_L}{R_L + R_{o2}}\right) \tag{1-21a}$$

$$A_{so} = \begin{pmatrix} \text{input} \\ \text{loading} \\ \text{factor} \end{pmatrix} \times \begin{pmatrix} \text{open-circuit} \\ \text{voltage} \\ \text{gain 1} \end{pmatrix} \times \begin{pmatrix} \text{interaction} \\ \text{factor} \\ \text{between} \\ \text{stages} \end{pmatrix}$$

$$\times \begin{pmatrix} \text{open-circuit} \\ \text{voltage} \\ \text{gain 2} \end{pmatrix} \times \begin{pmatrix} \text{output} \\ \text{loading} \\ \text{factor} \end{pmatrix} \tag{1-21b}$$

The input and output loading factors are the same forms considered earlier when a single amplifier stage was connected between a source and a load. The input loading factor involves the input impedance of the first stage, and the output loading factor involves the output impedance of the second stage. There are, of course, two open-circuit voltage gain factors appearing in this case. Finally, an interaction factor between the stages appears in the expression. This factor is a voltage divider relationship reflecting the fraction of the available output voltage of the first stage appearing across the input of the second stage under loaded conditions. If the circuit contained more than two stages, similar interaction factors would be required between all stages.

EXAMPLE 1–4   The linear amplifier of Example 1–3 is connected between a source and a load as shown in Figure 1–12(a). (a) Determine the net loaded voltage gain between the open-circuit source voltage and load resistance. (b) Given that the open-circuit source voltage is 0.2 V, determine the amplifier input voltage, open-circuit output voltage, and output voltage under loaded conditions.

(a)

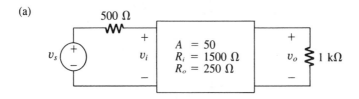

(b)

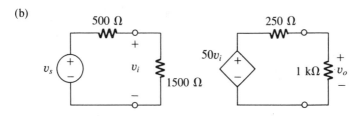

FIGURE 1–12
Circuit of Example 1–4.

Solution
(a) The equivalent circuit under connected conditions is shown in Figure 1–12(b). Using the simplified intuitive approach discussed in this section, the net voltage gain between source and load can be expressed as

$$A_{so} = \frac{v_o}{v_s} = \left(\frac{1500}{1500 + 500}\right) \times 50 \times \left(\frac{1000}{1000 + 250}\right) = 30 \qquad \textbf{(1–22)}$$

Observe that the loaded gain is considerably less than the open-circuit gain of 50.
   (b) The voltage v_i at the input terminals is determined by the voltage divider rule.

$$v_i = \frac{1500}{1500 + 500} \times v_s = 0.75 \times 0.2\,\text{V} = 0.15\,\text{V} \qquad \textbf{(1–23)}$$

Thus, the available open-circuit source voltage of 0.2 V is reduced to 0.15 V at the amplifier terminals by the interaction between R_s and R_i.
   The open-circuit output voltage is the voltage that would appear across the output if the 1-kΩ load resistor were removed. Let v_o' represent this voltage, and it is simply

$$v_o' = Av_i = 50 \times 0.15 = 7.5\,\text{V} \qquad \textbf{(1–24)}$$

The loaded output voltage v_o is the fraction of v_o' that appears across R_L as a result of the voltage division between R_o and R_L. We have

$$v_o = \frac{1000}{1000 + 250} \times v_o' = 0.8 \times 7.5 = 6\,\text{V} \qquad \textbf{(1–25)}$$

An alternate way to determine the output voltage under loaded conditions is to multiply the loaded gain A_{so} by the open-circuit source voltage. Thus,

$$v_o = A_{so}v_s = 30 \times 0.2 = 6 \text{ V} \tag{1-26}$$

which clearly agrees with the result of (1–25).

1–7   DECIBEL GAIN COMPUTATIONS

Decibel gain and loss relationships are widely used in analyzing and specifying linear amplifier circuits. Although most readers at this level are at least familiar with such computations, a review of some of the details will be worthwhile.

The original and most "proper" basis for decibel measurement is that of a power comparison. Consider the block diagram of a linear amplifier shown in Figure 1–13. The input signal is assumed to deliver a power P_i to the amplifier input, and the amplifier in turn delivers an output power P_o to the external load. The **power gain** G is defined as

$$G = \frac{P_o}{P_i} \tag{1-27}$$

Decibel Power Gain

The **decibel power gain** G_{dB} in decibels (abbreviated as dB) is defined by

$$G_{dB} = 10 \log_{10} G = 10 \log_{10} \frac{P_o}{P_i} \tag{1-28}$$

Assume that the power P_o is delivered to an external load resistance R_L, and assume that the input impedance is R_{in}.* The load and input power levels can then be expressed in terms of the output and input voltage levels as

$$P_o = \frac{v_o^2}{R_L} \tag{1-29}$$

and

$$P_i = \frac{v_i^2}{R_{in}} \tag{1-30}$$

Substitution of (1–29) and (1–30) in (1–28) yields

$$G_{dB} = 10 \log_{10} \frac{v_o^2/R_L}{v_i^2/R_{in}} \tag{1-31}$$

---

*Strictly speaking, v_o and v_i must be interpreted as rms, or effective values if P_o and P_i are specified as average power values.

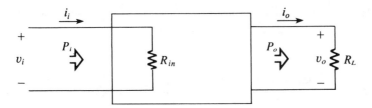

FIGURE 1–13
Power input and output comparison for a linear amplifier.

Before proceeding further with this analysis, we will pause briefly to review some pertinent logarithmic relationships. They are

$$\log_{10}x^2 = 2\log_{10}x \tag{1–32}$$

$$\log_{10}xy = \log_{10}x + \log_{10}y \tag{1–33}$$

$$\log_{10}\frac{1}{x} = -\log_{10}x \tag{1–34}$$

Assume momentarily that $R_L = R_{in}$. In this case, the resistance factors in (1–31) cancel, and with the use of (1–32), the following form for G_{dB} is obtained:

$$G_{dB} = 20\ \log_{10}\frac{v_o}{v_i} = 20\ \log_{10}A \tag{1–35}$$

where A is the voltage gain between the input terminals and the output load. (If A is negative, the logarithm of the magnitude is used.)

A similar form would be obtained for the current gain if the load power were expressed in terms of the load current and the input power were expressed in terms of the input current. *Thus, a factor of 10 appears in decibel computations involving power gain, and a factor of 20 appears when voltage and current gains are used.*

Decibel Voltage Gain

Strictly speaking, the form of (1–35) requires that the input and load resistances be equal in order to conform to the original definition of decibel measure on a power basis. However, it is common practice in the electronics industry to define decibel gain even when the impedances are not equal, and many electronic devices have specifications given in this way. It should be remembered in this case that the decibel ratio reflects not a power gain, but rather a voltage or current gain converted to a logarithmic basis.

Since many decibel computations in small signal amplifiers involve gains computed between points with different impedance levels, to avoid confusion with true decibel power levels, a quantity A_{dB} will be defined as follows:

$$A_{dB} = 20\ \log_{10}\frac{v_o}{v_i} = 20\ \log_{10}A \tag{1–36}$$

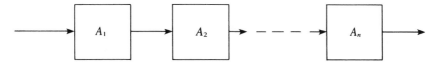

FIGURE 1–14
Cascade amplifier connection.

This definition agrees exactly with (1–35), but the use of the symbol A_{dB} indicates that this is a voltage gain converted to the appropriate logarithmic level for decibel comparison rather than a proper decibel power ratio. Thus, when the load and source resistances are equal, A_{dB} and G_{dB} mean exactly the same thing.

When two or more stages are connected in cascade, decibel measures are particularly convenient to use in dealing with the combination. Consider the cascade connection shown in Figure 1–14. It is assumed for this discussion that either there are no loading or interaction effects between different stages, or the loading and interaction effects are represented as equivalent gains (less than one in such cases). The net voltage gain A is determined in absolute form as the product of the individual gains; that is,

$$A = A_1 A_2 \cdots A_n \tag{1–37}$$

The decibel gain is

$$A_{dB} = 20 \log A = 20 \log(A_1 A_2 \cdots A_n) \tag{1–38}$$

Application of (1–33) to (1–38) yields

$$A_{dB} = 20 \log_{10} A_1 + 20 \log_{10} A_2 + \cdots + 20 \log_{10} A_n \tag{1–39a}$$

$$A_{dB} = A_{1dB} + A_{2dB} + \cdots + A_{ndB} \tag{1–39b}$$

where the terms on the right of (1–39b) refer to the corresponding decibel gains of the various stages. Thus, the net decibel gain is simply the sum of the individual decibel gains for the different stages.

When the output voltage is less than the input voltage, the decibel voltage gain is negative. Thus, if any of the various stages in Figure 1–14 represent interaction factors, whose voltage divider output/input ratios are usually less than unity, the decibel values must be treated as "negative gains" in the expression of (1–39b).

A negative decibel gain should not be confused with a negative linear gain. The former corresponds to the reduction of a signal level through the circuit, while the latter corresponds to an inversion of the polarity.

For voltage divider networks it is sometimes more convenient to work with losses than with gains. Let α_{dB} represent the decibel loss ratio for a system with input voltage v_i and output voltage v_o. The quantity is defined as

$$\alpha_{dB} = 20 \log_{10} \frac{v_i}{v_o} = 20 \log_{10} \frac{1}{A} \tag{1–40}$$

Employing (1–34), we can readily show that

$$\alpha_{dB} = -20\log_{10}\frac{v_o}{v_i} = -20\log_{10}A = -A_{dB} \qquad (1\text{–}41)$$

Thus a gain of -6 dB represents a loss of $+6$ dB, and vice versa. The use of the loss factor is very convenient in working with systems where the signal levels may go either up or down at different points. In such cases, it is less awkward to refer to a 20-dB loss than a -20-dB gain. Remember, however, that when working with a cascaded system such as the one shown in Figure 1–14, one should convert all decibel ratings shown to gains (either positive or negative) when using (1–39b).

Conversion from Decibels

A final point concerns the conversion from decibel levels back to actual gain levels. Consider (1–36) relating A_{dB} to A. Both sides are first divided by 20 as follows:

$$\frac{A_{dB}}{20} = \log_{10}A \qquad (1\text{–}42)$$

Both sides of (1–42) are equated as powers of 10, and the result is

$$A = 10^{A_{dB}/20} \qquad (1\text{–}43)$$

Operations of this form may be readily performed with calculators having a 10^x or a y^x function.

A newcomer to the electronics field might question the value of decibel specifications and computations, since the process represents what might be criticized as unnecessary computations. However, decibel forms have been used since the very early days of the electronics industry, and it is virtually impossible to deal with the complexity of device operation and specifications without using such forms. Actually, one of the strong reasons why decibel forms remain so imbedded is that they are very convenient to use once you become adept at doing so. It is possible to deal with large gains or losses very quickly and almost intuitively using decibel forms without resorting to unwieldy computations.

To assist the reader in developing a more intuitive feeling for decibel computations, a summary of some voltage gain values and their decibel equivalents is given in Table 1-1. One common slight approximation is made here to simplify the results. A voltage gain of 2 corresponds to a decibel gain of 6.02 dB, and a voltage gain of ½ corresponds to a decibel gain of -6.02 dB. However, where absolute precision is not required, these values are usually rounded to 6 dB and -6 dB, respectively. These rounded values will be used throughout the text in accordance with common practice.

EXAMPLE 1–5   Determine the decibel gain of the complete amplifier circuit of Example 1–4 by combining the decibel gain of the amplifier with the decibel losses of the input and output circuits.

TABLE 1–1
Voltage gains and their decibel equivalents

Positive dB Values		Negative dB Values	
A	A_{dB}	A	A_{dB}
$\sqrt{2} = 1.4142$	3 dB	$1/\sqrt{2} = 0.7071$	-3 dB
2	6 dB	$1/2 = 0.5$	-6 dB
$2\sqrt{2} = 2.8284$	9 dB	$1/2\sqrt{2} = 0.3536$	-9 dB
$\sqrt{10} = 3.1623$	10 dB	$1/\sqrt{10} = 0.3162$	-10 dB
4	12 dB	$1/4 = 0.25$	-12 dB
5	14 dB	$1/5 = 0.2$	-14 dB
8	18 dB	$1/8 = 0.125$	-18 dB
10	20 dB	$1/10 = 0.1$	-20 dB
20	26 dB	$1/20 = 0.05$	-26 dB
40	32 dB	$1/40 = 0.025$	-32 dB
50	34 dB	$1/50 = 0.02$	-34 dB
100	40 dB	$1/100 = 0.01$	-40 dB
1000	60 dB	$1/1000 = 0.001$	-60 dB
2^n	$6n$ dB	2^{-n}	$-6n$ dB
10^n	$20n$ dB	10^{-n}	$-20n$ dB

Solution
Refer to Figure 1–12 and the computations of Example 1–4. The amplifier has a voltage gain $A = 50$, and the corresponding decibel value is

$$A_{dB} = 20\log_{10}50 = 33.98\,\text{dB} \tag{1–44}$$

A simplified way to arrive at this value is to note that since 50 is ½ of 100, the decibel level is 6 dB below the decibel level corresponding to 100, which is 40 dB. This simplified approach leads to 34 dB, a value that is sufficiently close for most purposes.

    Let α_{idB} represent the loss factor in decibels for the input circuit, and let α_{odB} represent the corresponding factor for the output circuit. We have

$$\alpha_{idB} = 20\log_{10}\frac{v_s}{v_i} = 20\log_{10}\left(\frac{1500 + 500}{1500}\right) = 2.50\,\text{dB} \tag{1–45}$$

and

$$\alpha_{odB} = 20\log_{10}\frac{v_o'}{v_o} = 20\log_{10}\left(\frac{1000 + 250}{1000}\right) = 1.94\text{dB} \tag{1–46}$$

Since (1–45) and (1–46) both represent *losses,* the losses must be treated as negative gains when combined with the amplifier gain. Let $A_{dB}^{(so)}$ represent the net decibel gain between source and output. Combining the three individual values results in

$$A_{dB}^{(so)} = 33.98 - 2.50 - 1.94 = 29.54\,\text{dB} \tag{1–47}$$

For illustration, both decibel gain and loss calculations were made in this example. The complete analysis could be performed in terms of gain values if desired. With this approach, the "gain" values for the interaction functions would be defined in terms of the output-to-input voltage ratios. The absolute ratios of these quantities are less than unity, and the corresponding decibel "gains" are thus negative. The values of -2.50 dB and -1.94 dB would have been obtained directly as the respective "gains" with this latter approach.

As a check, the net decibel gain can be obtained in one step using the result of (1–22):

$$A_{dB}^{(so)} = 20\log_{10}A_{so} = 20\log_{10}30 = 29.54\,\text{dB} \qquad \textbf{(1–48)}$$

which agrees with (1–47).

1–8 FREQUENCY RESPONSE CONSIDERATIONS

Thus far in the text, all gains and all impedance values have been assumed to be constants, and no frequency-dependent parameters have been considered. Such assumptions are usually valid over a reasonable frequency range with most linear amplifier circuits. However, all linear circuits have frequency-limiting characteristics, and it is necessary to apply frequency response analysis before a full treatment is possible.

It is assumed that most readers are familiar with the details of steady-state ac circuit analysis using complex number techniques. Consequently, only a brief review and strengthening of certain critical ideas will be given here. Readers having little or no background in ac circuits should refer to one of the various general or ac circuit analysis books for a more basic treatment.

In steady-state ac circuit analysis, all instantaneous voltages and currents are assumed to be single-frequency sinusoidal functions. The effects of the circuit parameters are represented in terms of complex impedances. Circuit analysis may be performed algebraically, but all quantities are complex; that is, they have both real and imaginary parts (or magnitude and angle).

Phasors

Let $\overline{V}$ be the phasor representation of an instantaneous sinusoidal voltage $v(t)$, and let $\overline{I}$ be the phasor representation of a corresponding current $i(t)$. The polar forms of $\overline{V}$ and $\overline{I}$ are

$$\overline{V} = V \ \underline{/\theta} \qquad \textbf{(1–49)}$$

$$\overline{I} = I \ \underline{/\phi} \qquad \textbf{(1–50)}$$

where $V = |\overline{V}|$ is the magnitude of $\overline{V}$, and $I = |\overline{I}|$ is the magnitude of $\overline{I}$. The quantities θ and ϕ are the angles of the two respective phasors.

Impedance

The complex impedance $\overline{Z}$ is defined as

$$\overline{Z} = \frac{\overline{V}}{\overline{I}} \tag{1–51}$$

Since $\overline{Z}$ is complex, it has both real and imaginary parts and can be expressed as

$$\overline{Z} = R + jX \tag{1–52}$$

where R is the **resistance** and X is the **reactance.**

Alternately, a complex admittance $\overline{Y}$ can be expressed as

$$\overline{Y} = \frac{\overline{I}}{\overline{V}} = \frac{1}{\overline{Z}} \tag{1–53}$$

The rectangular form of $\overline{Y}$ is

$$\overline{Y} = G + jB \tag{1–54}$$

where G is the **conductance** and B is the **susceptance.**

The three basic circuit parameters are **resistance, capacitance,** and **inductance.** In steady-state ac circuit analysis, these three parameters are represented as complex impedances. The forms of the three impedances are given in Figure 1–15 as functions of the **radian frequency** . The radian frequency ω in radians/second (rad/s) is related to the **cyclic frequency** f in hertz (Hz) by

$$\omega = 2\pi f \tag{1–55}$$

Referring to Figure 1–15, observe that resistance is represented by an impedance

$$\overline{Z}_R = R \tag{1–56}$$

which is independent of frequency. Capacitance is represented by a complex impedance $\overline{Z}_C$ given by

$$\overline{Z}_C = \frac{1}{j\omega C} = \frac{-j}{\omega C} \tag{1–57}$$

Finally, inductance is represented by a complex impedance $\overline{Z}_L$ given by

$$\overline{Z}_L = j\omega L \tag{1–58}$$

To solve an ac circuit problem using phasor analysis, convert all sinusoidal voltages and currents to phasor forms and passive components to their complex impedance forms. Then apply algebraic procedures using complex numbers, and determine the desired phasor voltages and currents.

Most ac circuit analysis performed at a more elementary level is based on the assumption of a single-frequency sinusoid so that the radian frequency ω (or f) is a constant. In

Instantaneous form   Phasor form

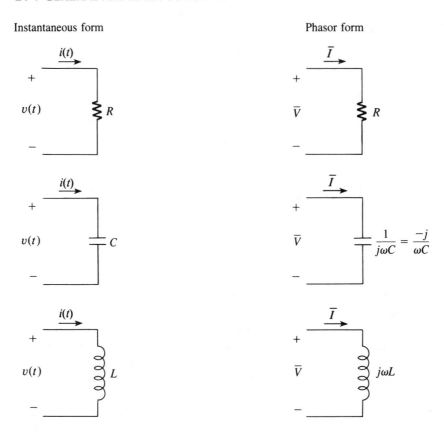

FIGURE 1–15
Passive circuit parameters and their steady-state phasor forms.

steady-state frequency response analysis, however, ω (or f) is considered to represent any arbitrary frequency at which the response is desired. Thus, ω is treated as a variable and the frequency-dependent form of the corresponding response is determined. In this manner, such questions as the gain variation of an amplifier as a function of frequency can be investigated.

Spectral Analysis

The concept of frequency response analysis is based on the *frequency spectrum* representation of signals. All analog signals can be thought of as a combination of sinusoidal components at different frequencies. The analytical basis for this concept is *Fourier analysis,* in which the frequencies comprising a signal may be determined. The corresponding distribution of the frequency components is called the *frequency spectrum.* This information is important in linear electronic circuit design so that a device having a frequency characteristic compatible with the signal spectrum can be selected. For example, audio signals such as music have spectral components ranging from well below 100 Hz to more than 15 kHz. Video waveforms, in contrast, have spectral components extending to the range of several megahertz.

Transfer Function

Consider a linear electrical circuit having one or more frequency-dependent parameters. Assume an instantaneous input voltage $v_i(t)$ and an instantaneous output voltage $v_o(t)$ as shown by the block diagram in Figure 1-16(a). In order to define an input-output function with frequency-dependent elements, the circuit is first converted to steady-state form. The input voltage is represented by a phasor $\overline{V}_i$, and the output voltage is represented by a phasor $\overline{V}_o$ as shown in Figure 1-16(b). Parameters within the circuit are represented by their impedance forms with $j\omega$ considered as a variable. The phasor output is determined in terms of the phasor input. A complex function $H(j\omega)$ can then be defined as

$$H(j\omega) = \frac{\overline{V}_o}{\overline{V}_i} \qquad (1\text{-}59)$$

The function $H(j\omega)$ is called the **steady-state transfer function** for the circuit. The steady-state transfer function is a generalization of the gain to the case where frequency-dependent elements are present, and it provides a mathematical basis for the output/input ratio. The argument $j\omega$ delineates the fact that this function varies with the frequency. Although $H(j\omega)$ is a complex function, it is not necessary to put a bar above it, since the argument $j\omega$ identifies the function as being complex.

Since $H(j\omega)$ is a complex function, it can be expressed in polar form as

$$H(j\omega) = M(\omega)\underline{/\theta(\omega)} \qquad (1\text{-}60)$$

where $M(\omega) = |H(j\omega)|$ is the magnitude or amplitude of the complex function, and $\theta(\omega)$ is the phase shift for the same function. The function $M(\omega)$ will be denoted as the **amplitude** (or **magnitude**) response in this text, and $\theta(\omega)$ will be denoted as the **phase response.** Both $M(\omega)$ and $\theta(\omega)$ are real, so the j is dropped from the argument in accordance with common practice. In actual practice, it is often convenient to express a given frequency response function in terms of the cyclic frequency $f = \omega/2\pi$, as will be demonstrated later. However, the basic mathematical forms of $M(\omega)$ and $\theta(\omega)$ will be retained in the functional notation.

In circuits where it is desired to emphasize the concept of amplifier gain, a symbol such as $A(j\omega)$ will be used to represent the frequency-dependent gain function. The more general symbol $H(j\omega)$ is used to emphasize that the input-output relationship may be different from that of a simple gain function.

Example 1-6 reviews some of the manipulations of ac circuit analysis. We will look at detailed examples of frequency response following explanation of the one-pole low-pass model in the next section.

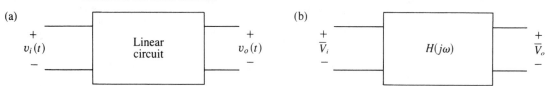

FIGURE 1-16
Development of steady-state transfer function concept.

EXAMPLE
1–6

As a review of ac circuit analysis procedures, convert the circuit of Figure 1–17(a) to steady-state phasor form if the frequency of the source is $f = 1$ kHz. Solve for the phasor values of the current and voltage across the capacitor.

Solution
The radian frequency ω is

$$\omega = 2\pi f = 2\pi \times 10^3 = 6283.2 \text{ rad/s} \qquad \textbf{(1–61)}$$

The inductive impedance $\overline{Z}_L$ is

$$\overline{Z}_L = j\omega L = j6283.2 \times 79.58 \times 10^{-3} = j500 \text{ } \Omega \qquad \textbf{(1–62)}$$

The capacitive impedance $\overline{Z}_C$ is

$$\overline{Z}_C = \frac{1}{j\omega C} = \frac{-j}{6283.2 \times 0.2274 \times 10^{-6}} = -j700 \text{ } \Omega \qquad \textbf{(1–63)}$$

The resistive impedance is, of course, $\overline{Z}_R = 150$. Assuming a sine function as the phase reference and utilizing its peak value as the phasor magnitude, the steady-state ac circuit model is shown in Figure 1–17(b).
    The impedance $\overline{Z}$ seen by the source is

$$\overline{Z} = j500 - j700 + 150 = 150 - j200 = 250\underline{/-53.13°} \text{ } \Omega \qquad \textbf{(1–64)}$$

where both rectangular and polar forms have been given.

FIGURE 1–17
Circuit of Example 1–6.

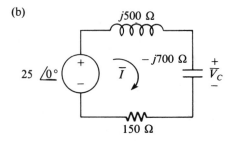

The phasor current $\bar{I}$ is

$$\bar{I} = \frac{25\underline{/0°}}{250\underline{/-53.13°}} = 0.1\underline{/53.13°}\text{A} \qquad (1\text{-}65)$$

The voltage across the capacitor $\bar{V}_C$ is

$$\bar{V}_C = \bar{Z}_C\bar{I} = (-j700)(0.1\underline{/53.13°}) = (700\underline{/-90°})(0.1\underline{/53.13°}) = 70\underline{/-36.87°}\text{ V} \qquad (1\text{-}66)$$

1-9   ONE-POLE LOW-PASS MODEL

In the previous section, the techniques for predicting the frequency response of a circuit were discussed. The steady-state transfer function, the amplitude response, and the phase response were defined. These techniques may be applied to virtually all linear electronic circuits.

A general method known as **Bode plot analysis** (named after Hendrik W. Bode) permits determination of amplitude and phase responses for rather complex circuits by some simplified graphical procedures. A complete treatment of Bode plot analysis is not within the objectives of this book, but most intermediate- and advanced-level circuit and network books have details on this procedure.

The most common frequency response form arising in linear integrated circuits is a function that could be appropriately labeled in Bode plot analysis as the **one-pole low-pass model.** Most operational amplifiers, for example, exhibit this characteristic over a wide frequency range. Because of its importance, this form will be developed and discussed in some detail in this section. In this manner, those readers not proficient in Bode plot analysis should acquire enough insight to deal with the one-pole low-pass form whenever it occurs. Other specific frequency response forms will be developed within the text as required. Readers who have proficiency in Bode plot analysis will be able to apply that background whenever frequency response considerations arise.

RC Low-Pass Circuit

To develop an intuitive feeling for the type of physical parameters that produce the response in question, the passive circuit of Figure 1-18(a) will be used as the starting point in the development. The input is $v_i(t)$, and the output is $v_o(t)$. The circuit is converted to steady-state phasor form in Figure 1-18(b). The instantaneous voltages are replaced by the phasors $\bar{V}_i$ and $\bar{V}_o$, and the capacitance C is replaced by the capacitive impedance $1/j\omega C$.

The circuit is analyzed most easily by the voltage divider rule. The output phasor $\bar{V}_o$ is readily determined as

$$\bar{V}_o = \frac{1/j\omega C}{R + (1/j\omega C)} \times \bar{V}_i \qquad (1\text{-}67\text{a})$$

$$= \frac{1}{1 + j\omega RC} \times \bar{V}_i \qquad (1\text{-}67\text{b})$$

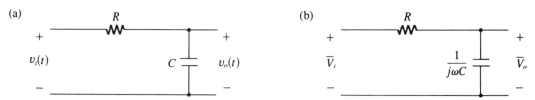

FIGURE 1–18
RC circuit used to develop the form of a one-pole low-pass model.

The steady-state transfer function is

$$H(j\omega) = \frac{\overline{V}_o}{\overline{V}_i} = \frac{1}{1 + j\omega RC} \tag{1-68}$$

The transfer function of (1–68) is a form of a one-pole low-pass function. It is recognized as such by the presence of a constant numerator (unity in this case) and a denominator of the form $1 + j\omega\tau$, where $\tau = RC$ is the time constant for the circuit.

The amplitude response corresponding to (1–68) is determined by computing the magnitude of $H(j\omega)$. The magnitude of the ratio of two complex numbers is the ratio of the two magnitudes. Thus, the amplitude response $M(\omega)$ is determined as

$$M(\omega) = \frac{1}{\sqrt{1 + (\omega RC)^2}} \tag{1-69}$$

The quantity $M(\omega)$ can be thought of in much the same manner as the gain A of the ideal amplifier, but in this case $M(\omega)$ is a function of the frequency of a signal.

The phase response is determined by computing the phase angle of $H(j\omega)$. The net phase angle corresponding to the ratio of two complex numbers is the numerator angle minus the denominator angle. The net phase response $\theta(\omega)$ is thus

$$\theta(\omega) = -\tan^{-1} \omega RC \tag{1-70}$$

Bode Plot

The graphical form of the amplitude response will be investigated next. In Bode plot analysis, it is standard practice to use a semilog scale and to plot the decibel form of the amplitude response on the linear scale as a function of frequency on the logarithmic scale. Let

$$M_{dB}(\omega) = 20 \log_{10} M(\omega) \tag{1-71}$$

in accordance with the earlier definition of decibel gain.

The form of $M_{dB}(\omega)$ on a semilog scale is given in Figure 1–19. The abscissa f/f_b (or ω/ω_b) needs some clarification. From (1–69), it can be determined that $M(\omega) = 1/\sqrt{2}$ when $\omega RC = 1$. This value corresponds to the point where $M_{dB}(\omega)$ −3.01 dB (or −3 dB rounded off). This frequency, a very convenient reference frequency in Bode plot analysis, is called the **break frequency** or **corner frequency** for reasons that will be clear shortly.

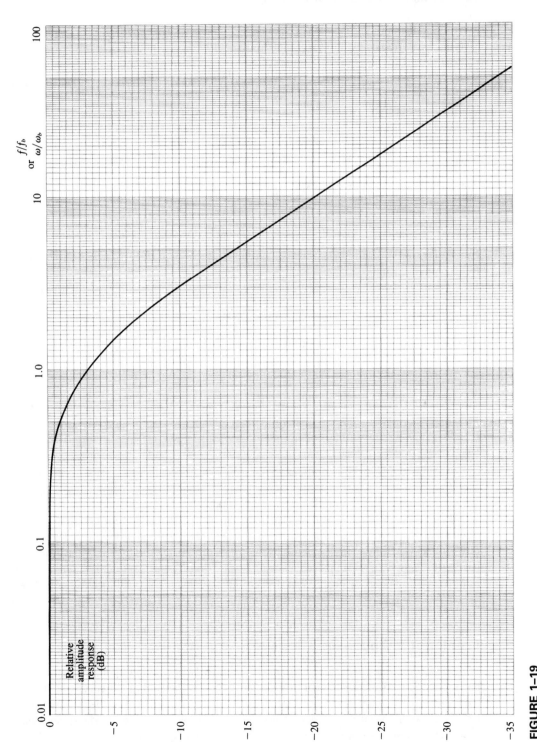

FIGURE 1–19
Relative decibel amplitude response of one-pole low-pass model.

Let (ω_b) represent the break frequency in radians/second, and let $f_b = \omega_b/2\pi$ represent the break frequency in hertz. Since $\omega_b RC = 1$ at this point, we have

$$\omega_b = \frac{1}{RC} \tag{1-72}$$

and

$$f_b = \frac{1}{2\pi RC} \tag{1-73}$$

The quantity ωRC can then be expressed in either of the forms

$$\omega RC = \frac{\omega}{\omega_b} = \frac{f}{f_b} \tag{1-74}$$

Using the ratios defined in (1–74), it is convenient to redefine $H(j\omega)$ in either of the forms

$$H(j\omega) = \frac{1}{1 + j\dfrac{\omega}{\omega_b}} = \frac{1}{1 + j\dfrac{f}{f_b}} \tag{1-75}$$

The magnitude and phase functions can be expressed as

$$M(\omega) = \frac{1}{\sqrt{1 + \left(\dfrac{\omega}{\omega_b}\right)^2}} = \frac{1}{\sqrt{1 + \left(\dfrac{f}{f_b}\right)^2}} \tag{1-76}$$

$$\theta(\omega) = -\tan^{-1}\frac{\omega}{\omega_b} = -\tan^{-1}\frac{f}{f_b} \tag{1-77}$$

In Figure 1–19, the amplitude response is nearly constant at a 0-dB level (corresponding to an absolute gain of unity) in the frequency range $f \gg f_b$. Conversely, in the frequency range $f \ll f_b$, the amplitude response approaches a straight line with a slope of -6 dB/octave or -20 dB/decade. A change in one octave corresponds to a doubling of the frequency; thus, each time the frequency is doubled, the response drops 6 dB. This property can be verified on Figure 1–19 in the upper frequency range where the response becomes a straight line. At the frequency $f/f_b = 1$ or $f = f_b$, the response is down by 3 dB as previously noted. A change in one decade corresponds to a tenfold increase in the frequency.

Break-point Approximations

With Bode plot analysis, the actual curve of Figure 1–19 is often approximated by the break-point approximation shown in Figure 1–20. With this simplified curve, the amplitude response is assumed to be constant at a 0-dB level at all frequencies in the range $f < f_b$. At $f = f_b$, the curve "breaks" at a slope of -6 dB/octave as shown. The significance of the terms *break frequency* and *corner frequency* is readily apparent.

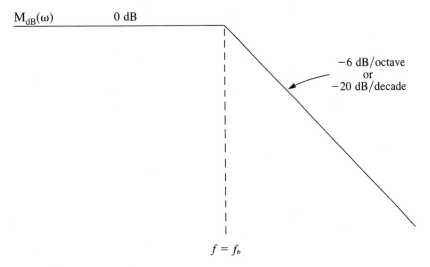

FIGURE 1–20
Break-point approximation for amplitude response of one-pole low-pass model.

While the break-point approximation is not exact, it is a quick, simplified way of sketching the rough form of an amplitude response. Further, it is relatively easy to "correct" the approximation by putting in the true values of the response when higher accuracy is required. For example, the true response is -3 dB at $f = f_b$, and this point represents the largest error in the break-point approximation. At $f = 0.1f_b$ and $f = 10f_b$, the true response is only 0.043 dB below the straight-line approximation.

The phase response $\theta(\omega)$ corresponding to the one-pole low-pass function is shown in Figure 1–21. At very low frequencies relative to the break frequency, there is a small phase lag (negative phase shift). As the frequency increases, the phase lag increases. At the break frequency $f = fb$, the phase shift is exactly $-45°$. As the frequency continues to increase, the phase shift eventually approaches $-90°$.

A break-point approximation for the phase response is not quite as simple as for the amplitude response. In fact, several variations appear in the literature. The most common one is shown in Figure 1–22. The approximation is made that the phase response is $0°$ up to $f = 0.1f_b$. At that point, a straight line intersecting $-45°$ at $f = fb$ is drawn, and the curve reaches $-90°$ at $f = 10f_b$. The value of $-90°$ is then assumed to hold for all higher frequencies. The worst errors occur at the two frequencies where the curve changes slope. At $f = 0.1f_b$, the actual phase shift is $-5.71°$; and at $f = 10f_b$, the actual phase shift is $-84.29°$. The error magnitude at both points is $5.71°$.

Extension to Amplifiers

The simple *RC* circuit has been used in the development because it displays the one-pole low-pass characteristic in the most simplified and intuitive way. As the frequency increases, the capacitive reactance decreases and more current is drawn from the source. A high-

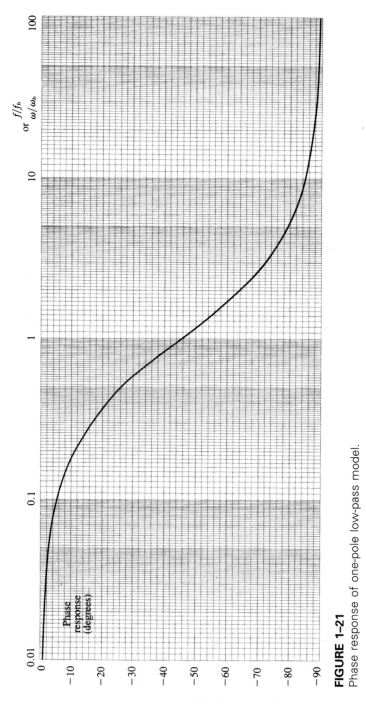

FIGURE 1-21
Phase response of one-pole low-pass model.

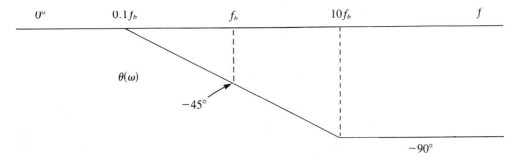

FIGURE 1–22
Break-point approximation for the phase response of one-pole low-pass model.

er portion of the input signal voltage is dropped across the series resistance, and a lower portion reaches the output. A "roll-off" effect results. Simultaneously, a larger phase shift occurs between input and output.

Many complex linear integrated circuits have frequency response functions of the general form of the RC circuit, at least over a major frequency range of interest. In such cases, it may not be possible actually to identify the values of R and C that cause the effect because these parameters are internal to the chip. Instead, the break frequency is identified from appropriate measurements or specifications.

Consider an amplifier whose frequency response is dominated by a one-pole low-pass model. The general form of the transfer function can be represented as

$$H(j\omega) = \frac{A_o}{1 + j\dfrac{\omega}{\omega_b}} = \frac{A_o}{1 + j\dfrac{f}{f_b}} \tag{1–78}$$

The form of (1–78) is the same as that of (1–75) except that the numerator is now A_o instead of unity, a result of the fact that amplification is assumed in the present case. Consequently, the amplitude response of (1–78) is of the same form as previously developed in this section except for one change. The maximum amplitude level is now A_o instead of 1, so the maximum decibel level is now $20 \log 10 A_o$ instead of 0 dB. As a result of the additive property of logarithms, the only effect is to shift the decibel level by a constant value. Indeed, Figures 1–19 and 1–20 may be interpreted for a function of the form of (1–78) as the decibel level *relative* to maximum. Thus, at $f = fb$, the response is 3 dB below the maximum gain. The phase response is unaffected by the constant A_o and is exactly the same as for the simple RC circuit, provided that A_o is positive.

The intuitive approach to the break-point approximation is based on first observing (1–78) and noting the values of A_o and f_b. The break-point approximation for the amplitude response is then constructed as explained earlier. Note, however, that the transfer function must be in the correct form of (1–78), including the constant of unity in the denominator. This process will be illustrated in the examples that follow.

EXAMPLE 1–7

The frequency response of a certain amplifier can be represented by a one-pole low-pass model. Some analysis on the amplifier results in the following transfer function:

$$H(j\omega) = \frac{1200}{20 + j10^{-3}\omega} \tag{1–79}$$

(a) Determine the low-frequency gain, the 3-dB break frequency in radians/second, and the corresponding value in hertz. (b) Sketch the break-point approximation for the amplitude and phase response functions.

Solution

(a) Although the constant numerator and the constant plus a $j\omega$ term in the denominator indicate a one-pole low-pass model, the function is not quite in the proper form. Inspection of (1–78) reveals that the constant in the denominator must have a unity value, a condition readily achieved by dividing both numerator and denominator by 20. Thus,

$$H(j\omega) = \frac{60}{1 + j5 \times 10^{-5}\omega} \tag{1–80}$$

Once the denominator has a constant term of unity as in (1–80), the value of ω_b, the break frequency in radians/second, is obtained by setting the factor of j to be unity; that is, $5 \times 10^{-5}\ \omega = 1$. The value of ω that satisfies this equation is ω_b. Thus,

$$5 \times 10^{-5}\omega_b = 1$$

or

$$\omega_b = \frac{1}{5 \times 10^{-5}} = 2 \times 10^4 \text{ rad/s} \tag{1–81}$$

The corresponding value of f_b, the break frequency in hertz, is

$$f_b = \frac{\omega_b}{2\pi} = \frac{2 \times 10^4}{2\pi} = 3183 \text{ Hz} \tag{1–82}$$

The forms ω/ω_b and f/f_b are then readily expressed as $\omega/\omega_b = \omega/2 \times 10^4$ and $f/f_b = f/3183$. The transfer function can be written in the form of (1–78) as

$$H(j\omega) = \frac{60}{1 + \dfrac{j\omega}{2 \times 10^4}} \tag{1–83a}$$

$$H(j\omega) = \frac{60}{1 + \dfrac{jf}{3183}} \tag{1–83b}$$

With some practice, it is possible to manipulate the denominator of (1–80) directly into the form of either (1–83a) or (1–83b) without performing all the steps in this example.

The low-frequency gain A_o is the constant numerator in the transfer function of either (1–83a) or (1–83b). This value is readily seen to be

$$A_o = 60 \qquad\qquad\qquad (1\text{–}84)$$

The corresponding decibel value is $20 \log_{10}60 = 35.56$ dB.

(b) From a knowledge of the low-frequency gain and the break frequency, the breakpoint approximations to the amplitude and phase functions are readily sketched as shown in Figures 1–23 and 1–24. While either ω or f can be used as the horizontal variable, f has been chosen here due to its more common practical usage. If ω had been used, the curves would have the same forms, but the break frequency would be 2×10^4 rad/s. Although these curves are approximations, they may be modified to determine exact values in accordance with the general procedures and exact curves given in this section.

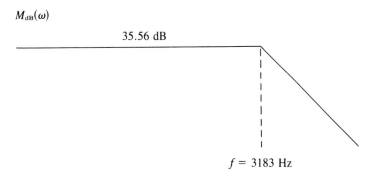

FIGURE 1–23
Bode plot amplitude response approximation of the amplifier in Example 1–7.

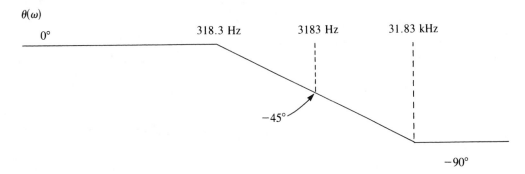

FIGURE 1–24
Bode plot phase response approximation of the amplifier in Example 1–7.

EXAMPLE 1–8

A certain amplifier has a frequency response of the one-pole low-pass type. Specifications indicate that the low-frequency gain is 40 dB, and the 3-dB break frequency is 200 kHz. Write an expression for H($j\omega$) both in terms of f and in terms of ω.

Solution

The first step is to convert the low-frequency decibel gain to an absolute gain. The value 40 dB is converted to $A_o = 100$. The values of A_o and f_b are then substituted in (1–78), and the transfer function is expressed in terms of cyclic frequency as

$$H(j\omega) = \frac{100}{1 + \dfrac{jf}{2 \times 10^5}} \qquad (1\text{–}85)$$

Since $f_b = 2 \times 10^5$ Hz, $\omega_b = 2\pi \times 2 \times 10^5 = 4\pi \times 10^5$ rad/s. The transfer function can then be expressed in terms of radian frequency as

$$H(j\omega) = \frac{100}{1 + \dfrac{j\omega}{4\pi \times 10^5}} \qquad (1\text{–}86)$$

1–10 MULTISIM EXAMPLES

The last section of each chapter will be devoted to a collection of Multisim examples, most of which involve adapting earlier analytical problems for computer solutions using Multisim. Each example is self-contained, so there will be very little general discussion preceding the examples in most cases.

The self-contained discussion within each example will assume some introductory familiarity with Multisim (or Electronics Workbench). Readers not having this background should first refer to Appendix E and read it carefully. That information, along with further details provided in each example, should be sufficient to deal with the examples in the text.

MULTISIM EXAMPLE 1–1

For the circuit of Example 1–4 (Figure 1–12), develop a Multisim model and use it to determine the various node voltages and branch currents in the circuit when the open-circuit source voltage is 0.2 V.

Solution

The circuit model of Figure 1–12(b) is constructed in the Multisim circuit window using the techniques of Appendix E, and the result is shown in Figure 1–25. The circuit contains five nodes, and they are numbered as 0 through 4. The input voltage is measured at node 1 and the output voltage is measured at node 4.

The resistors are obtained from the **Basic** parts bin, which is the second one from the top. It is noted that the two top buttons in this bin both represent resistors. The one on the left utilizes standard values, so if it is left-clicked, an additional window opens for specifying the value. The resistor button on the right is labeled with **virtual,** which is used to

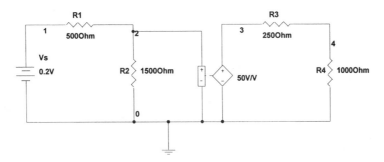

FIGURE 1–25
Circuit of Multisim Example 1–1.

indicate that any value can be set with this resistor. The value is set after the component is brought to the circuit area, dropped by one left-click, and selected for modification by an additional double left-click. While some of the resistance values in this particular circuit are standard values, for consistency, the virtual resistors were used for all cases.

Since there are no reactive components in the circuit, either a dc or an ac analysis could theoretically be employed. When such a choice is available, it is usually simpler to use dc, and this is the choice made in this case.

The input voltage source is obtained from the **Sources** bin (top one), and it has the appearance of a battery. Double-clicking on it opens a properties window, and the value is set to 0.2 V. The label is also changed to correspond to the text circuit.

A voltage-controlled voltage source (VCVS) is required, and it can also be obtained from the **Sources** parts bin. The left-hand terminals of the VCVS are connected to the control node numbers (2 and 0), with the upper node being the positive reference. Likewise, the right-hand terminals of the VCVS are the output nodes with the upper node being the positive reference. The voltage gain A is set at 50. While this value is actually dimensionless, Multisim lists the units as volts per volt (V/V), meaning the same thing.

The dc analysis is performed by first left-clicking on the **Analyses** button on the **Design Bar** (fifth button from right). A window providing different analysis options will then open, and a dc analysis is performed by left-clicking on the **DC Operating Point** button. An alternate path to get to the same point is to first left-click on **Simulate** on the **Menu Row,** followed by a left-click on **Analyses.**

When the **DC Operating Point** button is clicked, a new window with the title **DC Operating Point Analysis** opens, from which a number of choices can be made. The default tab position is with the upper tab set to **Output variables,** and the possible variables that can be tabulated are the node voltages and the currents through the voltage sources. For this circuit, there are six variables. The numbers 1 through 4 represent the four node voltages and the bottom two variables represent the currents through the two sources. The desired variables must be highlighted. Since more than one is to be read, hold the computer **Ctrl** button down while left-clicking on each of the six quantities. Next, left-click on the block entitled **Plot during simulation.** The numbers will then appear in the right-hand block. Finally, left-click on **Simulate** and the window shown in Figure 1–26 appears. This is a compilation of the output data.

DC Operating Point

1	200.00000m
2	150.00000m
3	7.50000
4	6.00000
e#branch	−6.00000m
vvs#branch	−100.00000u

FIGURE 1–26
Output data of Multisim Example 1–1.

The first four values in the table represent the four node voltages in volts. Note that the first two have the letter m at the end, representing millivolts. These values can be readily verified from the analysis of Example 1–4.

The negative signs for the two currents may seem strange and need some explanation. In the original SPICE program, currents could only be measured directly through voltage sources. The reference direction of conventional current flow for any SPICE component is into the positive terminal and out of the negative terminal. Since the actual currents in this circuit flow out of the positive source terminals, they appear with negative values. The one labeled **e#branch** is the controlled source current and it can be interpreted as a clockwise current of 6 mA. The second one, labeled **vvs#branch,** is the intput source current and the u refers to "micro." Thus, this current is 100 μA in a clockwise direction.

MULTISIM EXAMPLE 1–2

To illustrate the one-pole roll-off response, design a simple RC low-pass filter with a break frequency of 1 Hz. Plot the decibel amplitude and phase responses on semilog scales from 0.01 Hz to 100 Hz.

Solution
The transfer function for a one-pole RC low-pass circuit was derived in Section 1–9 and was given in (1–68) as

$$H(j\omega) = \frac{1}{1 + j\omega RC} \tag{1–87}$$

The break frequency in hertz was given in (1–73) as

$$f_b = \frac{1}{2\pi RC} \tag{1–88}$$

By specifying that $f_b = 1$ Hz, the product RC required to achieve this break is determined from (1–88) as

$$RC = \frac{1}{2\pi} \tag{1–89}$$

Since there are two variables, one must be arbitrarily selected, and the other can be determined from (1–89). Thus, there are an infinite number of possible solutions. As one possible solution, we will select $R = 10^6 \ \Omega = 1 \ M\Omega$ and $C = 0.159 \ \mu F$.

This circuit requires one source and two components. Since a frequency response is to be determined, an ac voltage source is obtained from the **Sources** bin. An ac voltage source displays a sine wave on the button along with a + sign. When it is opened in the circuit window, the default voltage and frequency values are 1 V and 1000 Hz. The voltage value is left at 1 V, but since a wide range of frequencies are to be employed, the label for the frequency is removed. (Actually, a frequency response analysis automatically sets the input voltage to 1 V irrespective of the value set for the source.) Finally, the title of the source is changed from v_1 to v.

Both the resistor and the capacitor are obtained from the **Basic** parts bin. As in the case of the resistor, the capacitor contains both a standard component (on the left) and a **virtual** component on the right. The latter capacitor is used in this case since the exact value desired is not a standard one. The final circuit is shown in Figure 1–27.

The frequency response analysis is performed by first left-clicking on the **Analyses** button on the **Design** bar (fifth button from right). A window providing the analysis options will then open, and an ac analysis is accomplished by a left-click on the **AC Analysis** button. An alternate path to get to the same point is to first left-click on **Simulate** on the **Menu Row,** followed by a left-click on **Analyses.**

When the **AC Analysis** button is clicked, a new window with the same title opens. The default tab position is for **Frequency Parameters.** The **Start frequency** is set at 0.01 Hz, and the **Stop frequency** is set at 100 Hz. The **Sweep type** is set as **decade.** This choice is the most convenient to use for a wide-range frequency sweep such as this. The **Number of points per decade** is arbitrarily set at 100. Finally, the **Vertical** scale is set to **Decibel.**

The next step is to click on the **Output variables** tab, and possible variables that can be tabulated are the two node voltages and the current through the voltage source. The desired output variable is the voltage at node 2, and thus the value 2 should be highlighted. Next, left-click on the block entitled **Plot during simulation.** The number will then appear in the right-hand block. Finally, left-click on **Simulate** and graphs of the amplitude and phase response functions will appear on the screen.

When the plots first appear, the titles at the top are the file name (**1-2** in this case) and **AC Analysis.** The top plot will initially be in the select mode, which means that the title and axes labels may be modified. A small tab on the left-hand side of the screen indi-

FIGURE 1–27

Circuit of Multisim Example 1–2.

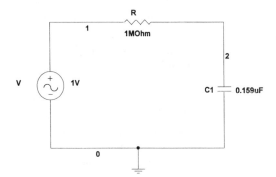

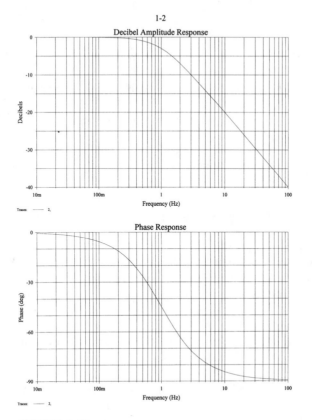

FIGURE 1–28

Frequency response functions of Multisim Example 1–2.

cates this. Modifications in labeling can be achieved by first left-clicking on the fifth button from the right and following the menu options given there. Initially, the default tab will indicate **General** and a new **Title** may be entered. For this particular plot, the title was changed to "Decibel Amplitude Response." Next, left-click on the **Left Axis** tab. In this case a **Label** block appears. At the time of this writing, the vertical axis appeared as **Voltage (V).** The units should be in decibels, so this represented a "bug" in Version 6.21. This should be corrected in later versions. The label was changed to "Decibels." Since the input voltage was 1 V, this may be interpreted as the amplitude or magnitude response in decibels. A grid may also be added by left-clicking on the tenth button from the right.

The bottom plot may be selected for label modification by left-clicking anywhere in the area. The procedures previously discussed may now be repeated for this plot as desired. This graph is given the title "Phase Response." The final response functions with proper labeling are shown in Figure 1–28. These curves obtained with Multisim may be compared with Figures 1–19 and 1–21, respectively.

SUMMARY

1. An ideal linear amplifier increases the level of a signal while preserving its exact shape.
2. An electronic signal source can be modeled in terms of an ideal voltage or current source in combination with a resistance (or possibly a complex impedance).
3. Four possible types of controlled (or dependent) sources may be used to describe control mechanisms in linear electronic circuits: voltage-controlled voltage source, voltage-controlled current source, current-controlled voltage source, and current-controlled current source.
4. A realistic linear amplifier at low frequencies may be described from a knowledge of three parameters: open-circuit voltage gain, input impedance, and output impedance.
5. The loaded gain of an amplifier with additional source and load impedances is determined by multiplying the unloaded gain by the input loading factor and the output loading factor.
6. The loaded gain of two or more amplifiers in cascade is determined by the process of (5) multiplied by the interaction factors between successive stages.
7. Decibel gain measures are used to express input/output ratios on a logarithmic basis.
8. Both decibel power and decibel voltage definitions are employed, and they reduce to the same values when input and load impedances are equal.
9. The decibel gain of a cascade of amplifier stages is determined by adding the individual decibel gains.
10. Steady-state ac analysis is performed by replacing all circuit elements by their complex impedance forms and replacing sinusoidal sources with phasors.
11. Impedance is the complex sum of resistance and reactance, and admittance is the complex sum of conductance and susceptance.
12. When the frequency is assumed to be a variable, the output/input ratio of a system can be described by a steady-state transfer function.
13. The magnitude of the steady-state transfer function as a function of frequency is called the magnitude or amplitude response, and the angle is called the phase response.
14. The frequency response of many amplifier circuits can be described over a wide frequency range by a function called the one-pole low-pass model.
15. Bode plot curves with straight line segments can be used to closely approximate the amplitude and phase curves for the one-pole low-pass model.

PROBLEMS

Drill Problems

1–1. For the waveforms of Figure 1–3 (Example 1–1), suppose that the vertical sensitivities of channels 1 and 2 had been 0.1 V/division and 5 V/division, respectively. Determine the voltage gain.

1–2. The input and output voltages of a certain amplifier are measured with an rms voltmeter, and the readings are 0.2 V and 4 V, respectively. Determine the voltage gain.

1–3. The open-circuit voltage of a certain transducer output is 50 V, and the equivalent output resistance is 10 kΩ. Determine two circuit models for the source.

1–4. The open-circuit voltage of a certain electronic device is 5 V. When a resistance of 8 kΩ is connected across the output, the voltage drops to 4 V. Assuming that the output impedance is resistive, determine two circuit models for the source.

1–5. An ideal VCVS of the form shown in Figure 1–7(a) has $A = 20$. Determine v_o if $v_i = 0.4$ V.

1–6. An ideal VCIS of the form shown in Figure 1–7(b) has $g_m = 40$ mS. Determine i_o if $v_i = 0.5$ V.

1–7. An ideal ICVS of the form shown in Figure 1–7(c) has $R_m = 2\ k\Omega$. Determine v_o if $i_i = 6$ mA.

1–8. An ideal ICIS of the form shown in Figure 1–7(d) has $\beta = 12$. Determine i_o if $i_i = 3$ mA.

1–9. An ideal VCVS of the form shown in Figure 1–7(a) produces an output voltage of V when the input voltage is 20 mV. Determine the voltage gain A.

1–10. An ideal VCIS of the form shown in Figure 1–7(b) produces an output current of 2 mA when the input voltage is 40 μV. Determine the transconductance g_m.

1–11. An ideal ICVS of the form shown in Figure 1–7(c) produces an output voltage of 8 V when the input current is 2 mA. Determine the transresistance R_m.

1–12. An ideal ICIS of the form shown in Figure 1–7(d) produces an output current of 12 mA when the input current is 60 μA. Determine the current gain β.

1–13. A certain linear amplifier has the following specifications:

$$\text{Input impedance} = 5\ k\Omega \qquad (\textit{resistive})$$
$$\text{Output impedance} = 100\ \Omega \qquad (\textit{resistive})$$
$$\text{Open-circuit voltage gain} = 200$$

Draw an equivalent circuit for the amplifier.

1–14. A certain linear amplifier has the following specifications:

$$\text{Input impedance} = 2\ k\Omega \qquad (\textit{resistive})$$
$$\text{Output impedance} = 200\ \Omega \qquad (\textit{resistive})$$
$$\text{Open-circuit voltage gain} = -60$$

Draw an equivalent circuit for the amplifier. (Note the sign of the gain.)

1–15. The equivalent circuit of a certain linear amplifier is shown in Figure P1–15. Determine the following: **(a)** input impedance; **(b)** output impedance, and **(c)** open-circuit voltage gain.

FIGURE P1–15

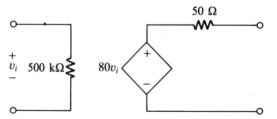

1–16. The equivalent circuit of a certain linear amplifier is shown in Figure P1–16. Determine the following: **(a)** input impedance; **(b)** output impedance; **(c)** open-circuit voltage gain. (Note the sign of the voltage gain. What does it imply?)

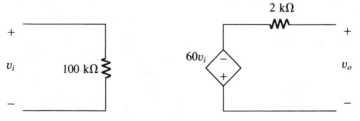

FIGURE P1–16

1–17. A signal source having an open-circuit voltage of 30 mV and an internal output resistance of 2.5 kΩ is connected to the input of the amplifier circuit of Problem 1–13, and a 300-Ω resistive load is connected across the output.

    a. Draw the complete equivalent circuit.

    b. Calculate the net loaded voltage gain A_{so} between the open-circuit source voltage and the load.

    c. Calculate the amplifier input and output voltages from the circuit model under loaded conditions.

    d. Use the result of part (b) to predict the output voltage directly, and compare it with the value obtained in part (c).

1–18. The amplifier of Problem 1–16 is connected in the circuit of Figure P1–18.

    a. Determine the net loaded voltage gain A_{so} between the open-circuit source voltage and load resistance under loaded conditions.

    b. If $v_s = 0.2$ V, what is the output loaded voltage v_o?

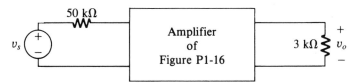

FIGURE P1–18

1–19. Two linear amplifier circuits are connected in cascade, and the source and load are connected as shown in Figure P1–19.

    a. Draw the complete equivalent circuit.

    b. Calculate the net loaded voltage gain A_{so} between the open-circuit source voltage and the load.

    c. For $v_s = 40$ mV, determine v_{i1}, v_{i2} and v_o from the circuit model.

    d. Use the result of part (b) to predict the output voltage directly, and compare it with the value obtained in part (c).

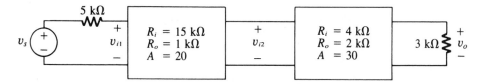

FIGURE P1–19

1–20. Three linear amplifier circuits are connected in cascade, and the source and load are connected as shown in Figure P1–20.

    a. Determine the net loaded voltage gain $A_{so} = v_o/v_s$.

    b. For $v_s = 20$ mV, determine v_o directly using the result of (a).

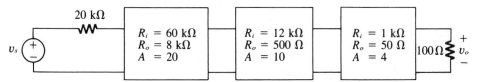

FIGURE P1–20

1–21. The following values represent the absolute voltage gains of some amplifiers (or possibly attenuators when the value is less than unity). Calculate the corresponding decibel value A_{dB} in each case. **a.** 1; **b.** 10; **c.** 100, **d.** 10^n, **e.** 10^{-n}.

1–22. The following values represent the absolute voltage gains of some amplifiers (or possibly attenuators when the value is less than unity). Calculate the corresponding decibel value A_{dB} in each case. **a.** 2; **b.** 4; **c.** 8; **d.** 600; **e.** 0.5.

1–23. The following values represent the decibel voltage gains of some amplifiers (or possibly attenuators when negative). Calculate the corresponding linear gain A in each case. **a.** 30 dB; **b.** 60 dB; **c.** 16 dB; **d.** 200 dB; **e.** −40 dB.

1–24. The following values represent the decibel voltage gains of some amplifiers (or possibly attenuators when negative). Calculate the corresponding linear voltage gain A in each case. **a.** 64 dB; **b.** 22 dB; **c.** $20n$ dB; **d.** $-20n$ dB; **e.** −100 dB.

1–25. The frequency response of a certain amplifier is dominated by a one-pole low-pass model. Some analysis on the amplifier indicates that the transfer function is

$$H(j\omega) = \frac{2000}{5 + j10^{-4}\omega}$$

a. Determine the low-frequency gain, the 3-dB break frequency in radians/second, and the corresponding value in hertz.

b. Sketch the break-point approximations for the amplitude and phase response functions.

1–26. The frequency response of a certain amplifier is dominated by a one-pole low-pass model. Some analysis on the amplifier indicates that the transfer function is

$$H(j\omega) = \frac{10^6}{5000 + jf}$$

a. Determine the low-frequency gain, the 3-dB break frequency in hertz, and the corresponding value in radians/second.

b. Sketch the break-point approximations for the amplitude and phase response functions.

1–27. A certain amplifier has a frequency response of the one-pole low-pass type. Specifications indicate that the low-frequency gain is 60 dB and the 3-dB break is 50 kHz. Write an expression for $H(j\omega)$ both in terms of f and in terms of ω.

1–28. A certain amplifier has a frequency response of the one-pole low-pass type. Specifications indicate that the low-frequency gain is 40 dB and the 3-dB break frequency is 20 kHz. Write an expression for $H(j\omega)$ both in terms of f and in terms of ω.

1–29. Consider the amplifier of Problem 1–27. *Calculate* the amplitude response $M_{dB}(\omega)$ and the phase response $\theta(\omega)$ at each of the following frequencies: **a.** 5 kHz; **b.** 25 kHz; **c.** 50 kHz; **d.** 100 kHz; **e.** 500 kHz. Use Figures 1–19 and 1–21 to check your results to within the accuracy of the curves. (Recall that the curve of Figure 1–19 is based on a 0-dB-level maximum reference, so a constant shift level must be assumed.)

1–30. Consider the amplifier of Problem 1–28. *Calculate* the amplitude response $M_{dB}(\omega)$ and the phase response $\theta(\omega)$ at each of the following frequencies: **a.** 5 kHz; **b.** 15 kHz; **c.** 20 kHz; **d.** 30 kHz; **e.** 100 kHz. Use Figures 1–19 and 1–21 to check your results to within the accuracy of the curves. (See note at end of Problem 1–29.)

Derivation Problems

1–31. Assume that an amplifier circuit is matched at both input and output; that is, the source internal resistance is equal to the input resistance, and the load resistance is equal to the output resistance. Show that the loaded gain is

$$A_{so} = \frac{A}{4}$$

1–32. For the one-pole low-pass magnitude response of (1–76), verify that for frequencies well above the break frequency: **a.** the response decreases by about 6 dB for each octave increase in frequency; and **b.** the response decreases by about 20 dB for each decade increase in frequency.

1–33. The Bode plot amplitude approximation for the one-pole low-pass model predicts a level of -20 dB relative to the maximum level at a frequency $10 f_b$.
   a. Verify that the actual level is 0.043 dB lower than the ideal -20-dB level.
   b. Verify that the actual level at a frequency $0.1 f_b$ is 0.043 dB below the ideal 0-dB level.

1–34. The Bode plot phase approximation for the one-plot low-pass model predicts phase shifts of $0°$ at $f = 0.1 f_b$ and $-90°$ at $f = 10 f_b$.
   a. Verify that the actual phase shift at $f = 0.1 f_b$ is $-5.71°$.
   b. Verify that the actual phase shift at $f = 10 f_b$ is $-84.29°$.

Design Problems

Design problems involving specific circuits will appear to a greater extent in Chapter 2 and later chapters. Only limited design-type exercises can be formulated from the material of this chapter, but here are a few.

1–35. A certain amplifier is to be selected to amplify the signal between a 1-kΩ resistance source and a 5-kΩ load. If the signal loss due to loading effects at each end is not to exceed 0.5 dB (that is, 1 dB total), determine **a.** the minimum value of input impedance, and **b.** the maximum value of output impedance.

1–36. A certain amplifier with an adjustable gain A is to be placed between a 500-Ω resistive source and a 2-kΩ resistive load. The input and output impedances of the amplifier are 1.5 kΩ and 500Ω, respectively. The open-circuit source voltage is 50 mV, and it is desired to establish 4 V across the load. Determine the required value of A.

1–37. Design an *RC* low-pass circuit to produce a 3-dB break frequency of 1 kHz.

1–38. Design an *RC* low-pass circuit to produce a phase shift of $-60°$ at 1 kHz.

2

IDEAL OPERATIONAL AMPLIFIER ANALYSIS AND DESIGN

OBJECTIVES

After completing this chapter, the reader should be able to:

- Discuss the general properties of an operational amplifier (op-amp) and its significance in modern electronics.
- Draw a diagram showing how two power supplies are used to provide power to an op-amp and indicate the reference ground terminal.
- Extract some of the basic op-amp characteristics, maximum ratings, and pin diagrams from data sheets.
- Estimate the op-amp saturation voltages from a knowledge of the power supply voltages.
- Draw the schematic symbol for an op-amp showing the three signal terminals, and draw the corresponding ideal circuit model.
- State the three assumptions used in analyzing a circuit containing an ideal op-amp and the three implications associated with the assumptions.
- Draw the circuit diagrams and derive the gains for an inverting amplifier and a noninverting amplifier.
- Draw the circuit diagram of a voltage follower circuit and derive the gain.
- Analyze an inverting or noninverting amplifier with load to determine all circuit voltages and currents.
- Design an inverting or noninverting amplifier circuit to meet simplified design objectives.
- Draw the circuit diagrams, analyze, and design three different types of voltage-controlled current sources, a current-controlled voltage source, and a current-controlled current source.
- Draw the circuit diagram, analyze, and design a linear combination circuit or a summing circuit.

- Design circuits using linear combination circuits and amplifiers to achieve arbitrary linear combinations of signals.
- Draw the circuit diagram, analyze, and design a closed-loop differential amplifier circuit.

2–1 INTRODUCTION

The most important single linear integrated circuit is the *operational amplifier.* Operational amplifiers are available as inexpensive circuit modules, and they are capable of performing a wide variety of linear and nonlinear signal-processing functions. We refer to the operational amplifier frequently throughout the book by the shortened form *op-amp.*

Both the analysis and the design of many operational amplifier circuits are amazingly simple and staightforward whenever such circuits can be assumed to be ideal. This chapter will emphasize this idealized approach, and the reader should be able to quickly gain skill and confidence in dealing with the resulting circuits. The more realistic imperfections and performance limitations thus will be delayed until the idealized approaches are clearly understood. We select actual operational amplifiers to approximate as closely as possible the ideal models, so the assumptions made here represent the desirable conditions in all circuits.

Along with the general methods of idealized analysis, many important basic circuits will be introduced. Included are the most common voltage and current amplifiers of various types, linear combination circuits (including the summing circuit), and closed-loop differential amplifier circuits. We place major emphasis throughout on the design of circuits to achieve a given set of specifications.

2–2 GENERAL APPROACH

The approach in this chapter and throughout most of the book is to focus on the operation and subsequent applications of the op-amp from the viewpoint of the *external* terminals. Ideal operation will be assumed in this chapter, but in subsequent chapters we will discuss various limitations as they affect external circuit behavior. This is the approach believed to be appropriate for the vast majority of op-amp users, since only a few find it necessary to fully understand the internal circuit details of the op-amp. In fact, users of this text in a formal electronics curriculum have probably studied the details of some of the circuits within the op-amp in a basic course in electronic devices and circuits.

Appendix D offers some background in the evolution of the modern op-amp from a basic differential amplifier. We provide a thorough analysis of a discrete component differential amplifier using small signal analysis and show some of the definitions used throughout the book as they relate to actual circuit details. Following the analysis of the differential amplifier stage, we present various circuit forms used on integrated circuit chips. Finally, the formulation of a complete op-amp is discussed. Thus, we offer Appendix D as a supplement to the text, as it provides additional material on the internal operation of an op-amp.

2–3 OPERATIONAL AMPLIFIER (OP-AMP)

The modern *operational amplifier* is a high-gain, integrated-circuit, direct-coupled amplifier capable of performing a large number of linear and nonlinear amplification and signal-processing functions. If there is any one device that could be called the building block of the analog circuit field, it would very likely be the operational amplifier.

Most analog low-power signal-processing functions below 1 MHz or so can be done more easily with op-amps than with any other devices. There is still a critical need for discrete component devices at higher frequencies and at higher power levels. However, at the time of this writing, op-amps are available that are actually usable to frequencies exceeding 50 MHz and power levels exceeding 50 W.

At the present time, the majority of applications employing op-amps tend to be in the frequency range below 1 MHz or so. With some of the least expensive op-amps, applications are often restricted to the range of a few kilohertz, but moderately expensive, high-performance types can routinely extend the range to above 1 MHz. Within the range under discussion, there are a myriad of different applications for which the op-amp is ideally suited.

The actual circuit diagram of a typical IC op-amp is shown in Figure 2–1. The obvious complexity of this diagram may be overwhelming; however, be assured that it is not necessary to understand everything that is happening on the chip to be able to use it properly. The actual circuit details were developed by the IC manufacturers over a period of time, and the purposes of all the individual components are not always clear without much research into the subject. We will learn how to employ the chip in various circuit applications and how to predict its performance characteristics from specified data.

Power Supply Connections

Most op-amps are powered from dual power supplies of opposite polarities. The most common values for modern IC op-amps are +15 V and −15 V, and many circuit development boards provide these standard voltages. However, some op-amps require voltages below these values, while others require higher voltages. It is even possible to power the op-amp with only one voltage level under special conditions (to be considered in Chapter 5). At this time, however, we will assume the popular dual symmetrical voltage connections.

The manner in which the power supply voltages for most op-amps are connected is illustrated in Figure 2–2 for the case of ±15 V. The positive terminal is connected to the positive supply voltage terminal, and the negative terminal is connected to the negative supply voltage terminal. These terminals are designated as V+ (or +V) and V− (or −V), respectively, on many pin connection diagrams.

The common ground point usually is not connected to the op-amp itself. Rather, the center point between the power supplies becomes the reference ground point, and all input signals to the op-amp have their common grounds connected to this point. Further, all output loads are connected between the op-amp signal output terminal (to be discussed later) and this common ground point.

In order to simplify detailed circuit diagrams, the power supply connections are usually omitted, and this practice will be followed extensively throughout this book. However, it should be clear to the reader that the power supply connections are required in all cases for the circuit to work, and their presence is understood.

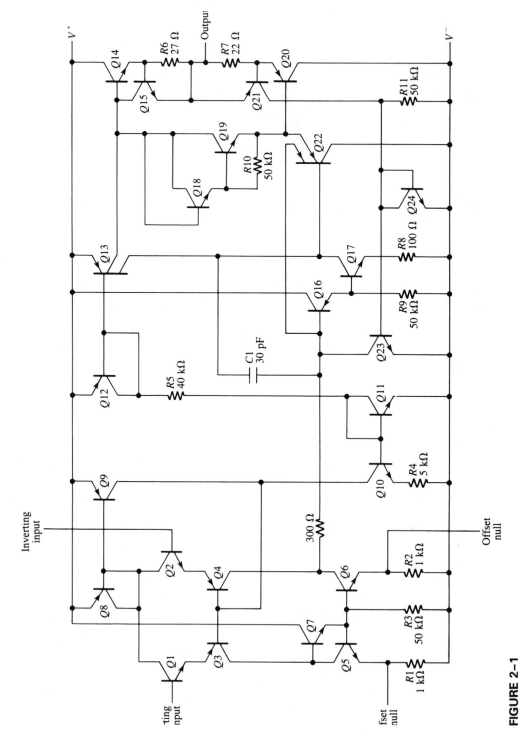

FIGURE 2–1

Schematic diagram of typical integrated circuit operational amplifier. (Courtesy of Fairchild Camera & Instrument Co.)

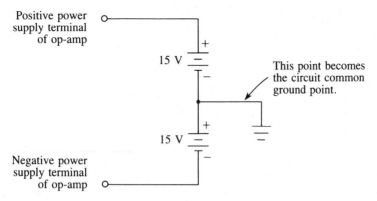

Positive power
supply terminal
of op-amp

15 V

This point becomes
the circuit common
ground point.

15 V

Negative power
supply terminal
of op-amp

FIGURE 2–2
Dual power supply connections.

While there are many different available op-amps, one particular type will be used for illustration and discussion at numerous points in the text. The op-amp of reference is the widely employed 741, which was introduced by Fairchild Semiconductor in 1968. Variations of the 741 are now produced by many different manufacturers. It is widely available, inexpensive, and internally stabilized, and it has internal short-circuit protection. While its performance characteristics are somewhat limited, as we will see later, it is capable of performing many general applications quite well.

741 Data

Refer to Appendix C for an overview of the 741 specifications. Note that Fairchild uses the designation "μA741." Note also that a number of different variations are indicated by one or more letters following the number. As a simplified approach, we will use the designation "741" throughout to refer to any of the different forms, except where it is necessary to recognize a specific property of a given form.

Do not be alarmed by the myriad of specifications given for the 741; much of this book is devoted to developing techniques for analyzing such data. A few of the basic specifications will be discussed here, and others will be investigated at various points in the text. Many of the rather detailed specifications are used in assessing the requirements for very demanding applications.

The various available packages and pin connection diagrams are delineated on the first page of the 741 data sheets. Observe that the pin numbers for different terminals of the 10-pin flatpack are different from those of the other packages.

The absolute maximum ratings for the 741 are given on the second page of the 741 data sheets. The maximum supply voltages for the 741A, 741, and 741E are given as ± 22 V, and the 741C has corresponding values of ± 18 V. The internal power dissipation ratings are a function of the type of package, and they range from 310 mW to 570 mW. Note that when the ambient temperature exceeds 70°C, it is necessary to derate or reduce these maximum ratings.

The differential input voltage rating of ± 30 V represents the maximum *difference* in voltages across the two input terminals, and the input voltage rating of ± 15 V represents the maximum levels at the two input terminals. The significance of the preceding two ratings will be clearer after an understanding of op-amp operation is acquired.

As previously noted, the most common values of op-amp dc power supply voltages are ± 15 V. However, it is actually possible to power the 741 with voltages well below these values if some degradation in performance is accepted. Observe the curves on the sixth page of the 741 data sheets in Appendix C, entitled "Open Loop Voltage Gain as a Function of Supply Voltage." (The two curves represent different versions of the 741.) The curves extend down to ± 2 V, implying that the op-amp actually will still work at these low power supply voltage levels, but, as we will see shortly, the dynamic range of the output will be severely limited. Observe also on the second page of the 741 data sheets that the absolute maximum input voltage rating must be reduced to the supply voltage levels as indicated in Note 2.

Saturation Voltages

The concept of dynamic range will be considered next. For the sake of discussion, assume power supply voltages of ± 15 V. For a purely resistive op-amp circuit, the signal at the output of the op-amp can never go above $+15$ V or below -15 V. As a matter of fact, it cannot reach these levels. As the extremes of the power supply levels are being approached by the internal circuitry, certain transistor saturation and junction drops appear between the power supply voltage and the signal output voltage terminal. The result is that the peak positive output reaches a value somewhat lower than $+15$ V, and the peak negative output reaches a value somewhat higher than -15 V. Once the limiting values are reached, the output establishes a signal saturation condition, which does not change until the controlling signal swings the signal output in the opposite direction. The signal output voltages under these conditions are called the output *saturation voltages.* In the most general case, the values may be different on positive and negative saturation, but for symmetrical power supply connections, the two magnitudes will be very nearly the same. Unless otherwise stated, we will assume that the two magnitudes are the same, and the symbols $\pm V_{\text{sat}}$ will be used to denote the two quantities.

In general, the magnitudes of the saturation voltages are about 2 V below the magnitudes of the dc power supply levels. This value can be inferred from the curves titled "Output Voltage Swing as a Function of Supply Voltage" shown on the sixth page of the 741 data sheets. For supply voltages of ± 15 V, the output peak-to-peak voltage is about 26 V for both curves. This value implies that $\pm V_{\text{sat}} \simeq \pm\ 13$ V. For linear operation, it is very important that the input signal not cause the output signal to reach saturation, or the output will be distorted.

Op-Amp Symbol and Circuit Model

The schematic symbol of an op-amp denoting *only* the signal terminals is shown in Figure 2–3(a). As previously noted, power supply connections are certainly required but are often omitted on circuit diagrams. One other set of terminals not shown is the offset null terminals, whose function will be considered in Section 3–6. The offset null terminals are left unconnected in many noncritical applications.

(a)

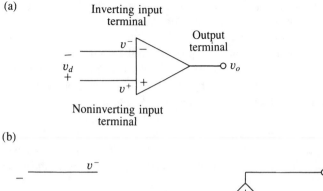

(b)

$$v_d = v^+ - v^- \qquad Av_d = A(v^+ - v^-)$$

$$A \to \infty$$

FIGURE 2–3
(a) Op-amp symbol, and (b) operational model of ideal op-amp.

The circuit model of an ideal op-amp is shown in Figure 2–3(b). Reference to both parts of Figure 2–3 can be made in the development that follows.

Signal Terminals and Voltages

Observe that there are two signal input terminals and one signal output terminal. The two input terminals are shown on the left, and the output terminal is shown on the right. The two input terminals are denoted as the *inverting* and *noninverting* terminals, respectively. The inverting terminal is universally indicated by a $(-)$ symbol, and the noninverting terminal is indicated by a $(+)$ symbol. However, these signs have nothing whatsoever to do with the polarities of possible voltages at the terminals, since the signal voltage at either terminal may be positive or negative with respect to ground. Rather, the signs refer to the polarities of the voltage gains obtained for signals at the two terminals, as we will see shortly.

2–4 IDEAL OPERATIONAL AMPLIFIER

In this section, we will begin the study of operational amplifier circuits utilizing an ideal model. In this manner, the reader will quickly see the typical operations that can be performed with op-amps, and the techniques for analyzing many different circuits will be established. The underlying theory that leads to the establishment of these results will be investigated in great detail in Chapters 3 and 4, and the limitations of the ideal assumptions will be established then.

While the ideal assumptions to be considered do not quite exist, actual operational amplifiers come very close to the ideal model when operated over the proper range of conditions for which intended. For a given application, the circuit designer will attempt to select an op-amp in which the imperfections do not degrade the response significantly from that obtained from the ideal model form. Thus, it is desired that an op-amp used for a particular application be as close as possible to the ideal.

Some of the assumptions made in the ideal model may seem a little strange and perhaps unjustified to the beginner. Bear in mind that a full development and justification will be made later. For the moment, the reader is asked to accept these results on faith and to start learning how to use them in solving problems.

First, a general assumption must be made that all circuits to be analyzed by this approach are operated as *stable, linear* circuits. The reader is not expected to know at this time when a given circuit meets these conditions. As the book progresses, the reader will gain considerable insight as to when this general condition is met. However, virtually all active circuits, if improperly designed or constructed, possess the potential for instability or for nonlinear operation, so the general assumption is always a point of vulnerability. It is sufficient to say that all circuits to be considered at this point in the book are capable of stable operation.

In view of the widespread practice of using $(-)$ and $(+)$ symbols to denote the inverting and noninverting terminals, respectively, we will denote the signal voltages at these two terminals as v^- and v^+, respectively, and both as measured with respect to ground. These symbols could create a point of confusion because the specification sheets often indicate the power supply connections as V^- and V^+. However, the power supply connections will rarely be shown on a diagram in which signals are being analyzed, so this convention should be understood. Further, the symbols v^- and v^+ are the most natural ways of denoting the two respective input voltages for clarity. Finally, the use of lowercase letters for most instantaneous signals and uppercase letters for power supply voltages should fully resolve the situation.

To summarize this notation, v^- is the voltage at the inverting terminal with respect to ground and v^+ is the voltage at the noninverting terminal with respect to ground, each of which may be either positive or negative with respect to ground. The output voltage will be denoted as v_o. In most circuit diagrams, the inverting terminal will be shown at the top, but an occasional awkward circuit diagram may be improved by reversing the positions.

Differential Input Voltage

The difference between the voltages at the noninverting input and the inverting input is defined as the ***differential input voltage***, v_d. Thus,

$$v_d = v^+ - v^- \tag{2-1}$$

The complex action of the operational amplifier results in the amplification of the differential voltage by a large gain factor A, as shown in Figure 2–3(b). A voltage-controlled voltage source with value $Av_d = A(v^+ - v^-)$ appears in the output of the op-amp. The output voltage v_o with respect to ground is then

$$v_o = Av_d = A(v^+ - v^-) \tag{2-2}$$

Later, A will be redefined as the open-loop differential gain and denoted as A_d. However, at this time, it will be referred to simply as the open-loop gain, and the symbol A will be used for simplicity.

Ideal Assumptions

The assumptions made in this ideal model are as follows:

1. The input impedance of the op-amp as viewed from the two input terminals is infinite, as indicated in Figure 2–3(b) by the open circuit across the two input terminals.
2. The output impedance of the op-amp viewed from the output terminal with respect to ground is zero. This condition is indicated in Figure 2–3(b); note that no Thevenin resistance appears in series with the dependent voltage source.
3. The open-loop gain A approaches an infinite value in the ideal case. The limiting case $A \rightarrow \infty$ is indicated below the dependent source in Figure 2–3(b), because it would appear awkward to actually show $A = \infty$ on the circuit diagram. However, the assumption $A = \infty$ will be made in many of the developments that follow.

Implications

Let us next consider the implications of these ideal assumptions. The implications will be considered in the same order as the assumptions.

1. The assumption of infinite input impedance implies that no current will flow either into or out of either input terminal of the op-amp. Thus, when several branches are connected to an input terminal of an op-amp, Kirchhoff's current law can be applied to the node involved, but zero current may be assumed for the op-amp input terminal current in the ideal case.
2. The assumption of zero output impedance implies that the voltage v_o at the output terminal does not change as the loading is varied. The ideal op-amp then produces the same output voltage irrespective of the current drawn into a load.
3. The implication of the third assumption is most significant and is best seen by first solving for v_d in (2–2). We have

$$v_d = v^+ - v^- = \frac{v_o}{A} \tag{2–3}$$

Consider now the general assumption that a circuit of interest is operated in a linear stable mode. Thus, v_o must be a finite voltage, which for the majority of small signal op-amps would be under 13 V or so. As $A \rightarrow \infty$ in (2–3),

$$\lim_{A \to \infty} v_d = \lim_{A \to \infty} \frac{v_o}{A} = 0 \tag{2–4}$$

where v_o is finite. This result indicates that *the differential voltage approaches zero*. An equivalent implication is

$$v^+ - v^- = 0 \tag{2–5}$$

or

$$v^+ = v^- \tag{2–6}$$

which indicates that *the voltages at the two input terminals are forced to be equal in the limit*.

TABLE 2–1
Ideal op-amp assumptions and their implications

Assumption	Implication
(1) $R_{in} = \infty$	Input current is zero
(2) $R_{out} = 0$	Output voltage is independent of load
(3) $A = \infty$	Differential input voltage is zero

The reader may be puzzled and asking, If the output voltage $v_o = Av_d$ and v_d is zero, how could v_o have a nonzero value? The answer is that v_d is not actually zero, but rather it is a very small value such that when multiplied by the very large value of gain produces a nonzero finite value for v_o. For example, typical values of A and v_d for stable operation are $A = 10^5$ and $v_d = 20\ \mu V$, which results in $v_o = Av_d = 10^5 \times 20 \times 10^{-6} = 2$ V, a reasonable value. Thus, v_d never quite reaches zero and A never quite reaches an infinite value, but the ideal assumptions $v_d = 0$ and $A = \infty$ are extremely convenient, as will be demonstrated shortly. A summary of the preceding assumptions is given in Table 2–1.

Although we will assume that the voltages at the two input terminals of an op-amp are the same, *the terminals must not be connected together* in a circuit. As indicated in the last paragraph, a small potential difference must exist in order for the circuit to function properly, and this situation would be impossible if the terminals were connected.

Now that the ideal assumptions for analyzing op-amp circuits have been established, several of the most important circuits will be investigated. The purposes are both to demonstrate the method of analysis and to begin the tabulation of a useful collection of practical circuits.

Inverting Amplifier

One of the most widely used single op-amp configurations is the ***inverting amplifier*** circuit, whose basic form is shown in Figure 2–4. This circuit represents one form of a voltage-controlled voltage source, as we will see shortly. In actual practice, a resistance is often connected between the noninverting input and ground to minimize certain effects due to bias currents, and this procedure will be considered later. However, the circuit will work without the resistance, so to keep the circuit initially as simple as possible, it will not be considered at this point.

FIGURE 2–4
Inverting amplifier circuit with voltages and currents labeled for analysis.

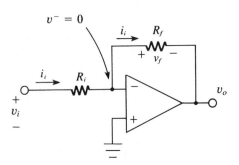

Assuming stable linear operation, the differential input voltage v_d is forced to be zero, or $v^- = v^+$. Since $v^+ = 0$ because the noninverting terminal is grounded, the inverting terminal is also forced to be at ground potential; that is, $v^- = 0$. This condition is referred to as a ***virtual ground,*** meaning that the inverting terminal is not really connected to ground, but it acts as if it were grounded from a voltage point of view. (However, the inverting terminal *must not* be connected to ground, since the small differential voltage must be maintained as previously noted.)

Since $v^- = 0$, the input voltage v_i appears across R_i. The input current i_i is then

$$i_i = \frac{v_i}{R_i} \tag{2-7}$$

Next, the assumption of zero amplifier input current is used. The current i_i flows to the junction point at the amplifier inverting input terminal. However, since no current can flow into or out of the amplifier, the current must flow through R_f. A voltage v_f is then produced across R_f, given by

$$v_f = R_f i_i = R_f \frac{v_i}{R_i} = \frac{R_f}{R_i} v_i \tag{2-8}$$

Since the inverting input of the op-amp is at a virtual ground, the output voltage is the voltage across R_f, but with the reference positive terminal on the left-hand side of R_f. This leads to

$$v_o = -v_f = -\frac{R_f}{R_i} v_i \tag{2-9}$$

Let A_{CL} represent the closed-loop voltage gain of the circuit. We have

$$A_{CL} = \frac{v_o}{v_i} = -\frac{R_f}{R_i} \tag{2-10}$$

Thus, the closed-loop voltage gain is a simple resistance ratio and is independent of the open-loop gain. Since stable resistance values may be established (either by manufacture or by measurement) to a high degree of accuracy, it is possible to implement very accurate amplifier gains by this technique.

The input impedance R_{in} of the circuit is readily determined by noting once again that the input voltage v_i appears across R_i. Thus

$$R_{in} = \frac{v_i}{i_i} = R_i \tag{2-11}$$

The reader should not confuse the input impedance of the op-amp itself, which has been assumed to be infinite, with the input impedance of the composite circuit, which is given by (2–11).

Some of the practical implications of the inverting amplifier will be made in the next section. Our major interest at this point is in developing analysis skills, so we will continue on to a different circuit.

Noninverting Amplifier

Another very important op-amp configuration is the ***noninverting amplifier*** circuit, whose basic form is shown in Figure 2–5. Like the inverting amplifier circuit, the noninverting amplifier is a form of a voltage-controlled voltage source. In actual practice, a resistance is often connected in series with the noninverting input to minimize the effects of bias current, and this procedure will be considered later. However, the circuit will work without the resistance, so it will not be considered at this point.

Observe that the input signal is applied directly to the noninverting input. Assuming stable linear operation, the differential input voltage is forced to be zero, and thus

$$v^- = v^+ = v_i \qquad (2\text{--}12)$$

This voltage appears across the resistance R_i, so the current i_i is readily determined as

$$i_i = \frac{v_i}{R_i} \qquad (2\text{--}13)$$

Since no current flows into or out of the op-amp inverting terminal, the current must be flowing through the resistance R_f from the op-amp output. A voltage v_f appears across R_f and is

$$v_f = R_f i_i = R_f \frac{v_i}{R_i} \qquad (2\text{--}14)$$

The voltage v_i with respect to ground, the voltage v_f, and the output voltage v_o constitute a closed loop, and a Kirchhoff voltage law equation could be written if desired. Either by this formal approach or by a more intuitive approach, it can be readily deduced that

$$v_o = v_i + v_f = v_i + \frac{R_f}{R_i} v_i = \left(1 + \frac{R_f}{R_i}\right) v_i \qquad (2\text{--}15)$$

The closed-loop voltage gain A_{CL} is then given by

$$A_{\mathrm{CL}} = \frac{v_o}{v_i} = 1 + \frac{R_f}{R_i} = \frac{R_i + R_f}{R_i} \qquad (2\text{--}16)$$

As in the case of the inverting amplifier, the closed-loop gain is a function of a resistance ratio and is independent of the open-loop gain.

The input impedance of the ideal noninverting amplifier is infinite, meaning that no current is drawn from the signal source in the ideal case.

FIGURE 2–5
Noninverting amplifier circuit with voltages and currents labeled for analysis.

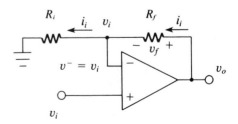

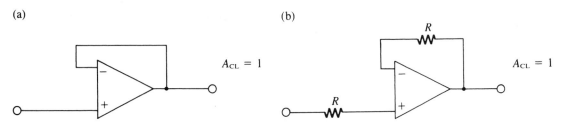

FIGURE 2–6
Voltage follower circuits.

Voltage Follower

A special case of the noninverting amplifier is the ***voltage follower circuit,*** whose basic form is shown in Figure 2–6(a). The closed-loop gain may be determined from (2–16) by considering the limiting case when $R_f = 0$ and $R_i = \infty$, and the result is

$$A_{\text{CL}} = 1 \qquad\qquad (2\text{–}17)$$

The voltage gain is thus unity, and the output voltage "follows" the input voltage.

  The reader's first reaction may be to question the value of the circuit, but the input impedance is infinite and the output impedance is zero in the ideal case. Thus, a definite power gain may be achieved with the circuit. Voltage followers are used where isolation between a source and a load are desired and where the exact level of the original voltage is to be maintained.

  As we will see later, there are situations that arise with the voltage follower circuit when a series resistance appears in the noninverting input (for example, signal source output resistance); and for compensation reasons, a resistance may be placed in the feedback path. These additional resistances are shown in Figure 2–6(b). The circuit, however, is still a voltage follower with unity gain. There is no signal current flowing into the op-amp input in the ideal case, so no drop appears across the series resistance, and the gain is not affected. Although the feedback resistance is no longer zero, we still have $R_i = \infty$, so (2–16) predicts a gain of unity.

EXAMPLE 2–1 A certain inverting amplifier circuit with an assumed ideal op-amp is shown in Figure 2–7(a). (a) Determine the closed-loop voltage gain $A_{\text{CL}} = v_o/v_i$. (b) Based on power supply voltages of ± 15 V, the output saturation voltages are assumed to be $\pm V_{\text{sat}} = \pm 13$ V. Determine the peak input signal voltage v_i (peak) for which linear operation is possible. (c) Determine the output voltage v_o for each of the following input voltages:

$$
\begin{aligned}
v_i &= 0 \text{ V} \\
&= -0.5 \text{ V} \\
&= 0.5 \text{ V} \\
&= 1 \text{ V} \\
&= -2 \text{ V}
\end{aligned}
$$

(a)

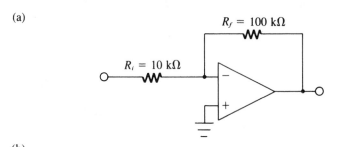

(b)

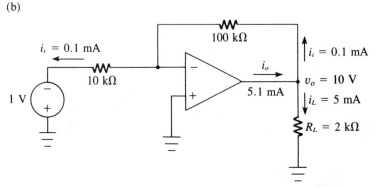

(c)

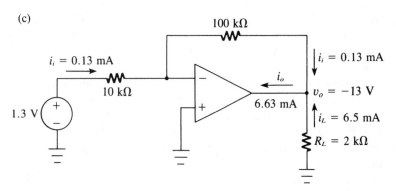

FIGURE 2–7
Circuit analysis of Example 2–1

(d) A load resistance $R_L = 2$ kΩ is connected between the output terminal and ground. Determine the op-amp output current for $v_i = -1$ V and for $v_i = 1.3$ V.

Solution

(a) The closed-loop gain is readily determined from the simple formula

$$A_{\text{CL}} = \frac{R_f}{R_i} = \frac{-100 \text{ k}\Omega}{10 \text{ k}\Omega} = -10 \tag{2-18}$$

in which the negative sign implies an inversion of the output with respect to the input.

(b) The peak magnitude of the input at which saturation is reached is given by

$$|v_i(\text{peak})| = \frac{V_{\text{sat}}}{|A_{\text{CL}}|} = \frac{13 \text{ V}}{10} = 1.3 \text{ V} \qquad (2\text{–}19)$$

Note that this result applies to either polarity of the input; that is, the input signal must lie in the range from -1.3 V to $+1.3$ V for linear operation.

(c) The various output voltages corresponding to different input voltages are determined by multiplying the gain by the input voltages:

$$v_o = A_{\text{CL}}v_i = -10v_i \qquad (2\text{–}20)$$

The results for the first four input voltages are as follows:

v_i	v_o
0	0
-0.5 V	5 V
0.5 V	-5 V
1 V	-10 V

Note that when the input is zero, there is no output. For other inputs, the output has a magnitude of ten times the input, but the sign is opposite. This is the characteristic of an inverting amplifier.

For $v_i = -2$ V, if one uses (2–20) as before, a predicted output of 20 V will result. This value is impossible since the saturation voltage is only 13 V. From part (b), it is clear that the peak input voltage for linear operation has been exceeded, and the linear gain equation of (2–20) is no longer valid. If one attempted to use the circuit as an amplifier under these conditions, the results would be most disappointing, since the signal would be clipped and distorted. Thus, $v_o \approx 13$ V for the condition given.

(d) With $R_L = 2$ kΩ and $v_i = -1$ V, the circuit conditions are as shown in Figure 2–7(b). The total op-amp output current is composed of two parts. One part is due to the actual load current i_L, which is

$$i_L = \frac{v_o}{R_L} \qquad (2\text{–}21)$$

Since the input voltage is $v_i = -1$ V, the output is $v_o = -10 \times -1$ V $= 10$ V, and i_L is

$$i_L = \frac{10 \text{ V}}{2 \text{ kΩ}} = 5 \text{ mA} \qquad (2\text{–}22)$$

An additional current i_i flows through the feedback network to ground. When the input voltage is negative, the positive direction of this current is to ground, and since v_i appears across R_i, we have

$$i_i = \frac{v_i}{R_i} = \frac{1 \text{ V}}{10 \text{ kΩ}} = 0.1 \text{ mA} \qquad (2\text{–}23)$$

The total op-amp output current i_o is

$$i_o = i_L + i_i = 5 + 0.1 = 5.1 \text{ mA} \qquad (2\text{--}24)$$

For $v_i = 1.3$ V, the circuit conditions are as shown in Figure 2–7(c). In this case, the positive directions of the two components of current are into the op-amp output terminal as shown. The reader may readily verify the values on the figure. The output current of the op-amp is

$$i_o = 6.63 \text{ mA} \qquad (2\text{--}25)$$

Borrowing terminology from digital electronics, the op-amp is said to be *sinking* current in this case.

From these values, it can be deduced that the two components of op-amp current are in the same direction for a given input, and their magnitudes increase as the input voltage magnitude increases. Thus, an estimate of the peak op-amp output current can be made by estimating the peak magnitude of the input voltage and using it to establish the two peak currents. If the input signal is symmetrical, the same magnitude of output current would occur on opposite halves of the cycle. For a nonsymmetrical input signal, the portion of the cycle having the largest magnitude would be used. It is important that the peak output current limit of the op-amp not be exceeded.

In this example, the current flowing through the feedback network is small compared with the load current as it should be in most circuits. If the resistance values are too small, this current can become excessive and overload the op-amp.

EXAMPLE 2–2

Assume that the op-amp of Example 2–1, with the same power supply voltages, is connected as in the noninverting configuration of Figure 2–8(a). Repeat the analysis of Example 2–1.

Solution

(a) The closed-loop gain is determined as

$$A_{\text{CL}} = 1 + \frac{R_f}{R_i} = 1 + \frac{90 \text{ k}\Omega}{10 \text{ k}\Omega} = 10 \qquad (2\text{--}26)$$

This gain has the same magnitude as the gain in Example 2–1. Lowering the feedback resistance from 100 kΩ to 90 kΩ produced the same gain for the noninverting amplifier as the gain magnitude for the inverting amplifier.

(b) Since the gain magnitudes are the same and the saturation voltages are the same, it follows that the peak value of the input voltage corresponding to saturation is the same as in Example 2–1; that is,

$$|v_i(\text{peak})| = 1.3 \text{ V} \qquad (2\text{--}27)$$

(c) The various output voltages are determined from the equation

$$v_o = A_{\text{CL}} v_i = 10 v_i \qquad (2\text{--}28)$$

FIGURE 2–8
Circuit of Example 2–2.

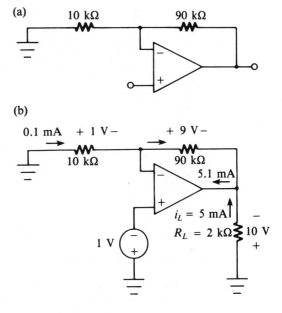

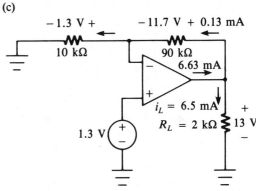

For the input voltages of Example 2–1, the output voltages are summarized as follows:

v_i	v_o
0	0
−0.5 V	−5 V
0.5 V	5 V
1 V	10 V
−2 V	−13 V (saturation)

The first four cases correspond to linear operation, and the magnitudes are the same as in Example 2–1. However, the outputs here are not inverted. For $v_i = -2$ V, the output is driven to saturation, but in this case, it is negative saturation.

(d) With $R_L = 2$ kΩ and $v_i = -1$ V, the circuit conditions are as shown in Figure 2–8(b). The value $v_i = -1$ V results in the same voltage across the 10-kΩ resistor, which means that the more positive terminal is at ground. A current given by 1 V/10 kΩ = 0.1 mA flows upward from ground through the 10-kΩ resistor and on through the 90-kΩ resistor. Across the latter resistor, a voltage given by 0.1×10^{-3} A $\times 90 \times 10^3$ Ω = 9 V is produced, with the positive terminal on the left. The output voltage is $-1 - 9 = -10$ V, which could, of course, have been predicted directly from the gain. A load current $i_L = 10$ V/2 kΩ = 5 mA flows upward from ground. The op-amp sinks a current $i_o = 5$ mA + 0.1 mA = 5.1 mA.

With $v_i = 1.3$ V, the conditions are shown in Figure 2–8(c). The various currents and voltages are determined in the same manner as for $v_i = -1$ V, except that polarities and directions are opposite. In this case, the op-amp output provides a net current $i_o = 6.63$ mA.

2–5 SIMPLIFIED DESIGN OF AMPLIFIER CIRCUITS

In this section, certain practical applications of the two amplifier circuits presented in the preceding section will be discussed. The designs of some representative circuits using these configurations will be illustrated. Remember that some limitations of op-amp circuits that must be considered in the most exact designs will be considered in later chapters. However, even with the limited amount of material covered thus far, it is possible to implement some useful and realistic circuit designs, provided that the circuits do not "push" the operating limits into the critical regions to be discussed later. As we progress through the text, the reader will acquire proficiency in making appropriate judgments of this type.

In designing an amplifier circuit, we select an operational amplifier and external resistances such that a specified gain is achieved. Another specification is that of frequency response, but we cannot consider it until Chapter 4 is completed. Indeed, the selection of a particular op-amp is most important in determining the bandwidth. For the moment, we will assume that whatever designs we develop will have adequate frequency response for the present applications.

The two amplifier circuits that we have considered thus far are summarized in Figure 2–9. The power supply connections have been omitted from the circuit diagrams.

Bias Compensating Resistance

In order to instill general design techniques from the beginning, the bias compensating resistor R_c has been included in both circuits. This resistor is selected because offsets due to bias currents are minimized if the effective dc resistances to ground at both terminals are equal. Ideal voltage sources are momentarily deenergized, as in applying superposition, when determining these equivalent resistances. Thus, when $v_i = 0$, the effective resistance to ground from the inverting terminal is $R_i \| R_f$ for the inverting circuit, so a compensating resistance $R_c = R_i \| R_f$ is placed between the noninverting terminal and ground as

(a) Inverting

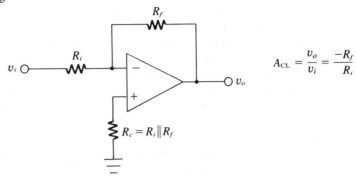

$$A_{CL} = \frac{v_o}{v_i} = \frac{-R_f}{R_i}$$

(b) Noninverting

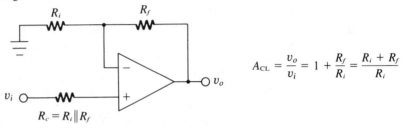

$$A_{CL} = \frac{v_o}{v_i} = 1 + \frac{R_f}{R_i} = \frac{R_i + R_f}{R_i}$$

FIGURE 2–9
Summary of voltage-controlled voltage source amplifier circuits.

noted. Similarly, a resistance of the same value is placed in series with the signal source in the noninverting circuit. (This resistance may represent all or a portion of the source resistance in this case.) This concept will be developed in much more detail in Chapter 3, but the resistance will be used automatically in a number of the circuits discussed in this chapter. Unlike R_i and R_f, the exact value of R_c is not critical, and the closest standard value is usually sufficient. However, to illustrate the principle clearly, the exact theoretical value will be labeled on circuit diagrams when it is used.*

Design Considerations

Both the inverting and noninverting amplifier circuits are examples of voltage-controlled voltage sources (VCVS). A required design often starts with the need to amplify the level of a given signal while maintaining the form of the signal preserved. It is assumed that the op-amp type and its bias supplies allow the necessary dynamic range for the required out-

---

*The compensating resistor does add some additional noise to the circuit, so it may be best not to use it in applications requiring absolute minimum noise levels.

put. For example, if the op-amp bias voltages are the standard values ± 15 V, one can expect the maximum output signal voltage range to be approximately ± 13 V or so. Thus, if one had a signal source having a range of ± 200 mV and applied it to a circuit having a gain of 100, but with the standard ± 15-V values, the user would be sadly disappointed because the required output swing would be $\pm 0.2 \times 100 = \pm 20$ V. In such a case, an op-amp with a higher voltage level would have to be found, and new power supply voltages would have to be obtained.

The gains of both the inverting and noniverting amplifier circuits are functions of the ratio R_f/R_i. Suppose that one desired $R_f/R_i = 10$. There are many combinations of R_f and R_i that produce a ratio of 10, so how does one know which values to use? The answer is that there is often a reasonable leeway on the part of the designer, and a number of different designs, all perfectly acceptable, may be achieved.

The following general comments concerning the resistance ratio should be noted. If a resistance level drops too low, the loading on the op-amp and/or the source may become excessive, and nonlinear operation (or worse) may result. In contrast, as the resistance level increases, the thermal noise produced by the resistors increases, and dc offsets due to bias currents may pose potential difficulties. While there are no optimum values for all cases, a reasonable operating range of resistance for most op-amp circuits is from around 1 kΩ to around 100 kΩ or so, with a large usage of resistances in the range of 10 kΩ to 100 kΩ. However, this is a rough guide rather than a restricting rule, and many exceptions to the pattern will be found.

Since the gains of both inverting and noninverting stages are functions of the R_f/R_i ratio, the question may arise as to whether the gain may be increased indefinitely by increasing this ratio. The answer is no for reasons that will become clear later. As the desired circuit gain (closed-loop gain) begins to approach the open-loop op-amp gain, the accuracy of the closed-loop gain expression is degraded. Second, the available bandwidth of the closed-loop response decreases as the gain increases. For this reason, closed-loop gains are established at values much smaller than the available values of open-loop gains.

One additional point concerns the input impedance. The input impedance of the inverting amplifier is R_i, so the possible loading effect of this resistance on the source must be considered. In contrast, the input impedance of the noninverting amplifier is ideally infinite, so no loading occurs.

After the circuit is connected, a check of the output current of the op-amp under worst-case conditions can be made. This process was illustrated in Examples 2–1 and 2–2.

Design Summary

A summary of some of the major considerations of the "simplified" design approach to op-amp small-signal amplifier circuits is as follows:

1. Be sure that the dynamic range of the op-amp, based on the power supply levels with some "backoff," will be sufficient for the desired output signal.

2. Whenever possible, choose resistance values in the range of 1 kΩ to 100 kΩ or so, except avoid resistances that load the output in the lower portion of the range. Values of 10 kΩ and greater are the most widely used for many applications.
3. Limit magnitudes of closed-loop gain to values much smaller than the available open-loop gain. (Typically, closed-loop gain magnitudes are kept under 100 or so.)
4. Check the maximum output current levels under worst-case conditions.

EXAMPLE 2-3
Using 5% resistance values, design an inverting VCVS amplifier to provide a voltage gain of −10. The input impedance must be no less than 10 kΩ to avoid loading the given signal source. The signal level will be sufficiently high so that thermal resistance noise and bias currents effects are not felt to be a critical problem, provided that resistance values less than 500 kΩ are used.

Solution
The desired gain is readily achieved by selecting $R_f/R_i = 10$, and there are a number of standard values in the given range that can achieve this ratio. Refer to Appendix A for a listing of the standard 5% resistance values. Since the minimum input impedance is 10 kΩ, the value of R_i cannot be less than this value. Thus, the lowest possible values that will satisfy the constraints are $R_i = 10$ kΩ and $R_f = 100$ kΩ. At the other extreme, possible standard values are $R_i = 47$ kΩ and $R_f = 470$ kΩ. The next standard values would cause R_f to exceed the specified range of 500 kΩ.

As an approximate mid-level set of values, we will somewhat arbitrarily select $R_i = 22$ kΩ and $R_f = 220$ kΩ; this design is shown in Figure 2–10. The ideal value of the compensating resistance is $R_c = R_i \| R_f = 22$ kΩ$\|220$ kΩ $= 20$ kΩ, which is a standard 5% value.

It should be understood that since 5% resistances were used in this design, the actual gain may vary somewhat from the design goal. If a more exact value of gain is needed, the two options are (a) to use precision resistors for R_i and R_f or (b) to use a combination of some fixed values and an adjustable resistance, so that a more exact value can be established by a calibration process. This problem was chosen to illustrate the ordinary application in which a precision gain is not needed, but in which standard, readily available values can be conveniently used.

FIGURE 2–10
Circuit of Example 2–3.

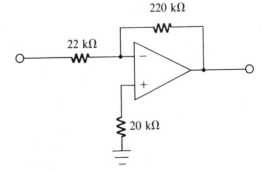

EXAMPLE 2-4

Using the standard 1% resistance values listed in Appendix A, design a VCVS noninverting amplifier circuit to provide a voltage gain of 18.

Solution

From the basic gain equation for the noninverting amplifier, we have the requirement that

$$A_{CL} = 18 = 1 + \frac{R_f}{R_i} \tag{2-29}$$

Solving for the R_f/R_i ratio, we have

$$\frac{R_f}{R_i} = 17 \tag{2-30}$$

Now we must find two resistances that have a 17-to-1 ratio. After some inspection of the values in Appendix A, we find that the values $R_i = 2$ kΩ and $R_f = 34$ kΩ are standard values that satisfy the constraint. Of course, $R_i = 20$ kΩ and $R_f = 340$ kΩ also satisfy the requirements; but the former values are likely to be more desirable, so that set will be selected. The required value of R_c is $R_c = R_i \| R_f = 2$ kΩ$\|34$ kΩ $= 1889$ Ω. The nearest standard value is 1870 Ω, so that value is used. The amplifier design is given in Figure 2-11.

This problem was formulated so as to yield a solution in which standard 1% resistance values could be used. A certain gain may be specified that is impossible to realize with standard values. In such a case, a variable resistance alone or a combination of a variable resistance and a fixed resistance may be used as either R_f or R_i. The variable resistance is then adjusted to yield the correct value.

In many applications, the actual level of the gain is not as critical as the knowledge of the exact value. In such cases, one can actually choose gain values that can be implemented with standard values.

FIGURE 2-11
Circuit of Example 2-4.

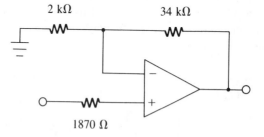

EXAMPLE 2-5

A naive circuit designer is given the task of designing an amplifier to produce a peak output signal voltage of 1 V from a transducer source, whose open-circuit voltage is 50 mV. It does not matter whether the output is inverted or not. The designer never quite understood Thevenin's theorem, so he neglects to obtain data on the output impedance of the source, which turns out to be 50-kΩ resistive. The designer produces the circuit shown in Figure 2-12. Calculate the actual output voltage obtained with the circuit.

Solution

The actual closed-loop voltage gain of the op-amp circuit is

$$A_{CL} = \frac{v_o}{v_i} = \frac{-100 \text{ k}\Omega}{5 \text{ k}\Omega} = -20 \tag{2-31}$$

which would produce the required output voltage *if* the input voltage to the circuit were 50 mV. However, since the designer failed to consider the 50-kΩ output impedance of the source, the relatively low input impedance of the op-amp circuit (5 kΩ) will excessively load the source.

When the source is connected, the actual input current is

$$i_i = \frac{50 \text{ mV}}{50 \text{ k}\Omega + 5 \text{ k}\Omega} = 0.909 \text{ }\mu A \tag{2-32}$$

The voltage across R_f is directly related to the output voltage, which is

$$v_o = -10^5 \times 0.909 \times 10^{-6} = -90.9 \text{ mV} \tag{2-33}$$

This value falls far short of the desired output voltage of 1 V! The actual gain referred to the open-circuit source voltage is −90.9 mV/50 mV = −1.818.

An alternate way to deduce the same result is to recognize that the source resistance becomes part of the effective R_i value. Let A'_{CL} represent the net gain on this basis, and we have

$$A'_{CL} = \frac{-R_f}{R_i + R_s} = \frac{-10^5}{50 \times 10^3 + 5 \times 10^3} = -1.818 \tag{2-34}$$

A final way to view the process is to interpret A_{CL} as the unloaded gain and to utilize the concept of loaded gain (A_{so}) between source and output as developed in Chapter 1. This approach yields

$$A_{so} = \frac{R_i}{R_i + R_s} \times A_{CL} = \frac{5 \times 10^3}{50 \times 10^3 + 5 \times 10^3} \times (-20) = -1.818 \tag{2-35}$$

which obviously agrees with the result of (2–34).

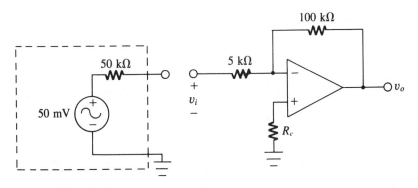

FIGURE 2-12
Circuit of Example 2–5.

This problem could be resolved by a new design in which the output impedance of the source is considered as a portion of the total input circuit resistance, with R_f selected to achieve the required gain. However, a better solution would probably be a noninverting amplifier, in which loading effects are minimal.

It is also recommended that the designer learn Thevenin's theorem!

EXAMPLE 2–6

An inverting amplifier with an adjustable gain is desired for a particular application. The desired adjustment range of the gain is from −2 to −12. If a 100-kΩ potentiometer is available, design a circuit to meet the requirements.

Solution

To vary the gain, a portion of either R_i or R_f must be adjustable. If R_i is adjustable, the input impedance will vary with the gain selected, so it seems preferable to vary R_f.

The proposed circuit diagram is shown in Figure 2–13. In general, $A_{\rm CL} = -R_f/R_i$, and the gain varies directly with R_f. When the potentiometer is at one end, its value is zero and $R_f = R_{f1}$. This corresponds to $A_{\rm CL} = -2$. When the potentiometer is at the other end, $R_f = R_{f1} + 10^4$, which corresponds to $A_{\rm CL} = -12$. The following two equations may thus be written:

$$\frac{R_{f1}}{R_i} = 2 \tag{2–36}$$

and

$$\frac{R_{f1} + 10^5}{R_i} = 12 \tag{2–37}$$

Simultaneous solution of these two equations yields

$$R_i = 10 \text{ k}\Omega \tag{2–38}$$

and

$$R_{f1} = 20 \text{ k}\Omega \tag{2–39}$$

FIGURE 2–13
Circuit of Example 2–6.

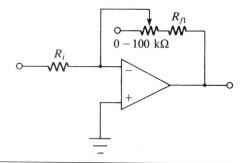

2–6 VOLTAGE-CONTROLLED CURRENT SOURCES

Thus far the emphasis has been on voltage-controlled voltage source amplifiers, since they are probably the most widely employed types of linear active circuits. Another useful linear amplifier circuit is the voltage-controlled current source (VCIS). The theoretical model was discussed in Section 1–4, and the reader may review the concept there if necessary.

There are several circuits that will perform the function, and we will consider three of the most common in this section. For clarity, the circuits will be classified as (a) floating-load inverting VCIS, (b) floating-load noninverting VCIS, and (c) grounded-load VCIS. The terms *inverting* and *noninverting* as used here are mainly for association with the corresponding VCVS circuits, since inversion and noninversion for current in a floating load are somewhat ambiguous.

Floating-Load Inverting VCIS

The floating-load inverting VCIS is shown in Figure 2–14. At first glance this circuit seems to be nothing more than the inverting VCVS amplifier circuit previously discussed, and for precisely this reason the adjective *inverting* was attached to the circuit. However, the difference lies in how the load is connected and how the function is interpreted. In the VCVS inverting amplifier, the feedback and input resistances are both fixed, and the output quantity of interest is the voltage to ground at the op-amp output terminal. In the VCIS circuit, the external load R_L is connected as the feedback resistance, and this load resistance may not be fixed. Observe that since the load is connected between two terminals of the op-amp, it is a *floating* load. This condition limits the application of this circuit to situations where no part of the load has to be connected to the actual circuit ground.

The input current i_i is established by the input controlling voltage v_i and the fixed resistance R. Assuming stable linear operation, the inverting terminal of the op-amp will be forced to be at ground potential, and the current i_i is

$$i_i = \frac{v_i}{R} \tag{2–40}$$

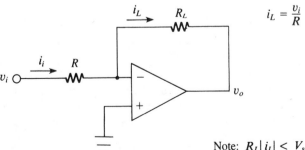

$$i_L = \frac{v_i}{R}$$

Note: $R_L|i_L| < V_{\text{sat}}$ for linear operation.

FIGURE 2–14
Floating-load inverting voltage-controlled current source.

This current must flow through R_L, and so $i_L = i_i$. Thus,

$$i_L = \frac{v_i}{R} \tag{2-41}$$

Observe that this load current is a function only of v_i and R and is completely *independent* of R_L. This is the required nature of a current source; that is, *the current is independent of the load resistance through which it flows.*

From earlier work, it can be recalled that a VCIS can be described by a transconductance g_m (in siemens, abbreviated S). The transconductance of this circuit is

$$g_m = \frac{1}{R} \tag{2-42}$$

The value of R is selected to provide the proper conversion between voltage input and current output.

This circuit will function as a linear VCIS for both positive and negative inputs, provided that operation is in the linear region. Since the focus is on the current through R_L, there is a tendency to overlook the fact that the voltage at the output terminal must remain below the saturation point. To avoid saturation, the following constraint is required for linear operation:

$$R_L|i_L| < V_{\text{sat}} \tag{2-43}$$

Floating-Load Noninverting VCIS

The floating-load noninverting VCIS is shown in Figure 2–15. This circuit obviously resembles the noinverting VCVS amplifier circuit, as implied by its name. However, as in the case of the previous current source analyzed, the load R_L is connected as the feedback resistance, and the desired output quantity is the current i_L through this resistance.

The load current i_L is the same as the current through R. Assuming stable linear operation, the potential at the inverting terminal is the same as v_i, and i_L is thus

$$i_L = \frac{v_i}{R} \tag{2-44}$$

The transconductance g_m is the same as for the previous current source:

$$g_m = \frac{1}{R} \tag{2-45}$$

The dynamic range of the noninverting circuit form is somewhat less than for the inverting circuit form because the inverting terminal is no longer at ground potential. To avoid saturation, the following constraint is required for linear operation:

$$(R + R_L)|i_L| < V_{\text{sat}} \tag{2-46}$$

The floating-load inverting VCIS has the advantage of a greater possible dynamic range of linear operation, while the floating-load noninverting VCIS has the advantage

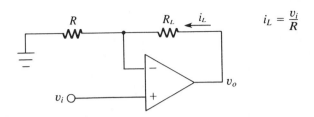

$$i_L = \frac{v_i}{R}$$

Note: $(R + R_L)|i_L| < V_{sat}$ for linear operation.

FIGURE 2–15
Floating-load noninverting VCIS.

of higher input impedance. The input impedance of the inverting circuit is $R_{in} = R$, while the input impedance of the noninverting circuit is theoretically (but not actually) infinite.

If the load resistance R_L is assumed to vary, it is not possible to provide a single value of compensation resistance R_c that will completely provide bias current compensation for either of the preceding circuits. This situation is similar to others where only partial compensation can be achieved with a single resistance. About the best that can be done is to base the selection of R_c on a nominal or mid-range expected value of the parallel combination of R and R_L.

Grounded-Load VCIS

A VCIS circuit in which the load is grounded is shown in Figure 2–16(a). Various quantities are labeled for analysis purposes in Figure 2–16(b). This circuit is often referred to as a *Howland current source*.

This circuit appears to be a bit more perplexing than others encountered thus far. However, the various assumptions given early in the chapter may be adapted to this circuit. Assume a load current i_L, a voltage i_L across R_L, and a voltage v_o at the output of the op-amp. First, a node voltage equation will be written at the junction between R_L and the two resistor branches with values R. Assuming zero current into the op-amp, and assuming positive currents leaving the node, we have

$$i_L + \frac{v_L - v_i}{R} + \frac{v_L - v_o}{R} = 0 \qquad (2\text{–}47)$$

Along the upper resistance path, the circuit is a simple voltage divider network, and the voltage at the inverting terminal is readily seen to be $v_o/2$. Since the voltages at the two op-amp input terminals are forced to be the same, we have

$$v_L = \frac{v_o}{2} \qquad (2\text{–}48)$$

(a)

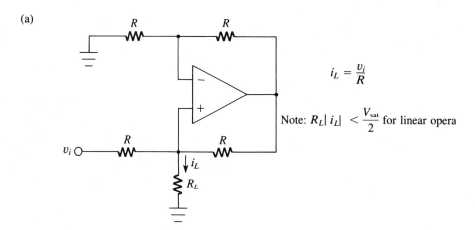

$$i_L = \frac{v_i}{R}$$

Note: $R_L |\, i_L | < \dfrac{V_{\text{sat}}}{2}$ for linear opera

(b)

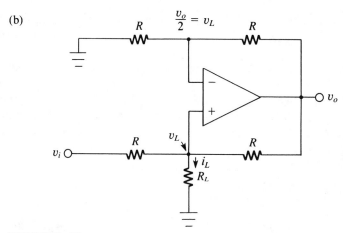

FIGURE 2–16

(a) Howland voltage-controlled current source for use with grounded load. (b) Various quantities used for analysis in the text labeled on the circuit.

Substituting the value of v_L from (2–48) in (2–47) and solving for i_L, we simplify the equation to

$$i_L = \frac{v_i}{R} \tag{2-49}$$

This result is fascinating because the terms involving v_o and v_L completely cancel, leaving i_L a function only of v_i and R! This is a good example of the numerous "tricks" possible with op-amp feedback circuits! As in the case of the floating current sources, the load current is completely independent of the load resistance and is a function of the controlling voltage as well as a fixed resistance R. The transconductance in this case is

$$g_m = \frac{1}{R} \tag{2-50}$$

One must be careful to ensure that the output voltage does not reach the saturation point. From (2–48), the output voltage is

$$v_o = 2v_L \qquad (2\text{--}51)$$

The following constraint is required for linear operation:

$$R_L i_L < \frac{V_{\text{sat}}}{2} \qquad (2\text{--}52)$$

Comparing (2–52) with (2–43), one can infer that if all other parameters are the same, the dynamic range of the grounded-load VCIS has half the value of the floating-load inverting VCIS. The factor of ½ in (2–52) is a result of the fact that the voltage across R_L can reach only half the level of the output voltage due to the voltage division required for symmetry. Thus, if $V_{\text{sat}} = 13$ V, the circuit will saturate when $R_L i_L = 6.5$ V.

As an additional point, it should be stressed that the four equal resistances in Figure 2–16 should be closely matched in order for this circuit to operate satisfactorily.

EXAMPLE 2–7

The circuit shown in Figure 2–17 is a simplified version of a circuit that can be used to implement a precision, high-impedance electronic dc voltmeter. The meter is a 0 μA–100 μA microammeter having an internal resistance of 2 kΩ. The desired dc voltage ranges are 0 V–0.1 V, 0 V–1 V, and 0 V–10 V. The microammeter scale will be relabeled as a voltage scale with maximum voltage corresponding to maximum current. Determine the values of the three resistances R_1, R_2, and R_3.

Solution

This circuit represents a good application of a VCIS. The meter responds to current, but the desired circuit function is to respond to voltage. The resistances are then chosen to provide the proper transconductance values so that maximum voltage at the input corresponds to maximum current through the meter.

Although given, the actual internal resistance of the meter is important only if it is large enough to cause the output to saturate over the given operating range. Since the max-

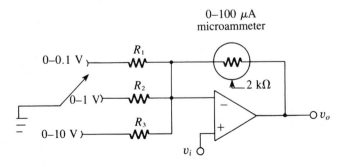

FIGURE 2–17
Simplified electronic voltmeter of Example 2–7.

imum current through the meter should be 100 μA, the maximum voltage drop across the meter is $(100 \times 10^{-6}$ A$) \times (2 \times 10^3$ $\Omega) = 0.2$ V. This voltage added to the maximum input voltage of 10 V results in a maximum output voltage of $v_o(\text{max}) = 10.2$ V, which is within the linear region for the normal range of bias supply voltages. However, these values illustrate that if a much higher input voltage scale were desired, a different circuit arrangement would be required.

In the required design, the three resistances R_1, R_2, and R_3 are chosen so that the maximum current of 100 μA flows when the maximum voltages of 0.1 V, 1 V, and 10 V, respectively, are considered. Thus,

$$R_1 = \frac{0.1 \text{ V}}{100 \ \mu\text{A}} = 1 \text{ k}\Omega \tag{2–53}$$

$$R_2 = \frac{1 \text{ V}}{100 \ \mu\text{A}} = 10 \text{ k}\Omega \tag{2–54}$$

$$R_3 = \frac{10 \text{ V}}{100 \ \mu\text{A}} = 100 \text{ k}\Omega \tag{2–55}$$

EXAMPLE 2–8

It is desired to perform some quality control measurements on a quantity of semiconductor diodes. Specifications call for a measurement of the forward-biased voltage drop when biased at a forward current of 5 mA. Design a circuit to accomplish this purpose so that no adjustment is required each time a new diode is connected.

Solution
Either of the current source circuits could be used, but the simplest one for this case is the floating-load inverting form, shown in Figure 2–18. The input circuit, consisting of the voltage source and the resistance R, is used to establish the required current of 5 mA. The diode is then inserted as the feedback element in the op-amp circuit. With the direction of the current as established in the figure, the output voltage is positive and is equal to the forward-biased diode voltage drop. Once the input circuit establishes the 5-mA current level, the current should ideally remain constant as different diodes are inserted. This result is in sharp contrast to a simple series connection of a diode, a resistor, and a battery, where the current would have to be readjusted each time a new diode is inserted.

There are an infinite number of combinations of E and R that would produce 5 mA, so the choice at this point would depend on the available resources. For example, if -15 V is available, $R = 3$ kΩ is the required value. To obtain the most exact value possible, one might use an existing negative voltage supply and then adjust a variable resistance to establish the current as closely as possible to 5 mA.

Since the op-amp inverting input terminal will assume a virtual ground level, the forward-biased diode voltage drop can be determined by measuring the op-amp output voltage with respect to the actual circuit ground. This procedure eliminates the necessity of connecting a voltmeter directly across the diode, which under some conditions could interfere with the normal function of the circuit.

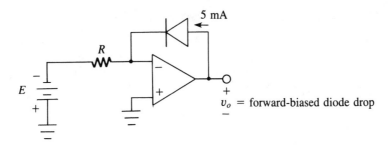

FIGURE 2-18
Measurement circuit of Example 2-8.

EXAMPLE 2-9

Design a voltage-controlled current source to provide a current of 0.5 mA through a variable resistive load in which one side is grounded. The current is to be established with a +15-V source. Determine also the range of the load resistance over which linear operation can be assumed if $V_{\text{sat}} = 13$ V.

Solution
The circuit to be employed is the Howland current source of Figure 2-16. The value of R required is determined from (2-49) to be

$$R = \frac{v_i}{i_L} = \frac{15 \text{ V}}{0.5 \times 10^{-3} \text{ A}} = 30 \times 10^3 \Omega = 30 \text{ k}\Omega \qquad (2\text{-}56)$$

Thus, four closely matched 30-kΩ resistors should be used in the circuit. The VCIS has a transductance of $g_m = 1/30 \times 10^3 = 33.3 \ \mu$S.

The range of load resistance is determined from (2-52). To determine the upper-bound value, the equation is established first as an equality; that is,

$$R_L i_L = \frac{V_{\text{sat}}}{2} = \frac{13}{2} = 6.5 \qquad (2\text{-}57)$$

We then determine R_L as

$$R_L = \frac{6.5}{i_L} = \frac{6.5}{0.5 \times 10^{-3}} = 13 \text{ k}\Omega \qquad (2\text{-}58)$$

Thus, the value of load resistance must be less than 13 kΩ to avoid saturation.

2-7 CURRENT-CONTROLLED SOURCES

In this section we will continue the development of possible op-amp circuits to implement the various voltage-current control functions by analyzing (a) a current-controlled voltage source and (b) a current-controlled current source. One example of each type will be considered.

Current-Controlled Voltage Source (ICVS)

The simplest implementation of an ICVS is shown in Figure 2–19. Since the inverting terminal of the op-amp appears as a virtual ground, current i_i flowing from the source "sees" an effective ground at that point. This current flows through the feedback resistance R, and the output voltage v_o is simply

$$v_o = -Ri_i \qquad (2\text{--}59)$$

The transresistance R_m (in ohms) is

$$R_m = R \qquad (2\text{--}60)$$

One could also associate the minus sign with the transresistance due to the inversion, but this will not be done here. The output voltage of this circuit is thus seen to be a constant times the input current.

Current-Controlled Current Source (ICIS)

An ICIS circuit is shown in Figure 2–20(a). Various quantities are labeled for analysis purposes in Figure 2–20(b). Assuming linear stable operation, the input current must flow through the resistance R_2 since no current flows into the amplifier terminals. A voltage v_2 across R_2 results from this current:

$$v_2 = R_2 i_i \qquad (2\text{--}61)$$

Since the inverting terminal is at a virtual ground, the voltage across R_1 is also v_2. A current then flows through R_1 from ground with the value

$$i_1 = \frac{v_2}{R_1} = \frac{R_2 i_i}{R_1} \qquad (2\text{--}62)$$

where the value of v_2 in terms of i_i in (2–61) was used.

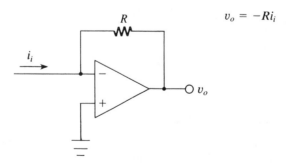

$$v_o = -Ri_i$$

FIGURE 2–19
Current-controlled voltage source circuit.

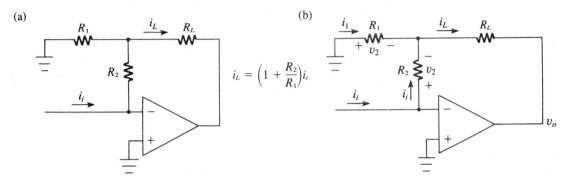

FIGURE 2-20
(a) Current-controlled current source circuit. (b) Various quantities used for analysis in the text labeled on the circuit.

The currents i_i and i_1 come together at the junction between R_1, R_2, and R_L. By Kirchhoff's current law, the current i_L is given by

$$i_L = i_i + i_i \tag{2-63}$$

Substitution of (2–62) in (2–63) and a slight rearrangement yield

$$i_L = \left(1 + \frac{R_2}{R_1}\right) i_i \tag{2-64}$$

The load current is seen to be a constant times the input current. The most important aspect of the circuit, however, is that *the load current is independent of the load resistance R_L.* Like the VCIS circuits of Section 2–6, the desired function is a current source, but in this case, the output current is controlled by the input current.

From the work of Section 1–4, the ICIS may be described by a current gain factor β. The value of β is readily determined from (2–64) as

$$\beta = 1 + \frac{R_2}{R_1} \tag{2-65}$$

The linear range of operation requires that the magnitude of the voltage at the output terminal be less than the saturation voltage. It can be shown that the output voltage magnitude $|v_o|$ is given by

$$|v_o| = \left[R_2 + R_L\left(1 + \frac{R_2}{R_1}\right)\right] |i_i| \tag{2-66}$$

The requirement of linear operation thus imposes the following constraint:

$$\left[R_2 + R_L\left(1 + \frac{R_2}{R_1}\right)\right] |i_i| < V_{\text{sat}} \tag{2-67}$$

EXAMPLE 2–10

It is desired to monitor the ac short-circuit collector signal current in a certain transistor circuit on an oscilloscope. The circuit shown in Figure 2–21 is proposed for this purpose. It is assumed that the capacitive reactance of the coupling capacitor is negligible in the frequency range of interest. Determine the value of R such that a peak short-circuit signal current of 1 mA produces an output peak voltage of 10 V.

Solution
This problem takes much longer to state than to solve, since R is readily determined as

$$R = \frac{10 \text{ V}}{1 \text{ mA}} = 10 \text{ k}\Omega \qquad (2\text{–}68)$$

However, much of the intent of this problem is to illustrate one of many different kinds of functions in which it is desired to convert a given current to a corresponding voltage in order to measure or monitor the current level. The oscilloscope is a voltage-measuring device, so the ICVS circuit performs the necessary conversion. On the oscilloscope, a voltage level of 10 V represents a small signal current level of 1 mA in the transistor circuit.

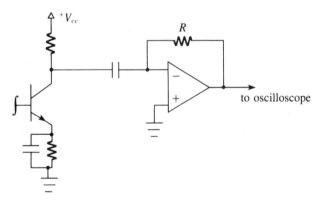

FIGURE 2–21
Circuit of Example 2–10.

EXAMPLE 2–11

It is desired to measure a direct current whose level is expected to be slightly less than 0.1 mA, but the only ammeter available has a full-scale value of 1 mA. The accuracy of the measurement would obviously be degraded considerably if the available meter were used directly, since deflection would be less than 10% of full scale. Design a circuit to multiply the current by a factor of 10 so that the given meter may be used more accurately. It is assumed that resistors will be either obtained or adjusted to meet whatever values are required. The internal resistance of the meter is 100 Ω.

FIGURE 2-22
Circuit of Example 2-11.

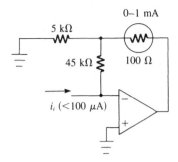

Solution

The given problem can be interpreted as the design of an ICIS having a current gain of $\beta = 10$. Thus, from (2–65),

$$1 + \frac{R_2}{R_1} = 10 \tag{2-69}$$

or

$$\frac{R_2}{R_1} = 9 \tag{2-70}$$

An inspection of the standard 1% resistance values of Appendix A reveals that no two standard values have an exact ratio of 9 to 1. It may be possible, however, to combine some of the values available to produce the required ratio. On the other hand, a variable resistance may be employed and adjusted to yield a correct value. For our purposes, we will simply provide the representative set of values $R_1 = 5$ kΩ and $R_2 = 45$ kΩ, and the circuit is as shown in Figure 2–22.

We need to check the maximum output voltage to ensure that saturation does not occur. From (2–66), $|v_o|$ is determined as

$$|v_o| = (45 \times 10^3 + 100 \times 10) \times 0.1 \times 10^{-3} = 4.6 \text{ V} \tag{2-71}$$

which is well within the linear range of the op-amp.

2-8 LINEAR COMBINATION CIRCUIT

The emphasis in this section will be directed toward the analysis and design of circuits that can be used to form a linear combination of several voltages. Assume that there are n voltages $v_1, v_2, \cdots, v_n$, which are to be combined to yield a voltage v_o in accordance with the equation

$$v_o = A_1 v_1 + A_2 v_2 + \cdots + A_n v_n \tag{2-72}$$

where a given A_k is a constant that may be positive or negative in the most general case. The voltage v_o in (2–72) is said to be a **linear combination** of the voltages v_1 through v_n.

Consider first the case where all A_k constants are negative. The circuit of Figure 2–23 may be used as a linear combination circuit in this case. Assuming stable linear operation, the circuit is readily analyzed by noting that the inverting terminal will be at ground potential, and the voltage across a given resistance R_k is the corresponding voltage v_k. Thus,

$$i_1 = \frac{v_1}{R_1} \qquad \text{(2–73a)}$$

$$i_2 = \frac{v_2}{R_2} \qquad \text{(2–73b)}$$

$$\vdots \qquad \vdots$$

$$i_n = \frac{v_n}{R_n} \qquad \text{(2–73c)}$$

By Kirchhoff's current law, the current i_f is

$$i_f = i_1 + i_2 + \cdots + i_n \qquad \text{(2–74a)}$$

$$i_f = -\frac{v_1}{R_1} + \frac{v_2}{R_2} + \cdots + \frac{v_n}{R_n} \qquad \text{(2–74b)}$$

The output voltage v_o is then

$$v_o = -i_f R_f \qquad \text{(2–75)}$$

Substituting the value of i_f from (2–74b) in (2–75), we obtain

$$v_o = -\frac{R_f}{R_1} v_1 - \frac{R_f}{R_2} v_2 - \cdots - \frac{R_f}{R_n} v_n \qquad \text{(2–76)}$$

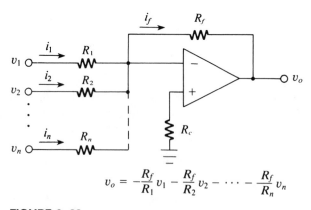

$$v_o = -\frac{R_f}{R_1} v_1 - \frac{R_f}{R_2} v_2 - \cdots - \frac{R_f}{R_n} v_n$$

FIGURE 2–23
Inverting linear combination circuit.

Comparing (2–76) with (2–72), one can readily observe that the linear combination function is achieved with all the constants negative. Although R_f is common in each of the gain constant factors, a different value of input resistance appears for each so that a given A_k is determined as

$$A_k = -\frac{R_f}{R_k} \tag{2–77}$$

✳ This circuit is most often called an **inverting summing circuit**. Strictly speaking, it does more than simply sum the signals, since arbitrary gain constants may be chosen for the n input signals. For this reason the term **linear combination circuit** has been preferred in this text.

If a simple addition and inversion are desired, the choice $R_k = R_f = R$ for all resistances is used. This **inverting addition circuit** is shown in Figure 2–24, and the output voltage v_o is

$$
\begin{aligned}
v_o &= v_1 - v_2 - \cdots - v_n \\
&= -(v_1 + v_2 + \cdots + v_n)
\end{aligned} \tag{2–78}
$$

☞ The reader may now wonder if the form of the noninverting amplifier circuit may be used to establish a noninverting linear combination circuit. The answer is yes, but the circuit is not nearly so convenient to design or implement as the inverting circuit. In the inverting circuit, all inputs are independent of each other as a result of the virtual ground at the inverting input. Thus, the individual gain constants in (2–76) are independent of resistances in other paths, and therefore inputs may be added or deleted at will without interaction with other inputs. Such is not the case with the noninverting linear combination circuit because all inputs interact with one another. For that reason, the noninverting linear combination circuit will not be considered in the text, but the analyses of some limited forms of this circuit are left as exercises for the reader (Problems 2–20, 2–21, 2–49, and 2–50).

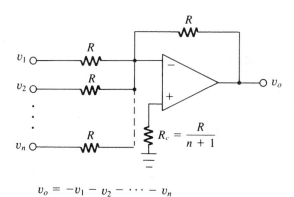

$$v_o = -v_1 - v_2 - \cdots - v_n$$

FIGURE 2–24
Inverting summing circuit.

Actually, in view of the very modest cost of general-purpose op-amps, the inversion limitation of the circuits of Figures 2–23 and 2–24 need not pose any special problems. For example, if all the constants in (2–72) are desired to be positive, a simple inverting circuit with a gain of −1 can be used in conjunction with the inverting linear combination circuit. If some of the gain constants are positive and some are negative, the use of an additional inverting combination circuit and/or simple inversion stages can usually create almost any reasonable combination. This process will be illustrated with the example that follows.

EXAMPLE 2–12

In a certain application, it is desired to combine two signals v_1 and v_2 to form a signal v_o according to the following combination:

$$v_o = -v_1 - 10v_2 \qquad (2\text{--}79)$$

The minimum input impedance for both signal inputs should be no less than 10 kΩ. Design a circuit to meet the requirements.

Solution
The inverting linear combination circuit of Figure 2–23 with two inputs will be used. The following two gain requirements must be met:

$$\frac{R_f}{R_1} = 1 \qquad (2\text{--}80)$$

and

$$\frac{R_f}{R_2} = 10 \qquad (2\text{--}81)$$

There are three resistances required, but only two rigid requirements to meet. However, the minimum acceptable input impedance establishes a third requirement.

For a given value of R_f, the largest gain corresponds to the smallest value of input resistance. We therefore select the input resistance of the signal requiring the largest gain to be at least 10 kΩ. Specifically, we will choose $R_2 = 10$ kΩ, although it could be selected as a larger value if desired. This choice of R_2 quickly leads to $R_f = 100$ kΩ from (2–81) and $R_1 = 100$ kΩ from (2–80). The circuit is shown in Figure 2–25.

FIGURE 2–25
Circuit of Example 2–12.

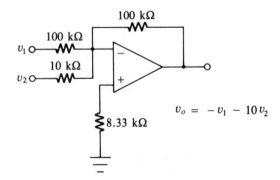

$$v_o = -v_1 - 10\,v_2$$

EXAMPLE 2-13

In a certain signal-processing system, it is necessary to combine and modify the levels of three signals, v_1, v_2, and v_3. The desired combination v_o is to be formed as follows:

$$v_o = 2v_1 + 5v_2 - 10v_3 \qquad (2\text{-}82)$$

Using linear combination circuits and/or inverting amplifiers, design a circuit to perform the given function.

Solution

If the signs of all the gain factors in (2–82) were negative, a single linear combination circuit would suffice. If all the signs were positive, a linear combination circuit followed by a simple inverting amplifier circuit would be appropriate. However, two of the three gain factors have positive signs and the third has a negative sign. Thus, the sign mixture necessitates a more elaborate scheme.

Since the gain factors for the first two terms have positive signs, an even number of inversions for those terms is required. However, the third gain factor has a negative sign, so an odd number of inversions is required for that term.

One solution is to employ a linear combination circuit, which will also invert, and to precede two of the inputs by inverting amplifiers to change the signs of v_1 and v_2 before adding to v_3. This solution is shown in Figure 2–26(a). However, one operational amplifier may be eliminated by first combining v_1 and v_2 in a linear combination circuit and then adding the combination to v_3. This solution is shown in Figure 2–26(b). The various gain factors could be distributed between the different amplifiers if desired. The primary criterion is that the net gain (including sign) for *each* path from input to output must be as specified. The reader may readily verify that this condition is met for the three paths starting with v_1, v_2, and v_3 for both solutions of Figure 2–26.

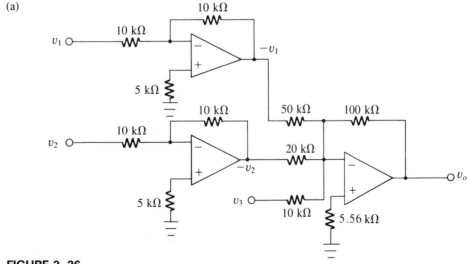

FIGURE 2-26
Two possible designs for the requirements of Example 2-13.

(b)

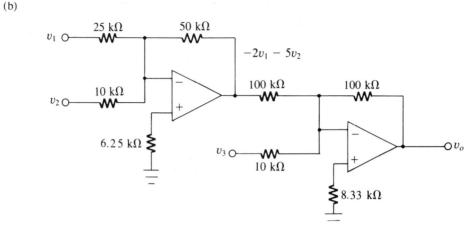

FIGURE 2–26 *(continued)*

2–9 CLOSED-LOOP DIFFERENTIAL AMPLIFIER

One linear combination circuit deserves special consideration because of its usefulness: the **closed-loop differential amplifier**, whose general form is shown in Figure 2–27(a). As we will see shortly, this circuit is capable of combining the signals v_1 and v_2 to yield an output v_o having the following form:

$$v_o = |A_1|v_1 - |A_2|v_2 \tag{2-83}$$

The circuit is most easily analyzed through the principle of superposition, and Figures 2–27(b) and (c) are used for this purpose. The effect of v_1 is considered in (b), and v_2 is assumed to be deenergized (replaced by a short) for this purpose. The signal v_1 is first attenuated by the R_1–R_2 circuit, and the voltage v^+ at the noninverting terminal is

$$v^+ = \frac{R_2}{R_1 + R_2} v_1 \tag{2-84}$$

The effect of v^+ on the output is that of a noninverting amplifier with v^+ as the effective signal. The component of the output v_o' due to v^+ is then

$$v_o' = \left(\frac{R_3 + R_4}{R_3}\right) v^+ \tag{2-85}$$

from the basic gain equation of a noninverting amplifier. Substitution of (2–84) in (2–85) yields

$$v_o' = \left(\frac{R_2}{R_1 + R_2}\right)\left(\frac{R_3 + R_4}{R_3}\right) v_1 \tag{2-86}$$

The effect of v_2 is considered in Figure 2–27(c), and v_1 is considered to be deenergized. The resistances R_1 and R_2 do not affect the presence of the virtual ground at the inverting terminal, since no signal current is assumed to flow through them. The resulting circuit is seen to be equivalent to an inverting amplifier form, and the component of the output v_o'' due to v_2 is readily expressed as

$$v_o'' = -\frac{R_4}{R_3} v_2 \qquad (2\text{-}87)$$

(a)

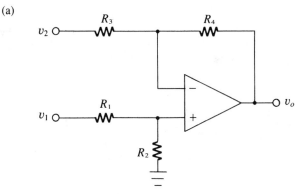

$$v_o = \left(\frac{R_2}{R_1 + R_2}\right)\left(\frac{R_3 + R_4}{R_3}\right)v_1 - \frac{R_4}{R_3} V_2$$

(b)

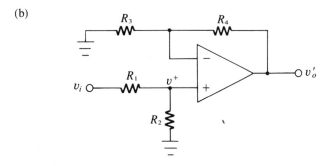

(c)

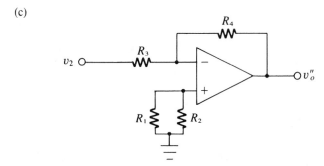

FIGURE 2-27
(a) Closed-loop differential amplifier circuit. When superposition is applied to the two separate inputs, the circuits of (b) and (c) are obtained.

By superposition,

$$v_o = v_o' + v_o'' \tag{2-88}$$

Substitution of (2–86) and (2–87) in (2–88) yields

$$v_o = \left(\frac{R_2}{R_1 + R_2}\right)\left(\frac{R_3 + R_4}{R_3}\right)v_1 - \left(\frac{R_4}{R_3}\right)v_2 \tag{2-89}$$

Comparing (2–89) with (2–83), we readily observe that the desired circuit function is obtained. Since there are two constants to be established and four available resistances, a number of possible combinations of different resistances may be determined for meeting required gain values.

Balanced Closed-Loop Differential Amplifier

The most important case of the closed-loop differential amplifier is the balanced circuit in which the two gain constants are equal and opposite; that is, $|A_1| = |A_2|$. Assume that the following constraints are satisfied:

$$\left.\begin{array}{l} R_1 = R \\ R_2 = AR_1 = AR \\ R_3 = R \\ R_4 = AR_3 = AR \end{array}\right\} \tag{2-90}$$

The resulting circuit is shown in Figure 2–28, and the output v_o in this case can be shown to be

$$v_o = A(v_1 - v_2) \tag{2-91}$$

where A is a positive constant. It can also be readily shown that the correct compensation for bias current is automatically satisfied with this circuit since the dc resistance to ground from both terminals is the same.

FIGURE 2–28
Closed-loop balanced differential amplifier circuit.

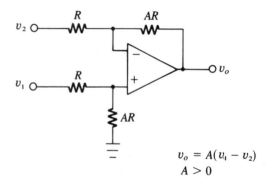

| EXAMPLE 2–14 | A circuit is desired for an instrumentation application in which the difference between two signals v_1 and v_2 is to be amplified by a factor of 10; that is, |

$$v_o = 10(v_1 - v_2) \qquad (2\text{–}92)$$

Design a circuit to perform the required function.

Solution

The design solution may be implemented with a balanced closed-loop differential amplifier, and the circuit form of Figure 2–28 may be adapted readily to the problem. Since no impedance specifications are given, arbitrary but reasonable values of 10 kΩ and 100 kΩ resistances are selected. The circuit design is shown in Figure 2–29.

FIGURE 2–29

Balanced differential amplifier design of Example 2–14.

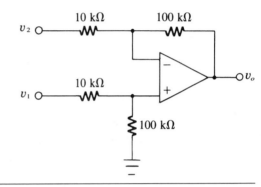

2–10 MULTISIM EXAMPLES

The examples presented in this section are devoted to the modeling and analysis of op-amp circuits. In Example 2–1, the nearly ideal model of an op-amp provided by Multisim will be employed. This model has only three terminals and is very easy to apply. It may be used in situations where ideal or nearly ideal operation can be assumed.

In Examples 2–2 through 2–7, the more realistic 741 model provided by Multisim will be employed. This model requires power supply connections and demonstrates realistic properties and limitations that would be encountered in an actual op-amp.

MULTISIM EXAMPLE 2–1

Using the **ideal** op-amp model, simulate the inverting amplifier circuit of Example 2–1 and perform a dc analysis on the circuit for an input voltage of 1 V.

Solution

As in the examples of Chapter 1, the input dc voltage source is obtained from the **Sources** bin and the resistors are obtained from the **Basic** bin. Since the two resistances are standard values, either the standard or the virtual resistors may be used.

All op-amp models are contained in the **Analog** parts labeled as **"ANA"**, which is the fifth bin from the top. When the arrow is brought over that bin, a new menu window opens showing several different classes of analog modules. The op-amp option is the upper-left button, and the word **OPAMP** appears when the arrow is placed above it. Left-click on this button, and a new window with the title **Component Browser** opens. A **Component List** menu appears on the window and various op-amp models may be selected. In this case, the first one indicated as **IDEAL_OPAMP** is selected. Left-click on **OK**, and the op-amp will be transferred to the circuit window in the same manner as other components and may then be dragged to the desired location. Left-click at the desired location and the op-amp will be left there. As in the case of other components, it may be moved later by selecting it.

This particular op-amp model is "nearly ideal" and may be assumed as ideal for most practical purposes. Basically, it is a three-terminal device having only the inverting and noninverting inputs and one output. Obviously, it is not very realistic, since it has no external power supply connections. However, it is useful for modeling in situations where ideal conditions can be assumed and where one does want to bother with the power supply connections of the more realistic models to be considered later.

The circuit involving the ideal op-amp is shown in Figure 2–30. Two points need to be made about the circuit that may be initially confusing to the reader.

The first point is that the op-amp model is arranged exactly as it comes out of the bin, that is with the noninverting input above the inverting input. This is contrary to the arrangement shown for most examples in this and other texts. If desired, we could left-click on **Edit**, followed by a left-click on **Flip Vertical**, and the arrangement would then be exactly as in the text. This will be done in some examples later in the book, but we hesitate to do it yet for the following reason: With the more realistic models to be considered later, this would require that the power supply connections be turned upside down. While this will be done in some later examples, the reader should learn to deal with both forms, since many "real-life" schematics are mixed with regard to these conventions.

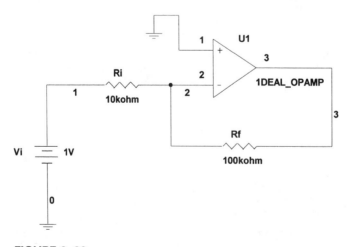

FIGURE 2–30
Circuit of Multisim Example 2–1.

DC Operating Point

2	9.99999u
3	-9.99989
1	1.00000
vvi#branch	-99.99900u

FIGURE 2–31
Output Data of Multisim Example 2–1.

The second point is that active devices have pin numbers that remain fixed on the schematic, irrespective of the numbers identified with nodes in the circuit. For example, there could be a pin number 2 for a device that could be connected to node 5.

To minimize any confusion surrounding these separate numbers, to the extent possible, the author will try to make node numbers the same as pin numbers. In Figure 2–30, it will be noted that pin 2 of the op-amp and node 2 of the circuit are the same, as well as pin 3 and node 3. The one place where this is not possible is a ground node. Thus, pin 1 of the op-amp is actually node 0 of the circuit. The reader need not necessarily be too concerned about doing it this way in building circuits on the screen as long as you remember that in performing analysis, it will be the node numbers for which data are provided in the various output windows.

To determine a dc analysis on the circuit, first left-click on the **Analysis** button and then left-click on **DC Operating Point.** The **DC Operating Point Analysis** window will then open and the variables to be measured are selected in the left-hand block. We will choose to display all of them so hold down the **Ctrl** key while all are highlighted. Left-click on **Plot during simulation** and then on **Simulate.**

A table displaying the three voltages and the current through the one voltage source is shown in Figure 2–31. Node 1 shows 1 V as expected. The output, node 3, shows -9.99989 V, which is about as close to the ideal value of -10 V as one could expect. Node 2 shows a value nearly equal to 10 μV, which is the magnitude of the output voltage divided by the open-loop gain. As explained in Multisim Example 1–1, the positive direction for current flow is into the positive terminal of a voltage source, so the negative value for the last entry is flowing out of the source as expected. Its value is very nearly 100 μA, which is 1 V/10 kΩ.

MULTISIM EXAMPLE 2–2

Using the 741 op-amp model, simulate the inverting amplifier circuit of Example 2–1 and perform a dc analysis on the circuit for an input voltage of 1 V.

Solution

This is essentially a repeat of the preceding example with the ideal model replaced by a more realistic model. The 741 model is obtained in the same fashion as the ideal model except that on the **Component List** menu, select **741** and left-click on **OK.** The actual schematic is much more involved and has the same pin numbers as the real 8-pin DIP package. The complete circuit containing the op-amp and external circuitry is shown in Figure 2–32.

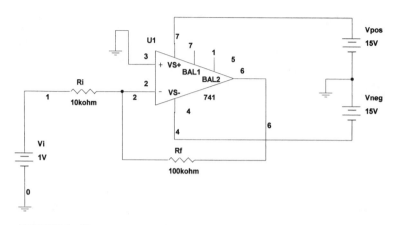

FIGURE 2–32
Circuit of Multisim Example 2–2.

The pin numbers closest to the op-amp are those of the package itself. As in the preceding example, the wiring was carefully selected in a manner to make as many circuit nodes as possible agree with the op-amp pin numbers. The one exception again is the noninverting input, which is pin 3 on the op-amp, but is actually connected to the circuit node 0.

A major difference between the wiring of the ideal op-amp and this more realistic component is that power supplies are required. Two dc power supplies (batteries) are connected on the right-hand side of the circuit with the midpoint grounded. Their **Reference ID** designations were set to the names shown, and each was set to a value of 15 V. The positive voltage is connected to pin 7 and the negative voltage is connected to pin 4. Pins 1 and 5 (offset null terminals) are left disconnected.

As in Multisim Example 2–1, a dc analysis was performed by using the **DC Operating Point Analysis**, and the tabulated results of all voltages and currents is shown in Figure 2–33. The voltages at nodes 1, 7, and 4 are obviously 1 V, 15 V, and −15 V, respectively. The output voltage (node 6) is −9.98120 V, which differs insignificantly from the ideal value of −10 V. In the preceding example, the value was −9.99989 V, which was

DC Operating Point

4	−15.00000
2	1.07936m
1	1.00000
vvi#branch	−99.89207u
vvpos#branch	−30.01499m
vvneg#branch	−30.01516m
7	15.00000
6	−9.98120

FIGURE 2–33
Output Data of Multisim Example 2–2.

a result of the almost ideal op-amp used. The voltage at the inverting input (node 2) is close to 1 mV.

Along with the current flowing out of the input source (approximately 100 μA), we also have the currents flowing in the power supply sources. Each has a value close to 30 mA and the direction in each case is out of the positive terminal of the source.

MULTISIM EXAMPLE 2–3

Using the **741** op-amp model, simulate the noninverting amplifier circuit of Example 2–2 and perform a dc analysis on the circuit for an input voltage of 1 V.

Solution

The circuit construction is very similar to that of the preceding example except that the input is applied in this case to the noninverting input. The final circuit is shown in Figure 2–34. Note that in this example none of the op-amp pins are grounded, so it was possible to force all the node numbers to coincide with the op-amp pin numbers. Once again, this step is not necessary, but the author has chosen to do it whenever possible to make the numbering system as clear as possible for the reader.

As in the preceding two examples, the **DC Operating Point Analysis** was performed and all node voltages and currents through voltage sources were read. The results are shown in Figure 2–35.

The voltages at nodes 3, 4, and 7 are 1 V, -15 V and 15 V as expected since these are source voltages with respect to ground. The op-amp output voltage (node 6) is 10.01584 V, which differs insignificantly from the ideal value of 10 V. The voltage at the inverting input (node 2) is 1.00095 V, which is extremely close to the voltage at the non-inverting input (1 V) as expected.

For the inverting amplifier considered in the first two examples, the input resistance was 10 kΩ, so the input current was readily predictable as 1 V/10 kΩ. For the non-inverting amplifier, the theoretical input current would be zero if the op-amp were perfect. Due to imperfections, we note a current of about 55 nA flowing out of the 1-V source. The two power supplies have currents of about 30 mA each.

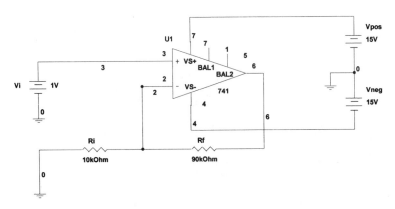

FIGURE 2–34
Circuit of Multisim Example 2–3.

DC Operating Point

4	-15.00000
2	1.00095
vvi#branch	-54.98716n
vvpos#branch	-30.01506m
vvneg#branch	-30.01516m
7	15.00000
6	10.01584
3	1.00000

FIGURE 2–35
Output Data of Example 2–3.

MULTISIM EXAMPLE 2–4

For the inverting amplifier circuit of Multisim Example 2–2, use the **DC Sweep** analysis command to plot an input-output characteristic curve as the input voltage is varied from −2 V to 2 V.

Solution

The circuit is essentially the same as in Multisim Example 2–2, and while it is unnecessary to build the circuit again, to illustrate certain points, it is shown again in Figure 2–36 with some slight modifications. A dc sweep involves selecting one or more dc sources, allowing them to change values over a specified range, and observing how one of the voltage or current variables changes. The range will be specified shortly in a separate window, so it does not matter what values have been set for the sources to be varied. In Figure 2–36, an arbitrary value of 0 V has been set, but it could be anything. Using the **Label** window of the **Label** tab in the properties window, an additional comment was inserted to alert the reader that the voltage is to be swept from −2 V to 2 V. This was done for convenience and is not really necessary.

To perform a dc sweep, first left-click on the **Analysis** button on the **Design Bar,** and when the menu opens, left-click on **DC Sweep**. A window with the title **Dc Sweep**

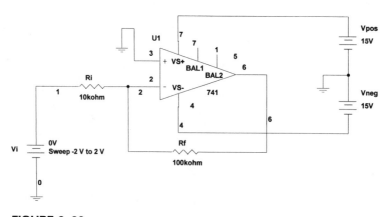

FIGURE 2–36
Circuit of Multisim Example 2–4.

Analysis will then open. The tab will initially be set at **Analysis Parameters** and the purpose here is to select up to two sources to vary. Only one source will be varied in this example and so highlight **vvi** beside **Source** under **Source 1**. The next three quantities to be set are the **Start value,** the **Stop value,** and the **Increment** (step size). The first two values are −2 and 2, respectively. An arbitrary choice of 0.1 V was made for the step size.

Selection of the output variables is achieved by a procedure similar to that of other analysis types considered earlier. Left-click on the **Output variables** tab and a new window will open to allow the appropriate variable to be identified. Only the op-amp output voltage will be selected in this case, so node number **6** should be highlighted. Left-click on **Plot during simulation** and the node number will be transferred to the right-hand side. Finally left-click on **Simulate** and the analysis will be performed.

The plot resulting from the simulation is shown in Figure 2–37, with some slight modifications. We have chosen to rename the horizontal and vertical axes as "Input Voltage (V)" and "Output Voltage (V)," respectively. Multisim uses the title "DC transfer characteristic," and it has been left that way. This is a term meaning the same as "Input-Output Characteristic," but we prefer the latter term so that it is not confused with the frequency-dependent type of transfer function considered in Chapter 1. A grid was also added.

Certain realistic properties of this op-amp model are apparent at the sides of the graph. The saturation voltages appear to be in the neighborhood of −14 V and 14 V. We could actually measure these voltages more accurately using **cursors** provided with Multisim, but we will postpone that process until later. Note that the slope of the curve in the active region is essentially −10, which is the expected voltage gain of the circuit.

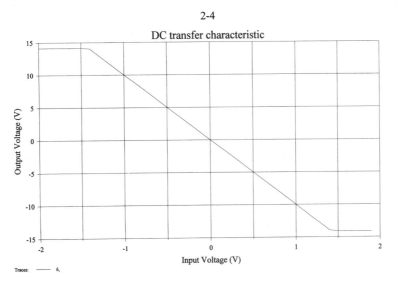

FIGURE 2–37

DC sweep of Multisim Example 2–4.

MULTISIM EXAMPLE 2-5

For the noninverting amplifier circuit of Multisim Example 2–3, use the **DC Sweep** analysis command to plot an input-output characteristic curve as the input voltage is varied from −2 V to 2 V.

Solution

This example follows exactly the same procedure as in Multisim Example 2–4 except that the noninverting amplifier of Multisim Example 2–3 is employed. The circuit with very slight changes for illustration is shown in Figure 2–38. Again, the source voltage is arbitrarily set at 0 and the sweep range has been labeled near the source.

Following the same sequence of steps as in Multisim Example 2–3, the resulting input-output characteristic curve is shown in Figure 2–39. Once again we see saturation occurring at about −14 V and 14 V. In this case, the characteristic curve in the active region has a positive slope of 10, representing the voltage gain.

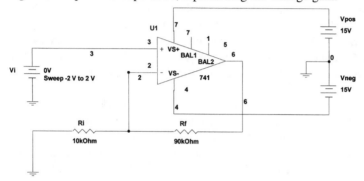

FIGURE 2–38
Circuit of Multisim Example 2–5.

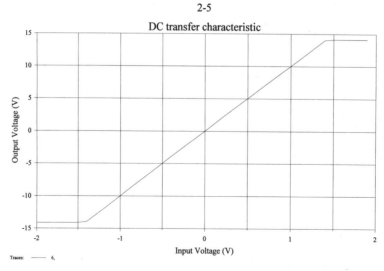

FIGURE 2–39
DC sweep of Multisim Example 2–5.

MULTISIM EXAMPLE 2–6

For the inverting amplifier circuit of Multisim Examples 2–2 and 2–4, use the **Transient Analysis** command to observe the input and output waveforms over two cycles when the input is a sinusoidal source having a frequency of 1 kHz and a peak amplitude of 1 V.

Solution

Transient analysis provides instantaneous plots on a point-by-point basis of various voltages and currents in a circuit. Therefore, the observed behavior is similar to an oscilloscopic view of the instantaneous variables over a specified time interval.

The other forms of analysis considered thus far are capable of being as accurate as the models employed in the analysis. That is not always the case with transient analysis. The actual behavior of energy storage elements in a circuit have derivative and/or integral relationships, and these can only be approximated by digital computational techniques. If everything works out well, the results can be quite accurate, but with complex transient phenomena, it takes a reasonable experience to make judgements on the validity of the results. This is a clear illustration of how the computer will never replace the skilled insight of the user who is well-versed in the fundamental understanding of how a circuit works. As we employ transient analysis throughout the text, various pitfalls will be noted to help the reader acquire sound judgement in the use of transient simulation.

The inverting amplifier circuit with the dc source replaced by a sinusoidal source is shown in Figure 2–40. When the basic form of a circuit is to be retained, but with only a few parts to be replaced, there is an easy way to perform this task without redrawing the circuit. First, select the component to be replaced by left-clicking on it. In this case it is the input DC source. Next, left-click on **Edit** in the **Menu** row. When its menu opens left-click on **Replace Part** and a window with the title **Component Browser** will open. Highlight **AC_Voltage_Source** and left-click on **OK**. The dc voltage source will then be replaced by an ac sinusoidal voltage source whose default values are 1 V and 1000 Hz. Since these are the values desired, no further modifications are required in this example. Both the peak voltage and the frequency can be changed by opening the properties window for the ac source.

To perform a transient analysis, first left-click on the **Analysis** button on the **Design Bar**, and then left-click on **Transient Analysis**. A window with the title **Transient Analysis** will then open and the tab will initially be set in the **Analysis Parameters** set-

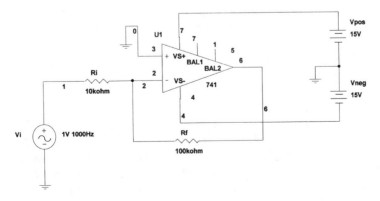

FIGURE 2–40
Circuit of Multisim Example 2–6.

ting. The starting time for the analysis is set in the **Start time** space, and the ending time is set in the **End time** space. For many transient problem the starting time will be assumed as $t = 0$, which is the value used in this example. Since the period corresponding to one cycle at 1 kHz is 1 ms, and two cycles are desired, set the **End time** to 0.002. (The time must be set in seconds.)

We now arrive at a point where the decision process may require some insight and some trial and error. The first block is entitled **Set maximum time step**. The default position is for this block to the checked, and it should be left in that position at this time.

The next three slots provide three different approaches to the transient analysis. The default option is the last one with the title **Generate time steps automatically**. This is the easiest one to use, since it requires no effort on the part of the user. However, the results may not always be accurate, as will be discussed later. Yet it is recommended that this option be used for the first run simply on a trial basis. For simple transient phenomena, particularly when the waveforms vary very little, this approach may suffice.

The option titled **Minimum number of time points** allows the user to specify a lower bound such that there will always be at least the number of points indicated. The author finds this option often useful when the input is a sine wave as will be illustrated shortly.

The option entitled **Maximum time step** permits the user to specify the maximum step between transient points. (Don't confuse this option with the checked box considered three paragraphs earlier.) When this option is being used, this value should be much less than the shortest time constant in the circuit.

After the parameters in this window are selected, left-click on the **Output variables** tab and then left-click on the **Output variables** tab at the top. A new window having the same form as several others already considered in other types of analysis will open. Under **Variables in circuit**, select the desired variables to be plotted, and left-click on **Plot during simulation**. In this example, the input voltage (node 1) and the output voltage (node 6) were selected. As in other forms of analysis, the variables will be transferred to the area on the right, which is called **Selected variables for display**. Finally, left-click on **Simulate**, and the analysis will be performed.

The two node voltages selected will appear on a graph in which the independent variable time is defined from 0 to 2 ms. When this particular analysis was performed with the **Generate time steps automatically** option, the waveforms appeared somewhat jagged at the peaks. A new run was obtained with the **Minimum number of time points** option, with the value set to 200 (or 100 points per cycle) the results were quite satisfactory and are shown in Figure 2–41. Several modifications of the graph were made to enhance the views. The input voltage is the solid curve having a peak value of 1 V and the output is the dashed curve having a peak value of 10 V, but in phase opposition to the input as expected.

In the version of 6.21 used by the author, the initial abbreviation for "second" appears as "S," which is the standard abbreviation for siemens (conductance). It was changed to "s" in this and subsequent examples in accordance with standard notation.

This first transient example illustrates how one must carefully evaluate the results of a transient analysis. This was easy to adjust in this case because we knew what the output was supposed to look like! In general, however, it is prudent to experiment with different approaches to see if consistent results are obtained.

2-6

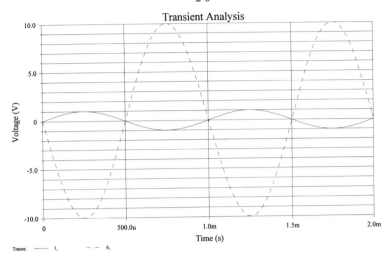

Transient Analysis

FIGURE 2–41
Transient response of Multisim Example 2–6.

MULTISIM EXAMPLE 2–7

For the noninverting amplifier circuit of Multisim Examples 2–3 and 2–5, use the **Transient Analysis** command to observe the input and output waveforms over two cycles when the input is a sinusoidal source having a frequency of 1 kHz and a peak amplitude of 1 V.

Solution
This example follows the same procedure as in Multisim Example 2–6 except that the non-inverting amplifier of Multisim Example 2–3 is employed. The circuit with the dc source replaced by the ac sinusoidal source is shown in Figure 2–42.

Following the same sequence of steps as in Multisim Example 2–6, plots of the input and output voltages are shown in Figure 2–43. In this case, the voltages are in phase as expected.

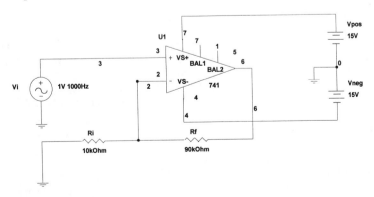

FIGURE 2–42
Circuit of Multisim Example 2–7.

2-7

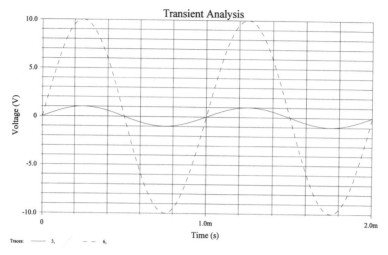

FIGURE 2–43

Transient response of Multisim Example 2–7.

SUMMARY

1. Operational amplifiers are available as high-gain direct-coupled linear integrated circuits and are capable of performing many signal-processing functions.

2. Most op-amps are powered by a series connection of two power supplies of opposite polarity, and the common ground point is established at the midpoint.

3. The op-amp saturation voltages typically occur about 2 volts back from the power supply levels; that is, for ±15-V power supplies, the saturation voltages are about ±13 V.

4. The three signal terminals of an op-amp are the noninverting input, the inverting input, and the output.

5. The three assumptions made for an ideal op-amp are infinite input impedance, zero output impedance, and infinite gain.

6. The three implications of the preceding assumptions are that the input signal current is zero, the output voltage is independent of the load, and the voltages at the two input terminals assume equal values.

7. The two basic types of voltage-controlled voltage sources are the inverting amplifier and the noninverting amplifier, whose closed-loop gains may be easily established by simple resistance ratios.

8. The voltage follower is a special case of a noninverting amplifier with a gain of unity, and it can be used to isolate a source from a load.

9. Operational amplifiers may be used to implement voltage-controlled current sources, current-controlled voltage sources, and current-controlled current sources.

10. Operational amplifiers may be used to form various linear combinations of signals, including a summing circuit.

11. A closed-loop differential amplifier, in which the output voltage is proportional to the difference between input signals, may be implemented with an op-amp.

PROBLEMS

Drill Problems

Note: Unless indicated otherwise, assume that the power supply voltages are ± 15 V and that the saturation voltages are $\pm V_{sat} = \pm 13$ V in all problems.

2–1. Consider the circuit of Figure P2–1 with $R_i = 10$ kΩ and $R_f = 50$ kΩ.
  a. Determine the closed-loop voltage gain.
  b. Determine the input impedance of the circuit.
  c. Determine the ideal output impedance of the circuit.
  d. Determine the peak input voltage v_i(peak) for which linear operation is possible.
  e. Determine the output voltage for each of the following values of input voltage: $v_i = 0$ V, -1 V, 2 V, -3 V, 4 V.

FIGURE P2–1

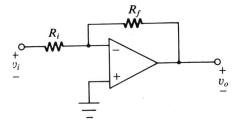

2–2. Consider the circuit of Figure P2–1 with $R_i = 1.2$ kΩ and $R_f = 18$ kΩ.
  a. Determine the closed-loop voltage gain.
  b. Determine the input impedance of the circuit.
  c. Determine the ideal output impedance of the circuit.
  d. Determine the peak input voltage v_i(peak) for which linear operation is possible.
  e. Determine the output voltage for each of the following values of input voltage: $v_i = 0$ V, -0.4 V, $+0.8$ V, -1.2 V, 1.6 V.

2–3. Consider the circuit of Figure P2–3 with $R_i = 10\Omega$ and $R_f = 50$ kΩ.
  a. Determine the closed-loop voltage gain.
  b. Determine the ideal input impedance of the circuit.
  c. Determine the ideal output impedance of the circuit.
  d. Determine the peak input voltage v_i(peak) for which linear operation is possible.
  e. Determine the output voltage for each of the following values of input voltage: $v_i = 0$ V, -1 V, 2 V, -3 V, 4 V.

FIGURE P2–3

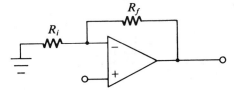

2–4. Consider the circuit of Figure P2–3 with $R_i = 12$ kΩ and $R_f = 180$ kΩ.
  a. Determine the closed-loop voltage gain.
  b. Determine the ideal input impedance of the circuit.
  c. Determine the ideal output impedance of the circuit.
  d. Determine the peak input voltage v_i(peak) for which linear operation is possible.
  e. Determine the output voltage for each of the following values of input voltage: $v_i = 0$ V, -0.4 V, $+0.8$ V, -1.2 V.

2–5. For the circuit of Problem 2–1 with $v_i = -2$ V, assume that an external load $R_L = 2$ kΩ is connected to the output. Determine the total op-amp output current.

2–6. For the circuit of Problem 2–2 with $v_i = -0.8$ V, assume that an external load $R_L = 2$ kΩ is connected to the output. Determine the total op-amp output current.

2–7. Consider the circuit of Figure P2–7 with $R = 1$ kΩ.

    a. Determine the transconductance.

    b. For $v_i = 10$ V, determine i_L for $R_L = 500$ Ω.

    c. Repeat (b) for $R_L = 1$ kΩ.

    d. Determine the maximum value of R_L for linear operation.

FIGURE P2–7

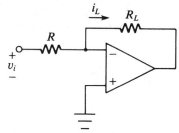

2–8. Consider the circuit of Figure P2–8 with $R = 5$ kΩ.

    a. Determine the transconductance.

    b. For $v_i = 2$ V, determine i_L for $R_L = 8$ kΩ.

    c. Repeat (b) for $R_L = 16$ kΩ.

    d. Determine the maximum value of R_L for linear operation.

FIGURE P2–8

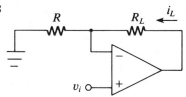

2–9. Consider the circuit of Figure P2–9 with $R = 2$ kΩ.

    a. Determine the transconductance.

    b. For $v_i = 6$ V, determine i_L for $R_L = 1.2$ kΩ.

    c. Determine the maximum value of R_L for linear operation.

FIGURE P2–9

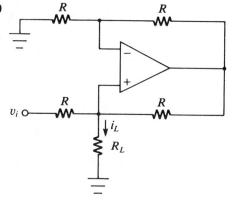

2–10. Consider the circuit of Figure P2–9 with $R = 20$ kΩ.
    a. Determine the transconductance.
    b. For $v_i = 8$ V, determine i_L for $R_L = 10$ kΩ.
    c. Determine the maximum value of R_L for linear operation.

2–11. Consider the circuit of Figure P2–11 with $R = 2$ kΩ.
    a. Determine the transresistance.
    b. Determine the output voltage for $i_i = 5$ mA.

FIGURE P2–11

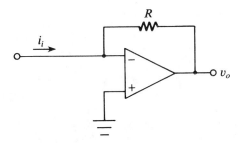

2–12. Consider the circuit of Figure P2–11 with $R = 2$ MΩ.
    a. Determine the transresistance.
    b. Determine the output voltage for $i_i = 5$ μA.

2–13. Consider the circuit of Figure P2–13 with $R_1 = 2$ kΩ and $R_2 = 8$ kΩ.
    a. Determine the value of β.
    b. For $i_i = 0.6$ mA and $R_L = 1$ kΩ, verify linear operation.
    c. Determine i_L for the condition of (b).

FIGURE P2–13

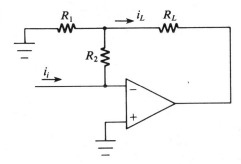

2–14. Consider the circuit of Figure P2–13 with $R_1 = 1$ kΩ and $R_2 = 15$ kΩ.
    a. Determine the value of β.
    b. For $i_i = 0.1$ mA and $R_L = 2$ kΩ, verify linear operation.
    c. Determine i_L for the condition of (b).

2–15. Consider the circuit of Figure P2–15 with $R_f = 100$ kΩ, $R_1 = 100$ kΩ, $R_2 = 50$ kΩ, and $R_3 = 25$ kΩ.
    a. Write an equation for v_o in terms of the three input voltages.
    b. Determine v_o for $v_1 = 10$ V, $v_2 = 3$ V, and $v_3 = -7$ V.
    c. Determine v_o for $v_1 = 8$ V, $v_2 = -4$ V, and $v_3 = 5$ V.

FIGURE P2–15

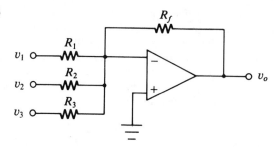

2–16. Consider the circuit of Figure P2–15 with $R_f = 100$ kΩ, $R_1 = 20$ kΩ, $R_2 = 10$ kΩ, and $R_3 = 5$ kΩ.

    a. Write an equation for v_o in terms of the three input voltages.

    b. Determine v_o for $v_1 = -2$ V, $v_2 = -1.5$ V, and $v_3 = 0.8$ V.

    c. Determine v_o for $v_1 = 2$ V, $v_2 = -1.5$ V, and $v_3 = -0.8$ V.

2–17. Consider the circuit of Figure P2–17 with $R_1 = 10$ kΩ and $R_2 = 20$ kΩ.

    a. Write an equation for v_o in terms of the two input voltages.

    b. Determine v_o for $v_1 = 3$ V and $v_2 = 3$ V.

    c. Determine v_o for $v_1 = 3$ V and $v_2 = -3$ V.

    d. Determine v_o for $v_1 = 5$ V and $v_2 = -5$ V.

FIGURE P2–17

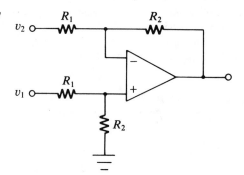

2–18. Consider the circuit of Figure P2–17 with $R_1 = 10$ kΩ and $R_2 = 50$ kΩ.

    a. Write an equation for v_o in terms of the two input voltages.

    b. Determine v_o for $v_1 = 4$ V and $v_2 = 2$ V.

    c. Determine v_o for $v_1 = 2$ V and $v_2 = 4$ V.

    d. Determine v_o for $v_1 = 4$ V and $v_2 = -2$ V.

Derivation Problems

2–19. An alternate form of a voltage-controlled current source for a grounded load is shown in Figure P2–19.

    a. Show that the load current is given by

$$i_L = -\frac{v_i}{R}$$

   b. Show that the requirement for linear operation is

$$|i_L|(R + 2R_L) < V_{sat}$$

FIGURE P2–19

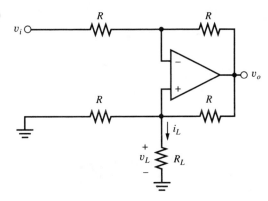

2–20. The circuit of Figure P2–20 is a two-input *noninverting linear combination circuit.* Show that the output voltage v_o is given by

$$v_o = \frac{(R_3 + R_4)}{R_3(R_1 + R_2)}(R_2 v_1 + R_1 v_2)$$

FIGURE P2–20

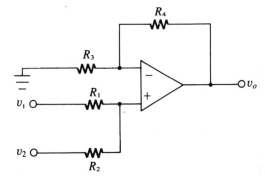

2–21. The circuit of Figure P2–21 is an *equal-gain noninverting linear combination circuit.* Show that for n inputs, $v_1, v_2, \cdots, v_n$, the output v_o is given by

$$v_o = \frac{(R_i + R_f)}{nR_i}(v_1 + v_2 + \cdots + v_n)$$

FIGURE P2–21

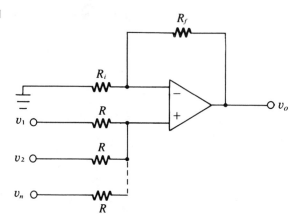

2–22. The circuit of Figure P2–22 permits a large gain to be achieved in an inverting circuit while maintaining a moderate to large input impedance and without requiring excessively large resistance values.

a. Show that the gain is

$$A_{CL} = \frac{v_o}{v_i} = -\left[\frac{R_3}{R_i}\left(1 + \frac{R_1}{R_2}\right) + \frac{R_1}{R_i}\right]$$

b. Evaluate the gain for $R_i = R_1 = R_3 = 10 \text{ k}\Omega$ and $R_2 = 100 \text{ }\Omega$.

FIGURE P2–22

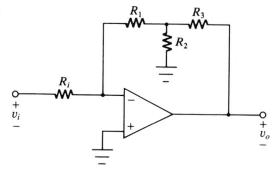

Design Problems

Note: In all design problems in which *standard* 1% or 5% resistance values are indicated, refer to Appendix A for the proper values. If standard values are not indicated, you may assume arbitrary values subject to the constraints given.

2–23. Design an inverting VCVS amplifier circuit using standard 1% resistance values to meet the following specifications:

a. magnitude of voltage gain = 20
b. input impedance $\geq 10 \text{ k}\Omega$
c. resistance values $\leq 500 \text{ k}\Omega$

2–24. Design an inverting VCVS circuit using standard 1% resistance values to meet the following specifications:
   a. magnitude of voltage gain $= 100$
   b. input impedance $\geq 5\ k\Omega$
   c. resistance values $\leq 1\ M\Omega$

2–25. Design a noninverting VCVS circuit using standard 1% resistance values to provide a voltage gain of 60. Use resistances in the range from $1\ k\Omega$ to $100\ k\Omega$.

2–26. Design a noninverting VCVS circuit using standard 1% resistance values to provide a voltage gain of 22. Use resistances in the range from $1\ k\Omega$ to $100\ k\Omega$.

2–27. Using standard 1% resistance values, design an inverting amplifier circuit to have a voltage gain whose magnitude is as close as possible to 25. Use resistances in the range from $10\ k\Omega$ to $500\ k\Omega$, and compute the "exact" gain achieved.

2–28. Repeat the design of Problem 2–27 for a noninverting amplifier.

2–29. An inverting amplifier having an adjustable gain is desired for a given application. The gain is to be adjustable from 0 to -20 by a single potentiometer. The input impedance must be at least $10\ k\Omega$ for all possible gain settings. Design the circuit and specify the range of the potentiometer.

2–30. The noninverting VCVS amplifier of Figure P2–30 is to be designed so that it can be adjusted from a gain of 10 to 25 by the potentiometer. The value of R_i is selected as $R_i = 1\ k\Omega$. Determine the value of R_f and the range of the potentiometer R_v.

FIGURE P2–30

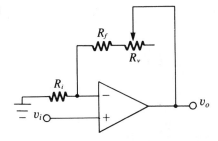

2–31. Design a floating-load, inverting voltage-controlled current source to have a trans-conductance $g_m = 10^{-4}$ S. Determine the maximum value of R_L for linear operation if i_L is set to 1 mA.

2–32. Repeat the design and analysis of Problem 2–31 given that the noninverting form of the floating-load VCIS is used.

2–33. Design a VCIS to produce a load current of 4 mA in a variable grounded load from a voltage source of 12 V. Determine the maximum value of the load resistance for linear operation.

2–34. Design a grounded-load VCIS to have a transconductance of 500 μS. For $v_i = 6$ V, determine the maximum value of the load resistance for linear operation.

2–35. Design a current-controlled voltage source to have a transresistance of $2\ k\Omega$. Determine the peak value $i_i(\text{peak})$ permitted for the input current for linear operation.

2–36. Design a current-controlled voltage source to have a transresistance of $10\ k\Omega$. Determine the peak value $i_i(\text{peak})$ permitted for the input current for linear operation.

2–37. Design a current-controlled current source using standard 1% resistors for $\beta = 15$. Use resistances in the range from $1\ k\Omega$ to $20\ k\Omega$.

2–38. Design a current-controlled current source using standard 1% resistors for $\beta = 2$. Use resistances in the range from $1\ k\Omega$ to $10\ k\Omega$.

2–39. Design a linear combination circuit to combine two signals as follows:

$$v_o = -4v_1 - 8v_2$$

The following specifications are imposed:
a. $R_{in} \geq 10$ kΩ at both inputs
b. all resistance values ≤ 100 kΩ

2–40. Design a linear combination circuit to combine three signals as follows:

$$v_o = -2v_1 - 8v_2 - v_3$$

The following specifications are imposed:
a. $R_{in} \geq 20$ kΩ at all inputs
b. all resistance values ≤ 200 kΩ

2–41. Design a balanced closed-loop differential circuit to combine two signals as follows:

$$v_o = 3(v_1 - v_2)$$

Use resistances in the range from 10 kΩ to 100 kΩ.

2–42. Design a balanced closed-loop differential circuit to combine two signals as follows:

$$v_o = 20(v_1 - v_2)$$

Use resistances in the range from 5 kΩ to 100 kΩ

2–43. In a certain signal-processing system, it is necessary to combine and modify the levels of four signals, v_1, v_2, v_3, and v_4. The desired combination v_o is to be formed as follows:

$$v_o = 2v_1 + 5v_2 - 10v_3 - 20v_4$$

The following requirements are imposed:
a. no more than 2 op-amps
b. $R_{in} \geq 10$ kΩ at all inputs
c. all resistance values ≤ 500 kΩ
Design a circuit to perform the function.
It is desired to add four signals to form a composite signal v_o as follows:

$$v_o = v_1 + v_2 + v_3 + v_4$$

2–44. The signals v_1, v_2, v_3, and v_4 (but not their negatives) are available. The following requirements are imposed:
a. no more than 2 op-amps
b. $R_{in} \geq 10$ kΩ at all inputs
Design a circuit to perform the function.

2–45. Design an op-amp circuit to perform the following linear operation:

$$v_o = 10v_1 - 5v_2 - 2v_3$$

The signals v_1, v_2, and v_3 (but not their negatives) are available. The following requirements are imposed:
a. no more than 2 op-amps
b. $R_{in} \geq 100$ kΩ at all external inputs
c. all resistance values ≤ 500 kΩ

2–46. Design an op-amp circuit to perform the following linear operation:

$$v_o = 80v_1 + 120v_2 - 200v_3$$

The signals v_1, v_2, and v_3 (but not their negatives) are available. The following requirements are imposed:

a. no more than 3 op-amps
b. $R_{in} \geq 100$ kΩ at all external inputs
c. all resistance values ≤ 500 kΩ

(*Hint:* Consider the possibility of one noninverting amplifier stage.)

2–47. In a certain audio recording system, it is desired to be able to mix two signals v_1 and v_2 and provide an adjustment for the level of each. The desired sum v_0 will be of the form

$$v_o = k_1 v_1 + k_2 v_2$$

where k_1 and k_2 are *each* adjustable from 0 to $+10$. Two 0-kΩ–100-kΩ adjustable potentiometers are available. Using no more than three op-amps, design a circuit to perform the desired function. The input impedance of each channel should be no less than 10 kΩ.

2–48. Using the closed-loop differential amplifier of Figure 2–27 but *without* the balanced condition, design a circuit to perform the following operation:

$$v_o = 2v_1 - 3v_2$$

2–49. Using the result of Problem 2–20, design a single-stage linear combination circuit to combine v_1 and v_2 to form v_0 as follows:

$$v_o = 5v_1 + 10v_2$$

Use resistances in the range of 10 kΩ to 200 kΩ.

2–50. Using the result of Problem 2–21, design a single-stage noninverting linear combination circuit to combine four signals v_1, v_2, v_3, and v_4 to form v_o as follows:

$$v_o = 10(v_1 + v_2 + v_3 + v_4)$$

Use resistances in the range of 10 kΩ to 500 kΩ.

LABORATORY EXERCISES

VCVS (Inverting Amplifier)
LE 2–1.

1. Connect a basic inverting amplifier circuit with $R_i = 10$ kΩ and $R_f = 100$ kΩ. You may omit the bias compensating resistor for this exercise.
2. Calculate the expected closed-loop voltage gain A_{CL}.
3. Indicate the expected value of the input impedance R_{in} of the composite circuit.
4. Apply a dc voltage of 1 V to the composite circuit input and measure the output voltage. Calculate the voltage gain and compare with the expected value.
5. Apply an ac sinusoidal voltage of 1 V peak. Observe the input and output voltages on a cathode-ray oscilloscope (CRO) noting their phase relationship. Measure the voltage gain again and compare with the expected value.
6. Measure the input impedance of the composite circuit using a voltage divider technique and compare with the expected value.
7. Increase the ac input voltage until saturation occurs on both peaks of the output voltage. Measure positive and negative saturation levels.

VCVS (Noninverting Amplifier)
LE 2–2.

1. Connect a basic noninverting amplifier circuit with $R_i = 10$ kΩ and $R_f = 100$ kΩ. You may omit the compensating resistor for this exercise.

Perform steps (2) through (5) and (7) of LE 2–1. You may *attempt* to perform the measurement of step (6), but don't waste too much time on it!

VCIS (Grounded Load)
LE 2–3.

1. Connect the grounded-load voltage-controlled current source of Figure 2–16(a) with $R = 10$ kΩ.
2. Calculate the expected value of the transconductance g_m.
3. Calculate the expected value of load current when $v_i = 1$ V.
4. Calculate the expected range of load resistance R_L over which the circuit will function properly.
5. Set the load resistance to some mid-range value based on step (4), and apply a dc voltage $v_i = 1$ V to the input. Measure the load current and compare with the expected value.
6. From the measured current, calculate g_m and compare with the expected value.
7. Vary the load resistance from zero to a level exceeding the range estimated in step (4), and observe the current variation.
8. Determine the value of load resistance at which the circuit ceases to function properly.

(Note: This circuit can serve as the current source input for the circuits of LE 2–4 and LE 2–5.)

ICVS
LE 2–4

1. Connect the current-controlled voltage source of Figure 2–19 with $R = 10$ kΩ.
2. Indicate the expected value of the transresistance R_m.
3. Calculate the expected value of output voltage when $i_i = 0.1$ mA.
4. Apply a dc input current $i_i = 0.1$ mA. Measure the output voltage and compare with the expected value (Note: The current source of LE 2–3 can serve as the input.)
5. From the measured output voltage, calculate R_m and compare with the expected value.

ICIS
LE 2–5.

1. Connect the current-controlled current source of Figure 2–19 with $R_1 = 1$ kΩ and $R_2 = 10$ kΩ.
2. Calculate the expected value of the current gain β.
3. Calculate the expected value of load current when $i_i = 0.1$ mA.
4. Calculate the expected range of load resistance R_L over which the circuit will function properly.
5. Set the load resistance to some mid-range value based on step (4), and apply a dc input current $i_i = 0.1$ mA to the input. Measure the load current and compare with the expected value. (Note: The current source of LE 2–3 can serve as the input.)
6. From the measured current, calculate β and compare with the expected value.
7. Vary the load resistance from zero to a level exceeding the range estimated in step (4), and observe the current variation.
8. Determine the value of load resistance at which the circuit ceases to function properly.

Linear Combination Circuit
LE 2–6.

1. Connect a linear combination circuit of the form shown in Figure 2–23 for two inputs with $R_1 = R_2 = R_f = 10\ \text{k}\Omega$.
2. Write an equation that expresses the expected output voltage v_o in terms of the two input voltages v_{i1} and v_{i2}.
3. Apply two convenient separate dc input voltages whose combined sum has a magnitude less than 13 V. Measure the output voltage and compare with the expected value.
4. Apply a negative dc voltage to one input and a 1-kHz ac sine wave to the other input. The sum of the dc voltage magnitude and the peak ac voltage should be less than 13 V. Observe the output and measure the maximum and minimum levels. Compare with expected values.

3

OPERATIONAL AMPLIFIER dc EFFECTS AND LIMITATIONS

OBJECTIVES

After completing this chapter, the reader should be able to:

- Draw a realistic low-frequency model of an op-amp showing *differential input resistance, output resistance,* and *open-loop gain,* indicate their typical values, and determine the values from op-amp data sheets.
- Draw the standard block diagram representation of a feedback amplifier and show how the noninverting amplifier fits the model.
- Calculate the actual values of the low-frequency closed loop gain, the input resistance, and the output resistance for a noninverting op-amp circuit and for an inverting op-amp circuit.
- Define the term *noise gain,* show how it relates to degradation in the ideal gain of an op-amp circuit, and determine the noise gain for a given op-amp configuration.
- Define the terms *input offset voltage, input bias current,* and *input offset current* and draw the equivalent circuit for determining their effects.
- Calculate the dc output offset voltages arising from both the input offset voltage and the input bias currents.
- Determine the value of compensation resistance that minimizes the effects of bias currents.
- Show how a potentiometer can be connected to an op-amp circuit so that the output dc offset can be adjusted to its minimum value.

3–1 INTRODUCTION

We have assumed ideal operational amplifiers for all circuits thus far. Before presenting further applications, we will investigate the validity of the ideal assumptions. None of these assumptions is completely true in practice, and to be fully competent in the analysis and design of op-amp circuits, one must be aware of the limitations. This chapter and Chapter 4 analyze in depth the various op-amp specifications and their effects on performance.

As we will see, most of the effects are difficult to analyze in their exact forms. However, we have developed a number of useful approximate techniques for estimating relative effects. For design purposes, the designer estimates the levels of the effects and then attempts to select an op-amp whose parameters are such that the op-amp may be assumed to be ideal for the desired conditions. Thus, one should always attempt to return to the ideal op-amp assumptions by proper design selections. This is possible only if one understands how to estimate the effects of the various limitations.

In this chapter we focus on various op-amp properties at dc and at very low frequencies, as suggested by the term *dc* in the title. The primary topics include the effects of finite open-loop gain, finite input resistance, nonzero output resistance, input offset voltage, and input bias currents.

We present a model for the op-amp displaying effects of finite open-loop gain, finite input resistance, and nonzero output resistance. From this model we will derive formulas for closed-loop gain, input resistance, and output resistance for both noninverting and inverting amplifiers.

Finally, we explore the significance of input offset voltage and input bias current parameters. A model for predicting the effects of these different parameters is given, and we discuss techniques for minimizing these effects.

3–2 LOW-FREQUENCY MODEL OF OP-AMP

The three basic assumptions employed in the idealized analysis of op-amp circuits in Chapter 2 will now be carefully investigated. A more realistic model applicable to dc and very low frequencies is shown in Figure 3–1. The resistance between the inverting and noninverting inputs is called the *differential input resistance* and is denoted as r_d. The *output resistance* is denoted as r_o. The open-circuit gain A should be properly called the *open-loop differential gain,* but until the common-mode gain is introduced in Section 4–6, we will simply call it the *open-loop gain* and omit any subscripts. The precise meaning of the adjective *differential* will be explained in Section 4–6.

Typical Parameter Values

Before we pursue any analysis, it may be helpful to check some typical values for these parameters. Refer to Appendix C for data on the 741 op-amp. The differential input resistance r_d is found under the heading "Input Resistance" on the third page of the 741 data sheets. A typical value is 2 MΩ, and the minimum value is given as 0.3 MΩ. On the eighth

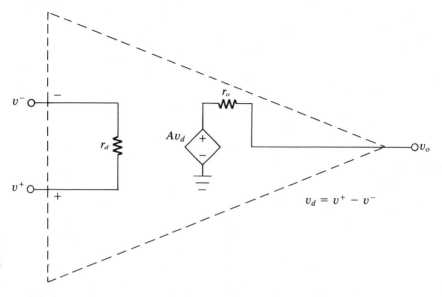

FIGURE 3–1
Realistic model of op-amp with finite input impedance and gain, and
nonzero output impedance.

page of the 741 data sheets, a curve titled "Input Resistance as a Function of Ambient Temperature" is given. In view of the variation with temperature, coupled with the variation of individual units, there is a great deal of uncertainty concerning the input resistance.

The output resistance is found under the heading "Output Resistance" on the third page of the 741 data sheets. A typical value is 75 Ω, and no minimum or maximum values are given.

The open-loop gain A is found under the heading "Large Signal Voltage Gain" on the third page of the 741 data sheets. A typical value for the 741 at dc is given as 200,000; and the minimum value is listed as 50,000. (The minimum value for the 741C is 20,000.) There is also an important variation of the gain as a function of frequency that will be postponed for consideration until Chapter 4. However, for the moment, remember that the single-value gain specification is the dc gain.

The preceding specifications given for the 741 are reasonably typical for bipolar junction transition (BJT) types of op-amps. The major difference with field effect transistor (FET) op-amps is that the input impedance can be much greater. It should be helpful to the reader to keep in mind the approximate range of these typical values for reference in the work that follows.

**EXAMPLE
3–1**

The dc and low-frequency model of a certain op-amp is shown in Figure 3–2. Determine the (a) differential input resistance, (b) output resistance, and (c) open-loop (differential) gain.

FIGURE 3–2
Circuit of Example 3–1.

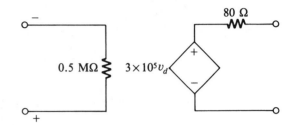

Solution
The three values are readily determined as follows:
(a) $r_d = 0.5 \text{ M}\Omega$
(b) $r_o = 80 \text{ }\Omega$
(c) $A = 3 \times 10^5$

3–3 NONINVERTING AMPLIFIER: GAIN, INPUT IMPEDANCE, AND OUTPUT IMPEDANCE

We will now consider the overall circuit effects when the realistic model introduced in Figure 3–1 is connected in the noninverting amplifier configuration. The noninverting form is selected because it is the easiest one to deal with in the analysis that follows. Further, the properties of other configurations may be deduced from the properties of the noninverting circuit.

Gain

Consider the closed-loop noninverting amplifier circuit of Figure 3–3. When all three non-ideal effects are considered at the same time, the resulting analysis gets very unwieldy, and one tends to get lost in the maze of equations. Based on typical values and the experiences of others who have analyzed the circuit in detail before, the effects of finite open-loop gain on closed-loop gain tend to be more profound in most regions of operation than the effects of finite input impedance and nonzero output impedance. Thus, as a starting point, we will assume that A is finite, but we will momentarily neglect the effects of r_d and r_o.

FIGURE 3–3
Noninverting op-amp circuit used in analysis of finite gain, input impedance, and output impedance effects.

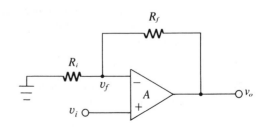

Specifically, it will be assumed that $v_o \simeq A(v_i - v_f)$ in Figure 3–3, even with a load connected to the output of the op-amp. This assumption is reasonable under many operating conditions because, as we will later show, the closed-loop output impedance is quite small, and very little internal drop should occur under most conditions.

As a second approximation, initially assume that any signal current flowing into the input terminals is small compared to the current through $R_i + R_f$. This assumption allows us to calculate the feedback voltage v_f in Figure 3–3 by the simple application of the voltage divider rule, which gives

$$v_f \simeq \frac{R_i}{R_i + R_f} v_o \tag{3–1}$$

Based on the preceding assumptions, a classical feedback amplifier block diagram can be constructed as shown in Figure 3–4. The open-loop forward gain is A, and the feedback ratio is β. The circle on the left represents an effective summing point in which the feedback signal v_f is subtracted from the input signal v_i to yield an "error" signal v_e as follows:

$$v_e = v_i - v_f \tag{3–2}$$

This error signal is actually the differential voltage v_d for this amplifier. However, in some block diagrams the effective error signal may be different, so a more general symbol is used. Note that the plus ($+$) and minus ($-$) symbols at the two inputs to the summing point define the signs of the linear combination, and the equals ($=$) symbol defines the output.

The error signal v_e is multiplied by the forward gain A to yield the output; that is,

$$v_o = Av_e = A(v_i - v_f) \tag{3–3}$$

The output voltage v_o is multiplied by the feedback factor β to produce the feedback signal v_f:

$$v_f = \beta v_o = \frac{R_i}{R_i + R_f} v_o \tag{3–4}$$

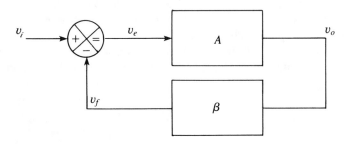

FIGURE 3–4
Classical feedback block diagram model used in analysis of noninverting amplifier.

where

$$\beta = \frac{R_i}{R_i + R_f} \tag{3-5}$$

is the fraction of the output signal fed back to the input.

When (3–4) is substituted in (3–3), the closed-loop gain A_{CL} in general terms becomes, after several manipulations,

$$A_{CL} = \frac{v_o}{v_i} = \frac{A}{1 + A\beta} \tag{3-6a}$$

or

$$A_{CL} = \frac{1/\beta}{1 + 1/A\beta} \tag{3-6b}$$

The results of (3–6a) and (3–6b) apply to any amplifier that can be represented by the model of Figure 3–4.

The quantity $A\beta$ appearing in the two forms of (3–6) is a very important parameter in feedback theory, and it is defined as

$$\text{Loop gain} = A\beta \tag{3-7}$$

Simply stated, the loop gain is the product of all gain factors around the loop with a subtraction circuit understood.

Assume now that the loop gain is very large; that is, $A\beta \gg 1$. In this case, the term $1/A\beta$ in (3–6b) is very small in comparison to unity, and the closed-loop gain becomes

$$A_{CL} \simeq \frac{1}{\beta} \quad \text{for } A\beta \gg 1 \tag{3-8}$$

This equation is one of the most important results of feedback theory, and it is the basis for a large body of linear circuit and system design. In general, A is the gain of an open-loop amplifier whose characteristics are often subject to wide fluctuations and uncertainties. However, β can be a transfer ratio of a simple resistance network whose properties can be precisely controlled to a high degree of certainty. By combining such a network with an amplifier and creating a large loop gain, the net gain can be made to be essentially a function of the stable resistance network alone. Since $\beta < 1$ in most amplifier applications, $1/\beta > 1$, and a net amplification results.

Let us now turn back to the specific circuit of interest, namely, the noninverting amplifier. Upon substitution of β from (3–5) in (3–6b), the closed-loop gain becomes

$$A_{CL} = \frac{(R_i + R_f)/R_i}{1 + \dfrac{R_i + R_f}{AR_i}} \tag{3-9}$$

The expression of (3–9) allows the actual closed-loop gain of the noninverting amplifier to be computed in terms of the open-loop gain. Observe that as $A \to \infty$ in (3–9), the closed-loop gain approaches the ideal limiting value $(R_i + R_f)/R_i$ as developed in Chapter 2. Observe also that since $\beta = R_i/(R_i + R_f)$, the ideal closed-loop gain is $A_{CL} = (R_i + R_f)/R_i = 1/\beta$ in accordance with (3–8).

From the denominator of (3–9), it is seen that the ideal gain approximation improves as A is made larger. However, as one attempts to establish a higher closed-loop gain, $(R_i + R_f)/R_i$ increases, and the loop gain decreases. As a result, there is more error in the closed-loop gain assumption. The trade-off involved will be investigated in Example 3–3 at the end of this section.

Input Impedance

We have established how finite open-loop gain affects the closed-loop gain; thus, the next step is to consider the actual closed-loop input impedance. The circuit shown in Figure 3–5 indicates the primary effects that must be considered in this analysis.

Assume a signal voltage v_i applied to the noninverting input terminal. Since the input impedance to the op-amp is no longer considered infinite, an input current i_i is assumed as shown. The voltage drop across r_d with the more positive terminal assumed at the bottom is $v_i - v_f$, so the current i_i can be expressed as

$$i_i = \frac{v_i - v_f}{r_d} \tag{3–10}$$

However, we will still assume that i_i is much smaller than the current through $R_i + R_f$, so the approximation of (3–1) and (3–4) will be assumed to be valid. Substitution of this approximation with β as a parameter in (3–10) yields

$$i_i = \frac{v_i - \beta v_o}{r_d} \tag{3–11}$$

The output voltage v_o can be expressed as

$$v_o = A(v_i - v_f) \tag{3–12}$$

in which the small internal drop due to the nonzero output impedance has been neglected.

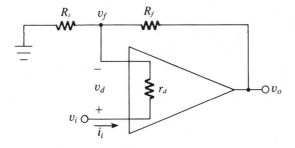

FIGURE 3–5
Model used to study finite input impedance of noninverting amplifier.

The differential voltage $v_i - v_f$ can be readily expressed in terms of the current i_i and the resistance r_d as

$$v_i - v_f = r_d i_i \tag{3-13}$$

Substitution of (3–13) in (3–12) yields

$$v_o = A r_d i_i \tag{3-14}$$

Substitution of (3–14) in (3–11) results in

$$i_i = \frac{v_i - A\beta r_d i_i}{r_d} \tag{3-15}$$

This equation may be readily manipulated to produce the ratio v_i/i_i, which is the input impedance R_{in} (resistive in this case):

$$R_{in} = \frac{v_i}{i_i} = (1 + A\beta)r_d \tag{3-16}$$

The result of (3–16) is quite significant in that it indicates that *the input impedance of the amplifier in the noninverting amplifier configuration is multiplied by* $1 + A\beta$. Since the loop gain is usually quite large in the normal range of stable closed-loop operation, the effective input impedance may be extremely large. For example, assume an open-loop differential input impedance of 500 kΩ, an open-loop dc gain of 10^5, and a β of 10^{-3}. For these typical values, the input impedance exceeds 50 MΩ. While some approximations were made in this development, the level of impedance obtained is sufficiently large to justify the assumption that the current i_i is quite small compared to the current through $R_i + R_f$ for most operating conditions.

In view of the typically large value of loop gain, the result of (3–16) is often approximated as

$$R_{in} \simeq \text{loop gain} \times \text{open-loop input impedance} \tag{3-17}$$

Output Impedance

The last effect that will be considered in this section is the nonzero output impedance. In order to determine this effect, one must use a form of Thevenin's theorem applicable to an active circuit with dependent sources. The circuit model used in this analysis is shown in Figure 3–6. The independent input signal v_i is assumed to be deenergized, so that point is connected to ground. The dependent source is $A(v^+ - v^-) = A(0 - v_f) = -Av_f$ as shown. A fictitious source v_x is connected to the output, and the associated current i_x is calculated. Observe that there are two components, i_1 and i_2.

If we again assume that the current flowing into the amplifier input terminals is small compared to the current through $R_i + R_f$, as was justified in the past few paragraphs, the current i_1 is simply

$$i_1 \simeq \frac{v_x}{R_i + R_f} \tag{3-18}$$

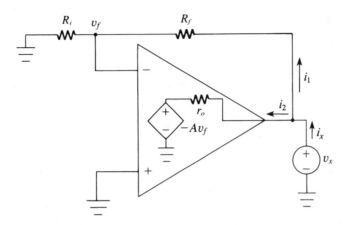

FIGURE 3-6
Model used to study nonzero output impedance of noninverting amplifier.

The current i_2 flowing into the amplifier is determined as

$$i_2 = \frac{v_x - (-Av_f)}{r_o} = \frac{v_x + Av_f}{r_o} \tag{3-19}$$

The voltage v_f may be readily expressed as

$$v_f = \beta v_x \tag{3-20}$$

Substitution of (3-20) into (3-19) yields

$$i_2 = \frac{v_x + A\beta v_x}{r_o} = \frac{(1 + A\beta)v_x}{r_o} \tag{3-21}$$

The numerator of (3-21) is typically much larger than the numerator of (3-18), but the denominator of (3-21) is typically much smaller than the denominator of (3-18). Both of these effects work together to make $i_2 \gg i_1$. Therefore, $i_x \approx i_2$, and the current i_1 will be ignored in the remainder of the analysis. Thus,

$$i_x \simeq \frac{(1 + A\beta)v_x}{r_o} \tag{3-22}$$

The output impedance R_{out} (resistive in this case) may then be determined from (3-22) as

$$R_{\text{out}} = \frac{v_x}{i_x} = \frac{r_o}{1 + A\beta} \tag{3-23}$$

The result of (3-23) is quite significant in that it indicates that *the output impedance of the amplifier in the noninverting amplifier configuration is divided by* $1 + A\beta$. Since the loop gain is usually quite large in the normal range of stable closed-loop operation, the effective output impedance may be extremely small. For example, assume an open-loop output impedance of 75 Ω, an open-loop dc gain of 10^5, and a β of 10^{-3}. For these typical values, the output impedance is less than 0.75 Ω. This small level of output impedance certainly seems to justify the assumption that the internal voltage drop is quite small for most conditions.

In view of the typically large value of loop gain, the result of (3–23) is often approximated as

$$R_{\text{out}} \simeq \frac{\text{open-loop output impedance}}{\text{loop gain}} \tag{3–24}$$

One practical comment on the very small effective output impedance should be made before proceeding further. For the purpose of discussion, assume a closed-loop value $R_{\text{out}} = 1\ \Omega$, and assume an open-circuit op-amp output signal voltage $v_o = 10$ V. Consider an external load requiring 5 mA connected to the op-amp output. Neglecting any current through the feedback circuit, the internal voltage drop is $\Delta v_o = 5$ mA $\times$ 1 Ω = 5 mV. The required load current is within the linear region of operation for most op-amps, and the effective internal resistance acts as the linear model predicts. However, consider the extreme possibility of reducing the effective load resistance down to the level that would suggest maximum power transfer. For such an absurd possibility, the load resistance would be 1 Ω, the current would be 10 V/2 Ω = 5 A, and the power delivered to the external load would be $(5\text{ A})^2 \times 1\ \Omega$ = 25 W! These results are obviously unrealistic for a small signal op-amp and were calculated without applying the realistic limitations on the linear model.

The fact is that the linear model displaying the very small output impedance is valid only if the loading in the circuit does not exceed the maximum that the internal circuitry is capable of bearing. Once the load current reaches a certain maximum level, nonlinear effects appear and change the situation drastically. Depending on the op-amp, either the short-circuit protection circuit may limit the current to a safe level or the op-amp may be burned out.

To summarize, the very low output impedance may be used appropriately in a linear model provided that the external load resistance is sufficiently large that the maximum current rating of the op-amp is not exceeded.

EXAMPLE 3–2

A certain op-amp has $A = 5 \times 10^4$, $r_d = 0.3$ MΩ, and $r_o = 120\ \Omega$. It is connected in a noninverting amplifier circuit with $R_i = 1$ kΩ and $R_f = 99$ kΩ. At dc and very low frequencies, calculate the values of (a) feedback factor β, (b) loop gain, (c) exact closed-loop gain, (d) closed-loop input impedance, and (e) closed-loop output impedance. (f) Use the result of (3–8) to determine the approximate gain and compare with the answer of (c).

Solution
(a) The value of β is

$$\beta = \frac{R_i}{R_i + R_f} = \frac{1\text{ k}\Omega}{1\text{ k}\Omega + 99\text{ k}\Omega} = 0.01 \tag{3–25}$$

(b) The loop gain is

$$\text{Loop gain} = A\beta = 5 \times 10^4 \times 0.01 = 500 \qquad (3\text{–}26)$$

(c) The exact closed-loop gain is

$$A_{CL} = \frac{A}{1 + A\beta} = \frac{5 \times 10^4}{1 + 500} = 99.8 \qquad (3\text{–}27)$$

(d) The closed-loop input impedance is

$$R_{in} = r_d(1 + A\beta) = 0.3 \times 10^6 \times 501 = 1.503 \times 10^8 \ \Omega \qquad (3\text{–}28)$$

(e) The closed-loop output impedance is

$$R_{out} = \frac{r_o}{1 + A\beta} = \frac{120}{501} = 0.24 \ \Omega \qquad (3\text{–}29)$$

(f) Using the approximation of (3–8), we have

$$A_{CL} \approx \frac{1}{\beta} = \frac{1}{0.01} = 100 \qquad (3\text{–}30)$$

This result differs from the result of (c) by about 0.2%.

　　Note also the very large value of input impedance and the very small value of output impedance. The assumption of infinite input impedance and zero output impedance for the noninverting amplifier certainly seems reasonable for most applications.

　　A useful rule of thumb for the *noninverting amplifier* can be stated as follows: *To achieve a desired dc and very low-frequency gain with an error not exceeding 1% of the value predicted by the ideal gain equation of Chapter 2, the open-loop gain should be at least 100 times as large as the desired closed-loop gain.*

EXAMPLE 3–3　The effects of finite open-loop dc gain and ideal closed-loop gain will be investigated in this exercise. Assume that the ideal closed-loop dc gain is first expressed by the ideal simple equation

$$A_{CL}(\text{ideal}) = \frac{R_i + R_f}{R_i} = 1 + \frac{R_f}{R_i} \qquad (3\text{–}31)$$

where the term $A_{CL}(\text{ideal})$ is defined momentarily to avoid confusion with the actual value A_{CL} that will be obtained from finite gain effects.

　　(a) Assume that $A = 10^5$ at dc. Calculate the closed-loop dc gain and error for each of the following ideal values of $A_{CL}(\text{ideal})$: 1, 100, 10^4.

　　(b) Next assume a lower-gain op-amp with $A = 10^4$ at dc. Calculate the closed-loop dc gain and error for each of the $A_{CL}(\text{ideal})$ values in (a).

Solution

The expression of (3–9) permits the actual gain to be computed as a function of the open-loop gain and pertinent resistance ratios. For the noninverting amplifier, this equation may be readily expressed in terms of $A_{CL}(\text{ideal})$ as

$$A_{CL} = \frac{A_{CL}(\text{ideal})}{1 + \dfrac{A_{CL}(\text{ideal})}{A}} \tag{3-32}$$

(a) For this part, $A = 10^5$, so (3-32) becomes

$$A_{CL} = \frac{A_{CL}(\text{ideal})}{1 + \dfrac{A_{CL}(\text{ideal})}{10^5}} \tag{3-33}$$

The percent error can be calculated for each case as

$$\text{Percent error} = \left(\frac{A_{CL}(\text{ideal}) - A_{CL}}{A_{CL}(\text{ideal})}\right) \times 100\% \tag{3-34}$$

The resulting values, rounded appropriately, are tabulated as follows:

A	$A_{CL}(\text{ideal})$	A_{CL}	Percent Error
10^5	1	1.000	0.001%
10^5	100	99.900	0.100%
10^5	10^4	9091	9.091%

(b) For this part, $A = 10^4$, and (3-32) becomes

$$A_{CL} = \frac{A_{CL}(\text{ideal})}{1 + A_{CL}(\text{ideal})/10^4} \tag{3-35}$$

The resulting values of A_{CL} at dc and the rounded percent error are tabulated as follows:

A	$A_{CL}(\text{ideal})$	A_{CL}	Percent Error
10^4	1	1.000	0.010%
10^4	100	99.010	0.990%
10^4	10^4	5000	50%

Noting the values of the closed-loop gain and the percent error for the two sets of data, one can observe the following trends:

1. For a given open-loop gain, the error increases as the desired closed-loop gain increases.
2. For a desired closed-loop gain, the error increases as the open-loop gain decreases.

By generalizing these results, we can infer an approximation for the small error case as follows:

$$\text{Percent error in dc gain} \simeq \frac{A_{CL}(\text{ideal})}{A} \times 100\% \tag{3-36}$$

This formula is very accurate for $A_{CL}(\text{ideal}) \leq A/100$, as can be verified from several of the preceding calculations. However, the accuracy of the formula decreases as $A_{CL}(\text{ideal})$ approaches A.

3-4 INVERTING AMPLIFIER: GAIN, INPUT IMPEDANCE, AND OUTPUT IMPEDANCE

In the last section, the effects of finite gain, finite input impedance, and nonzero output impedance were considered for the noninverting amplifier circuit. In this section, these same imperfections will be considered for the inverting amplifier configuration. It will be seen that some effects appear in essentially the same form as in the noninverting amplifier, but other effects are quite different.

Gain

As was the case for the noninverting amplifier, the effect of finite open-loop gain on closed-loop gain tends to be the most significant factor, so we will first focus on this phenomenon. The most obvious way of analyzing the inverting amplifier circuit is to write network equations directly for the circuit.

Consider the inverting amplifier circuit shown in Figure 3–7 with an assumed finite gain A for the amplifier. As previously indicated, the input impedance will be momentarily assumed as infinite, and the output impedance will be momentarily assumed as zero.

A conventional node voltage equation may be written at the inverting terminal. Summing currents leaving this node and assuming no current into the amplifier, we have

$$\frac{v_e - v_i}{R_i} + \frac{v_e - v_o}{R_f} = 0 \tag{3-37}$$

In order to obtain an expression relating v_i to v_o without other variables appearing, a second equation must be determined. Since the differential voltage is $v_d = v^+ - v^- = 0 - v_e = -v_e$, the output voltage v_o can be readily expressed as

$$v_o = -Av_e \tag{3-38}$$

FIGURE 3-7
Inverting op-amp circuit used in analysis of finite gain, input impedance, and output impedance effects.

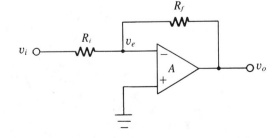

The variable v_e may then be eliminated between (3–37) and (3–38), and the closed-loop gain may be determined. In Problem 3–17, the reader will show that $A_{CL} = v_o/v_i$ is given by

$$A_{CL} = \frac{-R_f/R_i}{1 + \dfrac{(R_i + R_f)}{AR_i}} \tag{3–39}$$

Let us pause here and compare this result with the corresponding expression for the non-inverting amplifier, which was given by (3–9). The numerator of each expression represents the ideal closed-loop gain corresponding to infinite open-loop gain, and either expression reduces to the ideal form as $A \rightarrow \infty$. However, it is interesting to note that *the denominators of the two expressions are identical.* This property may puzzle some readers, and so we will investigate it further by deducing a feedback block diagram representation for the system.

The feedback block diagram model for the inverting amplifier is a bit more subtle than for the noninverting amplifier. For this reason we analyzed the circuit directly by writing the network equations first. However, a block diagram will help to clarify some important points about the circuit, such as the question raised in the last paragraph.

A major difference in the feedback forms for the two circuits is that the feedback for the noninverting amplifier is a form of *series* feedback, but the feedback for the inverting amplifier is a form of *shunt* feedback. The reader may not fully appreciate these terms without some background in feedback theory. In the case of series feedback, the input impedance is *increased*, as was evident for the noninverting amplifier in the last section. However, shunt feedback *reduces* the input impedance.

The result of this latter effect is that an attenuation factor must be applied to the input signal ahead of the summing point in the block diagram. The derivation of the transfer ratios within the block diagram is a little tricky. An intuitive approach to provide some justification for the results will be given here.

Refer to the block diagram of Figure 3–8 as it relates to the circuit of Figure 3–7. If we momentarily assume that $v_o = 0$ in Figure 3–7, the voltage v_i' that would appear at the inverting input terminal is determined by the voltage divider rule applied to the resistive network containing R_f and R_i; that is,

$$v_i' = \frac{R_f}{R_i + R_f} v_i \tag{3–40}$$

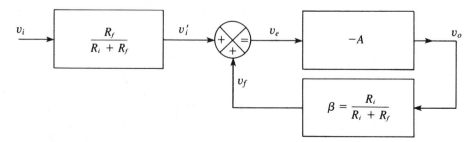

FIGURE 3–8
Feedback block diagram model adapted to inverting amplifier.

Next, we set $v_i = 0$, and determine the feedback voltage v_f established by the output voltage v_o. Again, the voltage divider rule is applied, and the feedback voltage v_f is

$$v_f = \frac{R_i}{R_i + R_f} v_o \tag{3–41}$$

The net "error" voltage v_e applied to the op-amp inverting input terminal is

$$v_e = v_i' + v_f \tag{3–42}$$

Finally, the output voltage is

$$v_o = -Av_e \tag{3–43}$$

The preceding four equations are then converted to block diagram forms as shown in Figure 3–8. When this diagram is compared with Figure 3–4 for the noninverting amplifier, several changes are evident. First, the additional voltage transfer ratio preceding the summing point in Figure 3–8 is present. Second, the summing block in Figure 3–8 is assumed to generate the **sum** of the modified input and feedback signals, while in Figure 3–4, the summing block generated the **difference** between the input and feedback signal. However, this point is largely a matter of convenience because either a *sum* or a *difference* operation could be used for either, provided that the signs of gain blocks are adjusted accordingly. In the noninverting case, the forward gain value was assumed to be positive, so the subtraction of the feedback signal in the block produced the correct overall sign required for negative feedback. However, in the inverting case, the forward gain value is assumed to be negative, so the summing is assumed to be positive.

The value of β for the inverting amplifier is determined from (3–41):

$$\beta = \frac{R_i}{R_i + R_f} \tag{3–44}$$

which is the same as for the noninverting circuit.

The net loop gain is determined as

$$\text{Loop gain} = A\beta = \frac{AR_i}{R_i + R_f} \tag{3–45}$$

This value is also the same as for the noninverting amplifier circuit. The conclusion is that *the loop gain for an inverting amplifier is the same as for a noninverting amplifier.*

An alternate expression for (3–39) in terms of the loop gain is

$$A_{\text{CL}} = \frac{-R_f/R_i}{1 + 1/A\beta} \tag{3–46}$$

Input Impedance

The next step in our analysis is to determine the actual input impedance of the inverting amplifier circuit. The circuit of Figure 3–9 shows the primary effects that must be consid-

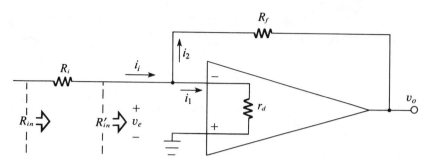

FIGURE 3–9
Model used to study input impedance of inverting amplifier.

ered in this analysis. In the ideal amplifier model that was developed in Chapter 2, the non-inverting terminal is forced to be at zero potential, so the ideal closed-loop input impedance at that point is zero.

The analysis is simplified by first disregarding R_i and computing the input impedance R'_{in} at the inverting terminal. The net circuit input impedance is then obtained very simply by adding R_i to the result.

Assume a voltage v_e and a current i_i. The current consists of two components i_1 and i_2. The component i_1 is the actual current flowing into the op-amp, and it can be readily determined as

$$i_i = \frac{v_e}{r_d} \tag{3–47}$$

The component i_2 can be expressed as

$$i_2 = \frac{v_e - v_o}{R_f} \tag{3–48}$$

Neglecting any voltage drop at the output due to the internal output impedance, v_o can be expressed as

$$v_o = -Av_e \tag{3–49}$$

Substitution of (3–49) in (3–48) yields

$$i_2 = \frac{v_e - (-Av_e)}{R_f} = \frac{(1 + A)v_e}{R_f} \tag{3–50}$$

The net current i_i is

$$i_i = i_1 + i_2 \tag{3–51}$$

In practice, the current i_2 of (3–50) is typically much larger than the current i_1 of (3–47). Thus, assume that $i_2 \gg i_1$ in which case $i_i \simeq i_2$ or

$$i_i \simeq \frac{(1 + A)v_e}{R_f} \qquad (3\text{–}52)$$

The input impedance R'_{in} at the inverting terminal is

$$R'_{in} = \frac{v_e}{i_i} = \frac{R_f}{1 + A} \qquad (3\text{–}53)$$

The net circuit input impedance R_{in} is then

$$R_{in} = \frac{v_i}{i_i} = R_i + \frac{R_f}{1 + A} \qquad (3\text{–}54)$$

A few brief comments will be made on this result. As $A \rightarrow \infty$, $R_{in} \simeq R_i$, which is the result developed in Chapter 2 for the ideal infinite gain op-amp. However, for finite gain A, the input impedance increases, because the potential at the inverting terminal increases to maintain the required differential input voltage. The most significant effect on the input impedance change is not a result of the finite input impedance of the op-amp, but rather it is caused by the external feedback resistance R_f.

Output Impedance

The final imperfection that will be considered for the inverting amplifier is the effect of nonzero output impedance. When the input signal is deenergized and a fictitious external source is applied to the output, the circuit model reduces to the same form as in Figure 3–6, which was used for the noninverting amplifier. Thus, the closed-loop output impedance R_{out} for the inverting amplifier circuit is the same as for the noninverting amplifier:

$$R_{out} = \frac{r_o}{1 + A\beta} \qquad (3\text{–}55)$$

EXAMPLE 3-4

The op-amp of Example 3–2 is used in an inverting amplifier with $R_i = 10$ kΩ and $R_f = 240$ kΩ. At dc and very low frequencies, calculate the values of (a) the feedback factor β, (b) loop gain, (c) exact closed-loop gain, (d) closed-loop input impedance and closed-loop output impedance. (e) Compare the results of (c) and (d) with the value based on an ideal op-amp.

Solution
From Example 3–2, the op-amp parameters are $A = 5 \times 10^4$, $r_d = 0.3$ MΩ, and $r_o = 120$ Ω.

(a) The value of β is

$$\beta = \frac{R_i}{R_i + R_f} = \frac{10\,\text{k}\Omega}{10\,\text{k}\Omega + 240\,\text{k}\Omega} = 0.04 \qquad (3\text{–}56)$$

(b) The loop gain is

$$\text{Loop gain} = A\beta = 5 \times 10^4 \times 0.04 = 2000 \tag{3-57}$$

(c) The closed-loop gain is

$$A_{CL} = \frac{\dfrac{-R_f}{R_i}}{1 + \dfrac{1}{A\beta}} = \frac{\dfrac{-240\,k\Omega}{10\,k\Omega}}{1 + \dfrac{1}{2000}} = -23.99 \tag{3-58}$$

(d) The closed-loop input and output impedances are

$$R_{in} = R_i + \frac{R_f}{1 + A} = 10\,k\Omega + \frac{240\,k\Omega}{1 + 5 \times 10^4} \tag{3-59}$$
$$= 10,000 + 4.8 = 10,004.8\,\Omega$$

$$R_{out} = \frac{r_o}{1 + A\beta} = \frac{120}{1 + 2000} = 0.06\,\Omega \tag{3-60}$$

(e) The assumption of an ideal op-amp results in $A_{CL} = -R_f/R_i = -240\,k\Omega/10\,k\Omega = -24$ and $R_{in} = R_i = 10,000\,\Omega$. These results are very close to the respective values of -23.99 and $10,004.8$ obtained in parts (c) and (d).

3-5 NOISE GAIN

The results of Equations (3-9) and (3-39) indicate basic forms for closed-loop gain as a function of open-loop gain for the specific amplifier circuits considered. However, there is a general pattern emerging that is applicable to more circuits than the two considered. The effects of finite gain can be represented for a wide variety of amplifier circuits by the form

$$A_{CL} = \frac{A_{CL}(\text{ideal})}{1 + (1/A\beta)} \tag{3-61}$$

where $A_{CL}(\text{ideal})$ is the ideal closed-loop gain, A_{CL} is the actual closed-loop gain, A is the open-loop gain, and β is the feedback factor. Observe that (3-9) and (3-39) both fit this general form. Observe also that $A_{CL} \simeq A_{CL}(\text{ideal})$ for $A\beta \gg 1$. Thus, as the loop gain increases, the actual closed-loop gain approaches the ideal limiting case.

Based on the analysis of many types of circuits fitting this model, it is convenient to introduce a new "gain factor" that delineates the effects of various imperfections, some of which have already been considered and some of which will be considered later. This gain factor, referred to as the **noise gain**, will be denoted by the symbol K_n. An equally appropriate description would be the *imperfection gain*, but we will use the established term *noise gain*.

The noise gain K_n is defined as

$$K_n = \frac{1}{\beta} \qquad \text{(3-62)}$$

where β is the voltage feedback factor of the closed-loop response. Substituting the definition of (3-62) in (3-61), we obtain

$$A_{CL} = \frac{A_{CL}(\text{ideal})}{1 + (K_n/A)} \qquad \text{(3-63)}$$

In this form, the opposite effects of K_n and A are most apparent. As A increases, K_n/A decreases, and A_{CL} more closely approaches $A_{CL}(\text{ideal})$. However, as K_n increases, K_n/A increases, and the difference between A_{CL} and $A_{CL}(\text{ideal})$ increases. Thus, larger values of K_n tend to accentuate imperfections in gain.

Determination of Noise Gain

A procedure will now be given for determining K_n for a variety of circuits. First, deenergize all external sources; that is, replace them by their internal impedances. Next, momentarily think of the circuit *as if it were a noninverting amplifier circuit*, and calculate the voltage gain *from the noninverting input terminal to the output terminal* on this basis. The resulting value of voltage gain is K_n.

Let us see if the noninverting and inverting amplifier circuits considered in the preceding two sections are satisfactorily analyzed by this procedure. First, consider the noninverting amplifier of Figure 3-3. In this case, the noise gain K_n is readily determined by the preceding procedure as

$$K_n = \frac{R_i + R_f}{R_i} \qquad \text{(3-64)}$$

Substituting (3-64) and the value of $A_{CL}(\text{ideal})$ in (3-63), we readily obtain the result of (3-9) as determined in Section 3-3.

Next, consider the inverting amplifier of Figure 3-7. The procedure indicates that we treat the circuit *as if* it were a noninverting amplifier. Thus, we momentarily ground the signal input and calculate the gain from the noninverting terminal to the output. The result is readily obtained as

$$K_n = \frac{R_i + R_f}{R_i} \qquad \text{(3-65)}$$

which is obviously the same as for the noninverting amplifier. Substituting (3-65) and the value of $A_{CL}(\text{ideal})$ in (3-63), we readily obtain the result of (3-39) as determined in Section 3-4. The noise gains for the basic inverting and noninverting amplifier circuits are identical, a further delineation of the similar feedback and loop gain forms.

Before proceeding, we should emphasize that *the noninverting amplifier has the property that the ideal closed-loop voltage gain and noise gain are identical.* However, this property is, in general, not true for other circuits. For example, the inverting amplifier has a noise gain given by $(R_i + R_f)/R_i$, but the magnitude of the ideal closed-loop gain

FIGURE 3–10

Circuit of Example 3–5.

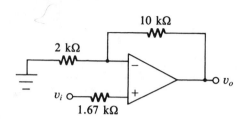

is R_f/R_i. The application of this concept to an entirely different type of circuit will be considered in Example 3–7.

EXAMPLE 3–5

Determine the noise gain for the noninverting amplifier circuit of Figure 3–10.

Solution

For the noninverting amplifier circuit, the noise gain is the same as the closed-loop gain, that is,

$$K_n = 1 + \frac{R_f}{R_i} = 1 + \frac{10 \text{ k}\Omega}{2 \text{ k}\Omega} = 6 \tag{3–66}$$

EXAMPLE 3–6

Determine the noise gain for the inverting amplifier circuit of Figure 3–11.

Solution

To determine the noise gain, we momentarily consider the circuit *as if* it were a noninverting amplifier. The input terminal on the left is, of course, grounded for this purpose, and the gain from the noninverting terminal to the output is

$$K_n = 1 + \frac{R_f}{R_i} = 1 + \frac{4 \text{ k}\Omega}{2 \text{ k}\Omega} = 3 \tag{3–67}$$

Note that the closed-loop gain is $A_{CL} = -4 \text{ k}\Omega/2 \text{ k}\Omega = -2$. Thus, the noise gain in this example is 50% larger than the magnitude of the signal gain.

FIGURE 3–11

Circuit of Example 3–6.

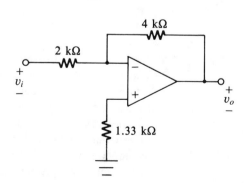

EXAMPLE 3–7

Consider the linear combination circuit in Figure 3–12(a). (a) Determine an ideal expression for v_o in terms of the three input voltages assuming infinite gain. (b) Determine the value of the noise gain. (c) Given that the open-loop dc gain of the amplifier is $A = 2 \times 10^4$, calculate the actual values of the dc gains for the three signals.

Solution

(a) The output voltage v_o may be expressed as

$$v_o = -A_1 v_1 - A_2 v_2 - A_3 v_3 \tag{3–68}$$

The ideal values of the three gain factors are readily determined as

$$A_1(\text{ideal}) = 240 \text{ k}\Omega/12 \text{ k}\Omega = 20$$

$$A_2(\text{ideal}) = 240 \text{ k}\Omega/6 \text{ k}\Omega = 40$$

$$A_3(\text{ideal}) = 240 \text{ k}\Omega/4 \text{ k}\Omega = 60$$

Thus, the ideal form for v_o is given as

$$v_o(\text{ideal}) = -20v_1 - 40v_2 - 60v_3 \tag{3–69}$$

(b) In order to determine the noise gain, we first deenergize the three voltage inputs and connect these points to ground. In Figure 3–12(b), the parallel equivalent resistance is an effective resistance R_i' whose value is

$$R_i' = 4 \text{ k}\Omega \| 6 \text{ k}\Omega \| 12 \text{ k}\Omega = 2 \text{ k}\Omega \tag{3–70}$$

Next we consider the gain as viewed from the noninverting terminal. The compensating resistance has no effect on the overall gain, so the noise gain is determined as

$$K_n = 1 + \frac{240 \text{ k}\Omega}{2 \text{ k}\Omega} = 121 \tag{3–71}$$

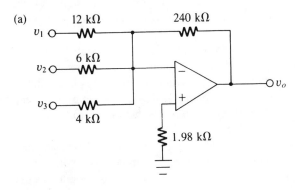

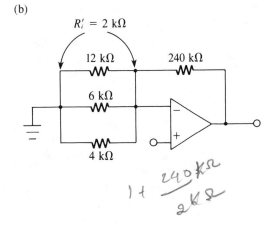

FIGURE 3–12
Circuit of Example 3–7 and noise gain computations.

(c) The actual dc gain of any of the three channels may be determined by the use of (3–63). The denominators will be identical, so if A_n(ideal) is the ideal gain of the nth input, the corresponding gain A_n is

$$A_n = \frac{A_n(\text{ideal})}{1 + \dfrac{K_n}{A}} = \frac{A_n(\text{ideal})}{1 + \dfrac{121}{2 \times 10^4}} = 0.994\, A_n(\text{ideal}) \qquad (3\text{–}72)$$

The actual values of three dc gains are computed as $A_1 = 19.88$, $A_2 = 39.76$, and $A_3 = 59.64$. These values may then be substituted in (3–69) as corrected values for the three gains.

Observe from the results of this problem that the noise gain is considerably larger than the gain for any of the three signals. This result is the sacrifice one makes for permitting a number of different signals to be combined in one circuit.

3–6 OFFSET VOLTAGE AND CURRENTS

According to the ideal gain model for an op-amp, the dc output voltage should be zero when the dc input voltage is zero. In practice, however, a dc offset voltage may be measured at the output terminal even when the dc input voltage is zero. With good circuit design, this voltage may be maintained to a very small level, and null adjustment terminals are available for most op-amps to reduce the offset to zero. However, improper circuit designs can lead to large offset levels, and even with the use of the nulling procedure, continual drifts in operating characteristics can render it impossible to maintain a perfect null in the output offset. Thus, it is important to understand the offset effects and how they are reduced or accentuated by different circuit design techniques.

First, it should be clear that offset effects are much more important in dc amplifiers than in ac amplifiers. If it is not necessary to amplify dc levels, coupling capacitors can be used at appropriate points to block dc levels and pass time-varying signals. In this case, the effects of the offsets are unimportant, unless, of course, one or more are large enough to shift operation to a point where saturation occurs at some point in the signal cycle. (Capacitive coupling at the input and output of amplifier circuits will be considered in conjunction with single dc supply operation in Section 5–8.)

In dc amplifiers, the relative importance of offsets depends on the signal levels. A 10-mV offset voltage may be relatively unimportant in a voltage follower circuit used to isolate a 12-V dc signal level. However, the same offset would be disastrous if the signal level were 10 mV. Thus, when one is working with small signal levels, op-amps with very low offset parameters are desired, and circuit design techniques that minimize the effects of the offsets should be used.

The total output offset voltage can be considered to be a function of two separate effects: (a) *input offset voltage* and (b) *input bias currents.* The input offset voltage is a phenomenon arising from the dc balance of the two inputs. The input bias currents are the actual currents that must flow into or out of the two input terminals to ensure proper operation of the solid-state circuitry. For bipolar junction transistors, the currents represent the actual bias currents for the two bases. Both input terminals must have a dc path to ground. Any dc connection to the output (e.g., voltage follower) or dc path through a grounded source meets this requirement.

Note that both of the two effects are referred to the *input*, even though the measured result is at the *output;* the reason will become clearer as we develop more details.

As previously noted, the total output offset voltage is a combination of these separate effects. We will see shortly, however, that the effects contribute to the output offset voltage in different ways, and for this reason they have been separated for specification purposes. Indeed, we will see that under certain circuit conditions, the output offset may be almost totally due to either one of the particular effects.

Offset Circuit Model

The most convenient way of analyzing offset effects is by an *offset equivalent circuit.* The basic form of the offset equivalent circuit is shown in Figure 3–13.

The circuit form as given could be interpreted as either an inverting or a noninverting amplifier configuration, depending on which of the two terminals on the left is used for an external signal and which is grounded. Thus, the results that will be developed here can be applied directly to either inverting or noninverting configurations.

Observe that one voltage source and two current sources appear in the circuit. These models are fictitious sources whose responses represent the actual effects produced by the offset phenomena. As we will see shortly, the effects may be predicted directly with circuit analysis using the offset source models.

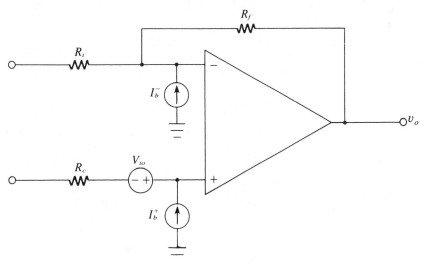

FIGURE 3-13
Op-amp equivalent circuit showing models of input offset voltage and input bias current effects.

The values of the three source models are defined as follows:

$$V_{io} = \text{input offset voltage}$$
$$I_b^+ = \text{input bias current at noninverting input}$$
$$I_b^- = \text{input bias current at inverting input}$$

A few comments about the polarities and directions are in order. The three preceding quantities are always specified as positive values. The actual directions of the two bias currents depend on the polarity of the transistor circuitry in the input stage. As we will see later, the magnitude, but not the polarity, of the output can be predicted with this model. Thus, as far as the magnitude is concerned, the directions of the current sources are *arbitrary* as long as *both have the same direction.* We will choose somewhat arbitrarily in this text to show the directions upward as in Figure 3–13, but some texts choose the opposite direction.*

The effect of the input offset voltage at the output can be either positive or negative at a given time, so the direction of V_{io} is arbitrary. The direction chosen in Figure 3–13 is the one that results in a positive voltage at the output, and this convention will be followed throughout the text. In addition, this voltage source can be placed at either of the two input terminals and the resulting magnitude effect is the same. However, it is somewhat simpler to perform the associated calculations when the source is placed at the noninverting terminal, and this convention will be used.

We are now ready to perform a circuit analysis to predict the effects of the three parameters in Figure 3–13. The basis for the analysis is the *principle of superposition,* which permits us to consider the effect of each source individually. While each individual source is being considered, all other sources are assumed to be deenergized. A voltage source is deenergized by replacing it by a short circuit, and a current source is deenergized by replacing it by an open circuit. Since we are considering the effects of offsets at this time, the input signal source is also deenergized. Thus, both input terminals in Figure 3–13 will be grounded for the analysis, and this condition assumes that the input signal source is an ideal voltage source. Otherwise, the internal source resistance should be added to either R_i or R_c, depending on the configuration.

Input Offset Voltage

The effect due to input offset voltage will be considered first. Let v_{o1} represent the output voltage due to the input offset voltage, and let $|v_{o1}|$ represent the corresponding magnitude. The circuit shown in Figure 3–14(a) is used for this analysis. Observe that I_b^+ and I_b^- have been deenergized (replaced by open circuits). Since no current is assumed to flow in either terminal in this model, the resistance R_c has no effect. The circuit thus reduces to a basic noninverting amplifier form, and the magnitude of the output voltage is

$$|v_{o1}| = \left(1 + \frac{R_f}{R_i}\right) V_{io} \tag{3–73}$$

The actual output voltage could be either this value or its negative.

---

*Strictly speaking, the current directions of Figure 3–13 are correct for PNP input circuitry.

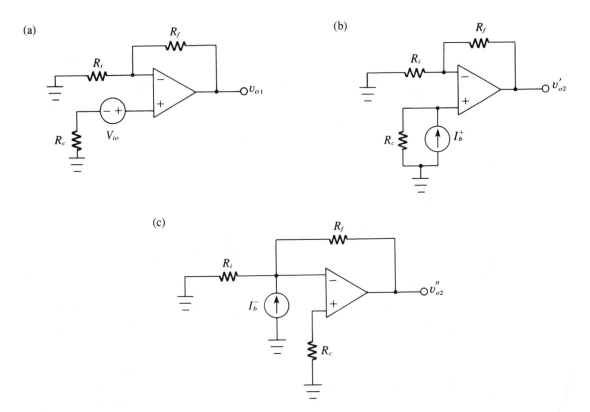

FIGURE 3-14
Superposition applied to op-amp equivalent circuit to obtain effects of offset and bias parameters.

Input Bias Currents

The effect due to I_b^+ will be considered next. Let v'_{o2} represent the component of the output voltage produced by I_b^+, and let $|v'_{o2}|$ represent its magnitude. The circuit shown in Figure 3–14(b) applies in this case. While the current I_b^+ actually represents the current flowing in the op-amp itself, the model may be treated as if the flow is external to the op-amp, with the op-amp itself then assumed to be ideal. The current I_b^+ can then be considered to create a voltage $R_c I_b^+$ at the noninverting terminal with respect to ground. This voltage is then multiplied by the noninverting gain for the amplifier, and the magnitude of the output is

$$|v'_{o2}| = R_c I_b^+ \left(1 + \frac{R_f}{R_i} \right) \tag{3-74}$$

The final effect will be the offset resulting from I_b^-, and the circuit used for this purpose is shown in Figure 3–14(c). *The direction of the offset produced by I_b^- will always be opposite to that of I_b^+.* Thus, if the offset produced by I_b^+ is positive, the offset produced by I_b^- is negative, and vice versa. Since we may not know which is which in a given case, to avoid sign difficulties in our analysis, it is convenient at this point to assume that the effect of I_b^+ given in (3–74) is positive in accordance with the assumed direction of I_b^+. Thus, the effect of I_b^- will be negative, and v_{o2}'' will be used to denote the corresponding negative output voltage.

From Figure 3–14(c), the virtual ground at the inverting terminal means that the voltage across R_i is zero. By Ohm's law, the current through R_i is then zero, and the net current source I_b^- may be assumed to flow through R_f. A voltage whose polarity is positive on the left appears across R_f, and the output voltage is

$$v_{o2}'' = -R_f I_b^-$$
(3–75)

The magnitude of the combined effects due to the two bias currents will be denoted as $|v_{o2}|$, and this voltage is determined by combining the results of (3–74) and (3–75). This value is

$$|v_{o2}| = \left| R_c \left(1 + \frac{R_f}{R_i} \right) I_b^+ - R_f I_b^- \right|$$
(3–76)

Note the magnitude bars around the entire expression. The effect of I_b^+ may be larger than that of I_b^- and vice versa, and the minus sign in the expression indicates the opposite sense of the two internal effects. The magnitude bars at the two ends of the expression, however, indicate that the final result is the magnitude of the net result.

Because of the minus sign in (3–76), the question arises as to whether it is possible to provide a cancellation of the terms inside the magnitude bars, thus forcing the net effect of the two bias currents to be zero. Before this question is answered, an initial assumption will be made that the two bias currents are approximately equal; that is, $I_b^+ = I_b^- = I_b$. This equality suggests identical transistors with the same operating point and gain characteristics at both inputs. Next, the equality of bias currents is substituted in (3–76) and the expression is equated to zero:

$$|v_{o2}| = \left| R_c \left(1 + \frac{R_f}{R_i} \right) I_b - R_f I_b \right| = 0$$
(3–77)

The reader is invited to show that the following result satisfies this equation:

$$R_c = \frac{R_i R_f}{R_i + R_f} = R_i \| R_f$$
(3–78)

Recall from Chapter 2 that this result is in total agreement with the design procedure established there. Further, we see that *if the two bias currents are identical, the component of the output offset produced by the input bias currents is zero.*

In practice, the two bias currents are never quite equal, but the value of compensating resistance given by (3–78) is still the optimum value. However, the output voltage is not zero in this case. In Problem 3–18, the reader will show that when the correct value of compensating resistance is used, the output offset voltage magnitude due to input bias currents is

$$|v_{o2}| = R_f |I_b^+ - I_b^-| \quad \text{for } R_c = R_i \| R_f$$
(3–79)

Observe that this expression will reduce to zero when $I_b^+ = I_b^-$, as previously noted.

Input Offset Current

Observe that the difference between I_b^+ and I_b^- appears in (3–79). When one is dealing with offset effects, this difference term appears so frequently that it is assigned a unique definition. Let I_{io} represent the **input offset current,** defined as

$$I_{io} = |I_b^+ - I_b^-| \qquad (3\text{–}80)$$

The input offset current is the magnitude of the difference between the bias currents at the two inputs. By the definition of I_{io}, the magnitude of output voltage offset produced by bias currents with an optimum value of compensating resistance is

$$|v_{o2}| = R_f I_{io} \quad \text{for } R_c = R_i \| R_f \qquad (3\text{–}81)$$

This expression replaces the more complex expression of (3–76) when the optimum compensating resistance is used.

General Summary

Reviewing our work to this point, the total output offset voltage results from a combination of input offset voltage and input bias currents. The component produced by input offset voltage was denoted as v_{o1}, and the component produced by input bias currents was denoted as v_{o2}. An expression for $|v_{o1}|$ was given by (3–73). An expression for $|v_{o2}|$ in the most general case was given by (3–76). However, when the optimum value of compensating resistance is used, $|v_{o2}|$ reduces to the much simpler expression of (3–81). The value of $|v_{o2}|$ in (3–81) is smaller than for the general case, so the optimum value of compensating resistance should be used when offset effects must be minimized. The preceding results are summarized in Table 3–1.

TABLE 3–1
Summary of offset and bias effects
(refer to Figure 3–13)

Output Offset Voltage Due to Input Offset Voltage
$|v_{o1}| = \left(1 + \dfrac{R_f}{R_i}\right) V_{io}$

Output Offset Voltage Due to Input Bias Currents	
General	$|v_{o2}| = \left| R_c\left(1 + \dfrac{R_f}{R_i}\right) I_b^+ - R_f I_b^- \right|$
When $R_c = R_i \| R_f$	$|v_{o2}| = R_f I_{io}$

Let v_o(offset) represent the total output offset voltage. The **worst-case** magnitude of the output offset can be expressed as

$$|v_o(\text{offset})| = |v_{o1}| + |v_{o2}| \qquad (3\text{–}82)$$

However, at a given time, the two components v_{o1} and v_{o2} may have the same or opposite signs, and either may be larger than the other. The result of (3–82) then represents an upper bound for the offset magnitude.

Since the input offset voltage and input bias current effects are different, it is important to understand the trends that accentuate and minimize the separate phenomena. To simplify the discussion that follows, we will assume that an optimum bias compensating resistance is used, in which case (3–81) applies for $|v_{o2}|$. However, the trend to be discussed applies even when the compensating resistance is not used, but the net effect will be even greater.

First, consider the effect of input offset voltage as given by (3–73). This expression applies to the amplifier form of Figure 3–13 irrespective of whether the amplifier is inverting or noninverting. The reader will recall an earlier point in the chapter in which the gain expression $1 + R_f/R_i$ occurred for either an inverting or a noninverting circuit. This concept was generalized by defining a quantity called the **noise gain** for the circuit. The same concept applies here with one possible qualification: Offset is a purely dc effect, so the **dc noise gain** is the quantity that applies. In most cases (all considered thus far in the book), the noise gain is a fixed value and is independent of frequency. However, to allow for a possible difference, let $K_n(\text{dc})$ represent the **dc noise gain.** In the general case, the output offset magnitude produced by input offset voltage is given by

$$|v_{o1}| = K_n(\text{dc}) \times V_{io} \qquad (3\text{–}83)$$

which includes the expression of (3–73) as a special case.

The procedure for determining noise gain here is virtually the same as that developed earlier. The circuit is momentarily considered as a noninverting amplifier with all other inputs deenergized, and the gain between the noninverting input terminal and the output is computed. The only difference in this case is that the **dc gain** is computed. Thus, if a capacitor appears in the circuit, it is represented by an open circuit for this purpose. This situation will be encountered later in the text.

It is readily concluded that the effect of the input offset voltage increases linearly with the noise gain of the circuit. Thus, a circuit with a very high signal gain and a circuit with many inputs (such as a linear combination circuit) may both have sizable effects produced by the input offset voltage. In a conventional amplifier, the output signal voltage also increases linearly with the gain, so the *relative* offset effect may not change as the gain increases.

Next, consider the effect of input bias currents when an optimum compensating resistance is used. In this case, the output voltage magnitude is given by (3–81). The voltage $|v_{o2}|$ in this case increases linearly with R_f. However, the gain of the circuit does not appear in the expression. Rather, this output offset effect is dependent only on R_f and I_{io}. Assuming that a given op-amp establishes I_{io}, the offset effect due to input bias currents may be minimized by keeping the resistance level reasonably low. In practice, there is a limit on the practical lower levels of resistances due to loading effects as discussed in Chapter 2. Further, in most cases, the effects of input offset voltage tend to be greater than input bias current effects, provided that a proper compensating resistance is used. Thus, the procedures established in Chapter 2 concerning acceptable resistance levels usually are adequate for this purpose. The effects of input bias currents become particularly troublesome, however, when resistance values of the order of 1 MΩ or more are used.

In summary of the ideas discussed thus far, the following patterns are noted:

1. At large values of voltage gain achieved with low or moderate values of resistance, the output offset voltage is produced almost entirely by the input offset voltage.
2. At low values of voltage gain achieved with very high resistance values, the output offset voltage is produced almost entirely by the input bias currents.

These trends may be used to devise measurement procedures for estimating the values of the offset parameters (see Problems 3–21 and 3–22). In the most general case, of course, the output is a combination of both effects.

Typical Offset Values

Before we consider any numerical examples, it will be helpful to note some typical values of the preceding parameters. Refer to Appendix C for data concerning the 741 op-amp. In the row entitled "Input Offset Voltage" on the third page of the 741 data sheets in Appendix C a typical value is listed as 1.0 mV, and the maximum value is listed as 5.0 mV. For the 741C op-amp, a typical value is listed as 2.0 mV, and the maximum value is listed as 6.0 mV.

In the analysis of bias currents, both I_b^+ and I_b^- were required. The value listed in specifications is the average of these two quantities, which will be denoted here as I_b. Thus, I_b is defined as

$$I_b = \frac{I_b^+ + I_b^-}{2} \tag{3–84}$$

In the row titled "Input Bias Current," a typical value is listed as 80 nA, and the maximum value is listed as 500 nA.

In the row titled "Input Offset Current," a typical value is listed as 20 nA, and the maximum value is listed as 200 nA. Observe that offset current is smaller than bias current, since the former represents the difference between the two values of the latter quantities. Amplifiers having more closely matched input circuits are capable of much smaller values of offset currents.

The preceding values are typical of general-purpose op-amps with bipolar junction transistor (BJT) circuitry, and they are suitable for applications where offsets are not particularly troublesome. However, op-amps are available with offset voltages in the microvolt range and offset currents in the picoampere range, the latter condition usually requiring field effect transistor (FET) circuits.

Null Circuit

Most op-amps have provisions for nulling or eliminating offset effects at a particular operating point when necessary. For example, refer to the seventh page of the 741 op-amp specifications in Appendix C. The variable resistance is adjusted so that the dc output level is zero. The op-amp should be connected in the actual configuration in which it is to be used. However, if one or more of the actual signals are dc, it may be necessary momentarily to replace the corresponding sources by resistances equal to the internal source resistances so that a signal level is not confused with an offset effect.

While the null circuit can eliminate offset at a particular time and under a particular set of conditions, drift will cause an offset to reappear as time passes. Consequently, if this approach is used, periodic checks and readjustments are required if the offset level is to be minimized.

EXAMPLE 3–8

Consider the noninverting amplifier circuit of Figure 3–15. Assume the typical offset and bias values for the 741 op-amp as given in Appendix C for the op-amp. (a) Determine the magnitude of the output voltage $|v_{o1}|$ produced by the input offset voltage. (b) With no compensating resistance ($R_c = 0$), determine the magnitude of the output voltage $|v_{o2}|$ produced by the input bias currents. (c) Determine the optimum value of R_c. (d) With R_c set at the value determined in part (c), determine the new value of $|v_{o2}|$. (The input signal terminal is assumed to be grounded through an ideal voltage source.)

Solution

The typical values of the offset and bias parameters for the 741 op-amp are determined from Appendix A as $V_{io} = 1$ mV, $I_b = 80$ nA, and $I_{io} = 20$ nA. The voltage gain of the circuit is readily determined as $A_{CL} = 1 + R_f/R_i = 1 + 10^6/10^4 = 101$. Since the circuit is a noninverting amplifier, the noise gain is equal to the closed-loop gain.

(a) The output offset magnitude $|v_{o1}|$ due to the input offset voltage is determined from either (3–73) or (3–83):

$$|v_{o1}| = 101 \times 1 \text{ mV} = 101 \text{ mV} \tag{3–85}$$

FIGURE 3–15
Circuit of Example 3–8.

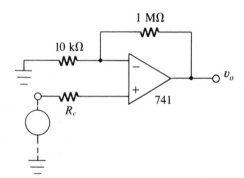

(b) There is some uncertainty concerning the exact values of I_b^+ and I_b^-, since only their average value I_b is specified. When no compensating resistance is used, a reasonable estimate is to assume that $I_b^+ = I_b^- = I_b$. The general equation for predicting the output offset magnitude produced by input bias currents as given by (3–76) is used in this part. Since $R_c = 0$, this expression reduces to

$$|v_{o2}| = |-R_f I_b^-| = R_f I_b = 10^6 \times 80 \times 10^{-9} = 80 \text{ mV} \qquad (3\text{–}86)$$

Reviewing parts (a) and (b), the worst-case magnitude of the total output offset voltage is $|v_o| = 101 + 80 = 181$ mV.

(c) The optimum value of the compensating resistance is

$$R_c = 10 \text{ k}\Omega \| 1 \text{ M}\Omega = \frac{10^4 \times 10^6}{10^4 + 10^6} = 9901 \ \Omega \qquad (3\text{–}87)$$

This is not a standard value, but very little difference in the results should occur by selecting $R_c = 10$ kΩ.

(d) When the proper value of compensating resistance is used, the assumption that $I_b^+ = I_b^-$ is no longer valid. Instead, the form of (3–81) utilizing the input offset current must be used. The corresponding output offset voltage magnitude is now

$$|v_{o2}| = R_f I_{io} = 10^6 \times 20 \times 10^{-9} = 20 \text{ mV} \qquad (3\text{–}88)$$

This value is much smaller than the result of (3–86), indicating the advantage of using the compensating resistance.

With the correct value of compensating resistance, the worst-case magnitude of the total output offset voltage is $101 + 20 = 121$ mV.

EXAMPLE 3–9   Consider the linear combination circuit of Example 3–7, and assume the typical offset and bias values for the 741 op-amp. (a) Determine the magnitude of the output voltage $|v_{o1}|$ produced by the input offset voltage. (b) Determine the magnitude of the output voltage $|v_{o2}|$ produced by the input bias currents. (The input signal terminals are assumed to be grounded through ideal voltage sources.)

Solution
The typical values of the offset and bias parameters are $V_{io} = 1$ mV, $I_b = 80$ nA, and $I_{io} = 20$ nA. The dc noise gain was determined in Example 3–7 to be $K_n(\text{dc}) = 121$.

(a) The dc output offset due to the input offset voltage is

$$|v_{o1}| = K_n(\text{dc}) \times V_{io} = 121 \times 1 \text{ mV} = 121 \text{ mV} \qquad (3\text{–}89)$$

(b) It can be readily verified in Figure 3–12 that the correct value of compensating resistance is used. For this condition, the dc output offset due to the input bias currents is

$$|v_{o2}| = R_f I_{io} = 240 \times 10^3 \times 20 \times 10^{-9} = 4.8 \text{ mV} \qquad (3\text{–}90)$$

3–7 MULTISIM EXAMPLES

The examples presented in this section are devoted to the modeling of various op-amp imperfections and limitations at dc. This modeling will include the effects of finite gain, finite input impedance, and nonzero output impedance. The Multisim 741 model will be utilized for all the circuits.

Along with basic measurements of the circuit variables, the transfer function command is utilized to study these phenomena. The transfer function operation permits a direct measurement of input impedance, circuit gain, and output impedance.

MULTISIM EXAMPLE 3–1

For the inverting amplifier of Figure 3–16, use the **Transfer Function** analysis command to determine the closed-loop gain, the input resistance, and the output resistance. (*Note:* This is not a particularly good design but is used to exaggerate some of the dc imperfections.)

Solution

The term **Transfer Function** as applied to Multisim, which is based on the original SPICE definition, applies only to dc. This is in sharp contrast to the widely employed frequency-dependent definition provided in Chapter 1, which is primarily dealt with through frequency response analysis.

Transfer Function analysis in Multisim allows the determination of voltage gain, input resistance, and output resistance for circuits containing one independent input source, and one or more dependent sources or active devices.

It is noted in Figure 3–16 that the input voltage is set at 0. Any small voltage that would not cause the amplifier to go into saturation could be used. Multisim actually determines a small-signal model for the circuit, and the parameters are measured as a small change is imposed on the input voltage. Therefore, a value of 0 will suffice, and it is used to remind us that we could not use any dc input voltage that would cause saturation to occur.

To perform the analysis, left-click on the **Analysis** button and then left-click on **Transfer Function**. A window with the title **Transfer Function Analysis** will then open.

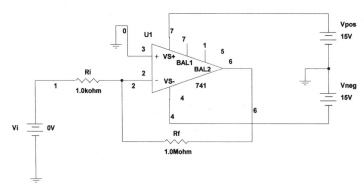

FIGURE 3–16
Circuit of Multisim Example 3–1.

For the **Input source**, slot, select **vvi**. (This is the input voltage.) Make sure that there is a check in the **Voltage** slot for **Output nodes/source**. Set **Output node** to 6 and **Output reference** to 0. (This is the default position for the latter node.) This means that the output variable is the voltage between node 6 and ground. Finally, left-click on **Simulate** and the table shown in Figure 3–17 will appear.

The entry "Transfer function" in the table refers to the voltage gain, which is about −995. The entry "vv1#Input impedance" refers to the input resistance and is about 1 kΩ, The entry "Output impedance at V(6,0" refers to the output resistance and is about 374 mΩ.

The theoretical values of voltage gain, input resistance, and output resistance may be calculated using (3–46), (3–54), and (3–55), respectively. A short table follows in which the theoretical and Multisim values are compared. The value of the 741 output resistance used is 75 Ω, which is the value provided for the Multisim model.

	Calculated Values	Multisim Values
Voltage Gain	−995.02	−995.01
Input Resistance (ohms)	1005	1005
Output Resistance (ohms)	0.3735	0.3741

One could hardly ask for any better agreement.

MULTISIM EXAMPLE 3-2

Use the procedure discussed in Problem 3–21 to estimate the input offset voltage for the Multisim 741 model.

Solution

The basic circuit diagram is given in Figure P3–21, and it is shown as a Multisim schematic diagram in Figure 3–18. As discussed in the procedure, the noise gain is chosen to have a relatively large value and is

$$K_n = 1 + \frac{R_f}{R_i} = 1 + \frac{10^4}{10} = 1001 \tag{3-91}$$

```
Transfer Function Analysis
───────────────────────────────────────────

Transfer function          -995.01166
vvi#Input impedance           1.00500k
Output impedance at V(6,0   374.09751m
```

FIGURE 3-17
Data for Example 3–1.

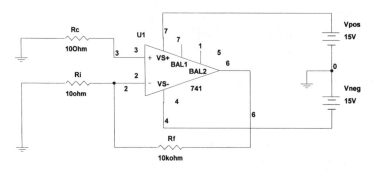

FIGURE 3–18
Circuit of Multisim Example 3–2.

Simultaneously, the resistance level is in a practical range for the measurement. Strictly speaking, the ideal value of the compensating resistance should be very slightly less than 10Ω, but the difference in the results should be negligible.

Note that there is no input signal source so the dc output will be a result of the input offset voltage and bias currents, with the latter contribution being very small compared to the former. To measure the dc output voltage, a **DC Operating Point** analysis was performed, and the only quantity of interest is the voltage at node 6. The result is shown in Figure 3–19 and is about 1.025 V. The input offset voltage is then estimated as

$$V_{io} = \frac{|v_o|}{1 + \dfrac{R_f}{R_i}} = \frac{|v_o|}{K_n} = \frac{1.025}{1001} = 1.024 \text{ mV} \tag{3–92}$$

This value compares very favorably with the "typical value" of 1 mV provided on the 741 data sheets in Appendix C.

MULTISIM EXAMPLE 3–3 Use the procedure discussed in Problem 3–22 to estimate the input bias currents and input offset current for the Multisim 741 model.

DC Operating Point

6	1.02538

FIGURE 3–19
Output Data of Multisim Example 3–2.

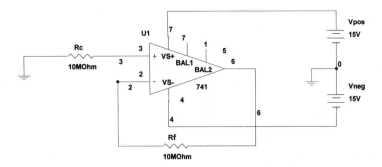

FIGURE 3–20
Circuit of Multisim Example 3–3.

Solution

The three circuit variations required are given in Figure P3–22, and all forms can be easily adapted from the Multisim schematic diagram of Figure 3–20. As given a value of 10 MΩ is used for both resistances and this corresponds to part (c) of Figure P3–22. In order to preserve the form of the circuit without nodes changing numbers, the "shorts" required in (a) and (b) can be easily achieved by simply changing the resistance values alternately to very small values (of the order of 1 Ω or so).

Three separate **DC Operating Point** analyses were performed in accordance with the procedure of Problem 3–22. The three measured values of the dc output voltages in the order of (a), (b), and (c) were −531.2363 mV, 693.3476 mV, and 153.79815 mV. With realistic rounding, the estimated values of the two bias currents are

$$I_b^+ = \frac{|v_{oa}|}{10^7} = \frac{0.531}{10^7} = 53.1 \text{ nA} \qquad \textbf{(3–93)}$$

$$I_b^- = \frac{|v_{ob}|}{10^7} = \frac{0.693}{10^7} = 69.3 \text{ nA} \qquad \textbf{(3–94)}$$

The estimated value of the input offset current is

$$I_{io} = \frac{|v_{oc}|}{10^7} = \frac{0.154}{10^7} = 15.4 \text{ nA} \qquad \textbf{(3–95)}$$

The "typical" value provided in the 741 data sheet for the bias current is 80 nA, and considering the wide possible range, the values of (3–93) and (3–94) appear reasonable. The "typical" value of input offset current is given as 20 nA, which compares reasonably well with the value of (3–95) when the wide range is considered.

Another check point is to take the difference between the two bias currents, which is 69.3 − 53.1 = 16.2 nA, which compares reasonably well with the measured value of input offset current. One could try to refine the measurement to compensate for the contributions from input offset voltage, but for estimation purposes, we will leave it at this point.

SUMMARY

1. A realistic low-frequency and dc signal model for an op-amp contains a finite differential input resistance, a nonzero output resistance, and a finite open-loop differential gain.
2. The loop gain is the product of the forward open-loop differential gain and the feedback factor.
3. As the loop gain increases, the input impedance of a noninverting amplifier increases and its output impedance decreases.
4. As the loop gain increases, the output impedance of an inverting amplifier decreases.
5. As the open-loop gain increases, the input impedance of an inverting amplifier decreases toward its limiting value of R_i.
6. As the loop gain increases, the gains for both noninverting and inverting amplifiers approach their ideal values.
7. The noise gain is a special defined parameter that is used to predict effects of various limitations of op-amps.
8. The noise gain is determined by deenergizing all external signal input terminals and determining the gain from the noninverting input to the output.
9. For a noninverting amplifier, the noise gain and desired closed-loop signal gains are equal, but in most circuits, the two quantities are different.
10. A dc offset voltage (usually a small value) will appear at the output of an op-amp even when the differential input voltage is zero and when the two power supply voltages are equal.
11. The dc offset may be insignificant in ac-coupled circuits or when the signal levels are large, but it poses a serious problem for applications involving small dc levels.
12. The output dc offset voltage can be considered to be due to two effects: input offset voltage and input bias currents.
13. The circuit model for predicting the offset effects utilizes a voltage source for representing the input offset voltage effect and two current sources for representing the bias current effects.
14. The effect of input offset voltage is most pronounced when the circuit has a large dc noise gain.
15. The effect of input bias currents is most pronounced when large resistance values are used in the circuit.
16. The effects of input bias current can be minimized by using a compensating resistor and in keeping the overall resistance levels to moderate values.
17. Most op-amps provide an additional provision for minimizing the output offset by providing special terminals to be connected through a potentiometer to a dc source.

PROBLEMS

Drill Problems

3–1. A certain op-amp has the following dc and very low-frequency parameters:

$$\text{dc open-loop gain} = 80,000$$
$$\text{differential input resistance} = 0.4 \text{ M}\Omega$$
$$\text{output resistance} = 50 \text{ }\Omega$$

It is used in a noninverting amplifier circuit with $R_i = 3 \text{ k}\Omega$ and $R_f = 47 \text{ k}\Omega$. At dc and low frequencies, calculate the values of the **a.** feedback factor, **b.** loop gain, **c.** exact closed-loop gain, **d.** closed-loop input impedance, and **e.** closed-loop output impedance. **f.** Use the result of (3–8) to determine the approximate gain and compare with the result of (c).

3–2. The op-amp of Problem 3–1 is used in a noninverting amplifier circuit with $R_i = 2$ kΩ and $R_f = 18$ kΩ. Repeat the calculations of Problem 3–1.

3–3. The op-amp of Problem 3–1 is used in an inverting amplifier circuit with $R_i = 30$ kΩ and $R_f = 270$ kΩ. Repeat the calculations of (a) through (e).

3–4. The op-amp of Problem 3–1 is used in an inverting amplifier circuit with $R_i = 20$ kΩ and $R_f = 100$ kΩ. Repeat the calculations of (a) through (e).

3–5. Consider a noninverting amplifier circuit with $R_i = 1$ kΩ and $R_f = 99$ kΩ. Calculate the exact closed-loop dc gain A_{CL} for each of the following open-loop op-amp dc gains:

a. $A = \infty$ **b.** $A = 10^5$ **c.** $A = 10^4$ **d.** $A = 1000$ **e.** $A = 100$

3–6. Consider a noninverting amplifier circuit with $R_i = 10$ kΩ and $R_f = 40$ kΩ. Calculate the exact closed-loop dc gain A_{CL} for each of the following open-loop op-amp dc gains:

a. $A = \infty$ **b.** $A = 10^5$ **c.** $A = 10^4$ **d.** $A = 1000$ **e.** $A = 100$

3–7. Consider an inverting amplifier circuit with $R_i = 10$ kΩ and $R_f = 510$ kΩ. Calculate the exact closed-loop dc gain A_{CL} for each of the following open-loop op-amp dc gains:

a. $A = \infty$ **b.** $A = 10^5$ **c.** $A = 10^4$ **d.** $A = 1000$ **e.** $A = 100$

3–8. Consider an inverting amplifier circuit with $R_i = 20$ kΩ and $R_f = 80$ kΩ. Calculate the exact closed-loop dc gain A_{CL} for each of the following open-loop op-amp dc gains:

a. $A = \infty$ **b.** $A = 10^5$ **c.** $A = 10^4$ **d.** $A = 1000$ **e.** $A = 100$

3–9. **a.** Determine the noise gain for the circuit of Figure P3–9.

b. Use the result to calculate the exact signal gain at dc and low frequencies if $A = 10^5$.

FIGURE P3–9

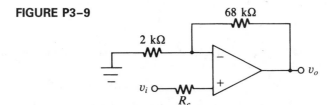

3–10. **a.** Determine the noise gain for the circuit of Figure P3–10.

b. Use the result to calculate the exact signal gain at dc and low frequencies if $A = 10^4$.

FIGURE P3–10

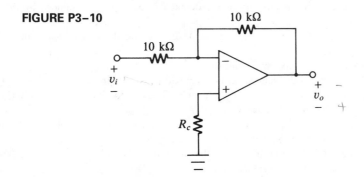

3–11. **a.** Determine the noise gain for the circuit of Figure P3–11.

b. Use the result to calculate the exact gain factors for the two signals if $A = 5 \times 10^4$.

FIGURE P3–11

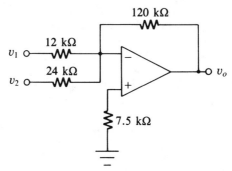

3–12. a. Determine the noise gain for the circuit of Figure P3–12.
   b. Use the result to calculate the exact gain factors for the two signals if $A = 2 \times 10^4$.

FIGURE P3–12

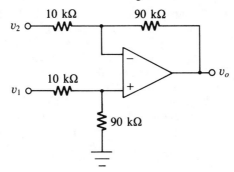

3–13. For the circuit of Problem 3–9 (Figure P3–9), assume that the op-amp bias and offset parameters are as follows:

$$\text{Input offset voltage} = 1.2 \text{ mV}$$
$$\text{Input bias current} = 60 \text{ nA}$$
$$\text{Input offset current} = 8 \text{ nA}$$

(You may now assume the ideal gain equation from Chapter 2.)
   a. Determine the magnitude of the output dc voltage $|v_{o1}|$ produced by the input offset voltage.
   b. With $R_c = 0$, determine the magnitude of the output dc voltage $|v_{o2}|$ produced by the input bias currents.
   c. Determine the optimum value of R_c.
   d. With R_c set at the value determined in part (c), determine the new value of $|v_{o2}|$.
3–14. Perform the analysis outlined in Problem 3–13 on the circuit of Problem 3–10 (Figure P3–10) assuming the op-amp parameters of Problem 3–13.
3–15. Perform the analysis outlined in Problem 3–13 on the circuit of Problem 3–11 (Figure P3–11) assuming the op-amp parameters of Problem 3–13.
3–16. Determine (a) $|v_{o1}|$ and (b) $|v_{o2}|$ for the circuit of Problem 3–12 (Figure P3-12) assuming the op-amp parameters of Problem 3–13.

Derivation Problems
3–17. Starting with (3–37) and (3–38), eliminate v_e and show that the closed-loop gain is given by (3–39).

3–18. Starting with (3–76), show that when the optimum value of compensating resistance given by (3–78) is used, the component of the output voltage due to bias currents reduces to (3–79).

3–19. Consider a linear combination circuit of the form developed in Section 2–8 whose output voltage is given by

$$v_o = -A_1 v_1 - A_2 v_2 + \cdots - A_n v_n$$

Show that the noise gain is given by

$$K_n = 1 + A_1 + A_2 + \cdots + A_n$$

3–20. Consider the balanced closed-loop differential amplifier of the form developed in Section 2–9 whose output voltage is given by

$$v_o = A(v_1 - v_2)$$

Show that the noise gain is given by

$$K_n = 1 + A$$

3–21. A procedure for estimating the input offset voltage V_{io} will be developed in this problem. Consider the circuit of Figure P3–21 with both inputs grounded. The resistance R_i and R_f are chosen to provide a relatively large dc noise gain, but the resistance levels are kept as small as practicable. With this strategy, the output offset component due to input offset voltage can be made to be considerably larger than the component due to bias currents. The latter component may then be neglected, in which case a reasonable estimate of the input offset voltage may be determined.

FIGURE P3–21

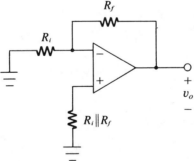

a. Based on the preceding discussion, let $|v_o|$ represent the magnitude of the measured dc output voltage. Assuming that $|v_o| \approx |v_{o1}|$, show that the input offset voltage is

$$V_{io} \approx \frac{|v_o|}{1 + \dfrac{R_f}{R_i}}$$

b. Assume that the maximum expected value of the input offset current is estimated as $I_{io}(\text{max})$. Calculate the output $|v_{o2}|$ due to this current and compare it with $|v_o|$ of part (a). Show that the maximum percent error in the estimate of (a) is about

$$\text{Percent error} \approx \frac{|v_{o2}|}{v_o} \times 100\%$$

c. For a particular op-amp, assume that $R_i = 100\ \Omega$ and $R_f = 10\ k\Omega$. The measured output dc voltage is $|v_o| = 130$ mV. Estimate the input offset voltage.

d. If the maximum expected value of input offset current is 200 nA, estimate the maximum percentage error in the estimate of (c).

3–22. A procedure for estimating the input bias and offset currents will be developed in this problem. Consider the three circuits of Figure P3–22. The resistance R is chosen to be sufficiently large that the contribution to the output from bias currents is considerably larger than the contribution from input offset voltage. The latter component may then be neglected, in which case a reasonable estimate of the bias currents may be determined. Note that the dc noise gain for each circuit is unity. Let $|v_{oa}|$, $|v_{ob}|$, and $|v_{oc}|$ represent the magnitudes of the three outputs, respectively.

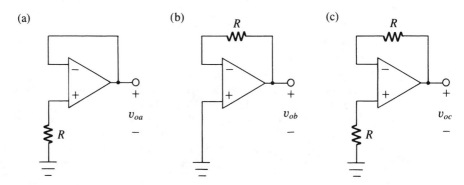

FIGURE P3–22

a. Show that

$$I_b^+ \approx \frac{|v_{oa}|}{R}$$

$$I_b^- \approx \frac{|v_{ob}|}{R}$$

$$I_{io} \approx \frac{|v_{oc}|}{R}$$

b. Assume that the maximum expected value of the input offset voltage is estimated as $V_{io}(\text{max})$. Calculate the output $|v_{o1}|$ due to this voltage and compare it with the three measured voltages of part (a). Normally, the maximum percent error would be that associated with I_{io}. Show that this error is about

$$\text{Percent error} \approx \frac{|v_{o1}|}{|v_{oc}|} \times 100\%$$

c. For a particular op-amp, assume that $R = 10\ M\Omega$. The three measured dc voltages of (a), (b), and (c) of Figure 3–22 are 700 mV, 580 mV, and 120 mV, respectively, Estimate the values of I_b^+, I_b^- and I_{io}.

d. If the maximum expected value of the input offset voltage is 3 mV, estimate the maximum percentage error in the estimate of (c).

3–23. The direct measurement of op-amp open-loop gain is very difficult due to the instability of the op-amp under open-loop conditions. In the frequency range above a few hundred hertz, the open-loop gain may be determined indirectly with any standard configuration by measuring the output voltage and the differential input voltage and dividing the former by the latter. However, at dc and very low frequencies, the open-loop gain is so large that it is impractical to measure the very small differential input voltage with most commonly available voltmeters. By cascading the op-amp with an attenuation network, the gain may be "degraded" by a sufficient amount to permit measurement of the differential input voltage. From a knowledge of the attenuation provided, a correction factor may be used to determine the open-loop gain of the amplifier.

Consider the circuit of Figure P3–23. The resistance R_1 and R_2 are selected to provide a large reduction in signal amplitude between v_a and the op-amp input. The resistances R_3 and R_4 are somewhat arbitrary, but are often chosen to be equal. The input voltage v_i is chosen to be large enough to provide measurable values of v_a and v_o, but small enough to avoid saturation.

FIGURE P3–23

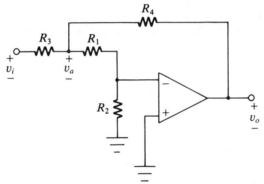

a. Assuming linear operation, show that the open-loop gain A is

$$A = \left(1 + \frac{R_1}{R_2}\right)\frac{v_o}{v_a}$$

b. For $R_1 = 10\ \text{k}\Omega$, $R_2 = 10\ \Omega$, and $R_3 = R_4 = 10\ \text{k}\Omega$, assume that two measured voltages are $v_a = 30\ \text{mV}$ and $v_o = 6\ \text{V}$. Estimate the open-loop gain A.

Note: This circuit as given is vulnerable to a possible large dc offset at the output. For the data of (b), it is assumed that the voltage values are either very low-frequency ac values, which may be readily separated from the dc levels, *or* that the values are *changes* as measured from the initial offset levels with signals removed. For further details on the offset effects, see Problem 3–24.

3–24. Consider the circuit of Figure P3–23.

a. Show that the dc output voltage $|v_{o1}|$ due to the input offset voltage is given by

$$|v_{o1}| = \left[\left(1 + \frac{R_1}{R_2}\right)\left(1 + \frac{R_4}{R_3}\right) + \frac{R_4}{R_2}\right]V_{io}$$

b. Show that the dc output voltage $|v_{o2}|$ due to the bias current I_b^- is given by

$$|v_{o2}| = \left[R_1 + R_4\left(1 + \frac{R_1}{R_3}\right)\right]I_b^-$$

c. For the measurement discussed in Problem 3–23, assume that $V_{io} = 1$ mV and $I_b^- = 100$ nA. Determine the values of $|v_{o1}|$ and $|v_{o2}|$.

Design Problems

This chapter has dealt with some of the limitations that must be considered in predicting op-amp circuit performance and in selecting op-amps to meet specific requirements. The "design" problems given here are not intended to represent complete designs, but rather to illustrate how the parameter values must often be selected to meet certain minimum conditions.

3–25. An op-amp is to be used at dc and very low frequencies. A closed-loop gain of 200 is required. Specifications indicate that the error due to finite open-loop gain cannot exceed 0.1%. Determine the minimum value of the dc open-loop gain required.

3–26. An op-amp is to be selected for an inverting amplifier to be used at dc and very low frequencies. A closed-loop nominal gain of -100 is desired. Specifications indicate that the error due to finite open-loop gain cannot exceed 0.1 dB. Determine the minimum value of the dc open-loop gain required.

3–27. Assume in the design of Problem 3–25 that the following two additional specifications are given:

$$\text{dc output due to input offset voltage} \leq 100 \text{ mV}$$

$$\text{dc output due to input offset current} \leq 5 \text{ mV}$$

a. Determine the maximum value of input offset voltage allowed for the op-amp.

b. When an op-amp is selected to meet the requirement of (a), assume that $I_{io} = 12$ nA. Calculate the maximum value of R_f permitted, assuming that a compensating resistance will be used.

3–28. Assume in the design of Problem 3–26 that the following two additional specifications are given:

$$\text{dc output due to input offset voltage} \leq 100 \text{ mV}$$

$$\text{dc output due to input offset current} \leq 5 \text{ mV}$$

a. Determine the maximum value of input offset voltage allowed for the op-amp.

b. If R_i is selected to meet a 10-kΩ requirement on input impedance, determine the maximum value of input offset current allowed for the op-amp, assuming that a compensating resistance will be used.

LABORATORY EXERCISES

Voltage Offset Null Circuit
LE 3–1.

In this exercise, investigate the use of a null circuit to reduce the output dc offset to its minimum possible value. Refer to the "Voltage Offset Null Circuit" given in the 741 op-amp data sheets of Appendix C. Although there is no specific closed-loop configuration shown, you will employ the circuit in combination with a closed-loop differential amplifier circuit as developed in Section 2–9. The reason for choosing this particular circuit to illustrate the null circuit is that the offset can introduce a major error when the difference between two nearly equal dc voltages is desired.

1. Connect the closed-loop difference amplifier of Figure 2–28 with $R = 10$ kΩ and $A = 1$.
2. Connect the "Voltage Offset Null Circuit" given in the 741 data sheets.
3. Connect the two external circuit inputs (v_1 and v_2) to ground and note the dc output voltage. As always, start on a high range, though you will likely need to reduce the voltage scale to a lower range.
4. Adjust the potentiometer until the dc output voltage magnitude is either zero or its minimum possible value. Record the minimum value of voltage attained.
5. Monitor the voltage for several minutes to see if there is any appreciable drift. If so, you will need to periodically readjust the null for applications in which the offset level is significant.

Note: This circuit forms the basis for Exercise LE 3–2.

Closed-Loop Differential Amplifier
LE 3–2.

Although this circuit was covered in Chapter 2, this exercise was delayed until offset effects were covered in Chapter 3. Exercise LE 3–1 is a prerequisite for the present exercise.

1. Complete the work of Exercise LE 3–1. Periodically in this exercise, you may find it necessary to disconnect the amplifier inputs from the signal sources, ground them, and readjust the output null condition.
2. Write an equation that expresses the expected output voltage v_o in terms of the two input voltages v_1 and v_2.
3. Calculate the ideal dc output voltage for each of the following six dc input combinations:
   a. $v_1 = 10$ V, $v_2 = 0$
   b. $v_1 = 0$, $v_2 = 10$ V
   c. $v_1 = v_2 = 10$ V
   d. $v_1 = 10$ mV, $v_2 = 0$
   e. $v_1 = 0$, $v_2 = 10$ mV
   f. $v_1 = v_2 = 10$ mV
4. Apply dc input voltages in the six combinations of step (3), and measure the six output voltages. In each case, compare the measured value to the expected value.

Measurement of dc Input Offset Voltage
LE 3–3.

1. Employ the procedure of Problem 3–21 to determine the input offset voltage v_{io} for a 741 op-amp. Set $R_i = 100$ Ω and $R_f = 10$ kΩ. Compare your result with 741 data in Appendix C.
2. Increase the value of R_f to 100 kΩ, and measure the output dc voltage again. Did it increase approximately by a factor of 10? If so, the effect of bias current is insignificant.

Measurement of dc Bias and Offset Currents
LE 3–4.

1. Employ the procedure of Problem 3–22 to determine the two bias currents I_b^+ and I_b^- and the input offset current I_{io}. Set $R = 560$ kΩ. Compare your result with 741 data in Appendix C.
2. Increase the value of R to 5.6 MΩ, and repeat step (a) again. Did the respective output voltages increase approximately by a factor 10? If so, the effect of input offset voltage on the measurement is insignificant.

4

OPERATIONAL AMPLIFIER ac EFFECTS AND LIMITATIONS

OBJECTIVES

After completing this chapter, the reader should be able to:

- Sketch the form of the open-loop gain of a typical op-amp.
- Define the *unity-gain frequency* or *gain-bandwidth product*.
- Determine the closed-loop 3-dB bandwidth of an op-amp circuit and the resulting rise time.
- Show graphically the trade-off between closed-loop gain and bandwidth on the open-loop gain curve of a noninverting amplifier.
- Define *slew rate* and discuss its significance.
- Determine the rise time resulting from the slew rate for a pulse-type waveform.
- Determine the slew rate limiting frequency for a complex waveform.
- Discuss the combined effects of finite bandwidth and slew rate, and indicate the conditions that accentuate either effect.
- Estimate the noise output of an op-amp circuit from noise specifications.
- Explain the difference between *differential gain* and *common-mode gain*, and show how to determine each from possible measurements.
- Define *common-mode rejection ratio* and discuss its significance.

4–1   INTRODUCTION

This chapter focuses on various op-amp limitations that either limit the frequency range of an op-amp or have frequency-dependent characteristics. This is suggested in the title of the chapter by the term *ac*. The primary topics covered are frequency response, slew rate, noise, and common-mode rejection.

In the preceding chapter, we considered the effect of finite op-amp open-loop gain on the closed-loop gain at dc and very low frequencies. In all cases, we assumed the open-loop gain to be the constant value specified for the op-amp at dc. In practice, the open-loop gain starts to decrease at a relatively low frequency, and this roll-off ultimately affects the closed-loop bandwidth. In this chapter, we derive the form of the closed-loop frequency response, as well as develop techniques for predicting the closed-loop 3-dB bandwidth. The noise gain parameter defined in Chapter 3 plays a key role in this process.

A second major effect that limits the bandwidth of an op-amp is the *slew rate*. We discuss this phenomenon in detail, and compare its effects to those of the bandwidth limitation.

Also, we survey sources of noise in op-amps, and we present ways of estimating overall noise levels.

Finally, we introduce the concept of common-mode gain and compare it to differential gain. Then we define the common-mode rejection ratio and discuss its significance.

4–2   CLOSED-LOOP FREQUENCY RESPONSE

We have observed how finite open-loop gain affects closed-loop gain in certain specific circuits as well as in general. However, the open-loop gain thus far has been considered to be a real number A. Specifically, the adjective "dc" was used extensively in previous developments and examples to delineate that a fixed value of A was being assumed.

The actual situation in op-amps is that A has a pronounced variation with frequency. The general behavior of A as a function of frequency for a large number of op-amps is illustrated in Figure 4–1. The abscissa is frequency, and the ordinate is the magnitude of the gain A. Both scales are assumed to be **logarithmic** in form. On some manufacturers' specification sheets, A is given in decibels (dB) on a linear scale, and frequency is given on a logarithmic scale. Such a plot will have the same shape as in Figure 4–1, but actual amplitude has been chosen for illustration since it is more convenient for this initial discussion.

At dc and very low frequencies, the magnitude of the gain is assumed to be $|A| = A_o$. However, the amplitude begins to drop at a relatively low frequency, as readily observed. We will define f_1 as the frequency at which the gain drops to $A_0/\sqrt{2}$ of its dc value. This value corresponds to a 3-dB drop in gain and is the open-loop break frequency as discussed in Section 1–9.

For stability purposes, the open-loop gain functions of most op-amps are deliberately established at -6 dB/octave or -20 dB/decade. Thus, in terms of absolute amplitude, the response on a log-log plot well above f_1 drops by a factor of 0.1 for a tenfold

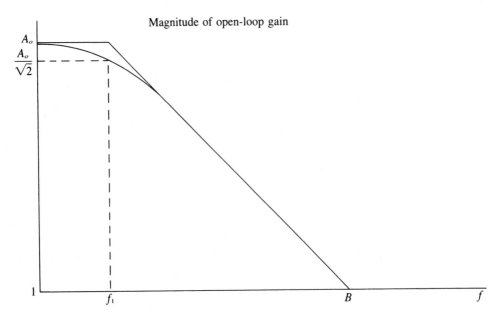

FIGURE 4–1
Form of the magnitude of open-loop gain for many op-amps as a function of frequency.
Both magnitude and frequency scales are assumed to be logarithmic.

increase in frequency. Eventually, the gain reaches the level $|A| = 1$. The frequency
at which this condition is met is called the ***unity-gain frequency,*** and it is denoted on
the curve as B. Above B, the slope of the gain function may be quite different from
-6 dB/octave.

The open-loop gain as a function of frequency may be closely represented over the
frequency range of primary interest by a one-pole low-pass function $A(j\omega)$ of the form

$$A(j\omega) = \frac{A_o}{1 + (jf/f_1)} \tag{4–1a}$$

or

$$A(j\omega) = \frac{A_o}{1 + (j\omega/\omega_1)} \tag{4–1b}$$

or

$$A(j\omega) = \frac{A_o\omega_1}{j\omega + \omega_1} \tag{4–1c}$$

The properties of the one-pole low-pass function were developed in Section 1–9, and
the results of that section can be applied here.

Gain-Bandwidth Relationship

Observe that at low frequencies, $A(j\omega)$ reduces to

$$A(j\omega) \approx A_o \quad \text{for} f \ll f_1 \tag{4-2}$$

as expected. However, in the frequency range well above f_1, the gain can be approximated as

$$A(j\omega) \approx \frac{A_o f_1}{jf} \quad \text{for} f \gg f_1 \tag{4-3}$$

which provides the -6-dB/octave roll-off expected.

Let $f = B$ and $\omega = 2\pi B$, and substitute these values in (4–3):

$$A(j2\pi B) = \frac{A_o f_1}{jB} = \frac{A_o f_1}{B} \underline{/-90^\circ} \tag{4-4}$$

However, we know by definition that the magnitude of the gain at this frequency is unity, so $A_o f_1/B = 1$. This requirement yields the following relationship:

$$B = A_o f_1 \tag{4-5}$$

Stated in words, **the unity-gain frequency is the product of the dc or low-frequency gain and the 3-dB frequency.** For this reason, the unity-gain frequency is also called the **gain-bandwidth product.** Both terms *unity-gain frequency* and *gain-bandwidth product* are used in specifications.

Frequency Response Data

We will pause at this point to investigate typical values of these parameters for a 741 op-amp. Refer to Appendix C for the appropriate data. On the seventh page of the 741 data sheets, a curve of the typical open-loop voltage gain as a function of frequency is shown. The very low-frequency gain is observed to be about $A_o \approx 2 \times 10^5$, and the unity-gain frequency is observed to be about $B = 1$ MHz. It is difficult to read f_1 exactly on the curve, but the use of (4–5) to determine f_1 leads to

$$f_1 = \frac{B}{A_o} = \frac{1 \text{ MHz}}{2 \times 10^5} = 5 \text{ Hz} \tag{4-6}$$

which is about as close as can be determined. These results clearly indicate that the amplitude response starts dropping at a very low frequency, and this effect must be considered in the complete analysis of an op-amp circuit.

The preceding development indicates that the open-loop frequency response is flat only for a very limited low-frequency region. This situation obviously would impose a very serious limitation on the use of the op-amp in an open-loop configuration. Fortunately, however, the bandwidth increases markedly when a closed-loop configuration is used, as we will see shortly.

Closed-Loop Bandwidth

Consider the expression of (3–63) giving the actual closed-loop gain as a function of the ideal closed-loop gain, the noise gain, and the open-loop gain. Next, substitute $A(j\omega)$ from (4–1a) as a frequency-dependent function for A in (3–63). The resulting closed-loop gain will be frequency dependent, and it will be designated as $A_{CL}(j\omega)$. In Problem 4–43, the reader will carry out the manipulations, and the form of the result can be expressed as

$$A_{CL}(j\omega) = \frac{A_{CL}(\text{ideal})/(1 + K_n/A_o)}{1 + j\dfrac{fK_n}{B(1 + K_n/A_o)}} \tag{4–7}$$

To simplify let

$$A_{CL}(\text{dc}) = \frac{A_{CL}(\text{ideal})}{1 + (K_n/A_o)} \tag{4–8}$$

where $A_{CL}(\text{dc})$ is the **actual dc gain** expressed as a function of the ideal closed-loop gain, the noise gain, and the dc open-loop gain. Basically, $A_{CL}(\text{dc})$ is the same form as (3–63), but with the stipulation that A in (3–63) is now the dc gain A_o. Actually, $K_n \ll A_o$ for most applications, so $A_{CL}(\text{dc}) \simeq A_{CL}(\text{ideal})$ in most cases. However, the corrected form of (4–8) will be retained for generality.

The 3-dB break frequency of the closed-loop response will be denoted as B_{CL}, and it can be determined from the denominator of (4–7). We will choose here to employ the approximation $K_n \ll A_o$, in which case $1 + (K_n/A_o) \simeq 1$. Utilizing this approximation and the form of (4–8), we can closely represent the closed-loop gain response $A_{CL}(j\omega)$ as

$$A_{CL}(j\omega) \approx \frac{A_{CL}(\text{dc})}{1 + j(f/B_{CL})} \tag{4–9}$$

where the closed-loop 3-dB frequency B_{CL} is closely approximated as

$$B_{CL} = \frac{B}{K_n} \tag{4–10}$$

The result of (4–10) is most significant, and it deserves some special consideration. In words, *the closed-loop 3-dB bandwidth is equal to the unity-gain frequency divided by the noise gain.* That is, the closed-loop bandwidth decreases as the noise gain increases. We thus see another way that the noise gain degrades the response of an op-amp circuit.

Graphical Interpretation

An interesting interpretation of the preceding concept may be developed by considering a noninverting op-amp circuit. Recall that for the noninverting op-amp circuit, $K_n = A_{CL}(\text{ideal}) = (R_i + R_f)/R_i$. Further, assume that $K_n \ll A_o$, so that $A_{CL}(\text{dc}) = A_{CL}(\text{ideal})$.

Refer to Figure 4–2 for the discussion that follows. The solid curve is the open-loop gain function. The dashed curve labeled $|A_{CL}^{(1)}|$ is a relatively large closed-loop gain. The 3-dB bandwidth of this response is $B_{CL}^{(1)}$, and it can be determined by noting the intersection of the constant dc gain level $A_{CL}^{(1)}$ (dc) with the open-loop gain function.

Next consider a desired smaller closed-loop gain, whose curve is labeled as $|A_{CL}^{(2)}|$. The 3-dB bandwidth of this response is $B_{CL}^{(2)}$, and the intersection of the constant dc gain level $A_{CL}^{(2)}$ (dc) with the open-loop gain function is observed to be at a higher frequency than for the first gain. The trade-off between bandwidth and gain is readily observed from this graphical interpretation.

The graphical interpretation can be adapted to the inverting amplifier configuration or to other possible circuits, but the noise gain level rather than the actual gain magnitude is used to establish the appropriate intersection level. In the noninverting amplifier, the actual gain and the noise gain are equal, and as a result the utility of the graphical approach for this circuit is enhanced.

Because of the simplicity of the relationship between noise gain and actual gain for the noninverting amplifier, the following simplified formula is found in the literature:

$$\text{Closed-loop bandwidth} = \frac{\text{unity-gain frequency}}{\text{closed-loop gain}} \Rightarrow \frac{B}{A_v} \quad \text{(4–11)}$$

Strictly speaking, this formula is correct only for the noninverting configuration, and our approach in this text will be to use the form of (4–10) employing the noise gain in general. However, this author has observed a widespread use of (4–11) for other configurations.

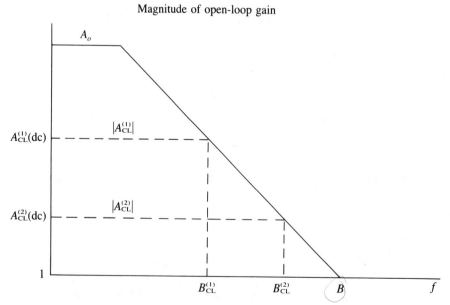

Magnitude of open-loop gain

FIGURE 4–2
Graphical interpretation of closed-loop bandwidth for a noninverting amplifier.

In circuits where the noise gain and actual gain are very close, the approximation may be quite good, but in other cases, significant error may occur.

Rise Time

An additional concept of importance concerns the relationship between the closed-loop bandwidth of an amplifier circuit and the rise time of a pulse type of waveform. When the amplifier is required to reproduce signals having nearly instantaneous step changes (for example, logic signals), it is quite important to maintain rise and fall times to tolerable levels.

In general, the rise time is inversely proportional to the closed-loop bandwidth. Thus, short rise times are achieved by amplifiers with higher bandwidths.

A quantitative relationship for rise time can be determined by first computing the response of the amplifier when excited by a step function (dc) input voltage of magnitude V_i as shown in Figure 4–3(a). Network theory can be used to show that the output voltage is of the form

$$v_o(t) = A_{CL}(dc)\, V_i(1 - e^{-2\pi B_{CL}t}) \qquad (4\text{–}12)$$

The form of this voltage is illustrated in Figure 4–3(b). The final value approaches $A_{CL}(dc)V_i$ as expected, since the dc gain is $A_{CL}(dc)$ and the input voltage magnitude is V_i. However, the finite bandwidth of the circuit keeps the output from reaching the final level instantaneously.

In view of the exponential increase, the output theoretically never quite reaches the limiting value. However, for virtually all practical purposes, the output can be considered to reach the final value in a time equal to about five time constants. The time constant in this case is $(2\pi B_{CL})^{-1}$.

A common means for specifying the effect involved is the rise time resulting from the finite closed-loop bandwidth. The standard IEEE definition of rise time is based on the time required for the output to change from 10% to 90% of the final level. The choice of these times eliminates the uncertainties at the beginning and final levels. It is left as an exercise for the reader (Problem 4–44) to show that the rise time for the function of (4–12) is given by

$$T_{CL} = \frac{0.35}{B_{CL}} \qquad (4\text{–}13)$$

This relationship is strictly correct only for the first-order form assumed for the gain function. However, in the most general case, $T_{CL} = K/B_{CL}$, where K is a constant depending on the nature of the amplifier frequency function. In most cases, K does not differ significantly from 0.35, so that value is often used for more complex situations.

As a final point, the concept of rise time applies equally well to fall time. Thus, if it takes a certain amount of time for a pulse to reach a final value at the output of an amplifier, it will take a comparable amount of time for the output to return to zero when the pulse is removed from the input.

(a)

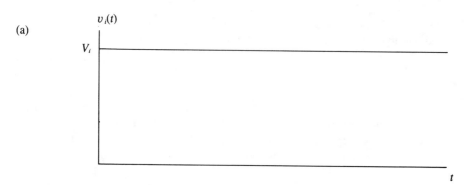

(b)

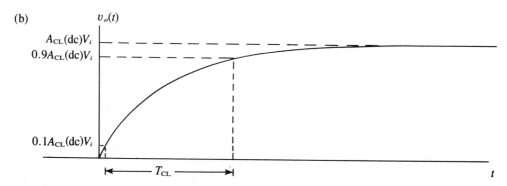

FIGURE 4–3
Step function input and corresponding output of closed-loop amplifier.

EXAMPLE 4–1 The unity-gain frequency of a certain 741 op-amp is $B = 1$ MHz. For a *noninverting* amplifier, calculate the closed-loop 3-dB bandwidth for each of the following ideal values of closed-loop gain: (a) 1000, (b) 100, (c) 10, and (d) 1.

Solution
The closed-loop bandwidth is determined from (4–10). Substituting $B = 1$ MHz, we have

$$B_{\mathrm{CL}} = \frac{10^6}{K_n} \qquad (4\text{–}14)$$

We recognize that $K_n = A_{\mathrm{CL}}(\text{ideal})$ for the noninverting amplifier. The values of B_{CL} are then calculated by substituting the four values of ideal closed-loop gain. The results are tabulated in the second column of Table 4–1. The values in other columns apply to later examples, and further discussion and comparison of these values will be made at the end of appropriate examples.

TABLE 4–1
Comparison of 3-dB closed-loop bandwidths and rise times for noninverting and inverting amplifier configurations with the 741 (See Examples 4–1, 4–2, and 4–3.)

Magnitude of Ideal Closed-Loop Gain	3-dB Bandwidth for Noninverting Amplifier	3-dB Bandwidth for Inverting Amplifier	Rise Time for Noninverting Amplifier	Rise Time for Inverting Amplifier
1000	1 kHz	999 Hz	0.35 ms	0.35 ms
100	10 kHz	9.9 kHz	35 μs	35.3 μs
10	100 kHz	90.91 kHz	3.5 μs	3.85 μs
1	1 MHz	500 kHz	0.35 μs	0.7 μs

EXAMPLE 4–2

Continuing in the same pattern as in Example 4–1, assume now an *inverting* amplifier configuration using the 741 op-amp. The four values of gain listed in Example 4–1 now represent *magnitudes*. Calculate the closed-loop 3-dB bandwidth for each of the four gain magnitudes, and compare their values with the results of Example 4–1.

Solution
We must first determine a relationship between the noise gain and the actual gain of an inverting amplifier. Recall that the noise gain of the inverting amplifier is the same as for the noninverting amplifier; that is,

$$K_n = 1 + \frac{R_f}{R_i} \tag{4–15}$$

However, the closed-loop ideal gain is

$$A_{CL}(\text{ideal}) = -\frac{R_f}{R_i} \tag{4–16}$$

Comparison of (4–15) and (4–16) leads to the conclusion that

$$K_n = 1 + |A_{CL}(\text{ideal})| \tag{4–17}$$

Thus, the noise gain is greater than the magnitude of the closed-loop gain for the inverting amplifier.

Substitution of (4–17) and $B = 1$ MHz into (4–10) leads to

$$B_{CL} = \frac{10^6}{1 + |A_{CL}(\text{ideal})|} \tag{4–18}$$

The values of B_{CL} are then calculated by substituting the four values of ideal closed-loop gain. The results are tabulated in the third column of Table 4–1.

It is interesting to compare the closed-loop bandwidths of inverting and noninverting amplifiers for fixed values of closed-loop gain. At large closed-loop gains, the two bandwidths are very nearly the same. However, as the desired closed-loop gain drops to much smaller values, the difference in bandwidth between the two configurations becomes quite significant. At the extreme of unity gain, the 3-dB bandwidth of the noninverting amplifier is twice the bandwidth of the inverting amplifier.

EXAMPLE 4-3

Compute the rise times associated with a pulse type of waveform for the various noninverting and inverting amplifier conditions of Examples 4–1 and 4–2.

Solution

For a given value of B_{CL}, the rise time T_{CL} can be readily calculated with (4–13). The various values of B_{CL} are tabulated in the second and third columns of Table 4–1 as developed in Examples 4–1 and 4–2. The corresponding values of rise time are tabulated in the fourth and fifth columns. Observe that as the bandwidth increases, corresponding to lower gains, the rise time decreases.

EXAMPLE 4-4

Consider the linear combination circuit of Example 3–7 as shown in Figure 3-12, and assume that the unity-gain frequency is $B = 1$ MHz. Compute the 3-dB closed-loop frequency.

Solution

The noise gain of the circuit was determined in Example 3–7 to be $K_n = 121$. Substituting this value and the value of B in (4–10), we obtain

$$B_{CL} = \frac{10^6}{121} = 8.264 \text{ kHz} \tag{4-19}$$

Observe that the noise gain is much greater than the gain for either of the input signals, and thus it degrades the frequency response.

EXAMPLE 4-5

The open-loop gain of a certain op-amp is shown by the solid curve in Figure 4–4. The op-amp is connected in a noninverting amplifier circuit. Draw the Bode plot of the closed-loop gain function for each of the following ideal values of closed-loop gain: (a) 100, (b) 10, and (c) 1. In each case, identify the 3-dB closed-loop bandwidth directly from the curve.

Solution

The desired closed-loop gain is drawn as a dashed horizontal line for each case, and the value of B_{CL} is noted at the intersection with the open-loop gain. The three curves are labeled as (a), (b), and (c), respectively. The closed-loop bandwidth values are readily observed as follows:

(a) $B_{CL} = 10$ kHz

(b) $B_{CL} = 100$ kHz

(c) $B_{CL} = 1$ MHz

$B_{CL} = \dfrac{B}{K_N} = \dfrac{10^6}{A_{CL}}$

These values could have been readily computed from (4–10), of course, but the graphical approach adds an additional point of interpretation.

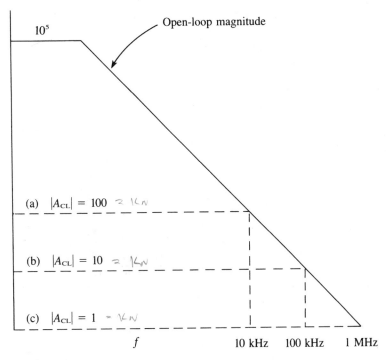

FIGURE 4–4
Open-loop and closed-loop frequency response curves of noninverting
amplifier of Example 4–5.

It should be noted that the curves are the straight-line Bode plot approximations. The actual frequency response curves are very close to the straight-line approximations at frequencies much lower and much higher than the break frequencies. The largest error is at the break frequency, where the actual curve is $1/\sqrt{2}$ times the straight-line approximation value (3 dB down).

4–3 SLEW RATE

Operational amplifiers typically have one or more capacitors contained within the circuitry for the purpose of stabilizing closed-loop operation. These capacitors must charge and discharge as the input signal varies. The effect is to impose a maximum rate at which the output signal may vary instantaneously. If the input signal exceeds a certain maximum rate of change, the output will simply fail to follow the instantaneous form of the input.

The process in which the output voltage of an op-amp can change only at a finite rate is referred to as the *slew rate limitation.* This effect is described quantitatively by specifying a *slew rate S* for the given op-amp. The basic unit for measuring slew rate is volts/second (V/s), and the value indicates how fast the output of the op-amp is capable of changing. In practice, the speeds are sufficiently high that most op-amp slew rate specifications are given in volts/microsecond (V/μs) for convenience.

Slew Rate Data

Before discussing practical interpretations of the quantity, a quick check of typical values is in order. In Appendix C, the row indicated as "Slew Rate" on the third page of the 741 data sheets provides the information desired, and a typical value is listed as 0.5 V/μs. As we will see later, this value strongly limits the application of the 741 op-amp in many circuits. Units with much higher slew rates are available but, of course, at a higher price than for the 741 op-amp.

Effect on Pulse Waveform

The effect of the slew rate on a pulse type of waveform will now be investigated. Let T_{SR} represent the rise time resulting from the slew rate phenomenon; this value is the minimum time required for the output voltage magnitude to change by V_o volts. Since slew rate is the rate of change in volts per unit time, the minimum rise time is determined as

$$T_{SR} = \frac{V_o}{S} \qquad (4\text{--}20)$$

This phenomenon is illustrated in Figure 4–5. If V_o is expressed in volts and S is expressed in volts/microsecond, T_{SR} will be expressed in microseconds. Conversely, if S is expressed in volts/second, T_{SR} will be expressed in seconds. If the op-amp output starts at zero, V_o is the magnitude of the final output voltage. However, if the output has to swing from one polarity to the opposite polarity, V_o could be interpreted as the peak-to-peak magnitude of the change in voltage, and T_{SR} could be interpreted as the time required for the change to occur.

Before proceeding further, we should clarify that the rise time due to the slew rate effect is a different phenomenon from the rise time due to the finite bandwidth as discussed

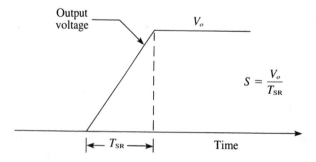

FIGURE 4–5
Slew rate for a pulse type of waveform and the associated rise time.

in Section 4–2. The two concepts are related through the complex interaction of the circuitry within the op-amp, but there are some significant differences in their effects. The rise time produced by the finite bandwidth is based on the exponential buildup of the output voltage. However, the rise time produced by the slew rate tends to be a straight-line or ramp variation as noted in Figure 4–5. For this reason, it is more convenient to define the associated rise time due to the slew rate effect as the total time required for the output to change, rather than the 10% to 90% time difference used for the more conventional rise time definition.

If the required op-amp output voltage is changing at a rate less than the slew rate, the amplifier will be able to "track" the signal as far as the slew rate is concerned. (The finite bandwidth may impose limitations.) Consider the waveforms shown in Figure 4–6, and assume that any degradation due to finite bandwidth is negligible. Let A_{CL} represent the closed-loop gain of the amplifier. The input waveform v_i shown in part (a) is such that the required output rate of change is less than the slew rate; that is, $A_{CL} V_i/t_i < S$. The output v_o shown in (b) then will be able to follow the input, and no degradation will result. For the same amplifier, however, assume that the input is the signal shown in (c), whose rise time is much shorter. In this case, $A_{CL} V_i/t_i > S$, and thus the required rate of change of the output exceeds the slew rate. The output therefore cannot follow the input, and the signal will be degraded as shown in (d). The actual rise time of the output will be T_{SR}, which is a function of the slew rate as given by (4–20).

In general, both finite bandwidth and slew rate affect the output rise time, and it is necessary to investigate the limitations of both phenomena when analyzing or designing a complete circuit. Section 4–4 is devoted to comparing both phenomena to determine their relative effects.

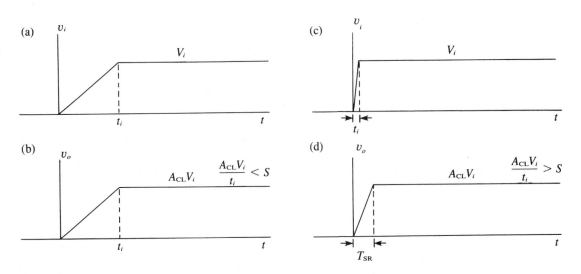

FIGURE 4–6
Slew rate effects for two different waveforms as a function of different rates of change of the input voltage.

Effect on Complex Waveform

Interpretation of rise time and slew rate of a pulse type of waveform is straightforward. However, a more difficult question concerns how the slew rate degrades a more complex waveform (for example, an audio waveform).

Since the slew rate determines the maximum rate of change permitted for the output voltage, an approach that has proved to be reasonable is to use a sinusoidal waveform as the basis of reference. The frequency at which the maximum rate of change of the sinusoid equals the slew rate is identified as the highest possible operating frequency. In this manner, a workable bandwidth is related to the slew rate. While a sinusoid is obviously not a complex waveform, this approach tends to give a worst-case bound, and the results have been found to be reasonable in practice.

To obtain a quantitative expression for the phenomenon involved, assume that the output v_o of an op-amp can be expressed in sinusoidal form as

$$v_o = V_o \sin \omega t \tag{4-21}$$

where V_o is the peak value and ω is the angular frequency (in radians/second).

The derivative or rate of change of the output can be expressed as

$$\frac{dv_o}{dt} = \omega V_o \cos \omega t \tag{4-22}$$

The maximum rate of change of the output voltage occurs when $\cos \omega t = 1$, corresponding to $t = 0$ and subsequent integer multiples of a period. We have

$$\left(\frac{dv_o}{dt}\right)_{max} = \omega V_o \tag{4-23}$$

In order for the op-amp to be able to follow the sinusoidal signal, the slew rate S must be greater than, or at a minimum equal to, the maximum rate of change of the output signal. Thus,

$$S \geq \left(\frac{dv_o}{dt}\right)_{max} \tag{4-24}$$

Let f_{SR} represent the highest frequency at which the op-amp can track the input sinusoid. The frequency is referred to as the **slew rate limiting frequency.** It is determined by setting $\omega = 2\pi f_{SR}$ in (4–23) and substituting that expression as an equality in (4–24):

$$2\pi f_{SR} V_o = S \tag{4-25}$$

or

$$f_{SR} = \frac{S}{2\pi V_o} \tag{4-26}$$

Thus, f_{SR} can be considered as an estimate of the highest operating frequency that can be processed without slew rate distortion by an op-amp.

As in the case of a pulse type of waveform, both finite closed-loop bandwidth and slew rate effects must be considered for a complete analysis. How the two effects together determine the workable bandwidth will be considered in Section 4–4.

EXAMPLE 4–6

A typical value of slew rate for the 741 op-amp is $S = 0.5$ V/μs. Assuming a pulse type of signal, determine the rise time due to the slew rate effect when the output voltage is required to change from zero to (a) 0.1 V, (b) 1 V, and (c) 10 V.

Solution
From (4–20), the rise time is readily determined as

$$T_{SR}(\mu s) = \frac{V_o}{0.5} \tag{4–27}$$

where we are emphasizing that the rise time is measured in microseconds since the slew rate in volts/microseconds has been substituted in (4–27). One can always resort to basic units if desired, and this procedure is recommended if one has doubts about working with modified units. In this case, the slew rate would be expressed as $S = 0.5 \times 10^6$ V/s.

Substituting the three values of V_o in (4–27), we readily determine the corresponding values of rise time, and the results are tabulated as follows:

V_o	(a) 0.1 V	(b) 1 V	(c) 10 V
T_{SR}	0.2 μs	2 μs	20 μs

Note that the rise time increases linearly with the peak value of the output voltage, as expected.

Observe that nothing was indicated about the gain of the amplifier circuit. Instead, only the peak value of the output voltage was stated. Consider, for example, the peak output of (c), which was 10 V. This value could be the output of a voltage follower having an input of 10 V, or it could be the output of an amplifier with a gain of 10 and an input of 1 V. The slew rate effect does not matter because the peak output voltage is the same in either case. Thus, *for a given peak output voltage, the rise time due to slew rate is independent of the gain of the amplifier.*

The rise time due to slew rate is in sharp contrast to the rise time due to the finite closed-loop bandwidth, since the closed-loop gain strongly affects the bandwidth and the corresponding rise time.

EXAMPLE 4–7

As in Example 4–6, assume a 741 op-amp with a slew rate S of 0.5 V/μs. However, now assume complex signals, and determine the highest possible operating frequency due to the slew rate effect for each of the following peak values of output voltage: (a) 0.1 V, (b) 1 V, and (c) 10 V.

Solution
From (4–26), the slew rate limiting frequency can be estimated as

$$f_{SR} = \frac{0.5 \times 10^6}{2\pi V_o} \tag{4–28}$$

where $S = 0.5 \times 10^6$ V/s has been substituted in (4–26). One could substitute $S = 0.5$ V/μs, and the frequency would then be expressed in megahertz. However, since units are changed somewhat in this computation, a return to basic units is probably the best approach.

Substituting the three values of V_o in (4–28), we readily determine the corresponding values of f_{SR}, and the results are tabulated as follows:

V_o	(a) 0.1 V	(b) 1 V	(c) 10 V
f_{SR}	796 kHz	79.6 kHz	7.96 kHz

Note that the highest operating frequency is inversely proportional to the peak value of the output. Thus, the sacrifice for operating an op-amp with a high peak output voltage is a degradation in frequency response.

Example 4–7 illustrates how important slew rate can be when the output voltage is large. From these results, it can be inferred that the 741 op-amp would be inadequate for audio applications whenever the peak output voltage is more than a few volts.

Note that the highest operating frequency due to slew rate depends on the peak output voltage but not on the gain itself. Thus, the discussion at the end of Example 4–6 referring to a pulse type of signal applies equally well to complex signals.

4–4 COMBINATION OF LINEAR BANDWIDTH AND SLEW RATE

In Section 4–2, the effects of the open-loop bandwidth in establishing closed-loop bandwidth and rise time were determined. In Section 4–3, the effects of slew rate in establishing the rise time and the highest operating frequency were determined. In each section, the particular phenomenon under investigation was assumed to be isolated from the other phenomenon in determining the limits of performance. In practice, both finite bandwidth and slew rate limitations are always present, and the overall performance may be worse than the effect predicted from either individual phenomenon. In this section, we will investigate regions of operation in which either effect may determine the overall limitation, and we will develop some insight as to the selection of desired op-amp parameters to minimize degradation in performance.

To establish a convenient framework of reference, we will refer to the finite closed-loop bandwidth phenomenon of Section 4–2 as the *finite bandwidth effect.* The slew rate phenomenon of Section 4–3 will be referred to as the *finite slew rate effect.* The results developed in these two sections will then be related to the overall situation as presented in this section.

As considered in the preceding two sections, it is convenient to classify the analysis on the basis of either (1) **complex signals** or (2) **pulse types of signals.** A given application will often determine which form is better for establishing a workable criterion. For example, if the op-amp is to be used to amplify an analog signal such as an audio signal or the instantaneous output of an analog transducer, a criterion based on complex signals will usually be better. On the other hand, if the op-amp is to be used in some digital or pulse application where square waves or pulse signals are to be processed, a criterion based on pulse types of signals will usually be better. In some situations, it might even be desirable to use both criteria for determining operating specifications.

Each of the two categories will now be investigated individually. Formulas developed in the preceding two sections will be used freely as required.

Complex Signals

Assume that the op-amp has a unity-gain frequency (gain-bandwidth product) B. The closed-loop 3-dB frequency B_{CL} as developed in Section 4–2 is given by

$$B_{CL} = \frac{B}{K_n} \tag{4–29}$$

where K_n is the noise gain for the particular configuration.

Let S represent the slew rate of the op-amp. The highest operating frequency f_{SR} at which the output can track the input as developed in Section 4–3 is given by

$$f_{SR} = \frac{S}{2\pi V_o} \tag{4–30}$$

where V_o is the peak value of the output.

Let f_H represent the highest frequency contained in the input analog signal. In order that the op-amp circuit be capable of processing the given complex signal without distortion, it is necessary that *both* of the following two criteria be satisfied:

$$B_{CL} \gg f_H \tag{4–31}$$
$$f_{SR} > f_H \tag{4–32}$$

Observe that the inequality of (4–31) indicates a strong inequality, that is much greater than. The reason is that at a frequency B_{CL}, the closed-loop response due to the finite bandwidth is already 3 dB down and thus may represent a serious degradation for a precision amplifier. (The extent of the error in magnitude will be discussed shortly.) However, the inequality of (4–32) indicates that the slew rate must be at least large enough to track the output signal at the highest frequency contained in the signal.

The extent to which the inequality of (4–31) is desired will now be investigated. From the work of Section 4–2, the closed-loop gain function of many op-amp circuits can be closely approximated over a wide frequency range by

$$A_{CL}(j\omega) = \frac{A_{CL}(\text{dc})}{1 + j\dfrac{f}{B_{CL}}} \tag{4–33}$$

The amplitude response $M(\omega)$ corresponding to (4–33) can be expressed as

$$M(\omega) = \frac{A_{CL}(\text{dc})}{\sqrt{1 + (f/B_{CL})^2}} \tag{4–34}$$

If $A_{CL}(dc)$ is positive (noninverting gain), the phase response $\theta(\omega)$ is

$$\theta(\omega) = -\tan^{-1}\left(\frac{f}{B_{CL}}\right) \tag{4–35}$$

When $A_{CL}(dc)$ is negative (inverting gain), an additional $180°$ phase shift should be added to (4–35).

The magnitude and phase functions of a one-pole low-pass function were developed in Section 1–9, and the results of that section may be adapted to the functions of (4–34) and (4–35). Specifically, Figure 1–19 may be used to predict the closed-loop decibel magnitude response of (4–34) relative to the peak level, and Figure 1–21 may be used to predict the closed-loop phase response. The break frequency f_b is interpreted as B_{CL} for this purpose.

To provide more accurate data than the curve permits, Table 4–2 lists specific gain levels relative to maximum gain and their corresponding frequencies. From the curves and the table, it is possible to determine the error in amplitude response for a given closed-loop bandwidth. Conversely, it is possible to determine the required minimum 3-dB bandwidth for a specified maximum error in amplitude or phase.

Since both (4–31) and (4–32) must be satisfied, it is important to understand the factors that contribute to both conditions. Since $B_{CL} = B/K_n$, the inequality of (4–31) is improved by either (1) increasing B or (2) decreasing K_n. The first condition requires the selection of an op-amp with a wider bandwidth, and the second condition normally requires designing for less gain in the given stage. Any trade-off between these two parameters that increases the inequality of (4–31) will flatten the pass-band response of the amplifier and decrease the error in gain due to the finite bandwidth effect.

The slew rate limiting frequency is $f_{SR} = S/(2\pi V_o)$. The inequality of (4–32) is increased by either (a) increasing the slew rate or (b) decreasing the peak output voltage. The first condition requires the selection of an op-amp with a higher slew rate, and the second condition requires maintaining the peak output voltage to a lower level.

TABLE 4–2
Closed-loop decibel gain levels relative to maximum gain and the corresponding frequencies for op-amp closed-loop gain

Gain Level Relative to Maximum Gain	f/B_{CL}
−0.01 dB	0.0480
−0.02 dB	0.0679
−0.025 dB	0.0760
−0.05 dB	0.1076
−0.1 dB	0.1526
−0.2 dB	0.2171
−0.25 dB	0.2434
−0.5 dB	0.3493
−1 dB	0.5088
−1.5 dB	0.6423
−2 dB	0.7648
−2.5 dB	0.8822
−3 dB	0.9976

The finite bandwidth limitation depends critically on the gain established for the circuit, but it is independent of the actual signal level within the linear region of operation. Conversely, the slew rate limitation depends critically on the output signal level, but it is independent of the gain established for the circuit.

The following general trends indicate the relative pattern and are worth remembering:

1. At very low signal levels, the finite bandwidth limitation *tends to be* the dominating factor in establishing the frequency range of the circuit.
2. At very high signal levels, the slew rate limitation *tends to be* the dominating factor in establishing the frequency range.

Pulse Types of Signals

While frequency response is the appropriate criterion for dealing with complex signals, rise time is usually the best criterion for pulse types of signals. The rise time T_{CL} due to the closed-loop bandwidth B_{CL} as developed in Section 4–2 is

$$T_{CL} = \frac{0.35}{B_{CL}} = \frac{0.35 \, K_n}{B} \tag{4-36}$$

The rise time T_{SR} due to the slew rate effect as determined in Section 4–3 is

$$T_{SR} = \frac{V_o}{S} \tag{4-37}$$

Recall that T_{SR} represents the total time for the signal to change at the output, while T_{CL} represents the standard definition of rise time between 10% and 90% points.

Let t_i represent the rise time of the input pulse-time waveform. Faithful reproduction of the pulse waveform requires that the rise time be preserved; that is, negligible additional rise time should be introduced by the amplifier. For correct reproduction, it is necessary that *both* of the following criteria be satisfied:

$$T_{CL} \ll t_1 \tag{4-38}$$

$$T_{SR} < t_1 \tag{4-39}$$

As the inequality of (4–38) increases, the effect of the additional rise time introduced by the amplifier becomes less significant. The actual rise time t_o of the pulse at the amplifier as a function of the input rise time and the rise time introduced by the finite gain effect of the amplifier can be approximated as

$$t_o \simeq \sqrt{T_{CL}^2 + t_i^2} \tag{4-40}$$

From this result, it can be shown that when $T_{CL} \leq 0.1 t_i$, the percentage increase in output rise time is less than 0.5%.

The general trends noted for complex signals apply to pulse types of signals as well. Thus, at very low output signal levels, finite bandwidth often tends to be the limiting factor for rise time, and at high output signal levels, slew rate often tends to produce the most significant effect.

LM318 Op-Amp

Although the 741 op-amp is capable of serving many general-purpose applications quite well, it is somewhat restricted in frequency and rise time as a result of slew rate and gain-bandwidth limitations. Other op-amps are available with both increased slew rate and gain-bandwidth specifications (at a higher price, of course). One particular op-amp that will be mentioned in this context is the LM118/LM218/LM318, and the specifications are given in Appendix C. The three numbers refer to the military, industrial, and commercial temperature specifications, respectively. For simplicity, subsequent references will be made simply to the "318" op-amp.

Inspection of the 318 data reveals that the minimum slew rate is 50 V/μs, and a typical value is 70 V/μs. These values exceed the corresponding value for the 741 op-amp by at least a 100-to-1 ratio. It is also possible to add external compensation to the 318 to boost the slew rate to more than 150 V/μs.

The unity-gain frequency of the 318 op-amp has a typical value of 15 MHz, which is about 15 times the corresponding value for the 741 op-amp.

EXAMPLE 4–8

Consider the noninverting amplifier circuit of Figure 4–7 utilizing a 741 op-amp. The circuit is to be used to amplify some complex analog signals. Investigate the frequency limits of operation when the input signal has a peak value of (a) 20 mV and (b) 500 mV. Assume that $B = 1$ MHz and $S = 0.5$ V/μs.

Solution

It is necessary to consider the effects of both finite bandwidth and slew rate. The closed-loop gain of the circuit is first determined to be

$$A_{CL} = 1 + \frac{24 \text{ k}\Omega}{1 \text{ k}\Omega} = 25 \tag{4-41}$$

The 3-dB closed-loop bandwidth is next calculated as

$$B_{CL} = \frac{B}{K_n} = \frac{1 \times 10^6}{25} = 40 \text{ kHz} \tag{4-42}$$

where it has been recognized that $K_n = A_{CL}$ for a noninverting amplifier. The value of B_{CL} is independent of the signal level, so this quantity can be used in both parts (a) and (b) that follow.

(a) Recall that the peak value of the **output** voltage determines the slew rate limitation. Since the peak value of the input voltage is 20 mV in the first case, the peak value of the output is $V_o = 25 \times 20$ mV $= 500$ mV $= 0.5$ V. The slew rate limiting frequency is

$$f_{SR} = \frac{S}{2\pi V_o} = \frac{0.5 \times 10^6}{2\pi \times 0.5} = 159.2 \text{ kHz} \tag{4-43}$$

Comparing (4–43) with (4–42), one can see that the value of B_{CL} sets the upper limit of frequency operation at this signal level. In other words, in order for *both* (4–31) and (4–32)

FIGURE 4–7
Circuit of Example 4–8.

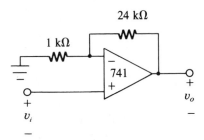

to be satisfied, all signal frequencies must be less than 40 kHz. Depending on the tolerance of gain error permitted in the passband, the highest frequency might be restricted to be much smaller than 40 kHz.

(b) The peak value of the input voltage in this case is 500 mV, so the peak value of the output voltage is $V_o = 25 \times 0.5$ V $= 12.5$ V. The highest operating frequency due to slew rate in this case is

$$f_{SR} = \frac{0.5 \times 10^6}{2\pi \times 12.5} = 6.366 \text{ kHz} \tag{4–44}$$

Comparing (4–44) with (4–42), one can see that the value of f_{SR} sets the upper frequency limit in this case. Thus, in order for *both* (4–31) and (4–32) to be satisfied, all signal frequencies must be less than 6.366 kHz.

Example 4–8 clearly illustrates the general trend discussed earlier. At the very low signal level, the upper frequency was limited by the finite closed-loop bandwidth, while at the high signal level, the upper frequency was limited by the slew rate. This is a general trend, and the conditions were selected to illuminate the trend. However, there are exceptions to the pattern, so the best approach is to check all criteria under worst-case conditions.

EXAMPLE 4–9

Consider the noninverting amplifier of Example 4–8 with the same parameters and peak input signal values. However, now the circuit is to be used with a pulse type of signal. Investigate the limiting values of rise time for the two cases. The two peak input values now represent the amplitudes of pulses.

Solution
The closed-loop bandwidth was calculated in Example 4–8 to be $B_{CL} = 40$ kHz. The corresponding rise time T_{CL} is

$$T_{CL} = \frac{0.35}{B_{CL}} = \frac{0.35}{40 \times 10^3} = 8.75 \ \mu s \tag{4–45}$$

The value of T_{CL} is independent of the signal level, so this quantity can be used in both parts (a) and (b) that follow.

(a) The peak value of the output pulse in this case is $V_o = 0.5$ V. The rise time T_{SR} due to the slew rate is

$$T_{SR} = \frac{V_o}{S} = \frac{0.5}{0.5 \times 10^6} = 1 \ \mu s \qquad (4\text{–}46)$$

Comparing (4–46) with (4–45), one can see that the value of T_{CL} sets the minimum input signal rise time in this case. In other words, in order for *both* (4–38) and (4–39) to be satisfied, the rise time of any input pulse must be greater than 8.75 μs in order that no degradation occur. Depending on the tolerance of rise time degradation permitted, the input rise time might be restricted to be much larger than 8.75 μs.

(b) The peak value of the output pulse in this case is 12.5 V, and T_{SR} is given by

$$T_{SR} = \frac{12.5}{0.5 \times 10^6} = 25 \ \mu s \qquad (4\text{–}47)$$

Comparing (4–47) with (4–45), one can see that the value of T_{SR} sets the minimum input signal rise time in this case. Thus, in order for *both* (4–38) and (4–39) to be satisfied, the rise time of any input pulse must be greater than 25 μs in order that no degradation in rise time occur.

The trends for the amplifier with pulse types of signals are compatible with the trends for complex signals. Thus, at the very low input signal level, the rise time was limited by the finite bandwidth, while at the high signal level, the rise time was limited by the slew rate.

EXAMPLE 4–10

An op-amp is to be selected for an audio application. The noninverting configuration is to be used, and a gain of 50 is desired. The peak input signal is expected never to exceed 0.2 V. Specifications require that the gain not vary more than 0.25 dB over the frequency range from near dc to 20 kHz. Determine the minimum values of (a) unity gain frequency and (b) slew rate for the op-amp selected.

Solution
From Table 4–2 we can see that a 0.25-dB change in the amplitude response occurs for $f/B_{CL} = 0.2434$. Since the highest frequency in the signal is 20 kHz, the maximum change of 0.25 dB in the gain is assured by setting

$$\frac{20 \times 10^3}{B_{CL}} = 0.2434$$

or

$$B_{CL} = 82.17 \ kHz \qquad (4\text{–}48)$$

This result indicates that in order to keep the amplitude response variation within the 0.25-dB range, it is necessary that the 3-dB frequency be set to a frequency much higher than the audio frequency upper limit of 20 kHz.

Since $B_{CL} = B/K_n$ and $K_n = A_{CL}$ for a noninverting amplifier, the minimum value of B is determined as follows:

$$\frac{B}{K_n} = \frac{B}{50} = 82.17 \text{ kHz} \tag{4-49}$$

or

$$B = 4.109 \text{ MHz} \tag{4-50}$$

The peak output voltage is $V_o = 50 \times 0.2 \text{ V} = 10 \text{ V}$. Since the slew rate limiting frequency is $f_{SR} = S/(2\pi V_o)$, the minimum slew rate is determined by setting $f_{SR} = 20$ kHz and solving for S:

$$20 \times 10^3 = \frac{S}{2\pi \times 10} \tag{4-51}$$

or

$$S = 1.257 \times 10^6 \text{ V/s} = 1.257 \text{ V/}\mu\text{s} \tag{4-52}$$

In order to meet the specifications, the op-amp selected must have minimum values of $B = 4.109 \text{ MHz}$ and $S = 1.257 \text{ V/}\mu\text{s}$. Since op-amp parameters vary from one unit to another, a conservative approach would suggest selecting an op-amp having parameters above these minimum values in order to provide some leeway. A 741 op-amp does not meet the minimum specifications in this example.

4-5 NOISE IN OPERATIONAL AMPLIFIERS

Noise originating in an op-amp is a rather complex phenomenon, and a complete analysis is quite involved. Detailed models have been developed for isolating the various noise sources and predicting their relative levels. However, such details are probably best left to the specialists in this area. The vast majority of IC users usually settle for some reasonable estimate of the overall combined noise level.

The approach taken here will be first to explain in qualitative terms the major noise effects that appear. The use of manufacturers' specifications to approximate the overall noise level will then be emphasized. This approach should be sufficient for most practical conventional circuit designs.

Noise Classifications

Internal noise within integrated circuits can be roughly divided by its origin into *thermal noise, shot noise, flicker noise,* and *popcorn noise. Thermal noise* is the result of random motion of the charge carriers in a resistance. The random motion as well as the associated noise increases as the temperature increases.

Shot noise arises from the discrete nature of current flow in electronic devices. *Flicker noise* (also called 1/*f* noise) is a somewhat vaguely understood form of noise occurring in active devices at very low frequencies. *Popcorn noise* (also called *burst noise*) results from a momentary change in input bias current, usually occurring below 100 Hz, and is caused by imperfect semiconductor surface conditions.

The various noise sources combine to produce a net total noise, and it is somewhat difficult to separate the relative contributions previously defined. However, it is convenient to separate the resulting effects according to their frequency distribution. For example, both thermal and shot noise are forms of *white noise,* and their effects are usually lumped together. *White noise* is any noise whose frequency spectrum is constant over a very wide bandwidth. The adjective *white* has its origin in the associated analogy with white light, which can be considered to be composed of all colors in the light spectrum. If a tunable bandpass filter were connected to the output of a white noise source, equal noise power would be measured in any constant bandwidth over which the filter were tuned.

Flicker noise follows a 1/*f* frequency variation; that is, the relative noise level increases significantly at very low frequencies. However, flicker noise is usually negligible above 1 kHz or so.

Noise Data

We can observe how noise specifications are provided by referring to the eighth page of the 741 op-amp specifications in Appendix C. Note the three graphs entitled "Input Noise Voltage as a Function of Frequency," "Input Noise Current as a Function of Frequency," and "Broadband Noise for Various Bandwidths." Observe on the first two graphs how the noise increases at low frequencies. The presence of the 1/*f* or flicker noise causes this pronounced increase. Above 1 kHz or so, the noise curves flatten out and become white noise functions.

The presence of both input noise voltage and input noise curves represents the approach that is taken when a complete detailed analysis is performed. In such an analysis, the noise effects are separated into voltage and current forms in much the same way as was done for offset effects.

We will limit the noise analysis in this text to an overall simplified approach using the curves labeled "Broadband Noise for Various Bandwidths." On these curves, the effects of the noise voltage and current contributions, as well as the associated thermal noise contribution of the source resistance, have been computed and are shown graphically for several specified values of bandwidth. The abscissa is the source resistance, and the ordinate is the rms noise voltage referred to the input.

Each of the three curves corresponds to a specific bandwidth. Observe that these curves, as well as the noise voltage and curves previously discussed, do not provide any data concerning the noise below 10 Hz. Thus, the bandwidths given consider the combined noise from 10 Hz to the upper frequency limits. Observe that the noise voltage increases as the bandwidth increases. Pure thermal noise voltage increases with the square root of bandwidth. One can observe this general trend by noting that when the upper frequency increases from 1 kHz to 100 kHz (a 100-to-1 increase), the noise voltage increase is not drastically different from 10 to 1. However, the presence of flicker

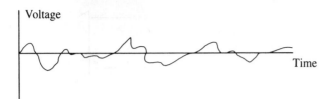

FIGURE 4–8
Random noise waveform.

noise in the combined effect results in some deviation from this square-root relationship of ideal thermal noise.

Observe that at relatively small values of source resistance, the rms noise voltage is relatively constant. In this region, the noise introduced by the op-amp is much greater than the thermal noise introduced by the input resistance, and there is very little change as the input resistance changes. Eventually, however, thermal noise introduced by the input resistance starts to add enough additional noise to cause a noticeable increase in the total noise. Depending on the bandwidth, the noise introduced by the input resistance begins to display a noticeable effect around 10 kΩ or so. By the point where the source resistance is 100 kΩ, the noise voltage is increasing rapidly and would continue increasing in the range where the curves could be extended. These curves indicate the desirability of maintaining circuit resistances to moderate levels in accordance with the guidelines of Chapter 2, particularly where the noise level is critical.

The extent to which the noise level is critical depends on the relative signal level. At signal levels of a few volts or so, there are usually no problems, and even "poor" circuit designs, from a noise point of view, may be totally adequate. Conversely, at very small signal levels, even the best circuit designs may be inadequate if the op-amp is too noisy. The only solution in this case might be to find an op-amp with ultra-low noise specifications.

The actual noise is a random process whose instantaneous value can be described only in statistical terms. The random erratic behavior of this noise is illustrated in Figure 4–8. This sort of waveform can be seen on an oscilloscope by observing the output of the op-amp with no signal applied. The best way to specify such a noise is by means of the **root-mean-square** value of the voltage, which is the equivalent power-producing measure of the waveform. The actual peak value of the noise will exceed the rms value very frequently and will occasionally exceed the rms voltage by a very large ratio. As a practical general rule, the peak value will remain in the range of about three times the rms value for about 99.7% of the time. Thus, if the rms voltage is 10 μV, the instantaneous noise will vary from about -30 μV to $+30$ μV almost all the time.

Noise Circuit Model

The rms noise voltage obtained from the curves under discussion represents the combined effect of all noise contributions referred to the **input** of the op-amp. This effect is best represented by an rms noise voltage V_{ni} at the noninverting input as shown in Figure 4–9.

FIGURE 4–9
Circuit model used to represent op-amp
noise. (V_{ni} is rms noise voltage referred
to input.)

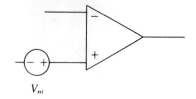

$$V_{ni}$$

Let V_{no} represent the rms noise voltage at the output of the op-amp. This voltage is given by

$$V_{no} = K_n V_{ni} \tag{4–53}$$

where K_n is the noise gain of the particular amplifier configuration. In the case of the noninverting amplifier, the noise gain is equal to the closed-loop voltage gain of the circuit as previously noted.

EXAMPLE 4–11

Consider the noninverting amplifier circuit shown in Figure 4–10. It is desired to estimate the total noise level in the bandwidth from 10 Hz to 1 kHz. (a) Determine the total rms noise referred to the input. (b) Determine the corresponding output noise level.

Solution
(a) It will be assumed that the major contribution to external noise is from the 1 kΩ resistance.

From the curve "Broadband Noise for Various Bandwidths" for the 741 op-amp, the input rms noise for a source resistance of 1 kΩ is estimated to be

$$V_{ni} = 1 \ \mu V \tag{4–54}$$

(b) The noise gain for this circuit is equal to the closed-loop voltage gain and is readily determined to be $K_n = 1001$. The rms output noise voltage is

$$V_{no} = 1001 \times 1 \ \mu V \approx 1 \ mV \tag{4–55}$$

FIGURE 4–10
Circuit of Example 4–11.

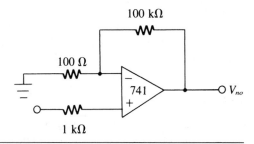

4–6 COMMON-MODE REJECTION

There are two different types of gain functions associated with a differential amplifier:
(1) *differential gain* and (2) *common-mode gain.* The open-loop gain function used extensively earlier in the book is actually the differential gain. The concept of common-mode
gain will be introduced in this section, and its significance will be discussed.

To define the two separate gain concepts, refer to Figure 4–11. Consider first the
measurement shown in part (a). This circuit is said to be excited in the *differential
mode.* A signal voltage v_d is applied between the inverting and noninverting terminals,
and the corresponding output voltage v_{od} is measured. The *differential gain* A_d is then
defined as

$$A_d = \frac{v_{od}}{v_d} \tag{4–56}$$

This definition is in agreement with our earlier interpretation of the open-loop gain of an
op-amp. The symbol A has been used exclusively up to this point, but the subscript d is
added here to distinguish differential gain from common-mode gain.

Consider next the measurement shown in Figure 4–11(b), in which a voltage v_{cm} is
applied simultaneously to both inputs of the op-amp. This circuit is said to be excited in
the *common mode*. From the basis of the ideal model developed in Chapter 2, the op-amp
output should be zero. However, a small signal voltage v_{ocm} will appear at the output under
this condition. This small voltage results from slight differences in gain between the two
inputs.

Let A_{cm} represent the common-mode gain of the amplifier, which is defined as

$$A_{cm} = \frac{v_{ocm}}{v_{cm}} \tag{4–57}$$

We thus see that a differential amplifier can be described by two gain functions, one of
which is based on a differential input signal, and the other of which is based on a common-
mode input signal.

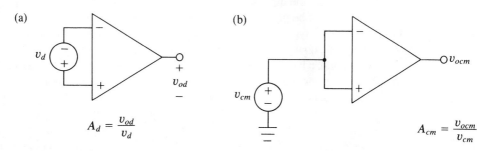

FIGURE 4–11
Circuits used to define concepts of (a) differential gain and (b) common-mode gain.

For an ideal amplifier, $A_{cm} = 0$. In any high-quality amplifier, $A_d \gg A_{cm}$. The extent of this inequality can be expressed by a quantity called the ***common-mode rejection ratio*** (CMRR), which is defined as

$$\text{CMRR} = \frac{A_d}{A_{cm}} \tag{4-58}$$

This quantity is commonly stated in decibel form as

$$\text{CMRR(dB)} = 20 \log_{10} \text{CMRR} = 20 \log_{10} \frac{A_d}{A_{cm}} \tag{4-59}$$

Note that the ideal op-amp would have an infinite common-mode rejection ratio.

Common-Mode Rejection Data

A check of some typical values is in order. Refer to the 741 op-amp data in Appendix C. On the line titled "Common Mode Rejection Ratio," on the third page of the 741 data sheets, a typical value is given as 90 dB, and the minimum value is listed as 70 dB. The absolute values corresponding to these decibel values are 3.16×10^4 and 3.16×10^3, respectively.

Although not specifically stated, the preceding values are the dc and low-frequency values for a common-mode rejection ratio. The frequency-dependent nature is observed from the graph titled "Common Mode Rejection as a Function of Frequency" on the sixth page of the 741 data sheets. On this curve, the common-mode rejection ratio is seen to be constant at 90 dB to just above 100 Hz, at which point it starts to decrease.

We have defined the two gain functions and the common-mode rejection ratio, but the puzzled reader may ask, "So what?" In other words, what is the significance of these quantities on the operating performance of an op-amp?

Significance of CMRR

A development of the quantitative significance of CMRR as it relates to the instrumentation amplifier will be performed in Section 5–2. The application of that particular amplifier represents the type of situation in which common-mode rejection is very important. At this point, some qualitative comments are in order.

In general, a large value of CMRR is desirable in two types of situations: (1) circuits in which large common-mode undesirable signals are present, and (2) circuits used to establish a delicate balance between two signals.

The first condition is typical of small signal instrumentation systems operating in noisy environments with long leads and/or where ground loops exist. With proper physical placement, much of the noise and stray pickup tends to be of a common-mode type. An amplifier with a large value of CMRR tends to suppress these spurious components. The second situation is typical of the bridge type of amplifier, where an absolute null is required to perform a precise measurement.

FIGURE 4–12
Circuit model used to represent
common-mode signal effect.

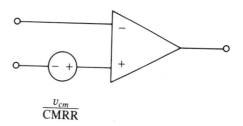

$$\frac{v_{cm}}{\text{CMRR}}$$

As we progress through the book, we will revert to the term ***closed-loop gain*** and the symbol A for differential gain in situations where common-mode rejection is not under consideration. Where common-mode rejection is important, we will use the notation and terminology established in this section.

CMRR Circuit Model

Although common-mode rejection ratio is defined in terms of open-loop gains, it is much easier to work with its effect on a closed-loop basis. In a stable closed-loop configuration, the effect can be represented in terms of the gain ratios, and the output level can be predicted without an explicit knowledge of the actual open-loop gains.

A circuit model that can be used for this purpose is shown in Figure 4–12. The voltage v_{cm} is the common-mode voltage at the two op-amp inputs. A fictitious voltage source is placed at the noninverting input whose value is v_{cm}/CMRR. In general, the effect at the output may have either polarity, so the polarity chosen here is the one that results in a positive value at the output. Further, the source could be placed at either input, but it is easier to work with at the noninverting input.

The reader will recognize this form as the same used to model the effects of input offset voltage and noise. Consequently, the effect at the output can be determined by multiplying this voltage by the noise gain of the circuit used. Once again, the significance of noise gain is evident.

**EXAMPLE
4–12**

A certain 741 op-amp has a common-mode rejection ratio of 80 dB. If the differential gain is 106 dB, determine the common-mode gain.

Solution

A common-mode rejection ratio of 80 dB corresponds to $\text{CMRR} = 10^4$. A differential gain of 106 dB corresponds, with very slight rounding, to $A_d = 2 \times 10^5$. Since

$$\text{CMRR} = \frac{A_d}{A_{cm}} \tag{4–60}$$

then

$$A_{cm} = \frac{A_d}{\text{CMRR}} = \frac{2 \times 10^5}{10^4} = 20 \tag{4–61}$$

4–7 MULTISIM EXAMPLES

The examples presented in this section are devoted to the study of finite bandwidth and slew rate using the Multisim 741 model. Both frequency response and transient analysis are employed. The effects of the limitations will be demonstrated for both sinusoidal and pulse waveforms.

MULTISIM EXAMPLE 4–1

Using the 741 model, design a simple noninverting amplifier circuit with a gain of 10, and measure its frequency response.

Solution

A gain of 10 is easily achieved with values of $R_i = 10$ kΩ and $R_f = 90$ kΩ. The schematic diagram as created by Multisim is shown in Figure 4–13. For simplicity in the design, the compensating resistance has been eliminated.

The sinusoidal source is used as the input and its peak value is set at 1 V. (Actually, with an ac frequency response analysis, Multisim automatically changes the input level to 1 V irrespective of its actual level, but we might as well set it at that level from the beginning.) The labeling of "1 kHz − 10 MHz" was achieved from the source properties window by first left-clicking on **Label** and then typing the information in the **Label** slot.

To perform the frequency response, first left-click on the **Analyses** button and then left-click on **AC Analysis**. A window with the title **AC Analysis** will then open and the default tab position will be at **Frequency Parameters**. The **Start Frequency** is set to 1 kHz and the **Stop Frequency** is set to 10 MHz. These choices will provide four decades of appropriate frequency coverage. The **Sweep type** is chosen as **Decade,** and for good resolution, the **Number of points per decade** was chosen as 100. The **Vertical Scale** is chosen as **Decibel.**

A left-click on the **Output variables** tab allows the selection of the various circuit variables. Only the voltage at node 6 (op-amp output terminal) is of interest, so left-click on 6 and then left-click on **Plot during simulation.** Finally, left-click on **Simulate** and the amplitude and phase functions shown in Figure 4–14 will appear. A grid has been added and some changes in the labels have been made for clarity.

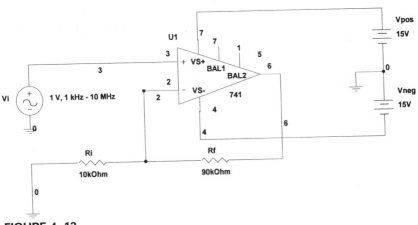

FIGURE 4–13
Circuit of Multisim Example 4–1.

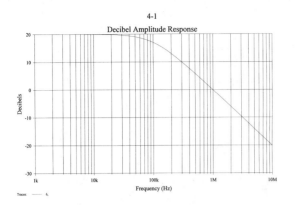
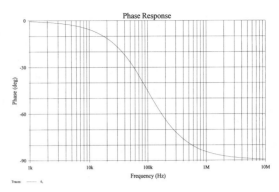

FIGURE 4–14
Frequency response of Multisim Example 4–1.

The noise gain of the circuit is 10, and based on an assumed unity-gain frequency of 1 MHz for the 741, the theoretical value of the 3-dB closed-loop bandwidth should be

$$B_{CL} = \frac{1\ MHz}{10} = 100\ kHz \qquad \text{(4–62)}$$

Since the low-frequency gain is 10, which corresponds to 20 dB, the actual response at 100 kHz should be 17 dB. The curve shows a very strong agreement with this value.

MULTISIM EXAMPLE 4–2

For the noninverting amplifier of Multisim Example 4–1, change the voltage gain to unity (voltage follower) and perform a frequency response analysis.

Solution
The Multisim circuit schematic is shown in Figure 4–15 and was directly adapted from Figure 4–13. The resistance R_i was deleted and replaced by an open circuit. One could delete R_f and replace it by a short circuit. However, to keep the node numbering system the

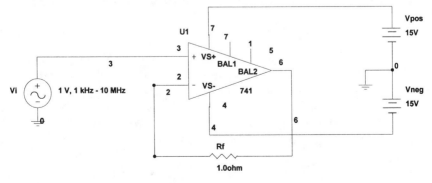

FIGURE 4–15
Circuit of Multisim Example 4–2.

FIGURE 4–16
Frequency response of Multisim Example 4–2.

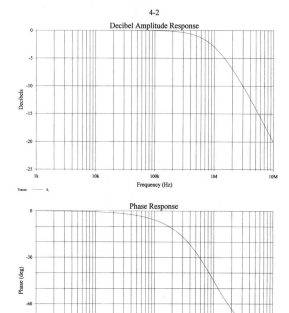

same, its value was changed to 1 Ω. (Theoretically, the gain should be unity for any finite value of R_f if R_i is replaced by an open circuit.)

The frequency response analysis is performed in the same manner as in Multisim Example 4–1, and the results are shown in Figure 4–16. Since the absolute voltage gain is unity, the decibel gain is 0 dB at low frequencies as expected. The 3-dB frequency is the same as the op-amp unity gain frequency and is at 1 MHz as expected.

MULTISIM EXAMPLE 4–3

Using the 741 model, design a simple inverting amplifier with a gain of −1 and measure its frequency response.

Solution

The Multisim circuit schematic is shown in Figure 4–17, and two 10-kΩ resistors are used to establish the gain. As in previous examples, a compensating resistance is not used. The frequency range of the source has been labeled on the diagram for clarity.

The frequency response is achieved by the same procedure as in Multisim Example 4–1, and the results are shown in Figure 4–18. For an inverting amplifier with a gain of −1, the noise gain is 2, so the closed-loop bandwidth should be

$$B_{\text{CL}} = \frac{1 \text{ MHz}}{2} = 500 \text{ kHz} \tag{4-63}$$

The measured amplitude response is in agreement with this value.

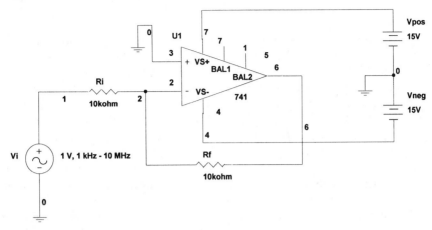

FIGURE 4–17
Circuit of Multisim Example 4–3.

FIGURE 4–18
Frequency response of Multisim Example 4–3.

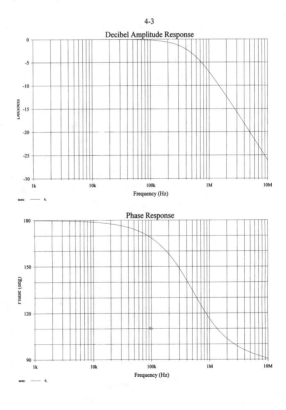

MULTISIM EXAMPLE 4–4 Example 4–8 in the text considered various effects of finite bandwidth and finite slew rate on the circuit of Figure 4–7 with a sinusoidal input. In this example and Multisim Examples 4–5 and 4–6, the effects will be observed on an instantaneous basis. In this example, assume a sinusoidal input of 20 mV and perform a **Transient Analysis** at (a) 1 kHz and (b) 40 kHz.

Solution

The Multisim circuit schematic is shown in Figure 4–19, and no compensating resistance is employed. Using the sinusoidal source properties window, the source is set to a peak value of 20 mV. The two frequencies have been labeled on the circuit for clarity but only one can be set at a time. Thus, two separate runs are required for the plots.

(a) In this first part, the frequency is set at 1 kHz. It was determined in Example 4–8 that the closed-loop bandwidth is 40 kHz and the slew rate limiting frequency for this amplitude is 159.2 kHz. Therefore, the amplifier should function very well at 1 kHz for the input level of 20 mV.

To perform the analysis, left-click on the **Analyses** button and then left-click on **Transient Analysis.** A window with the title **Transient Analysis** will then open and the tab will initially be set in the **Analysis Parameters** setting. The **Start time** is set to 0 and the **End time** will be set to 0.002 or 2 ms. This will provide a display of two cycles at 1 kHz. The analysis was performed with the **Minimum number of time points** activated and set to 200. The only variable of interest is the output voltage of the op-amp, which is node 6.

A plot of the instantaneous output is shown in Figure 4–20. A slight upward shift of the curve will be noted. This is due to the effect of the input offset voltage, whose effect is noticeable at the small signal level and large noise gain. Assuming an input offset voltage of about 1 mV and a noise gain of 25, a shift of about 25 mV occurs, and this is about the level noted. This could be eliminated by using the offset null terminals of the op-amp but we will not bother for the purpose intended. After subtracting out this offset, the peak value of the output is 500 mV, which is the expected value based on an input voltage of 20 mV and a gain of 25. Note also that the output appears to be very sinusoidal in form, since no slew rate effects should appear for the given situation.

(b) In this part, the input signal frequency is increased to 40 kHz while the peak value remains at 20 mV. The period corresponding to 40 kHz is $1/(40 \times 10^{-3}) = 25 \ \mu s$. The ending time is set to 50 μs, corresponding to two cycles again. After observing some "jagged" segments with 200 points, the **Minimum number of time points** was increased to 1000. The result of the simulation is shown in Figure 4–21.

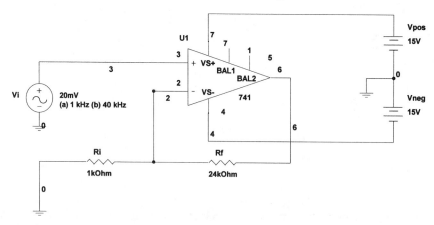

FIGURE 4–19
Circuit of Multisim Example 4–4.

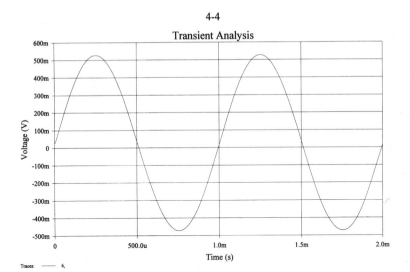

FIGURE 4–20
Transient response of Multisim Example 4–4 at 1 kHz.

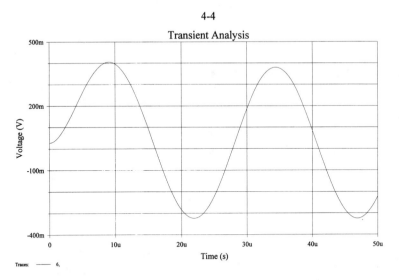

FIGURE 4–21
Transient response of Multisim Example 4–4 at 40 kHz.

Because of the much higher frequency, there is a noticeable initial transient effect, but the response quickly settles into a steady-state sinusoidal form. There is no evidence of slew rate effects for the conditions used, but the peak value has been decreased somewhat. When the offset effect is considered, the output peak value is around 70% of the low-frequency peak value, which is to be expected since the 3-dB frequency is 40 kHz.

MULTISIM
EXAMPLE
4–5

For the circuit of Multisim Example 4–4, increase the peak value of the sinusoidal input to 0.5 V and repeat the analysis for frequencies of (a) 1 kHz and (b) 10 kHz.

Solution

The Multisim circuit schematic is shown in Figure 4–22 with the sinusoidal input source peak value increased to 0.5 V. For clarity, the two frequencies have been labeled on the circuit but two separate runs are required.

(a) In this first part, the frequency is set at 1 kHz. Since the closed-loop bandwidth is independent of the signal level, it is still 40 kHz. However, for a peak input value of 0.5

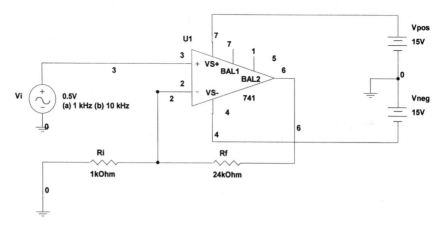

FIGURE 4–22
Circuit of Multisim Example 4–5.

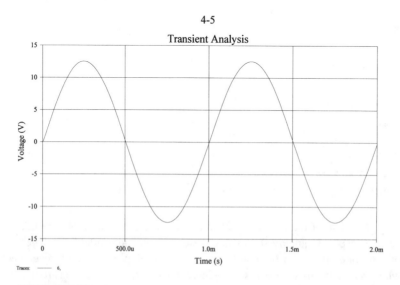

FIGURE 4–23
Transient response of Multisim Example 4–5 at 1 kHz.

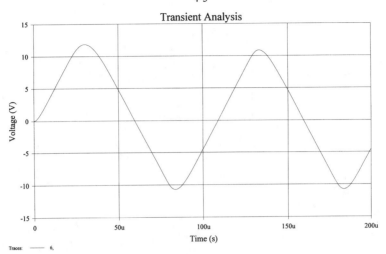

FIGURE 4–24
Transient response of Multisim Example 4–5 at 10 kHz.

V, the slew rate limiting frequency was determined in Example 4–8 to be 6.366 kHz. This should not present any problem at 1 kHz.

A transient analysis was performed by the same procedure as in Multisim Example 4–4(a). The result is shown in Figure 4–23. Note that at this signal level, the effect of the input offset voltage is negligible. Moreover, the output peak voltage is about 12.5 V, which is expected for an input signal level of 0.5 V and a gain of 25. The response also appears to have the proper sinusoidal shape.

(b) In this part, the frequency is increased to 10 kHz. This is below the 3-dB frequency of 40 kHz but is above the slew-rate limiting frequency of 6.366 kHz. The result for this run over two cycles with the **Minimum number of time points** set to 1000 is shown in Figure 4–24. The peak value has not been affected greatly but the shape of the function deviates somewhat from a pure sinusoid. The effect of slew rate limiting has now entered the picture and significant non-linear effects are now evident.

It should be emphasized that the frequency response analysis of the first three examples in this chapter do not bring out slew rate limiting effects. Slew rate is a nonlinear effect that can be exhibited only through an instantaneous transient analysis.

MULTISIM EXAMPLE 4–6

For the circuit of Multisim Examples 4–4 and 4–5, and with a peak value of the sinusoidal input set to 0.5 V, repeat the analysis with the frequency set to 100 kHz.

Solution
The Multisim circuit schematic is shown in Figure 4–25. The frequency is now well above both the 3-dB closed loop bandwidth of 40 kHz and the slew rate limiting frequency of 6.366 kHz.

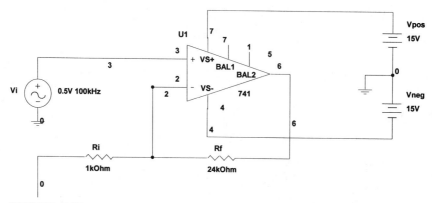

FIGURE 4–25
Circuit of Multisim Example 4–6.

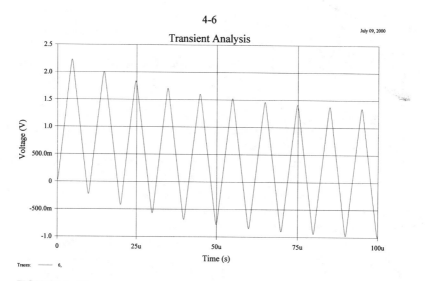

FIGURE 4–26
Transient response of Multisim Example 4–6 at 100 kHz.

To more precisely show the transient effect, the response in this case was run for 10 cycles with the **Minimum number of time points** set to **5000**. The result is shown in Figure 4–26. Note that the peak amplitude is reduced from 12.5 V to a little more than 1 V after the initial transient settles. Note also that the signal looks more like a triangular wave than a sine wave. The slew rate limiting effect has essentially taken over and is now the dominating effect.

MULTISIM EXAMPLE 4–7

Example 4–9 in the text considered various effects of finite bandwidth and finite slew rate on the circuit of Figure 4–7 with a pulse type input. In this example and Multisim Examples 4–8 and 4–9, the effects will be observed on an instantaneous basis. In this example, assume a pulse train oscillating between 0 V and 20 mV and perform a **Transient Analysis** at (a) 1 kHz and (b) 40 kHz.

Solution

The Multisim circuit schematic is shown in Figure 4–27. This circuit is adapted from the one in the preceding example by replacing the sinusoidal source with a pulse source. To perform this task, double left-click on the source, and when the **AC Voltage** properties window opens, left-click on **Replace**. A window with the title **Component Browser** will then open. In the **Component Family** slot, move through the options until you reach **PULSE_VOLTAGE_SOURCE** and then left-click **OK**. The sinusoidal source will then be replaced by a pulse source and a new window with the title **Pulse Voltage** will open. The default position for the tab should be at **Value**. If not, left-click on this tab. Various pulse parameters must now be set.

Some of the pulse parameters can probably be left at their initial values, but all should be checked. The pulse source allows two constant values within a cycle, and these are denoted as **Initial Value** and **Pulsed Value**. The two values should be set at 0 V and 20 mV. Leave the tolerance slot unchecked at this time. However, the additional five slots should be checked.

The definitions of these are as follows:

Delay Time: Initial time delay before **Pulsed Value** is first set

Rise Time: Time for leading edge of pulse to increase from initial value to pulse value

Fall Time: Time for trailing edge of pulse to decrease from pulse value to initial value

Pulse Width: Time width of pulse

Period: Time width of period (reciprocal of frequency)

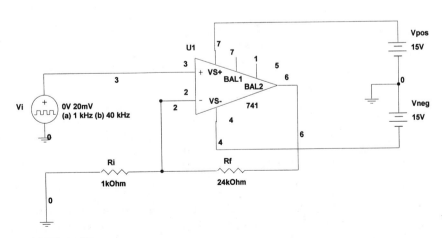

FIGURE 4–27

Circuit of Multisim Example 4–7.

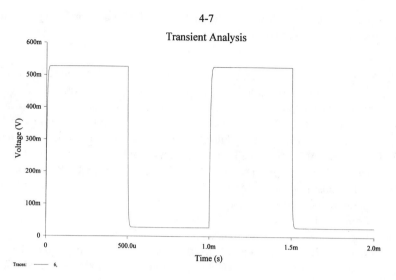

FIGURE 4–28

Transient response of Multisim Example 4–7 at 1 kHz.

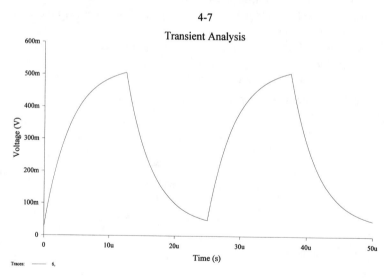

FIGURE 4–29

Transient response of Multisim Example 4–7 at 40 kHz.

(a) In this part, the frequency is 1 kHz so set **Period** to 1 ms. For a 50% duty cycle, set **Pulse Width** to 0.5 ms. Set **Delay Time** to 0. The **Rise Time** and **Fall Time** could be set to any times that are small compared with the pulse width and period. A value of 1 ns for each should suffice.

The transient analysis for two cycles is shown in Figure 4–28. As was the case with a sinusoidal input, there is a small positive shift due to the input offset voltage and it is of the order of 25 mV. The effect of the finite closed-loop bandwidth also introduces some rise and fall times, but for many applications, this level of degradation may be acceptable

(b) In this part, the input signal frequency is increased to 40 kHz while the pulse value oscillates from 0 to 20 mV. The period is now 25 μs and the pulse width is 12.5 μs. The **Minimum number of time points** was set to 1000. The results of the simulation are shown in Figure 4–29.

The effects of finite bandwidth are obviously much more pronounced at a frequency of 40 kHz than was the case at 1 kHz. The pulse is considerably rounded and does not quite reach its final level before the end of a half-cycle. As in several other cases, the effect of input offset voltage has pushed the level up slightly.

MULTISIM EXAMPLE 4–8

For the circuit of Multisim Example 4–7, increase the peak value of the pulse input to 0.5 V and repeat the analysis for frequencies of (a) 1 kHz and (b) 10 kHz.

Solution
The Multisim circuit schematic is shown in Figure 4–30 with the pulse input source peak value increased to 0.5 V. For clarity, the two frequencies have been labeled on the circuit but two separate runs are required.

(a) In this first part, the frequency is set at 1 kHz. Since the closed-loop bandwidth is independent of the signal level, it is still 40 kHz. However, for a peak input value of 0.5 V, the slew rate limiting frequency was determined in Example 4–8 to be 6.366 kHz, which theoretically should not present any problem at 1 kHz.

A transient analysis was performed by the same procedure as utilized in earlier examples with the **Minimum number of time points** set to 1000. The result is shown in

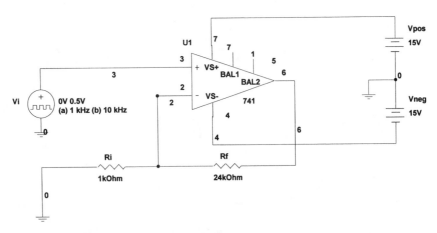

FIGURE 4–30
Circuit of Multisim Example 4–8.

Figure 4–31. When this result is compared with that of Figure 4–28, which was obtained at the same frequency, but at a smaller signal level, there appears to be a slight increase in the rise time. The formula for the slew rate limiting frequency was an approximation based on a sinusoidal input, and when the function is more complex, slew rate effects may "creep in" below the theoretical slew rate limiting frequency.

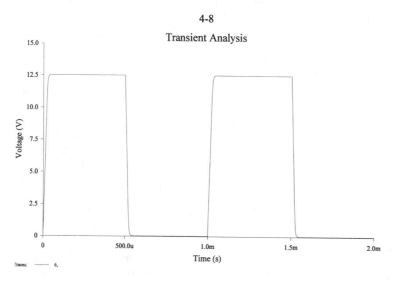

FIGURE 4–31
Transient response of Multisim Example 4–8 at 1 kHz.

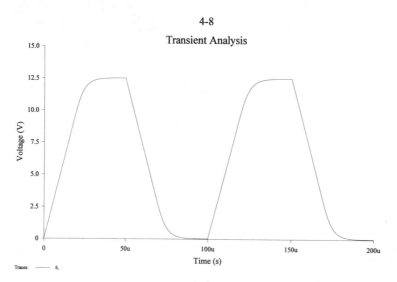

FIGURE 4–32
Transient response of Multisim Example 4–8 at 10 kHz.

(b) In this part, the frequency is increased to 10 kHz. This is below the 3-dB frequency of 40 kHz but is above the slew-rate limiting frequency of 6.366 kHz. The result for this run is shown in Figure 4–32 with the **Minimum number of time points** set to 1000. The peak value has not been affected greatly, but the shape of the function during the leading and trailing edges assumes a ramplike nature due to the slew rate effect.

SUMMARY

1. Most op-amps display a very high gain at dc, but the gain starts decreasing at a relatively low frequency.
2. The gain roll-off for most op-amps follows the one-pole low-pass form and decreases at 6 dB/octave or 20 dB/decade.
3. The gain reaches a level of unity at a frequency called the unity-gain frequency.
4. The unity-gain frequency is also called the gain-bandwidth product, since its value is equal to the product of dc gain and 3-dB open-loop bandwidth.
5. The closed-loop 3-dB bandwidth is equal to the unity-gain frequency divided by the noise gain.
6. The rise time is inversely proportional to the closed-loop 3-dB bandwidth.
7. The process limiting the maximum rate at which an op-amp output voltage can change is the slew rate limitation.
8. The slew rate effect imposes a rise time on a pulse-type signal or distorts a complex signal if the output attempts to change at a rate exceeding the slew rate.
9. Degradation for either a pulse-type signal or a complex waveform results from the combined effects of finite bandwidth and slew rate.
10. As a general rule, the effect of finite bandwidth tends to be the dominating factor at low output signal levels and the effect of slew rate tends to be the dominating factor at high output signal levels.
11. Overall noise in an op-amp results from the combined effects of thermal noise, shot noise, flicker noise, and popcorn noise.
12. The output noise level can be established from the manufacturer's noise data and the circuit noise gain.
13. Differential gain is based on applying an input voltage between the two input terminals and measuring the output voltage.
14. Common-mode gain is based on applying the same voltage to both input terminals and measuring the output voltage.
15. Common-mode rejection ratio is the ratio of differential gain to common-mode gain.
16. A high common-mode rejection ratio is desirable when a signal contains large spurious common-mode components.

PROBLEMS

Drill Problems

4–1. A certain op-amp has a unity-gain frequency of 2 MHz. For a *noninverting* amplifier, calculate the closed-loop 3-dB bandwidth for each of the following ideal values of closed-loop gain:
   a. 500   **b.** 50   **c.** 5   **d.** 1

4–2. A certain wideband op-amp has a unity-gain frequency of 12 MHz. For a *noninverting* amplifier, calculate the closed-loop 3-dB bandwidth for each of the following ideal values of closed-loop gain:

    a. 1000   **b.** 100   **c.** 10   **d.** 1

4–3. In Problem 4–1, assume that the gain values listed represent the *magnitudes* for an inverting amplifier. Calculate the closed-loop 3-dB bandwidth for each case.

4–4. In Problem 4–2, assume that the gain values listed represent the *magnitudes* for an inverting amplifier. Calculate the closed-loop 3-dB bandwidth for each case.

4–5. Calculate the rise time T_{CL} associated with each gain value in Problem 4–1.

4–6. Calculate the rise time T_{CL} associated with each gain value in Problem 4–2.

4–7. Calculate the rise time T_{CL} associated with each gain value in Problem 4–3.

4–8. Calculate the rise time T_{CL} associated with each gain value in Problem 4–4.

4–9. Consider the circuit of Figure P4–9 with $R_i = 10\ \mathrm{k\Omega}$ and $R_f = 70\ \mathrm{k\Omega}$. If the unity-gain frequency of the op-amp is 1.6 MHz, determine the closed-loop 3-dB bandwidth.

FIGURE P4–9

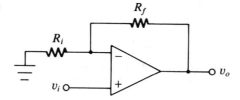

4–10. Consider the circuit of Figure P4–9 with $R_i = 2.2\ \mathrm{k\Omega}$ and $R_f = 47\ \mathrm{k\Omega}$. If the unity-gain frequency of the op-amp is 1.5 MHz, determine the closed-loop bandwidth.

4–11. Consider the circuit of Figure P4–11 with $R_i = 10\ \mathrm{k\Omega}$ and $R_f = 30\ \mathrm{k\Omega}$. If the unity-gain frequency of the op-amp is 1.2 MHz, determine the closed-loop 3-dB bandwidth.

FIGURE P4–11

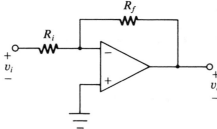

4–12. Consider the circuit of Figure P4–11 with $R_i = 15\ \mathrm{k\Omega}$ and $R_f = 120\ \mathrm{k\Omega}$. If the unity-gain frequency of the op-amp is 2 MHz, determine the closed-loop 3-dB bandwidth.

4–13. Calculate the rise time T_{CL} for the circuit of Problem 4–9.

4–14. Calculate the rise time T_{CL} for the circuit of Problem 4–10.

4–15. Calculate the rise time T_{CL} for the circuit of Problem 4–11.

4–16. Calculate the rise time T_{CL} for the circuit of Problem 4–12.

4–17. Consider the circuit of Figure P4–17 with $R_f = 100\ \mathrm{k\Omega}$, $R_1 = 10\ \mathrm{k\Omega}$, $R_2 = 20\ \mathrm{k\Omega}$, and $R_3 = 25\ \mathrm{k\Omega}$. If the unity-gain frequency of the op-amp is 1 MHz, determine the closed-loop 3-dB bandwidth.

FIGURE P4-17

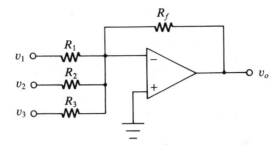

4-18. Consider the circuit of Figure P4-17 with $R_f = 200$ kΩ, $R_1 = 20$ kΩ, $R_2 = 10$ kΩ, and $R_3 = 5$ kΩ. If the unity-gain frequency of the op-amp is 1 MHz, determine the closed-loop 3-dB bandwidth.

4-19. Consider the circuit of Figure P4-19 with $R_1 = R_2 = 10$ kΩ. If the unity-gain frequency of the op-amp is 1 MHz, determine the closed-loop 3-dB bandwidth.

FIGURE P4-19

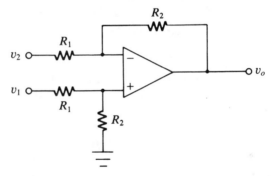

4-20. Consider the circuit of Figure P4-19 with $R_1 = 10$ kΩ and $R_2 = 100$ kΩ. If the unity-gain frequency of the op-amp is 2 MHz, determine the closed-loop 3-dB bandwidth.

4-21. Consider the circuit of Figure P4-9 with $R_i = 10$ kΩ and $R_f = 70$ kΩ. Assume that the op-amp slew rate is 0.8 V/μs. Calculate the rise time T_{SR} due to the slew rate when the input is a pulse that changes from zero to each of the following values:
   a. 10 mV   **b.** 0.1 V   **c.** 1 V

4-22. Consider the circuit of Figure P4-9 with $R_i = 2.2$ kΩ and $R_f = 47$ kΩ. Assume that the op-amp slew rate is 0.6 V/μs. Calculate the rise time T_{SR} due to the slew rate when the input is a pulse that changes from zero to each of the following values:
   a. 10 mV   **b.** 0.2 V   **c.** 0.5 V

4-23. Consider the circuit of Figure P4-11 with $R_i = 10$ kΩ and $R_f = 30$ kΩ. Assume that the op-amp slew rate is 0.5 V/μs. Calculate the rise time T_{SR} due to the slew rate when the input is a pulse that changes from zero to each of the following values:
   a. 0.2 V   **b.** 1 V   **c.** 3 V

4-24. Consider the circuit of Figure P4-11 with $R_i = 15$ kΩ and $R_f = 120$ kΩ. Assume that the op-amp slew rate is 1.2 V/μs. Calculate the rise time T_{SR} due to the slew rate when the input is a pulse that changes from zero to each of the following values:
   a. 0.1 V   **b.** 0.5 V   **c.** 1.5 V

4-25. Consider the circuit of Problem 4-21, but assume complex signals with peak values equal to the three values given. Calculate the slew rate limiting frequency in each case.

4–26. Consider the circuit of Problem 4–22, but assume complex signals with peak values equal to the three values given. Calculate the slew rate limiting frequency in each case.

4–27. Consider the circuit of Problem 4–23, but assume complex signals with peak values equal to the three values given. Calculate the slew rate limiting frequency in each case.

4–28. Consider the circuit of Problem 4–24, but assume complex signals with peak values equal to the three values given. Calculate the slew rate limiting frequency in each case.

4–29. Consider the noninverting amplifier circuit of Figure P4–29. It is desired to estimate the total noise level in the bandwidth from 10 Hz to 10 kHz.

    a. Determine the total rms noise referred to the input.

    b. Determine the corresponding output noise level.

FIGURE P4–29

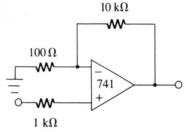

4–30. Consider the noninverting amplifier circuit of Figure P4–30. It is desired to estimate the total noise level in the bandwidth from 10 Hz to 100 kHz.

    a. Determine the total rms noise referred to the input.

    b. Determine the corresponding output noise level.

FIGURE P4–30

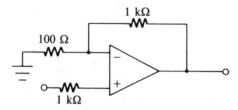

4–31. A certain op-amp has a common-mode rejection ratio of 86 dB. If the differential gain is 94 dB, determine the common-mode gain.

4–32. A certain op-amp has a differential gain of 112 dB and a common-mode gain of 14 dB. Determine the common-mode rejection ratio in dB.

4–33. Assume that a certain 741 op-amp has a differential gain of 100 dB at some very low frequency. Calculate the value of the common-mode gain corresponding to the minimum value of CMRR of 70 dB.

4–34. Repeat the analysis of Problem 4–33 based on the maximum value of CMRR of 90 dB.

4–35. For the op-amp of Problem 4–33, assume that there is a common-mode input voltage of 0.2 V. Determine the common-mode output voltage.

4–36. For the op-amp of Problem 4–34, assume that there is a common-mode input voltage of 0.2 V. Determine the common-mode output voltage.

Derivation Problems

4–37. Show that the 3-dB bandwidth of the voltage follower is simply

$$B_{CL} = B$$

4–38. Show that the 3-dB bandwidth of an inverting amplifier with $A_{CL} = -1$ is

$$B_{CL} = \frac{B}{2}$$

4–39. Consider a linear combination circuit whose output is given by

$$v_o = -A_1 v_1 - A_2 v_2 - \cdots - A_n v_n$$

Show that the 3-dB bandwidth is

$$B_{CL} = \frac{B}{1 + A_1 + A_2 + \cdots + A_n}$$

4–40. Consider a closed-loop balanced differential amplifier whose output is given by

$$v_o = A(v_1 - v_2)$$

Show that the 3-dB bandwidth is

$$B_{CL} = \frac{B}{1 + A}$$

4–41. a. Assume that it is desired to determine experimentally the 3-dB frequency B_{CL} of some op-amp closed-loop configuration. In order to perform the measurement, it is essential that the signal level be sufficiently small that the slew rate effect does not interfere. Let A_{CL} represent the closed-loop gain, and let V_o represent the peak value of the op-amp output. For a sinusoidal test signal, show that the slew rate effect should not interfere with the measurement if

$$V_o < \frac{S}{2\pi B_{CL}}$$

Since B_{CL} is the quantity being measured, it is necessary that a rough estimate of the value be known in advance, so as to satisfy the above equation. If there is a large uncertainty in the value, the voltage level should be decreased to satisfy the equation based on the largest possible value of B_{CL}.

b. For a 741 op-amp with $S = 0.5$ V/μs, estimate the upper bound of V_o based on the assumption that $B = 1$ MHz and the particular configuration giving the largest closed-loop bandwidth.

4–42. a. To measure slew rate, an op-amp is often connected in a voltage follower configuration and driven by a square wave of sufficient peak-to-peak value to swing the output between near-saturation levels. The rise time (or fall time) is then measured. It is assumed that the rise time of the input square wave and the rise time due to the finite bandwidth of the op-amp are both small compared with the slew rate rise time. Assume that the voltage changes by ΔV in a time ΔT. Show that the slew rate is

$$S = \frac{\Delta V}{\Delta T}$$

b. For a certain amplifier driven by a square wave, assume that the output changes from -12 V to $+12$ V in a time of 8 μs. Determine the slew rate.

4–43. Consider the expression of (3–63) giving the actual closed-loop gain as a function of the ideal closed-loop gain, the noise gain, and the open-loop gain. Substitute the expression of (4–1a) and show that after subsequent manipulations, the result of (4–7) is obtained.

4–44. Starting with (4–12), calculate the time t_1 for the voltage to reach 10% of its final value and the time t_2 for the voltage to reach 90% of its final value. Show that $T_{CL} = t_2 - t_1 = 0.35/B_{CL}$.

Design Problems

4–45. A noninverting amplifier with a gain of 25 is to be designed for use in an audio application. Specifications require that the gain vary no more than 1 dB from near dc to 20 kHz. The peak input signal is expected never to exceed 0.5 V. Determine the minimum values of **a.** unity-gain frequency and **b.** slew rate for the op-amp selected.

4–46. A noninverting amplifier with a gain of 50 is to be designed for use with complex signals. The peak value of the input signal is estimated to be 250 mV, and its frequency range is from near dc to 15 kHz. Assume that the following criteria are used: (1) closed-loop 3-dB bandwidth $\geq 4 \times$ highest signal frequency and (2) slew rate limiting frequency $\geq 2 \times$ highest signal frequency. Determine the minimum required values of the op-amp specifications for **a.** unity-gain frequency and **b.** slew rate.

4–47. An inverting amplifier with a gain of -8 is to be designed for use in a special type of video amplifier. Specifications require that the gain vary no more than 2 dB from near dc to 1 MHz. The peak input signal is estimated as 0.8 V. Determine the minimum values of **a.** unity-gain frequency and **b.** slew rate for the op-amp selected. **c.** Would the LM118/218/318 op-amp be satisfactory?

4–48. A linear combination circuit is to be designed to serve as a mixer for three signals. The function performed is

$$v_o = -A_1 v_1 - A_2 v_2 - A_3 v_3$$

where each value of A is to be adjustable from 0 to 10. Specifications require that the gain vary no more than 0.1 dB from near dc to 20 kHz. While the peak values of the input signals are not known accurately, an operator will always adjust the gains so that the output signal never quite reaches the saturation voltages of ± 13 V. Determine the minimum values of **a.** unity-gain frequency and **b.** slew rate for the op-amp selected.

4–49. A noninverting amplifier with a gain of 50 is to be designed to process pulse signals. The peak value of the input is estimated as 250 mV. The following specifications are imposed: (1) rise time resulting from finite bandwidth ≤ 1 μs and (2) rise time resulting from slew rate ≤ 2 μs. Determine the minimum required values of the op-amp specifications for **a.** unity-gain frequency and **b.** slew rate.

4–50. An inverting amplifier with a gain of -8 is to be designed to process a pulse train. The peak value of the input is estimated as 1.5 V. The following specifications are imposed: (1) rise time resulting from finite bandwidth ≤ 100 ns and (2) rise time resulting from slew rate ≤ 50 ns. Determine the minimum values of the op-amp specifications for **a.** unity-gain frequency and **b.** slew rate.

LABORATORY EXERCISES

Closed-Loop Bandwidth ($A_{CL} = -100$)
LE 4–1.

1. Connect an inverting amplifier circuit with $R_i = 1$ kΩ and $R_f = 100$ kΩ. You may omit the bias compensation resistor for this exercise.

2. Calculate the expected values of noise gain K_n and closed-loop 3-dB bandwidth B_{CL}.

3. Apply a variable-frequency sine-wave generator, and measure the 3-dB bandwidth. Compare with the expected value.

   Note: In order to avoid slew rate effects, it is necessary to maintain the output voltage below a certain level. See Problem 4–41.

Closed-Loop Bandwidth ($A_{CL} = -10$)
LE 4–2.

Repeat LE 4–1 for $R_i = 10$ kΩ and $R_f = 100$ kΩ.

Closed-Loop Bandwidth ($A_{CL} = -1$)
LE 4–3.

Repeat LE 4–1 for $R_i = 10$ kΩ and $R_f = 10$ kΩ.

Closed-Loop Bandwidth ($A_{CL} = +1$)
LE 4–4.

Repeat LE 4–1 for the case of a noninverting voltage follower.

Slew Rate (Square Wave)
LE 4–5.

1. Connect a noninverting voltage follower circuit.
2. Apply a variable frequency square wave as the input to the circuit and observe both input and output on the CRO at 1 kHz. If the square wave is symmetrical about zero volts, adjust its amplitude until the op-amp output is almost reaching the two saturation levels. If the square wave only has one polarity, adjust the amplitude until the output is nearly saturated on one side.
3. Observe the input and output as the frequency is varied from about 100 Hz to above 100 kHz.
4. Select a frequency at which the rise time can be easily measured, and measure this quantity along with the pertinent voltage change. Calculate the slew rate S, and compare with 741 op-amp data in Appendix C.

Slew Rate (Sinusoid)
LE 4–6.

1. Connect a noninverting voltage follower circuit.
2. Based on 741 op-amp data, calculate the slew rate limiting frequency f_{SR} based on a 10-V peak sinusoidal output.
3. Apply a variable frequency 10-V peak sine wave to the circuit with the frequency initially set to a lower value than f_{SR} calculated in step (2). Monitor the input and output waveforms.
4. Observe the input and output as the frequency is varied from about 100 Hz to well above 10 kHz.
5. Identify as best you can the frequency at which slew rate distortion starts to occur. Compare the measured value with the expected value of step (2). (*Note:* This frequency is difficult to measure precisely, since distortion is not always apparent.)

5

LINEAR OPERATIONAL AMPLIFIER CIRCUITS

OBJECTIVES

After completing this chapter, the reader should be able to:

- Analyze an instrumentation amplifier circuit and discuss its advantages as compared to other basic amplifier circuits.
- Discuss the nature of common-mode spurious noise components and show how the instrumentation amplifier suppresses them.
- Analyze the operation of a true integrator circuit and explain the difficulties that arise from implementation with a general-purpose op-amp.
- Analyze the operation of a true differentiator circuit and discuss the difficulties with high-frequency noise.
- Show the circuit form and determine the transfer function of a general inverting op-amp circuit with arbitrary impedances.
- Analyze or design an ac integrator circuit and determine the output waveform of an ac integrator circuit used for waveshaping.
- Analyze or design a low-frequency differentiator circuit and determine the output waveform of a low-frequency differentiator circuit used for waveshaping.
- Analyze or design an all-phase lag circuit and all-pass phase lead circuit.
- Show how an op-amp can be connected to operate with a single power supply, and design the circuitry for single power supply operation to meet prescribed low-frequency requirements.

5–1 INTRODUCTION

A collection of important linear circuits was introduced in Chapter 2, and our method of analysis at that point was based on the ideal op-amp model. In Chapters 3 and 4, we discussed the various limitations and imperfections that appear in realistic devices. One must be aware of these limitations in designing or utilizing all the circuits of Chapter 2 as well as other circuits we will develop throughout the book. In this chapter, we will return for the most part to the ideal model assumption of Chapter 2, but the additional "sophistication" acquired in Chapters 3 and 4 will allow us to challenge the assumptions when required.

In one sense, the circuits introduced in this chapter represent a continuation of the linear circuits of Chapter 2, but many are more susceptible to the limitations discussed in Chapters 3 and 4. With the principal exception of the instrumentation amplifier, most of the circuits in this chapter have deliberate frequency-dependent properties incorporated in the design. Such designs usually employ capacitors along with resistors in the network external to the op-amp.

The first circuit we consider is the so-called instrumentation amplifier. The instrumentation amplifier is a precision circuit that can be used to amplify small signals in a noisy environment. A particular asset of this circuit is its large common-mode rejection ratio, and we establish the significance of this parameter in discussing its applications.

We discuss the purpose of integration and differentiation, and give the forms of ideal integrator and differentiator circuits using op-amps. This is followed by the analysis and presentation of design procedures for the ac integrator and low-frequency differentiator circuits. These latter circuits may be used with more general-purpose op-amps for performing certain types of differentiation and integration operations.

We also cover the analysis and design of phase shift circuits. Then we consider both phase lag and phase lead circuits. Such circuits allow arbitrary phase shifts to be established at a specific frequency for signals.

Lastly, we present circuits for biasing an op-amp with a single power supply. These circuits are included in this chapter because it is necessary to include frequency-dependent factors in designing the circuits.

✳5–2 INSTRUMENTATION AMPLIFIER

A common signal-processing requirement is the forming of the difference between two signals and amplifying the result by an arbitrary gain level. The *closed-loop differential amplifier* (or *difference amplifier* as it is also called) introduced in Section 2–9 is capable of performing this operation in many applications where accuracy requirements are moderate. The reader wishing to review the analysis and operation of that circuit should refer to Section 2–9.

Some limitations of the basic closed-loop differential amplifier are as follows:

✳1. The input impedances at the two signal inputs are finite, forcing the necessity to drive the two inputs with nearly ideal voltage sources when delicate balancing is required. Indeed, the input impedance at the noninverting signal input is a function of the signal level at the inverting signal input.

2. Common-mode rejection is a critical function of the external resistances connected to the circuit, and variations in the four resistance values will degrade the common-mode rejection.

3. In order to adjust the gain of the circuit, it is necessary to adjust two resistances, further complicating the balancing requirements.

A circuit that performs the required function while eliminating or at least minimizing the preceding difficulties is the ***instrumentation amplifier***, whose basic form is shown in Figure 5–1. Such circuits are available as packaged units, and most general users obtain them in this fashion. The fixed resistance values are established by the manufacturer at a high degree of accuracy, and the gains of the separate signal paths are closely matched. This delicate balancing, along with the fact that the op-amps are high-quality types, results in a very large value of common-mode rejection ratio (typically 120 dB or more).

Observe that the two signal inputs are applied to noninverting op-amp input terminals, resulting in high input impedances for the two signals. The output stage is a closed-loop differential amplifier circuit of the form previously considered. As we will see shortly, the two gains for the separate signal paths are both adjusted by the single resistance R_x.

We will now perform an analysis of the instrumentation amplifier. Consider the circuit diagram of Figure 5–2, on which various variables pertinent to the analysis have been labeled. The ideal assumptions considered in Chapter 2 will be used.

The starting point in the analysis is to recognize that under closed-loop stable conditions, the voltage at the inverting input terminal for each input op-amp will equal the voltage at the noninverting terminal. Since the inverting terminals are connected to the two ends of R_x, the voltages at the respective ends of R_x are equal to the corresponding input voltages as shown.

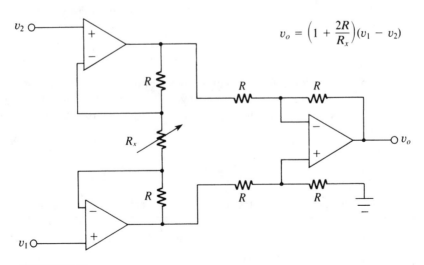

$$v_o = \left(1 + \frac{2R}{R_x}\right)(v_1 - v_2)$$

FIGURE 5–1
Instrumentation amplifier circuit.

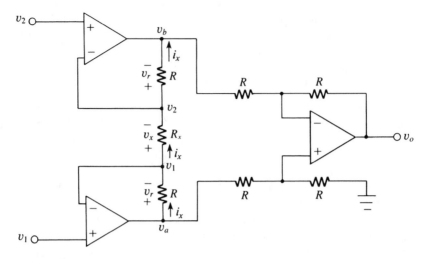

FIGURE 5–2
Instrumentation amplifier labeled for analysis of input stages.

We will arbitrarily consider v_1 as the more positive reference of the two inputs (although which variable is chosen is irrelevant as long as the remaining analysis is consistent with the initial assumption). The voltage across R_x is denoted as v_x, and it is seen to be

$$v_x = v_1 - v_2 \tag{5-1}$$

This voltage causes a current i_x to flow through R_x, and the value of i_x is

$$i_x = \frac{v_x}{R_x} = \frac{v_1 - v_2}{R_x} \tag{5-2}$$

Since no current flows in the op-amp input terminals, the current i_x must be flowing upward from the lower op-amp and flowing out of R_x on toward the upper op-amp. Since the upper and lower resistances are equal, the voltage v_r across each is the same, and this value is

$$v_r = Ri_x = \frac{R(v_1 - v_2)}{R_x} \tag{5-3}$$

The voltage v_a at the output of the lower op-amp is

$$v_a = v_1 + v_r \tag{5-4}$$

The voltage v_b at the output of the upper op-amp is

$$v_b = v_2 - v_r \tag{5-5}$$

The two voltages v_a and v_b act as two input signals to the closed-loop differential amplifier stage on the right. Since that portion of the circuit was analyzed in detail in Section 2-9, we will use the results of that section to deduce that

$$v_o = v_a - v_b = v_1 + v_r - (v_2 - v_r)$$
$$= v_1 - v_2 + 2v_r \tag{5-6}$$

The value of v_r from (5-3) can then be substituted in (5-6), and after some rearrangement of terms, the following result is obtained:

$$v_o = \left(1 + 2\frac{R}{R_x}\right)(v_1 - v_2) \tag{5-7}$$

The output voltage is seen to be a constant gain factor times the difference between the two input voltages, which is the same form as encountered for the closed-loop differential amplifier. However, for the instrumentation amplifier, the net gain for both signals can be varied by simply adjusting the value of R_x. Note that the gain decreases as R_x is increased.

Linear operation necessitates that all three op-amps operate in their linear regions. Because of the differencing operation, it is easy to overlook the possibility of saturation at the two input stages. The conditions for linear operation will now be investigated.

The output requirement is the most obvious and will be considered first. From (5-7), linear operation results when

$$\left(1 + \frac{2R}{R_x}\right)|v_1 - v_2| < V_{\text{sat}} \tag{5-8}$$

where the saturation voltages are assumed to be $\pm V_{\text{sat}}$. Note that since there are two inputs, it is necessary that (5-8) be satisfied for the worst-case combination of v_1 and v_2 occurring together.

The output of each of the two input stages must be within the linear region of operation. The pertinent voltages are v_a and v_b as given by (5-4) and (5-5), respectively. When v_r from (5-3) is substituted in (5-4) and (5-5), the following two inequalities result:

$$\left|\left(1 + \frac{R}{R_x}\right)v_1 - \frac{R}{R_x}v_2\right| < V_{\text{sat}} \tag{5-9}$$

$$\left|\left(1 + \frac{R}{R_x}\right)v_2 - \frac{R}{R_x}v_1\right| < V_{\text{sat}} \tag{5-10}$$

For linear operation, it is necessary that (5-8), (5-9), and (5-10) be satisfied.

Common-Mode Spurious Signals

The concept of common-mode rejection was introduced in the preceding chapter, but the full significance of the concept may not be clear. Now that we have introduced the instrumentation amplifier, it is appropriate to discuss common-mode problems in more detail and show the superiority of the instrumentation amplifier in dealing with such phenomena.

Exact models depicting the behavior of spurious pickup and other forms of noise are difficult to formulate, but we can use a somewhat simplified situation to illustrate the nature of undesired noise and how it can be reduced. Refer to Figure 5–3(a) containing a grounded signal source v_i. A cable is required to connect the signal to some amplifier at a different location. Stray pickup induces an undesired "noise" voltage n in each lead, as shown in Figure 5–3(b). If the leads are very close together, the voltages induced in the two sides will be nearly equal and are assumed to be exactly equal for this analysis. The resistances R_w represent wire resistance.

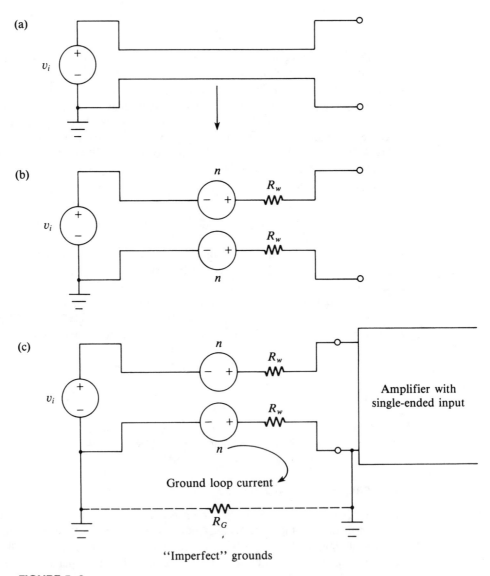

FIGURE 5–3
Illustration of common-mode noise pickup and ground-loop current.

Assume that the output of the cable is applied to one of the basic amplifier circuits with a grounded input, as shown in Figure 5–3(c). If the two grounds were perfect, the "best" situation would result in the upper noise component adding directly to the signal voltage at the amplifier input. More likely, however, is the additional presence of a *ground loop*. This phenomenon results from the presence of imperfect grounds from which there are small potential differences. When a circuit in such a situation is grounded at two or more points, as is often required for safety, a complete loop results, and a ground-loop current flows. This creates an additional undesired component at the input to the amplifier.

Use of Instrumentation Amplifier

The preceding problem can be reduced significantly by using an instrumentation amplifier as shown in Figure 5–4. The instrumentation amplifier utilizes a balanced input, of course, so neither input is grounded at the amplifier. Instead, the input signal appears as a differential input; that is, $v_d = v_i - v_2$. Let v_{od} represent the differential output, and let A represent the closed-loop differential gain established for the circuit. The desired output is then

$$v_{od} = A(v_1 - v_2) = Av_i \tag{5–11}$$

The noise component n appears as a common-mode input, that is, $v_{cm} = n$. Let A_{cm} represent the closed-loop common-mode gain of the complete circuit, and let v_{ocm} represent the common-mode output. This quantity is

$$v_{ocm} = A_{cm}n \tag{5–12}$$

The ratio of the differential output to the common-mode output is

$$\frac{v_{od}}{v_{ocm}} = \frac{Av_i}{A_{cm}n} \tag{5–13}$$

For the instrumentation amplifier, the common-mode rejection ratio CMRR is defined as

$$\text{CMRR} = \frac{A}{A_{cm}} \tag{5–14}$$

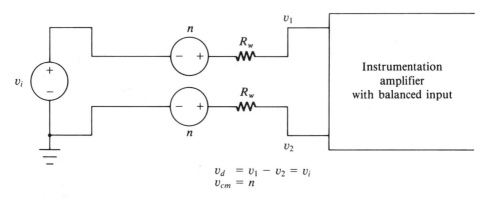

$$v_d = v_1 - v_2 = v_i$$
$$v_{cm} = n$$

FIGURE 5–4
Signal with spurious pickup applied to instrumentation amplifier with balanced input.

Substitution of this definition in (5–13) results in

$$\frac{v_{od}}{v_{ocm}} = \text{CMRR} \times \frac{v_i}{n} \tag{5–15}$$

An interesting interpretation of (5–15) is that v_i/n can be considered as an input signal voltage-to-noise voltage ratio and v_{od}/v_{ocm} can be interpreted as an output signal voltage-to-noise voltage ratio. The output signal-to-noise ratio is thus CMRR times the input signal-to-noise ratio.

EXAMPLE 5–1

A certain instrumentation amplifier of the form shown in Figure 5–1 has $R = 10$ $k\Omega$, and R_x is variable. Determine the value of R_x required if the desired output is to be

$$v_o = 10(v_1 - v_2) \tag{5–16}$$

Solution

Comparing (5–16) with the general form for the circuit as given by (5–7), we readily deduce that

$$1 + \frac{2R}{R_x} = 10 \tag{5–17}$$

Substituting $R = 10 \, k\Omega$, we obtain

$$R_x = 2222 \, \Omega \tag{5–18}$$

EXAMPLE 5–2

Consider the instrumentation amplifier of Example 5–1, with the gain value established according to (5–16). Check for linear operation, and determine the output for each of the following combinations of v_1 and v_2:
(a) $v_1 = 0.8$ V, $v_2 = 0.3$ V
(b) $v_1 = 0.8$ V, $v_2 = -0.3$ V
Assume that $\pm V_{\text{sat}} = \pm 13$ V for all op-amps.

Solution

We will check for linear operation in the two input stages using (5–9) and (5–10) first. If the outcomes of these tests are good, we will calculate the possible output voltage using (5–16). If this value is within the linear operating range, true linear operation is assured, and the value calculated with (5–16) will obviously be v_o.

(a) $v_1 = 0.8$ V, $v_2 = 0.3$ V. Application of (5–9) and (5–10) for these conditions yields

$$\left|\left(1 + \frac{10^4}{2222}\right)(0.8) - \frac{10^4}{2222}(0.3)\right| = 3.05 < 13 \tag{5–19}$$

$$\left|\left(1 + \frac{10^4}{2222}\right)(0.3) - \frac{10^4}{2222}(0.8)\right| = 1.95 < 13 \tag{5–20}$$

The two preceding results indicate that the first two stages are operating in their linear regions. We next calculate

$$v_o = 10(0.8 - 0.3) = 5 \text{ V} \qquad (5\text{--}21)$$

Since this value is within the linear region, it is the true output voltage.

(b) $v_1 = 0.8$ V, $v_2 = -0.3$ V. Application of (5–9) and (5–10) in this case yields

$$\left|\left(1 + \frac{10^4}{2222}\right)(0.8) - \frac{10^4}{2222}(-0.3)\right| = 5.75 < 13 \qquad (5\text{--}22)$$

$$\left|\left(1 + \frac{10^4}{2222}\right)(-0.3) - \frac{10^4}{2222}(0.8)\right| = 5.25 < 13 \qquad (5\text{--}23)$$

The first two stages are again linear, and we then calculate

$$v_o = 10[0.8 - (-0.3)] = 11 \text{ V} \qquad (5\text{--}24)$$

This value is within the linear region, and it is the true output voltage.

EXAMPLE 5–3

A certain instrumentation amplifier has a gain of 40 dB and a common-mode rejection ratio of 100 dB. It is used in a noisy environment in which the signal has a level of 50 mV, and the common-mode noise level is 100 mV. Determine (a) common-mode gain, (b) signal output, (c) noise output, and (d) output signal voltage-to-noise voltage ratio.

Solution

The specified gain is always the differential gain. A value of 40 dB corresponds to $A = 100$. A common-mode rejection ratio of 100 dB corresponds to CMRR $= 10^5$.

(a) The common-mode gain A_{cm} is

$$A_{cm} = \frac{A}{\text{CMRR}} = \frac{100}{10^5} = 10^{-3} \qquad (5\text{--}25)$$

(b) The signal output v_{od} is

$$v_{od} = Av_i = 100 \times 0.05 = 5 \text{ V} \qquad (5\text{--}26)$$

(c) The common-mode noise output v_{ocm} is

$$v_{ocm} = A_{cm}n = 10^{-3} \times 0.1 = 10^{-4} \text{ V} \qquad (5\text{--}27)$$

(d) The output signal-to-noise ratio is

$$\frac{v_{od}}{v_{ocm}} = \frac{5}{10^{-4}} = 50{,}000 \qquad (5\text{--}28)$$

Alternately, the input signal-to-noise ratio could have been first determined as $v_i/n = 50 \text{ mV}/100 \text{ mV} = 0.5$. The output signal-to-noise ratio could then have been determined from (5–15) as

$$\frac{v_{od}}{v_{ocm}} = 10^5 \times 0.5 = 50{,}000 \qquad (5\text{–}29)$$

5–3 INTEGRATOR AND DIFFERENTIATOR CIRCUITS

This section introduces circuits that perform the mathematical processes of differentiation and integration. These operations arise frequently in signal-processing functions. Both differentiation and integration change the shapes of the waveforms involved in accordance with the associated mathematical operations.

True Integrator Circuit

A block diagram of a hypothetical integrator circuit is shown in Figure 5–5. The input voltage is $v_i(t)$ and the output voltage is $v_o(t)$. With an ideal integration operation, the output $v_o(t)$ can be expressed as

$$v_o(t) = \int_0^t v_i(t)\, dt + v_o(0) \qquad (5\text{–}30)$$

where $v_o(0)$ is the initial value of the output (at $t = 0$). The integral is a form known in calculus as the **definite integral**, indicating an integration between two limits. The lower limit is $t = 0$, at which time the integration is started, and the upper limit is an arbitrary time t.

Integration is the process of area accumulation. The physical meaning of the integral sign is that all area under the curve of $v_i(t)$ is determined up to the time t. Since integration is an accumulative process, this accumulated area must be added to the initial value $v_o(0)$ in accordance with (5–30). If $v_o(0) = 0$, that is, if the initial value of the output voltage is zero, the integrator is said to be **initially relaxed.**

Now consider the capacitor shown in Figure 5–6. The instantaneous voltage across the capacitor is related to the instantaneous current flow into the capacitor by

$$v_c(t) = \frac{1}{C} \int_0^t i_c(t)\, dt + v_c(0) \qquad (5\text{–}31)$$

where $v_c(0)$ is the initial value. The voltage across the capacitor is thus proportional to the integral of the current, and this equation has the same form as (5–30). However, in (5–30)

FIGURE 5–5
Block diagram of hypothetical integrator.

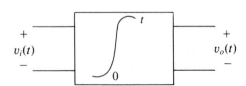

FIGURE 5–6

Capacitor with instantaneous voltage and current assumed.

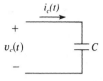

the variables were input and output voltages, while in (5–31), the variables are current and voltage associated with a capacitor.

The capacitor variables may be converted to appropriate input-output voltage variables by placing the capacitor as the feedback element in an op-amp circuit as shown in Figure 5–7. The circuit is one form of a voltage-controlled current source, and the input voltage $v_i(t)$ controls the capacitor current $i_c(t)$ as given by

$$i_c(t) = \frac{v_i(t)}{R} \tag{5-32}$$

This value of $i_c(t)$ can then be substituted in (5–31), in which case $v_c(t)$ is expressed directly in terms of $v_i(t)$. However, since the positive reference terminal of $v_c(t)$ is on the left, the output voltage is $v_o(t) = -v_c(t)$. Note also that $v_c(0) = -v_o(0)$. The net result for the output voltage is then

$$v_o(t) = \frac{-1}{RC} \int_0^t v_i(t)\,dt + v_o(0) \tag{5-33}$$

The only difference between (5–33) and (5–30) is the presence of the inversion sign and the constant $1/RC$. However, neither of these properties should cause any special difficulties, since additional inversion and/or gain could be added if required. The important property is that the output voltage is **proportional** to the integral of the input voltage, and this property is sufficient in most applications of the integrator.

The integrator circuit of Figure 5–7 is used extensively in analog computers for the solution of differential equations and in the simulation of physical systems. In such systems, the capacitor is charged to the desired initial value, and this voltage is maintained by auxiliary circuitry until the solution is initiated at $t = 0$.

To distinguish this integrator circuit from other integrator circuit forms, we will refer to it as the **true integrator circuit.**

FIGURE 5–7

True integrator circuit.

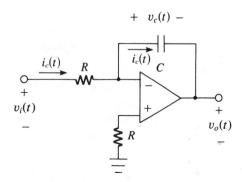

Difficulty with True Integrator Circuits

Operational amplifiers used in true integrator circuits must have extremely low dc offset voltage and bias currents. Recall that offset effects are equivalent to the presence of small dc sources at the input. A true integrator changes a constant dc voltage to a linearly increasing or decreasing ramp voltage. Depending on which direction dominates, the effect of offset and bias levels is that the output will slowly move toward saturation even with no signal present. For this reason, general-purpose op-amps such as the 741 are virtually unusable in true integration circuits. However, such op-amps may be used in the "ac" integrator circuit, which will be discussed in Section 5–5.

The result of this difficulty is that only op-amps having ultra-low offset voltage and bias currents are usable for true integrator circuits. One special type called the *chopper-stabilized* amplifier uses a mechanical switching process to correct continually for offset and bias effects. Most op-amps used in analog computers are the chopper-stabilized types.

True Differentiator Circuit

A block diagram of a hypothetical differentiator circuit is shown in Figure 5–8. The output voltage $v_o(t)$ can be expressed as

$$v_o(t) = \frac{dv_i(t)}{dt} \qquad (5\text{–}34)$$

The operation indicated by (5–34) is the mathematical process of differentiation, and in mathematical terms, $v_o(t)$ is the **first derivative** of $v_i(t)$; that is, $v_o(t)$ is equal to the rate of change (or slope) of $v_i(t)$. Thus, when $v_i(t)$ is changing rapidly, $v_o(t)$ has a large magnitude, but when $v_i(t)$ is changing very slowly, $v_o(t)$ is very small. When $v_i(t)$ is a constant (dc), $v_o(t) = 0$. In contrast to integration, there is no accumulative effect in differentiation; that is, the output at a given time is not affected by the "history" of the input, but is a function only of the rate of change of the input at a given time.

Refer to the capacitor shown in Figure 5–6. The current flow into the capacitor is related to the voltage across it by

$$i_c(t) = C \frac{dv_c(t)}{dt} \qquad (5\text{–}35)$$

This relationship is the inverse of (5–31) and can be obtained from the latter relationship by differentiating both sides.

The current flow into the capacitor is proportional to the derivative of the voltage, and this equation has the same form as (5–34). However, just as in the case of the integra-

FIGURE 5–8

Block diagram of hypothetical differentiator.

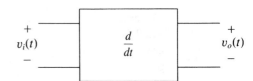

FIGURE 5–9
True differentiator circuit.

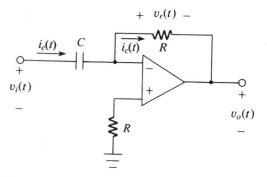

tor, one of the variables is a current, and it is desired to convert this current to a voltage before a workable circuit is achieved.

The op-amp differentiator circuit is shown in Figure 5–9. In this case, the capacitor is placed as the input element, and since the inverting terminal assumes ground potential, the capacitive current $i_c(t)$ is

$$i_c(t) = C \frac{dv_i(t)}{dt} \tag{5–36}$$

in accordance with (5–35) since $v_c = v_i$. This current flows through R and produces a voltage $v_r(t)$ across it with the reference positive terminal on the left. The output voltage $v_o(t)$ is the negative of this voltage and is

$$v_o(t) = -v_r(t) = -Ri_c(t) \tag{5–37}$$

Substituting $i_c(t)$ from (5–36) in (5–37), we obtain

$$v_o(t) = -RC \frac{dv_i(t)}{dt} \tag{5–38}$$

The only difference between (5–38) and (5–34) is the presence of the inversion sign and the constant RC. However, inversion and/or gain could be used if desired. The output is proportional to the derivative of the input, and this relationship is the fundamental property sought.

To distinguish this differentiator circuit from other differentiator circuit forms, we will refer to it as the ***true differentiator circuit.***

Difficulty with True Differentiation

The preceding development of the true differentiator was done to enhance the reader's understanding of the basic process, and the results are correct as far as the ideal models are concerned. In practice, however, true differentiator circuits are not used very often. Noise, which is always present in electronic circuits, is accentuated strongly by the differentiation process. Noise tends to have sudden abrupt changes and displays "spikes." Since the output of a true differentiator is proportional to the rate of change of the input, these sudden changes in noise result in pronounced output noise. Thus, a true differentiator circuit is very noisy, and its usefulness is somewhat limited. This difficulty can be alleviated by the low-frequency differentiator, to be considered in Section 5–6.

Comparison of Integration and Differentiation

A brief, qualitative comparison of the integration and differentiation processes as they affect signals will now be made. The process of integration involves accumulation of area, and sudden changes in the input signal are suppressed. Thus, an effective smoothing of the signal is achieved. It will be shown in Section 5–5 that integration can be viewed as one form of *low-pass filtering*.

In contrast to integration, differentiation results in the accentuation of sudden changes in the input signal. Constants or slowly changing signals are suppressed by a differentiator. It will be shown in Section 5–6 that differentiation can be viewed as a form of *high-pass filtering*.

EXAMPLE 5–4

A certain "true integrator circuit" produces an output voltage $v_o(t)$ given by

$$v_o(t) = \int_0^t v_i(t)\, dt + v_o(0) \tag{5–39}$$

where $v_i(t)$ is the input voltage and $v_o(0)$ is the initial value of the output voltage. Starting at $t = 0$ and with $v_o(0) = 0$, the periodic voltage waveform of Figure 5–10(a) is applied to the input. Sketch the output voltage and label significant levels.

Solution
The definite integral is simply the net accumulation of area as time passes. For a function consisting of straight-line segments, the definite integral may be determined most easily by a graphical approach. Each of several segments will be considered individually. Refer to the waveforms of both (a) and (b) of Figure 5–10 in the development that follows.

0 < t < 1.5 s
The input v_i in this interval is a positive constant, so the *area increases linearly with time.* To obtain a key point on the output curve, consider $t = 1.5$ s. The output voltage at this time is the area under the curve from $t = 0$ to $t = 1.5$ s, that is,

$$v_o(1.5) = 4 \times 1.5 = 6\text{ V} \tag{5–40}$$

1.5 < t < 4.5 s
The input v_i in this interval is a negative constant, so the *area decreases linearly with time.* However, the area always changes from the level reached prior to our point of consideration, so it decreases from the level of 6 V. A key point is $t = 4.5$, at which time the output voltage can be expressed as

$$v_o(4.5) = -4 \times 3 + v_o(1.5) = -12 + 6 = -6\text{ V} \tag{5–41}$$

Thus, the negative area from $t = 1.5$ to $t = 4.5$ causes the output voltage to decrease linearly from 6 V through 0 to −6 V.

(a)

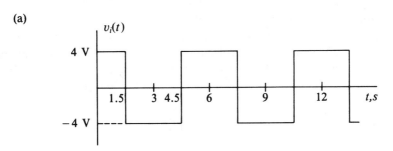

(b)

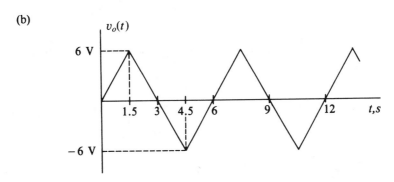

FIGURE 5-10
Waveforms of Example 5-4.

4.5 < t < 7.5 s
The input v_i in this interval is a positive constant, so the area increases linearly with time again. At $t = 6$, the output voltage crosses zero again, which corresponds to the end of one complete cycle. The overall pattern is thus clear and may be sketched for subsequent cycles as shown.

An important conclusion to draw from Example 5–4 is that *the integral of a symmetrical square wave is a triangular wave.* For this conclusion to be valid with a true integrator, it is necessary to restrict the square wave to a symmetrical form having no dc component. The preceding illustrates an important waveshaping operation that you will encounter in several places throughout the book.

EXAMPLE 5–5
A certain "true differentiator circuit" produces an output voltage $v_o(t)$ given by

$$v_o(t) = \frac{dv_i(t)}{dt} \tag{5-42}$$

where $v_i(t)$ is the input voltage. The voltage of Figure 5–11(a) is applied as the input. Sketch the output voltage and label significant levels.

Solution

The derivative is simply the slope of the function. For a function consisting of straight-line segments, the derivative may be determined most easily by a graphical approach. The function v_i will be divided into three segments for the analysis that follows. Refer to the waveforms (a) and (b) of Figure 5–11.

$0 < t < 0.5\,s$

The *change* in v_i is 6 V, and the *change* in t is 0.5 s. Thus,

$$v_o = \frac{6}{0.5} = 12 \text{ V} \tag{5–43}$$

$0.5 < t < 4.5\,s$

During this interval, there is *no change* in the dependent variable, and the slope is zero. Thus,

$$v_o = 0 \tag{5–44}$$

(a)

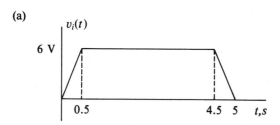

(b)

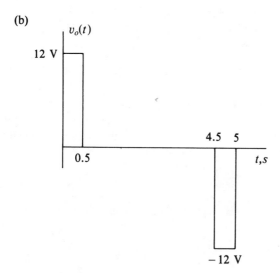

FIGURE 5–11
Waveforms of Example 5–5.

4.5 < t < 5 s

During this interval, the change in v_i is -6 V, while the change in t is 0.5 s. The slope is thus negative, and

$$v_o = \frac{-6}{0.5} = -12 \text{ V} \qquad (5\text{--}45)$$

This example illustrates an important property: *a differentiator circuit can be used to detect sudden changes in a signal* and *to generate trigger voltages based on these changes.* When the waveform of Figure 5–11(a) undergoes a positive change, the output voltage is a positive pulse. Conversely, when the input waveform undergoes a transition in the negative direction, the output voltage is a negative pulse.

As the transition becomes sharper (i.e., occurs in a shorter time), the magnitude of the output voltage increases. This property is exploited in a number of circuits utilizing the differentiator circuit.

5–4   GENERAL INVERTING GAIN CIRCUIT

So far, all op-amp amplifier circuits were designed to have constant gain values established by using resistors in the external network. In Chapter 4, gain variation as a function of frequency was discussed; this property was inherent, resulting from the frequency response limitation of the op-amp rather than a desired aspect of the design.

In the preceding section, true integrator and differentiator circuits were discussed. From the qualitative discussion of those circuits, one can infer that both have gain functions that vary in a pronounced manner with frequency. In those cases, however, the frequency variation is desired and is part of the design objective. The two circuits have been represented thus far in terms of their ideal mathematical operations, a natural process in view of the desired circuit objectives. However, both circuits can also be viewed in terms of their frequency response functions, as will be seen in the next two sections. In this manner, it is easier to consider modifications and variations on the basic operations than would be possible with the pure mathematical descriptions.

Circuits that are designed specifically to create frequency-dependent gain functions utilize reactive elements in addition to resistance. In theory, both capacitance and inductance could be used. In practice, however, capacitors tend to be much more ideal in the frequency range for which op-amp circuits are most often used. Recall from the preceding section that both the true integrator and the differentiator circuits used only capacitance as the reactive element. In general, virtually any realizable frequency-shaping function can theoretically be created with resistance, capacitance, and op-amps. Occasionally, inductors are used in some special op-amp circuits which are usable at high frequencies.

One form of an op-amp circuit widely used for certain frequency-dependent operations is the inverting circuit of Figure 5–12. In this case, the input and feedback circuits are shown as blocks, which represent complex impedances. The steady-state impedance of the input block is $\overline{Z}_i$, and the corresponding impedance of the feedback block is $\overline{Z}_f$. Each of the blocks may contain just one element, or each may contain a combination of two or more elements.

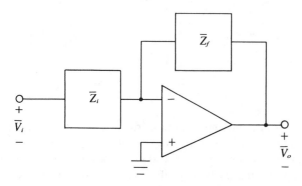

FIGURE 5–12
Inverting op-amp circuit form with arbitrary impedances.

In general, a bias compensating resistance would also be used between the noninverting terminal and ground, but this element has been eliminated here because it does not contribute to the signal analysis.

Since the blocks are impedances, it is necessary to use the transfer function concept to obtain an algebraic relationship between the input and output. The input/output ratio is determined by changing R_i and R_f in the basic gain equation for an inverting amplifier to $\overline{Z}_i$ and $\overline{Z}_f$, respectively. The transfer function $H(j\omega)$ is then readily expressed as

$$H(j\omega) = \frac{\overline{V}_o}{\overline{V}_i} = -\frac{\overline{Z}_f}{\overline{Z}_i} \tag{5–46}$$

When $\overline{Z}_f = R_f$ and $\overline{Z}_i = R_i$, this function obviously reduces back to the inverting amplifier gain, in which case $H(j\omega)$ becomes A_{CL}, the closed-loop gain.

The noninverting amplifier form can also be adapted to this general form. However, the noninverting form is not as useful for operations such as differentiation and integration, so we will restrict our consideration at this point to the inverting form.

5–5　ac INTEGRATOR

The true integrator considered in Section 5–3 does not operate satisfactorily with general-purpose op-amps due to the integration effects of dc offset and bias parameters. However, in this section a circuit will be described that permits integration of the time-variable portion of the input signal using general-purpose op-amps such as the 741. The circuit will be denoted in this text as the *ac integrator*.

Frequency Response of True Integrator

The most convenient way to introduce the ac integration circuit is through frequency response and impedance considerations utilizing the material of the last section. To that end, let us first establish the frequency response of the true integrator shown in Figure 5–13(a).

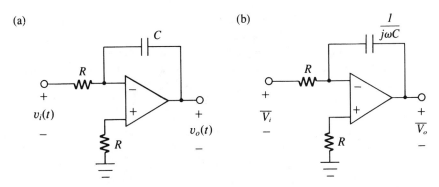

FIGURE 5–13
True integrator circuit and its steady-state phasor model.

(The circuit is assumed to be initially relaxed.) The corresponding steady-state phasor form for the circuit is shown in Figure 5–13(b). The impedances of the input and feedback components are

$$\overline{Z}_i = R \tag{5–47}$$

and

$$\overline{Z}_f = \frac{1}{j\omega C} \tag{5–48}$$

From (5–46), the transfer function of the true integrator is

$$H(j\omega) = \frac{-1/j\omega C}{R} = \frac{-1}{j\omega RC} \tag{5–49}$$

Comparing (5–49) and (5–33) with $v_o(0) = 0$, we note that integration in the time domain assumes the form of multiplication by $1/j\omega$ in the transfer function. Indeed, this property has a more formal mathematical basis in circuit and system theory. Thus, when the form $1/j\omega$ appears in a transfer function, it is an indication of true integration for the circuit.

The amplitude response $M(\omega)$ for the true integrator is readily determined from the magnitude of (5–49) as

$$M(\omega) = \frac{1}{\omega RC} \tag{5–50}$$

This function is very large at low frequencies but decreases as the frequency increases. This point is in agreement with the observation of Section 5–3 that true integration is a form of low-pass filtering. (If true low-pass filtering were desired, much better low-pass filters are available, as will be considered in Chapter 8. However, the point here is that integration displays a frequency response characteristic having a general low-pass quality.)

The theoretical value of the amplitude response of a true integrator is infinite at dc. This observation is a different way of deducing the difficulty with dc offsets and bias parameters. The effect of such parameters is to cause the output to saturate. If the output of the op-amp were not limited by finite saturation, it would continue to increase or decrease indefinitely in accordance with the prediction of the frequency response. The frequency response concept assumes a perfectly linear circuit and does not recognize such finite effects as saturation.

On a Bode plot with frequency on a logarithmic scale and decibels on a linear scale, the frequency response of the ideal integrator is a straight line with a negative slope of -6 dB/octave. The reader may momentarily look ahead to Figure 5–15. The straight-line characteristic with the negative slope is that of the true integrator.

Development of ac Integrator

Since the primary difficulty arises at dc, it may be circumvented by dropping the gain to a finite value at dc. This drop is achieved by placing a resistance R_f in parallel with the capacitor as shown in Figure 5–14(a), and this circuit is the **ac integrator**. The input resistance is now denoted as R_i. Since a capacitor is an open circuit at dc, the circuit reduces to a simple inverting amplifier with gain $-R_f/R_i$ at dc. However, as the frequency increases, the magnitude of the capacitive reactance decreases and more of the signal current is shunted through the capacitor. The operation of the circuit should eventually approach that of the true integrator as the frequency increases.

Frequency Response of ac Integrator

The ac integrator circuit will now be analyzed by the steady-state phasor approach. The circuit is first converted to steady-state form as shown in Figure 5–14(b). The feedback impedance is the parallel combination of R_f and $1/j\omega C$, which is

$$\overline{Z}_f = R_f \| \frac{1}{j\omega C} = \frac{R_f \times (1/j\omega C)}{R_f + (1/j\omega C)} = \frac{R_f}{1 + j\omega R_f C} \tag{5-51}$$

The input impedance is simply

$$\overline{Z}_i = R_i \tag{5-52}$$

The transfer function is then

$$H(j\omega) = \frac{-R_f/R_i}{1 + j\omega R_f C} = \frac{-R_f/R_i}{1 + j\omega \tau_f} \tag{5-53}$$

where $\tau_f = R_f C$ is the time constant of the feedback circuit.

The transfer function of (5–53) is the one-pole low-pass form introduced in Chapter 1 and considered again in Chapter 4 in the analysis of op-amp closed-loop frequency response. However, the present goal is to approximate integration, so our focus on the circuit is from a different point of view.

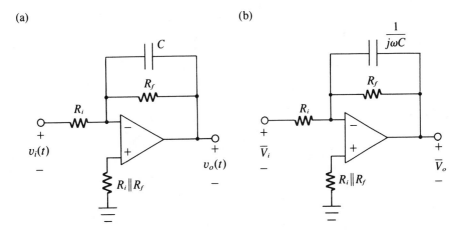

FIGURE 5–14
ac integrator circuit and its steady-state phasor model.

The amplitude response corresponding to (5–53) is

$$M(\omega) = \frac{R_f/R_i}{\sqrt{1 + (\omega R_f C)^2}} = \frac{R_f/R_i}{\sqrt{1 + (\omega \tau_f)^2}} \tag{5–54}$$

The decibel form of the Bode plot approximation of (5–54) is shown in Figure 5–15 along with the corresponding response of the true integrator. On this plot, it is assumed that $R_i = R$; that is, the input resistances for both circuits are equal.

The break frequency f_b for (5–54) is

$$f_b = \frac{1}{2\pi R_f C} = \frac{1}{2\pi \tau_f} \tag{5–55}$$

For $f \ll f_b$, $M(\omega) \simeq R_f/R_i$, which means that the circuit is acting as a constant gain amplifier as expected. However, for $f \gg f_b$, (5–54) can be approximated as

$$M(\omega) \simeq \frac{R_f/R_i}{\omega R_f C} = \frac{1}{\omega R_i C} \quad \text{for } f \gg f_b \tag{5–56}$$

Comparing (5–56) with (5–50), we see that the amplitude response of the ac integrator closely approximates that of the true integrator in the frequency range well above the break frequency. For $R_i = R$, the break-point approximation and the true integrator curves coincide in this frequency range.

Assume that the lowest frequency of the time-varying portion of some signal of interest is f_L. Assume that the integrator has been designed to satisfy the inequality $f_b \ll f_L$. Let $v_i'(t)$ represent the time-varying portion of the input signal, and let $v_o'(t)$ represent the time-varying portion of the output signal. The ac integrator circuit then performs the signal-processing operation

$$v_o'(t) = -\frac{1}{R_i C} \int v_i'(t) \, dt \tag{5–57}$$

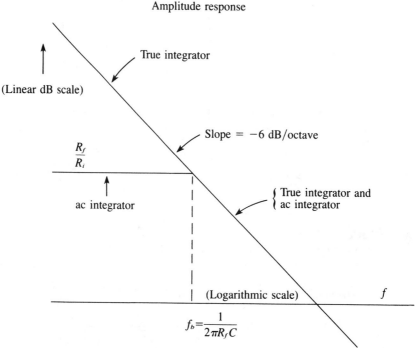

FIGURE 5–15
Bode break-point approximations for amplitude responses of true integrator and ac integrator.

Design Procedure

In the design of an ac integrator, there are three parameters to be selected: R_i, R_f, and C. Assume that the desired processing function is specified as

$$v_o'(t) = -K \int v_i'(t)\, dt \qquad (5\text{–}58)$$

where K is positive, and the inversion has been recognized by the negative sign. Assume that the lowest frequency f_L (other than dc) contained in the signal spectrum is known. Let $T_L = 1/f_L$ represent the period of the lowest frequency component. If the signal is periodic, f_L is the repetition rate of the composite waveform and T_L is the corresponding period. The specification of the gain constant K and the frequency f_L (or period T_L) is actually two specifications, and since there are three parameters, one may be selected arbitrarily. Often it is more convenient to select the capacitance value and adjust resistance values to match. The procedure given here is based on that assumption. If this approach is not suitable, one can modify it in accordance with the equations developed earlier.

The following procedure for designing an ac integrator circuit should suffice for many routine purposes:

1. Select a fixed value of C. Do not use an electrolytic capacitor unless the voltage across the capacitor always has one polarity. High-quality, stable capacitors made of such materials as polystyrene or Mylar® are recommended.

2. From the desired value of K, determine R_i by comparing (5–58) and (5–57). An expression for R_i is then determined as

$$R_i = \frac{1}{CK} \qquad (5\text{–}59)$$

3. The degree to which the ac integrator approaches ideal integration of the time-varying portion of the signal is determined either by the selection of the feedback time constant $\tau_f = R_f C$ appearing in the denominator of (5–53) or by the selection of the break frequency f_b as given by (5–55). The larger the value of τ_f or the smaller the value of f_b, the closer the integrator approaches ideal integration. However, as exact integration is approached, element values become larger and noise gain problems become greater. For precision integrator circuits, try either $\tau_f = 10T_L$ or $f_b = 0.01f_L$ as a starting point. In less demanding requirements, τ_f may be reduced or f_b may be increased. (The two conditions $\tau_f = 10T_L$ and $f_b = 0.01f_L$ do not yield identical results, but they are rounded general rules that produce results reasonably close to each other.)

   If the time constant τ_f is selected, the value of R_f is computed as

$$R_f = \frac{\tau_f}{C} \qquad (5\text{–}60)$$

If the break frequency f_b is selected, the value of R_f is computed as

$$R_f = \frac{1}{2\pi f_b C} \qquad (5\text{–}61)$$

Since these two equations for R_f are based on arbitrary rules of thumb and do not yield the same results, a reasonable adjustment (for example, a closest standard resistance value) may be acceptable for R_f when using either formula.

4. If either or both of the resistance values are excessively large or excessively small, a different value of capacitance should be selected, and the procedure repeated. If the values of resistance are initially too large, increase the value of capacitance on the next trial, and vice versa.

Waveshaping Applications

One useful application of the ac integrator is in waveshaping circuits for periodic waveforms. For example, a square wave can be converted to a triangular waveform and a triangular waveform can be converted to a parabolic waveform.

With some care, the standard procedures for graphical integration can be used for predicting the shape and magnitude of output signals in certain cases. Assume for this analysis that the input waveform $v_i(t)$ is *periodic* and that it can be represented as

$$v_i(t) = v_i'(t) + V_i(\text{dc}) \tag{5-62}$$

where $v_i'(t)$ is the ac (time-varying) portion of the input and $V_i(\text{dc})$ is the dc value of the input. One can determine the dc value by evaluating the net area in one cycle and dividing by the period:

$$V_i(\text{dc}) = \frac{\text{net area under curve in 1 cycle}}{T}$$

or

$$V_i(\text{dc}) = \frac{1}{T} \int_0^T v_i(t)\, dt \tag{5-63}$$

The time-varying portion of the signal is then determined by subtracting the dc level:

$$v_i'(t) = v_i(t) - V_i(\text{dc}) \tag{5-64}$$

The latter operation is simply shifting the signal up or down to remove the dc component.

By definition, the net area of $v_i'(t)$ integrated over one cycle is zero. (Why?) Thus any positive change in the output voltage during a portion of a cycle must be balanced by a corresponding negative change during the remainder of the cycle. For this analysis, assume that all the positive area occurs in a single, distinct portion of the cycle separate from the negative area. The results may be adapted to other cases by the basic techniques of calculus, but the restriction applies to most common cases of interest, and it simplifies subsequent discussions. The peak-to-peak output voltage $V_o(\text{pp})$ is then determined as the maximum change within a cycle produced by the integration process:

$$V_o(\text{pp}) = K \times |\text{net area with same sign}| \tag{5-65}$$

where the area involved is the net magnitude of the area of $v_i'(t)$ for *either* the positive or the negative portion of the cycle. Either area may be used, since the magnitudes should be the same, and $V_o(\text{pp})$ is thus a positive value. Note that if an input is given as $v_i'(t)$ with a dc component, it is necessary first to determine $v_i'(t)$ from (5–64) before the area can be computed.

The complete effect of the ac integrator on a composite signal $v_i'(t)$ can be determined by representing $v_i'(t)$ in the form of (5–62) and noting the different effects on the two components. Assuming that all frequency components of $v_i'(t)$ are well above the break frequency, the ac portion of the signal is integrated according to (5–57). However, the dc component is simply multiplied by the gain constant $-R_f/R_i$ of an inverting amplifier. Thus, the net output voltage $v_o'(t)$ can be expressed as

$$v_o(t) = v_o'(t) + V_o(\text{dc}) \tag{5-66a}$$

$$v_o(t) = -\frac{1}{R_iC} \int v_i'(t)\, dt - \frac{R_f}{R_i} V_i(\text{dc}) \tag{5-66b}$$

where the dc level of the output $V_o(dc)$ is observed to be

$$V_o(dc) = -\frac{R_f}{R_i} V_i(dc) \tag{5–67}$$

Some care must be exercised with this circuit when a dc component is present in order to remain in the linear region. As the ratio R_f/R_i increases, the magnitude of the output dc level increases, and this level plus the peak level of the output time-varying component should not reach saturation.

EXAMPLE 5–6

An ac precision integrator is desired for a particular application to perform the operation

$$v_o'(t) = -1000 \int v_i'(t) \, dt \tag{5–68}$$

where the primes indicate the ac portions of the respective functions. The lowest frequency (other than a possible dc component) of the input signal is estimated to be 1 kHz. Determine a suitable design.

Solution
As a starting point, the capacitance will be selected somewhat arbitrarily as $C = 0.1 \ \mu F$. Since $K = 1000 = 1/R_iC$, the value of R_i is then calculated as

$$R_i = \frac{1}{1000 \times 0.1 \times 10^{-6}} = 10 \ \text{k}\Omega \tag{5–69}$$

The time constant criterion will be used. Since $f_L = 1$ kHz, $T_L = 1/10^3 = 1$ ms. The value of τ_f will then be selected as $\tau_f = 10T_L = 10$ ms. The value of R_f is then determined as

$$R_f = \frac{\tau_f}{C} = \frac{10 \times 10^{-3}}{0.1 \times 10^{-6}} = 100 \ \text{k}\Omega \tag{5–70}$$

This value is a standard value, and the circuit is shown in Figure 5–16. Observe that a bias compensating resistance has been added. The reader is invited to show that the

FIGURE 5–16
Integrator circuit design of Example 5–6.

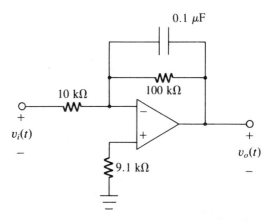

break frequency is $f_b = 15.92$ Hz. If the break-frequency criterion had been used, the design would have been based on an initial selection of $f_b = 10$ Hz, which would have yielded a different value of R_f. The reader should thus note that since T_f or f_b is somewhat arbitrary, there is no "correct" value for R_f. However, the values of R_i and C determine the actual gain constant for the circuit, so they must be selected with great care.

EXAMPLE 5–7   The square wave of Figure 5–17(a) is applied to the input of the ac integrator designed in Example 5–6. After steady-state conditions are established, determine the peak-to-peak output voltage $V_o(\text{pp})$.

Solution
For a general periodic waveform, the theory of Fourier series indicates that the steady-state frequency components will be at dc, the repetition rate of the waveform (fundamental fre-

(a)

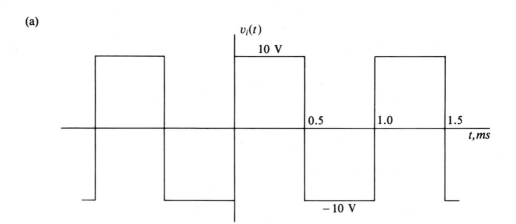

(b)

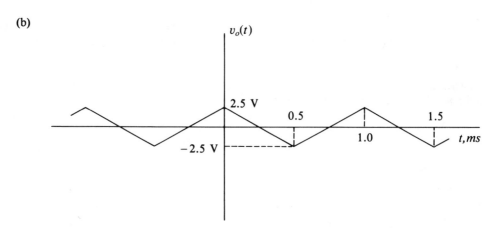

FIGURE 5–17
Waveforms of Examples 5–7 and 5–8.

quency), and at integer multiples of the repetition rate. For this waveform, the dc component is zero, since the net positive area in one cycle is equal to the magnitude of the net negative area. The period is $T = 1$ ms, and the frequency f of the square wave is $f = 1/T = 1/0.001 = 1$ kHz. Thus, the lowest frequency component of the square wave is 1 kHz, and all other frequencies will be integer multiples of that frequency. Since the circuit was designed to integrate components at 1 kHz and higher, the circuit is assumed to integrate the signal.

The positive half-cycle will be used as the basis for area evaluation. The net area in that half-cycle is

$$\text{Area} = 10 \text{ V} \times 0.5 \times 10^{-3} \text{ s} = 5 \times 10^{-3} \text{ V} \cdot \text{s} \qquad (5\text{--}71)$$

The peak-to-peak output voltage is determined from (5–65) as

$$V_o(\text{pp}) = 1000 \times 5 \times 10^{-3} = 5 \text{ V} \qquad (5\text{--}72)$$

EXAMPLE 5-8

For the square wave and integrator of Example 5–7, plot the output voltage $v_o(t)$ after steady-state conditions are reached.

Solution

The basic integral of a square wave is a triangular waveform. The area under the curve in one half-cycle times the gain constant represents the change (peak-to-peak value) of the triangular waveform over one half-cycle as determined in (5–72). Since this circuit also inverts, the sense of the area is reversed; that is, a *positive* area for the input results in a *decrease* in the output, and vice versa.

The dc value of the input voltage is zero, so the output voltage will not have a dc component. This means that the output will have equal magnitude of area above and below the time axis. Consequently, the output waveform will oscillate between 2.5 V and −2.5 V as shown in Figure 5–17(b).

This circuit form is useful for waveshaping purposes as suggested by this example. The input square wave has been converted to a triangular wave by the integration process.

Care must be exercised with this type of circuit if the input waveform contains a dc component. The dc gain of this particular circuit is $-100 \text{ k}\Omega / 10 \text{ k}\Omega = -10$. Consequently, any input dc level is multiplied in magnitude by this value. For example, if the input had contained a dc level of 1.05 V, the dc level of the output would be at −10.5 V, and on the negative side of the output cycle, the output would reach a level of $-10.5 - 2.5 = -13$ V, which is the approximate negative saturation level.

5-6 LOW-FREQUENCY DIFFERENTIATOR

The true differentiator considered in Section 5–3 tends to be quite noisy when used with a wide-band op-amp because of the pronounced increase in gain as the frequency increases. In this section a circuit will be described that provides a differentiation effect over a specified low-frequency range but that limits the increase in gain at higher frequencies. This circuit will be denoted in this book as the *low-frequency differentiator*.

Frequency Response of True Differentiator

As in the case of the integrator, the operation of the differentiator will be established through steady-state frequency response and impedance considerations. First, consider the true differentiator shown in Figure 5–18(a). The corresponding phasor form for the circuit is shown in Figure 5–18(b). The impedance of the input and feedback components are

$$\overline{Z}_i = \frac{1}{j\omega C} \tag{5–73}$$

$$\overline{Z}_f = R \tag{5–74}$$

From (5–46), the transfer function of the true differentiator is

$$H(j\omega) = \frac{-R}{1/j\omega C} = -j\omega RC \tag{5–75}$$

Comparing (5–75) and (5–38), we note that differentiation in the time domain assumes the form of multiplication by $j\omega$ in the transfer function. This operation has a more formal mathematical basis in circuit and system theory. Thus, when the form $j\omega$ appears in a transfer function, it is an indication of true differentiation for the system.

The amplitude response $M(\omega)$ for the true differentiator is readily determined from the magnitude of (5–75) as

$$M(\omega) = \omega RC \tag{5–76}$$

This function is very small at low frequencies, but increases linearly as the frequency increases. This point is in agreement with the observation of Section 5–3 that true differentiation is a form of high-pass filtering. The differentiator response is not suitable for most high-pass applications since the response is not flat, but it does display a high-pass type of effect.

The frequency response of the ideal differentiator is a straight line having a positive slope of +6 dB/octave. The reader may momentarily look ahead to Figure 5–20 to observe the frequency response of the ideal differentiator.

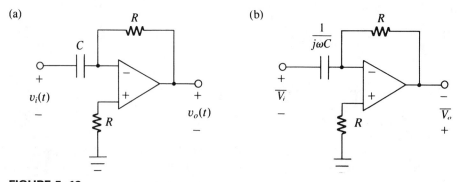

FIGURE 5–18
True differentiator circuit and its steady-state phasor model.

Development of Low-Frequency Differentiator

The primary difficulty with the true differentiator is the pronounced increase in gain at higher frequencies. This problem may be circumvented for many applications by placing a resistance R_i in series with the capacitor as shown in Figure 5–19(a), and this circuit is the *low-frequency differentiator*. The feedback resistance is now denoted as R_f. Since a capacitor is a short circuit at high frequencies, the circuit reduces to a simple inverting amplifier with gain $-R_f/R_i$ in that frequency range. However, at lower frequencies, the magnitude of the capacitive reactance increases so that its effect overshadows the effect of the series resistance. In this frequency range, the circuit will act as a differentiator.

Frequency Response of Low-Frequency Differentiator

The low-frequency differentiator circuit will now be analyzed by means of the steady-state phasor approach form as shown in Figure 5–19(b). The input impedance is the series combination of R_i and $1/j\omega C$, which is

$$\overline{Z}_i = R_i + \frac{1}{j\omega C} = \frac{1 + j\omega R_i C}{j\omega C} \tag{5–77}$$

The feedback impedance is simply

$$\overline{Z}_f = R_f \tag{5–78}$$

(The effect of the possible capacitor in shunt with R_f, whose purpose will be discussed later, has been ignored in this analysis.) The transfer function is

$$H(j\omega) = \frac{-j\omega R_f C}{1 + j\omega R_i C} = \frac{-j\omega R_f C}{1 + j\omega \tau_i} \tag{5–79}$$

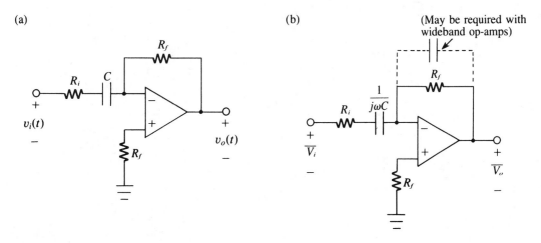

FIGURE 5–19
Low-frequency differentiator and its steady-state phasor model.

where $\tau_i = R_iC$ is the time constant of the input circuit. The transfer function of (5–79) is a one-pole high-pass form, which has not been encountered yet in the text. The amplitude response corresponding to (5–79) can be expressed as

$$M(\omega) = \frac{\omega R_f C}{\sqrt{1 + (\omega R_i C)^2}} = \frac{\omega R_f C}{\sqrt{1 + (\omega \tau_i)^2}} \tag{5–80}$$

The decibel form of the Bode plot approximation of (5–80) is shown in Figure 5–20 along with the corresponding response of the true differentiator. On this plot, it is assumed that $R_f = R$; that is, the feedback resistances for both circuits are equal.

The break frequency f_b for (5–80) is

$$f_b = \frac{1}{2\pi R_i C} = \frac{1}{2\pi \tau_i} \tag{5–81}$$

For $f \gg f_b$, $M(\omega)$ in (5–80) reduces to

$$M(\omega) \simeq \frac{\omega R_f C}{\sqrt{(\omega R_i C)^2}} = \frac{R_f}{R_i} \quad \text{for } f \gg f_b \tag{5–82}$$

This is the high-frequency range in which the circuit reduces to an inverting amplifier. In contrast, for $f \ll f_b$, (5–80) reduces to

$$M(\omega) \simeq \omega R_f C \quad \text{for } f \ll f_b \tag{5–83}$$

Comparing (5–83) with (5–76), we see that the amplitude response of the low-frequency differentiator closely approximates that of the true differentiator well below the break frequency. For $R_f = R$, the break-point approximation and the true differentiator curves coincide in this frequency range.

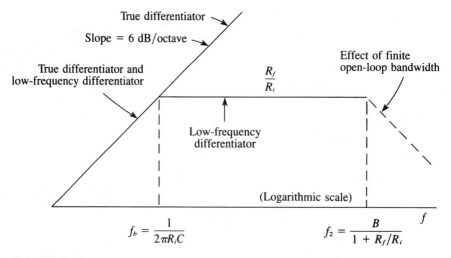

FIGURE 5–20
Bode break-point approximations for amplitude responses of true differentiator and low-frequency differentiator.

The analysis thus far has been based on the assumption of an ideal op-amp. When the finite open-loop bandwidth of the op-amp is considered, an additional alteration in frequency response occurs. Recall from Chapter 4 that the op-amp open-loop frequency response and the closed-loop response converge together above the closed-loop 3-dB frequency. This phenomenon is illustrated by the dashed line on the right-hand side of the response of Figure 5–20. The value of this frequency f_2 is given by

$$f_2 = \frac{B}{1 + R_f/R_i} \tag{5–84}$$

in accordance with the work of Chapter 4.

Assume that the highest frequency of some signal $v_i(t)$ of interest is f_H. Assume that the differentiator has been designed to satisfy the inequality $f_b \gg f_H$. The output signal $v_o(t)$ of the low-frequency differentiator then performs the signal-processing operation

$$v_o(t) = -R_f C \frac{dv_i(t)}{dt} \tag{5–85}$$

Design Procedure

In the design of a low-frequency differentiator, there are three parameters to be selected: R_i, R_f, and C. Assume that the desired processing function is specified as

$$v_o(t) = -K \frac{dv_i(t)}{dt} \tag{5–86}$$

where K is positive, and the inversion has been recognized by the negative sign. Assume that the highest frequency f_H contained in the signal spectrum is known. Let $T_H = 1/f_H$ represent the period of the highest-frequency component. Specification of the gain constant K and the frequency f_H (or period T_H) is actually two specifications, and since there are three parameters, one may be selected arbitrarily. Often it is more convenient to select the capacitance value and adjust resistance values to match. The procedure to be given here is based on that assumption. However, this procedure can be modified if necessary by returning to the equations developed in this section.

The following procedure for designing a low-frequency differentiator circuit should suffice for many routine purposes:

1. Select a fixed value of C. Do not use an electrolytic capacitor unless the voltage across the capacitor always has one polarity.
2. From the desired value of K, determine R_f by comparing (5–86) with (5–85). An expression for R_f is then determined as

$$R_f = \frac{K}{C} \tag{5–87}$$

3. The degree to which the low-frequency differentiator approaches ideal differentiation is determined either by the selection of the input time constant τ_i or the selection of the corresponding break frequency f_b. The smaller the value of τ_i or the

larger the value of f_b, the closer the differentiator approaches ideal differentiation. However, there is a limit on the frequency range in which differentiation can be achieved due to the inherent bandwidth of the op-amp as indicated by f_2 in Figure 5–20. Furthermore, even with wide-band op-amps, it may be desirable to add a small shunt capacitor around R_f to produce a gradual roll-off effect to reduce noise and other high-frequency difficulties. Finally, the frequency f_H may be difficult to estimate. Because of these various factors, the degree of precision in the practical differentiator is usually more limited than in the integrator, and good general rules on the selection of τ_i or f_b are more difficult to specify. For a periodic signal with frequency f or period $T = 1/f$, try $\tau_i = 0.01T$ or $f_b = 10f$. If the time constant τ_i is selected, the value of R_i is computed as

$$R_i = \frac{\tau_i}{C} \qquad (5\text{–}88)$$

If the break frequency f_b is selected, the value of R_i is computed as

$$R_i = \frac{1}{2\pi f_b C} \qquad (5\text{–}89)$$

The values of R_i calculated by (5–88) and (5–89) may differ in view of the arbitrary nature of the assumptions employed. In most cases, a reasonable adjustment (for example, a nearest standard value) may be acceptable for R_i.

4. If either or both of the resistance values are excessively large or excessively small, a different value of capacitance should be selected and the procedure can be repeated. If the values of resistance are initially too large, increase the value of capacitance on the next trial, and vice versa.

Waveshaping Applications

As in the case of the ac integrator circuit, the low-frequency differentiator circuit can be used in waveshaping circuits for periodic waveforms. For example, a triangular waveform can be converted to a square wave, and a square wave can be converted to a periodic train of narrow "spikes."

The standard procedures for graphical differentiation may be used for predicting the shape and magnitude of the output signal of a low-frequency differentiator in many cases. Assume for this analysis that the input waveform $v_i(t)$ is periodic. Assuming that the differentiator property is valid for all frequency components representing the input signal, the output voltage $v_o(t)$ is determined by a direct application of (5–85) or (5–86). When the input voltage is composed of straight-line segments, the output voltage can be expressed as

$$v_o(t) = -K \times (\text{slope of input voltage}) \qquad (5\text{–}90)$$

The peak-to-peak output voltage $V_o(\text{pp})$ can be expressed as

$$V_o(\text{pp}) = K \times (m^+ + |m^-|) \qquad (5\text{–}91)$$

where m^+ is the maximum positive slope and m^- is the maximum negative slope. Note that the *magnitude* of this latter slope is added to the positive slope m^+ to determine the peak-to-peak value.

FIGURE 5–21
Circuit of Example 5–9.

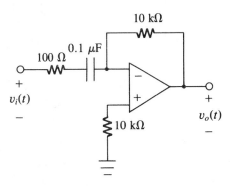

EXAMPLE 5–9

A low-frequency differentiator is desired for a particular application to perform the operation

$$v_o(t) = -0.001 \frac{dv_i(t)}{dt} \tag{5–92}$$

Based on a periodic signal with a frequency of 1 kHz, determine a suitable design.

Solution
As a starting point, the capacitance will be selected somewhat arbitrarily as $C = 0.1\ \mu\text{F}$. Since $K = 0.001 = R_f C$ the value of R_f is selected as

$$R_f = \frac{0.001}{0.1 \times 10^{-6}} = 10\ \text{k}\Omega \tag{5–93}$$

The time constant criterion will be used. Since $f = 1$ kHz, $T = 1/10^3 = 1$ ms. The value of τ_i will then be selected as $\tau_i = 0.01T = 0.01$ ms. The value of R_i is then determined as

$$R_i = \frac{\tau_i}{C} = \frac{10^{-5}}{0.1 \times 10^{-6}} = 100\ \Omega \tag{5–94}$$

This value is a standard value, and the circuit is shown in Figure 5–21. Observe that a bias compensating resistance equal to R_f has been added. The reader is invited to show that the break frequency is $f_b = 15.915$ kHz based on the time constant selected. However, if a 741 op-amp were used in this circuit, the closed-loop 3-dB frequency would actually be $f_2 = 1$ MHz/101 $\simeq$ 10 kHz, which would limit the attainable bandwidth.

EXAMPLE 5–10

The triangular waveform of Figure 5–22(a) is applied to the input of the low-frequency differentiator designed in Example 5–9. Plot the form of the output voltage after steady-state conditions are reached. Assume that all significant frequency components of the triangular waveform are within the differentiator range of the circuit.

Solution
Referring to Figure 5–17(b), note that the input waveform of the differentiator is exactly the same as the output of the integrator of Example 5–8. Assuming an ideal differentiator, we determine the derivative of each segment of the input signal by evaluating the slope over that segment.

In the first half-cycle, the slope m of the input is

$$m = \frac{-2.5 - 2.5}{0.5 \times 10^{-3}} = -10 \times 10^3 \text{ V/s} \qquad \text{(5–95)}$$

Multiplication by the gain constant of -0.001 yields $v_o = -0.001 \times (-10 \times 10^3) = 10$ V for that interval, as illustrated in Figure 5–22(b).

The slope in the second half-cycle is

$$m = \frac{2.5 - (-2.5)}{0.5 \times 10^{-3}} = 10 \times 10^3 \text{ V/s} \qquad \text{(5–96)}$$

The output voltage in this interval is then $v_o = -10$ V. The peak-to-peak value of the output voltage is thus $V_o(\text{pp}) = 20$ V. The waveform is shown in Figure 5–22(b).

Some rounding at the edges has been included to indicate the typical behavior of a realistic differentiator circuit. This rounding is due to roll-off at very high frequencies.

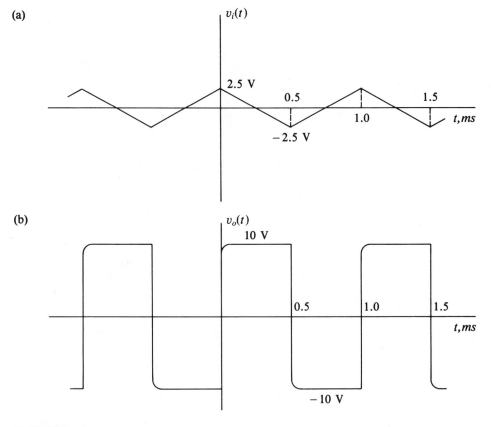

FIGURE 5–22
Waveforms of Example 5–10.

The process illustrated between the preceding several examples is that integration of a square wave produces a triangular wave, but subsequent differentiation of the triangular wave produces a square wave again. In these examples, the actual levels of the final square wave correspond to the original levels. This is true because the gain constant of the integrator was -1000, and the gain constant of the differentiator was -0.001, which results in $(-1000) \times (-0.001) = 1$.

5–7 PHASE SHIFT CIRCUITS

A common operation required in electronic circuit design is the process of producing a required phase shift at a given frequency. All reactive networks are capable of producing phase shifts, but the majority of such circuits have amplitude response functions that vary with frequency, which limits their usefulness to carefully controlled situations.

In this section two circuits will be considered that permit the phase shift to be adjusted over a wide range, but in which the amplitude response remains constant. These circuits are examples of a special class of circuits called **all-pass networks**. The circuits to be considered will be denoted as the **all-pass phase lag** circuit and the **all-pass phase lead** circuit. The names suggest the forms of the phase shift functions that can be realized, as will be seen shortly. The two circuits will be considered separately.

All-Pass Phase Lag Circuit

The form of the all-pass phase lag circuit is shown in Figure 5–23. The circuit can be analyzed most readily by noting that since the input signal v_i is connected to both signal paths, the circuit is equivalent to one having two identical inputs. The circuit converted to this form, along with the frequency domain representation, is shown in Figure 5–24(a). In this form, the circuit has the same structure as the closed-loop differential amplifier. However, the lower path is a frequency-dependent process as a result of the capacitance. Superposition will be used for the analysis. Let

$$\overline{V}_o = \overline{V}'_o + \overline{V}''_o \tag{5–97}$$

where $\overline{V}'_o$ is produced by the lower equivalent source, and $\overline{V}''_o$ is produced by the upper equivalent source.

FIGURE 5–23
All-pass phase lag circuit.

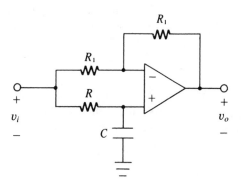

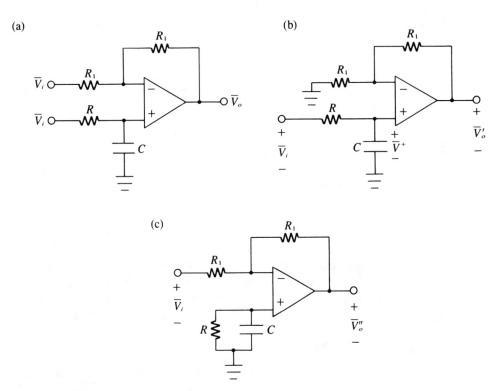

FIGURE 5–24
Analysis of all-pass phase lag circuit using superposition.

Consider first the effect due to the lower source, and thus the upper source is de-energized; that is, it is replaced by a short circuit as shown in Figure 5–24(b). The voltage $\overline{V}^+$ at the noninverting input is then determined by the voltage divider rule as

$$\overline{V}^+ = \frac{1/j\omega C}{R+1/j\omega C} \times \overline{V}_i = \frac{\overline{V}_i}{1 + j\omega RC} \tag{5–98}$$

From the noninverting terminal on to the output, the circuit is just a noninverting amplifier, so the output component $\overline{V}'_o$ is

$$\overline{V}'_o = \left(1 + \frac{R_1}{R_1}\right) \times \overline{V}^+ = 2\overline{V}^+ = \frac{2\overline{V}_i}{1 + j\omega RC} \tag{5–99}$$

The effect due to the upper source is determined by first deenergizing the lower source as shown in Figure 5–24(c). This circuit is simply an inverting amplifier, so the output component $\overline{V}''_o$ is

$$\overline{V}''_o = \frac{-R_1}{R_1} \overline{V}_i = -\overline{V}_i \tag{5–100}$$

Substituting (5–99) and (5–100) in (5–97), we find that the net output is

$$\overline{V}_o = \frac{2\,\overline{V}_i}{1 + j\omega RC} - \overline{V}_i = \frac{1 - j\omega RC}{1 + j\omega RC}\,\overline{V}_i \tag{5-101}$$

The transfer function $H(j\omega)$ is

$$H(j\omega) = \frac{\overline{V}_o}{\overline{V}_i} = \frac{1 - j\omega RC}{1 + j\omega RC} \tag{5-102}$$

The amplitude response $M(\omega)$ corresponding to (5–102) is

$$M(\omega) = \frac{\sqrt{1 + (\omega RC)^2}}{\sqrt{1 + (\omega RC)^2}} = 1 \tag{5-103}$$

The phase response $\theta(\omega)$ is

$$\theta(\omega) = -\tan^{-1}\omega RC - \tan^{-1}\omega RC = -2\tan^{-1}\omega RC \tag{5-104}$$

The forms of the amplitude and phase functions are shown in Figure 5–25.

The amplitude response is observed to have a constant value of unity at all frequencies. The description as an *all-pass* circuit is thus quite appropriate. Actually, the finite open-loop bandwidth of the op-amp causes the amplitude response eventually to drop in accordance with the analysis of Chapter 4, so an op-amp having sufficient bandwidth must be selected in a given application.

From (5–102) and Figure 5–25(b), the phase shift of the circuit is lagging and varies from 0° to −180° over an infinite frequency range. Note that the phase shift is a function of the product ωRC. For a given frequency, the RC product can be determined to provide a given phase shift. Conversely, for a given RC product, a frequency can be determined at which the phase shift will be a specified value. By varying one of the component values, one can adjust the phase shift at a given frequency.

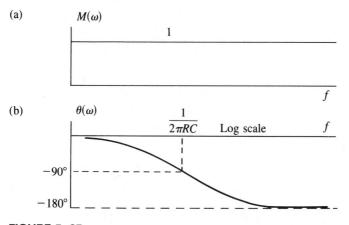

FIGURE 5–25
Amplitude and phase of all-pass phase lag circuit.

FIGURE 5–26
All-pass phase lead circuit.

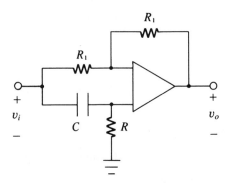

All-Pass Phase Lead Circuit

The form of the all-pass phase lead circuit is shown in Figure 5–26. An analysis of this circuit follows the same procedure as for the phase lag circuit, and the details will be left as an exercise (Problem 5–28). However, the results will be summarized here.

The transfer function for the phase lead circuit is

$$H(j\omega) = \frac{-1 + j\omega RC}{1 + j\omega RC} \qquad \textbf{(5–105)}$$

The amplitude response is the same as for the phase lag circuit:

$$M(\omega) = 1 \qquad \textbf{(5–106)}$$

The phase response for this circuit is

$$\theta(\omega) = 180° - 2\,\tan^{-1}\omega RC \qquad \textbf{(5–107)}$$

The form of the phase shift is shown in Figure 5–27. From (5–107) and this figure, we observe that the phase shift is leading and varies from 180° to 0° over an infinite frequency range. Observe also that the general shape and slope of the phase shift for the phase lead circuit are the same as for the phase lag circuit. However, the beginning and end points are quite different.

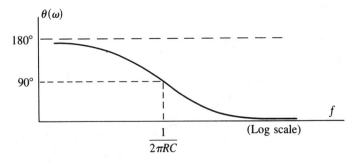

FIGURE 5–27
Phase of all-pass lead circuit. (Amplitude is same as in phase lag circuit.)

FIGURE 5-28
Circuit of Example 5-11.

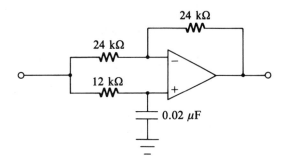

EXAMPLE 5-11

For the circuit of Figure 5–28, determine (a) amplitude response $M(\omega)$, (b) phase response $\theta(\omega)$, and (c) phase shift at 1 kHz.

Solution

By comparing the specific circuit of Figure 5–28 to the general form of Figure 5–23, it is deduced that the circuit is an all-pass phase lag circuit with $R_1 = 24$ kΩ, $R = 12$ kΩ, and $C = 0.02$ μF.

(a) The ideal amplitude response is

$$M(\omega) = 1 \tag{5-108}$$

Finite bandwidth will, of course, eventually cause a roll-off effect.

(b) The phase response is

$$\theta(\omega) = -2 \tan^{-1}\omega RC = -2 \tan^{-1}2\pi f \times 12 \times 10^3 \times 0.02 \\ \times 10^{-6} = -2 \tan^{-1}0.00151f \tag{5-109}$$

(c) The phase shift at the specific frequency of 1 kHz is

$$\theta = -2 \tan^{-1}0.00151 \times 10^3 = -2 \tan^{-1}1.51 = -2 \times 56.45° \\ = -112.9° \tag{5-110}$$

EXAMPLE 5-12

Design an all-pass phase lag circuit to produce a phase shift of $-135°$ at a frequency of 1 kHz.

Solution

Refer to the basic form of the phase lag circuit in Figure 5–23 and the equation for the corresponding phase lag as given by (5–104). The choice of R_1 is arbitrary, so the common value $R_1 = 10$ kΩ will be selected. We next equate the desired phase shift equal to the function of (5–104) with $\omega = 2\pi \times 10^3$ rad/s. The phase shift is considered as $-135°$ in the substitution, since it is a phase lag. The details follow:

$$-2 \tan^{-1} 2\pi \times 10^3 RC = -135°$$
$$\tan^{-1} 2\pi \times 10^3 RC = 67.5° \tag{5-111}$$
$$2\pi \times 10^3 RC = \tan 67.5° = 2.414$$

FIGURE 5-29
Circuit of Example 5-12.

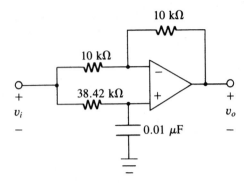

The required value of the RC product is thus

$$RC = 384.2 \times 10^{-6} \qquad \text{(5–112)}$$

Theoretically, there is an infinite number of values of R and C that will satisfy the required product. Since it is usually easier to adjust R than C in the given frequency range, the best approach is to select a standard value of C and determine the value of R required. As a reasonable choice, $C = 0.01$ μF will be selected. The corresponding value of R is determined from (5–112) to be $R = 38.420$ kΩ. An adjustable resistance could be used to establish the required value. Actually, since the capacitance value will have some error, the best procedure is to "tweak" the circuit to the correct phase shift by a measurement process with an adjustable resistance. In this manner, the resistance can correct for an error in the capacitance value. The circuit design is shown in Figure 5–29.

5-8 SINGLE POWER SUPPLY OPERATION

Operational amplifiers are directly coupled devices (that is, they have no series capacitors) and are capable of amplifying signal frequency components all the way down to dc. In order that the output voltage be zero when no input voltage is applied, most op-amps utilize two power supplies as discussed in Chapter 2. (A few single power supply, special-purpose op-amps having zero output voltage with no signal applied have appeared on the market, but the vast majority conform to the standard pattern noted earlier.)

Recall from Chapter 2 that the standard bias power supply connections involved a positive voltage connected to one op-amp terminal and a negative voltage applied to the other terminal. (Refer to Figure 2–2 if necessary.) The midpoint between the power supplies is established as the common ground, and this terminal is not connected to the op-amp itself. Rather, all input and output voltages are referred to this common ground reference. For most op-amps, a symmetrical bias arrangement (for example, ±15 V) results in nearly zero output voltage with no input. Small deviations from this zero level are likely a result of input offset voltage and bias currents as discussed in Chapter 3, and means are available for nullifying, or at least minimizing, these effects.

For the actual internal operation of the op-amp, a single power supply connected between the positive bias and the negative bias terminals (indicated here as $+V$ and $-V$) will suffice. However, the major difficulty is in establishing the common ground reference point for input and output voltages. There is no common point between power supplies when only one supply is used, so an artificial means must be used for establishing a common ground reference.

Operation with a single power supply is not too practical with general-purpose op-amps if dc response is required because of the difficulty associated with the undesired dc levels in the circuit. However, where the response at dc and very low frequencies can be sacrificed, practical single power supply circuits can be readily implemented. For example, many audio application requirements can be met with this approach. The technique involves coupling capacitors at both the input and output to remove the dc levels. In a sense, this type of operation has a strong similarity to traditional transistor ac-coupled amplifiers where a single power supply and coupling capacitors (plus bypass capacitors) are used.

The circuits that will be considered in subsequent developments will be established to allow equal peak signal levels on either side of the reference operating points. This situation is the most common and would likely be employed if no special information on special signal character were available. However, it should be clear from the development how the designs could be modified to produce a nonsymmetrical signal condition if desired.

Inverting Amplifier Connection

The first circuit to be considered is the ***inverting amplifier ac*** connection shown in Figure 5–30. A single power supply with voltage V_B is connected between the bias terminals indicated as $+V$ and $-V$. The $-V$ terminal is arbitrarily established as the ground reference, but the circuit may be readily modified to establish the other power supply terminal as the ground reference.

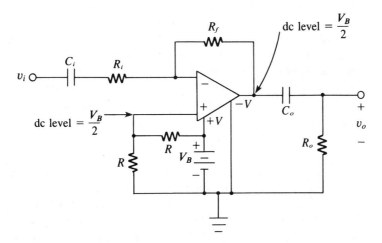

FIGURE 5–30
Inverting amplifier connected for single power supply, ac operation.

FIGURE 5–31
dc and ideal signal equivalent circuits for inverting ac amplifier.

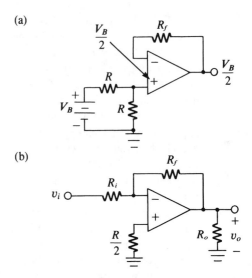

(a)

(b)

This circuit is best explained by observing both the dc circuit and the signal equivalent circuit. The essential elements of the dc circuit are shown in Figure 5–31(a). The voltage divider consisting of the two equal resistances of value R establishes a dc voltage $V_B/2$ at the noninverting op-amp input terminal. The op-amp functions as a voltage follower for this dc level, and thus the op-amp dc output is also $V_B/2$. Note that the capacitor C_i is absolutely essential on the inverting side. Without it, the circuit would attempt to multiply the dc level by $1 + R_f/R_i$, which would cause saturation under most conditions.

Input signals are coupled to the noninverting input resistance R_i through the capacitance C_i. Such time-varying signals cause a signal current to flow alternately up and down through R_f, and the voltage at the inverting input terminal varies instantaneously about its dc level. Feedback forces the op-amp output to vary in direct proportion about its dc level. The resulting signal component at the op-amp output is coupled to the load resistance R_o by the output capacitor C_o, which removes the dc level.

For the frequency range in which the capacitive reactances are negligible, the signal circuit is shown in Figure 5–31(b). The gain A_{CL} in this frequency range is simply

$$A_{CL} = \frac{v_o}{v_i} = -\frac{R_f}{R_i} \qquad (5\text{–}113)$$

Thus, the signal component at the output is inverted with respect to the signal at the input, which is a basic property of an inverting amplifier.

As the frequency decreases below a certain range, the series reactance of the input capacitance C_i increases, thus decreasing the net gain. Simultaneously, the series reactance of the output capacitance C_o increases, causing a further loss of gain. Thus, both capacitors contribute to a low-frequency roll-off of the gain. At the extreme point of zero frequency (dc), the net signal gain is zero, as already noted.

Selecting Capacitances

To maintain a relatively flat response over the desired band of frequencies, the capacitive reactances of the two capacitors must each be small compared to the corresponding series resistances at the lowest signal frequency. Let f_L represent the lowest frequency of interest. We then require that

$$\frac{1}{2\pi f_L C_i} \ll R_i \qquad\qquad (5\text{--}114)$$

and

$$\frac{1}{2\pi f_L C_o} \ll R_o \qquad\qquad (5\text{--}115)$$

Solving (5–114) and (5–115) for the corresponding capacitance inequalities, we have

$$C_i \gg \frac{1}{2\pi f_L R_i} \qquad\qquad (5\text{--}116)$$

and

$$C_o \gg \frac{1}{2\pi f_L R_o} \qquad\qquad (5\text{--}117)$$

Typical choices on the inequality range are ten or greater, but there is no "exact" value that will work in all cases. Bode plot analysis (or computer programs) may be used to study the actual amplitude response roll-off on a more exact basis. Note that using larger resistances for R_i and R_o will allow use of smaller capacitances for a given inequality ratio.

Saturation Levels

With the same assumptions on saturation made in Chapter 2, the linear range of op-amp output voltages varies from about 2 V to about $V_B - 2$ V. For example, if a single 15-V bias supply is used, the linear range of op-amp output voltage is from about 2 V to about 13 V, corresponding to a peak-to-peak dynamic range of 11 V. Obviously, for the same peak-to-peak range as in dual power supply operation, the voltage of the single supply should equal the sum of the two separate dual supply voltages.

Noninverting Amplifier Connection

A noninverting amplifier connected for single power supply, ac operation is shown in Figure 5–32. The dc equivalent circuit is the same as for the inverting amplifier and will not be repeated. [Refer to Figure 5–31(a)]. The output dc level is again $(V_B)/2$.

Operation of the noninverting circuit is very similar to that of the inverting amplifier, except that the signal is couple to the noninverting input through the capacitor C. In the

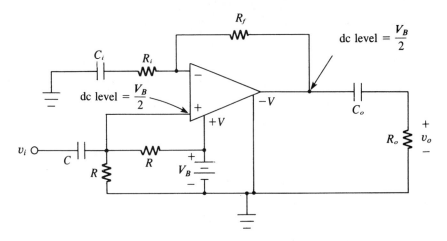

FIGURE 5–32
Noninverting amplifier connected for single power supply, ac operation.

frequency range where all capacitive reactances are negligible, the signal equivalent circuit is as shown in Figure 5–33. The gain A_{CL} is readily written as

$$A_{CL} = 1 + \frac{R_f}{R_i} \tag{5-118}$$

Three capacitors are required with the noninverting configuration, compared with two for the inverting circuit. Consequently, three reactances must be selected in this case. The portions of the circuit affecting C_i and C_o are the same forms as for the inverting amplifier, so the inequalities of (5–114) and (5–115) leading to (5–116) and (5–117) can be stated here as well. The additional inequality relates to the input branch consisting primarily of C and the two resistances of value R.

The effective resistance for the signal path is $R/2$, so the reactance of C must satisfy

$$\frac{1}{2\pi f_L C} \ll \frac{R}{2} \tag{5-119}$$

FIGURE 5–33
Ideal signal equivalent circuit for inverting
ac amplifier.

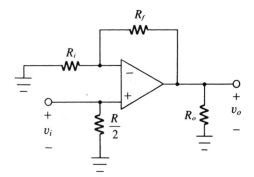

which leads to

$$C \gg \frac{1}{\pi f_L R} \tag{5-120}$$

The discussion earlier in this section of saturation levels, as considered for the inverting amplifier, applies to the noninverting amplifier as well.

One point about ac-coupled circuits is that dc offset voltage and current effects are removed by the output capacitor and *usually* do not cause any problems. Occasionally, however, if operation approaches one of the peak limits of the output, an offset could cause one limit of the output signal to saturate. This situation would indicate a rather marginal design, of course.

EXAMPLE 5–13

An ac-coupled inverting amplifier circuit is to be designed for an application in which a single 24-V power supply (negative terminal grounded) is available. The desired magnitude of the gain is 15. To minimize loading on the source, the input impedance must be at least 10 kΩ. The output will be coupled to a circuit having a resistive input impedance of 2.5 kΩ. As an arbitrary low-frequency design criterion, the reactances of all capacitors are required to be no greater than one-tenth the values of the corresponding series resistances at 50 Hz. (a) Specify components for the design. (b) What are the approximate linear limits of the input and output voltages?

Solution

(a) The design will have the same form as the circuit of Figure 5–30 with $V_B = 24$ V, so direct reference to the terminology of that circuit will be made in the development that follows. The minimum value of the input impedance is R_i, and this value occurs in the frequency range where the reactance of C_i is negligible. As a starting point, $R_i = 10$ kΩ will be selected, since it satisfies the input impedance requirement. The gain magnitude of 15 is then achieved by selecting $R_f = 150$ kΩ. The factor of 10 for the reactance of C_i at 50 Hz applied to either (5–114) or (5–116) yields

$$C_i \geq \frac{10}{2\pi f_L R_i} \tag{5-121}$$

With $f_L = 50$ Hz and $R_i = 10$ kΩ substituted in (5–121), we obtain

$$C_i \geq 3.183 \ \mu\text{F} \tag{5-122}$$

A reasonable choice in a standard value would be a 3.3-μF capacitor. The resistances R_i and R_f could be increased by some factor, and C_i could be reduced by the same factor if desired.

Since the output is to be connected to a fixed resistive load, we can interpret that resistance to be R_o, which establishes $R_o = 2.5$ kΩ. Unlike the input circuit, where the impedance level could be chosen somewhat arbitrarily, the output circuit impedance level is established by the external load. The factor of 10 for the reactance of C_o at 50 Hz applied to either (5–115) or (5–117) yields

$$C_o \geq \frac{10}{2\pi f_L R_o} \tag{5-123}$$

Substituting $f_L = 50$ Hz and $R_o = 2.5$ kΩ in (5–123), we obtain

$$C_o \geq 12.7 \ \mu\text{F} \qquad\qquad\qquad (5\text{–}124)$$

A reasonable choice in a standard value would be a 15-μF capacitor.

The only remaining components to be selected are the power supply voltage divider resistors with value R.

The actual value of R is not critical as long as both resistances are equal. In the possible event that bias currents could have noticeable effects, it seems reasonable to select R to minimize these effects. The dc path from the inverting input is 150 kΩ, and this value suggests $R = 300$ kΩ for the two resistors. The final design values are

$$R_i = 10 \text{ k}\Omega$$
$$R_f = 150 \text{ k}\Omega$$
$$R\ = 300 \text{ k}\Omega \quad \text{(two values)}$$
$$R_o = 2.5 \text{ k}\Omega \quad \text{(external load)}$$
$$C_i = 3.3 \ \mu\text{F}$$
$$C_o = 15 \ \mu\text{F}$$

Electrolytic capacitors will likely be required, so proper polarities must be observed.

(b) Allowing an approximate 2-V "backoff" from end limits, the op-amp linear swing is from about 2 V to about 22 V, corresponding to a peak-to-peak range of 20 V. After the dc level is removed, the output linear voltage swing is from about -10 V to $+10$ V. Since the output signal voltage magnitude is 15 times the input voltage magnitude, the corresponding input voltage swing is from -0.667 V to 0.667 V.

5–9 MULTISIM EXAMPLES

The examples in this section are employed to illustrate the operation of several of the circuits considered in this chapter. The circuits considered are the ac integrator, the low-frequency differentiator, and the all-pass phase lag circuit.

MULTISIM EXAMPLE 5–1

Using the 741 model, develop a Multisim model for the ac integrator circuit designed in Example 5–6, and use it to analyze the response to the square wave of Example 5–8. Perform two separate analyses on the circuit: (a) during startup with zero initial voltage on the capacitor and (b) after steady-state conditions have been reached.

Solution

The circuit in the text was shown in Figure 5–16 and has been reconstructed as a Multisim schematic in Figure 5–34, but with the compensating resistance eliminated. The reader will note a change in the manner in which the power supply connections have been made here relative to the approach used earlier in the text. The form shown here will be utilized extensively through the remainder of the book.

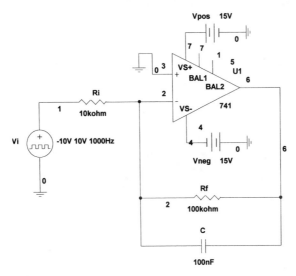

FIGURE 5-34
Circuit of Multisim Example 5-1.

The input source is the **PULSE-VOLTAGE-SOURCE**, whose format was first discussed in Multisim Example 4–7. However, in the present example, the **Initial Voltage** is set to −10 V and the **Pulsed Value** is set to 10 V. Since a frequency of 1 kHz is desired, the **Pulse Width** is set to 0.5 ms and the **Period** is set to 1 ms. The **Rise Time** and **Fall Time** values are arbitrarily set to 1 ns.

(a) In this run, we will start at $t = 0$ and set the initial voltage on the capacitor to 0. After the **Transient Analysis** window is opened, set the **Initial conditions** option to **Set to zero.** After some experimentation, it was decided to run from a **Start time** of 0 to an **End time** of 20 ms (0.02 s), corresponding to 20 cycles. The **Maximum time step** was set to 10 μs. Plots of both the input voltage and the output voltage are shown in Figure 5–35. The dashed line represents the input square wave and the solid line represents the output voltage. Note that there is an exponential upward trend as the output moves from its initial form toward its final form.

(b) As a means of estimating when final steady-state conditions should be reached, the time constant of the feedback resistance and capacitance is used as a basis. This value was selected in Example 5–6 as 10 ms, although it could be quickly calculated again as $RC = 10^5 \times 10^{-7} = 10$ ms. We will assume that the steady-state will be reached in 5 time constants or 50 ms. Therefore, the **Start time** is set to 50 ms (0.05 s). We will choose now to show two cycles, and the **End time** is set to 52 ms (0.052 s). It should be emphasized that these are the time limits of the display, not the time limits of the computation. The actual computation begins at $t = 0$.

The results of this run are shown in Figure 5–36. These curves are seen to be in agreement with the predicted response curves of Figure 5–17.

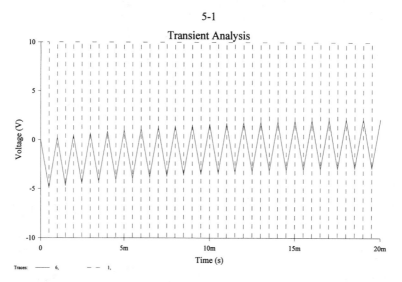

FIGURE 5–35
Waveforms of Multisim Example 5–1 during initial transient interval.

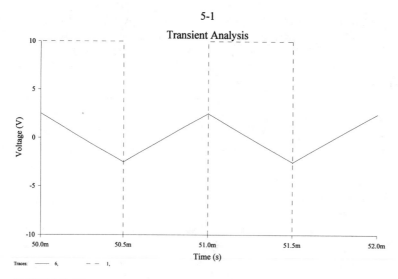

FIGURE 5–36
Waveforms of Multisim Example 5–1 after steady-state conditions are reached

MULTISIM EXAMPLE 5–2

Using the 741 model, develop a Multisim model for the low-frequency differentiator circuit designed in Example 5–9 and use it to analyze the response to the triangular waveform of Example 5–10.

Solution

The circuit in the text was shown in Figure 5–21 and it has been reconstructed as a Multisim schematic in Figure 5–37. Although the pulse voltage source could be used by increasing the rise and fall times to create a triangular effect, we will illustrate still another waveform generator in this example. Near the bottom of the **Sources** bin is one called the **PIECEWISE_LINEAR_VOLTAGE_SOURCE,** which we will hereafter refer to as the **PWL** source. This source can be used to establish a waveform created by a series of straight-line segments. The data required is a pair of values at each point providing the time and the voltage corresponding to the point.

A double left-click on the source causes a window with the title **PWL Voltage** to open. Left-click on the **Value** tab to open the data field. Although one could use an external data file, the default is the **Enter Points** option. Two columns with titles **Time** and **Voltage** appear on the right. Along each row enter the time in seconds and the voltage in volts representing the points at which the line segments join.

For the triangular waveform of Figure 5–22(a), the points for two cycles are entered as follows:

0	2.5
0.5e−3	−2.5
1e−3	2.5
1.5e−3	−2.5
2e−3	2.5

After entering all the data, left-click on **OK.**

A **Transient Analysis** is performed and the input and the waveforms are shown in Figure 5–38. While the +10-V and −10-V flat areas of the curve are in agreement, there is a pronounced ringing effect that occurs, and it even appears to momentarily drive the op-amp into saturation beginning with the second half-cycle of the input. Differentiation can be a troublesome process to achieve with a practical circuit and the limitations of the model may also contribute.

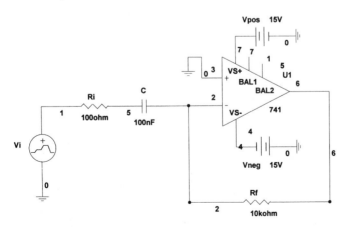

FIGURE 5-37

Circuit of Multisim Example 5-2.

When a 2-nF capacitor was placed in parallel with the feedback resistance, the response of Figure 5–39 was achieved. The ringing and saturation effects are eliminated at the expense of increased rise time.

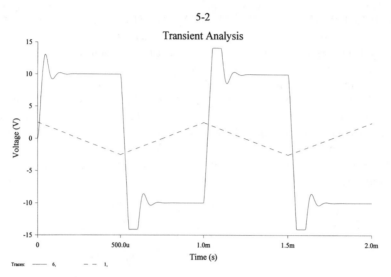

FIGURE 5–38
Waveforms of Multisim Example 5–2 based on schematic of Figure 5–37.

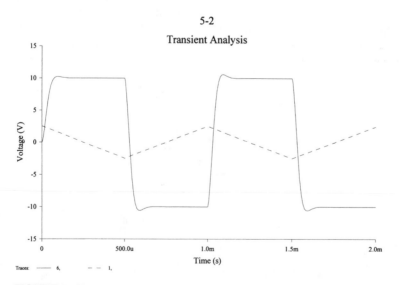

FIGURE 5–39
Waveforms of Multisim Example 5–2 with 2-nF capacitor placed in parallel will feedback resistor.

MULTISIM EXAMPLE 5–3

Using the 741 model, develop a Multisim model for the all-pass phase lag circuit designed in Example 5–12. Perform a frequency response analysis over the frequency range from 100 Hz to 10 kHz.

Solution

The circuit in the text was shown in Figure 5–29 and has been reconstructed as a Multisim schematic in Figure 5–40. The input is now a sinusoidal voltage source and an **AC Analysis** is required. The results of the analysis from 100 Hz to 10 kHz are shown in Figure 5–41. The **Range** of the amplitude response graph was changed from its initial values to a range of -2×10^{-3} dB to 0 dB.

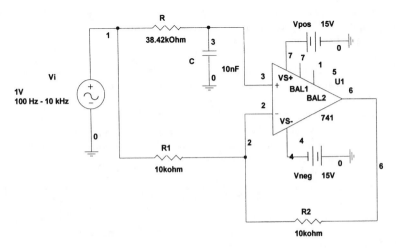

FIGURE 5–40
Circuit of Multisim Example 5–3.

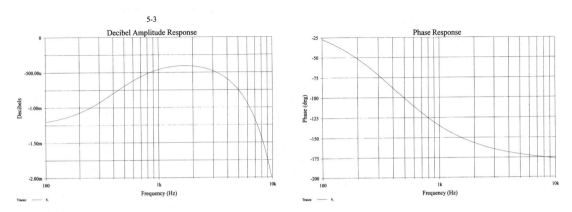

FIGURE 5–41
Frequency response of Multisim Example 5–3.

It is immediately observed that the amplitude response is not exactly a constant as expected and is a little below 0 dB. However, one must judge the response on a relative basis. Over the two decades, the response varies only by less than 2 mdB (2×10^{-3} dB) and is down from the level of 0 dB by about 1 mdB (1×10^{-3} dB). For most practical purposes, these deviations would be negligible.

The phase response curve follows the expected pattern and has the required phase shift at 1 kHz.

SUMMARY

1. An instrumentation amplifier is a high-quality unit having differential balanced inputs, high input impedances, and a high common-mode rejection ratio.
2. The gain for both inputs of an instrumentation amplifier may be changed by adjusting one resistance.
3. An instrumentation amplifier reduces significantly the effects of common-mode noise pickup.
4. A true integrator circuit produces an output proportional to the area underneath the curve of the input signal.
5. Op-amps used to implement true integrators must have ultra-low offset and bias parameters; general-purpose op-amps are usually unsatisfactory for this purpose.
6. Circuits that integrate only the ac components of the input signal ("ac" integrators) may be satisfactorily implemented with general-purpose op-amps.
7. A true differentiator circuit produces an output proportional to the slope of the input signal.
8. True differentiator circuits tend to be noisy due to the accentuation of high-frequency components of the signal.
9. Circuits that differentiate only the low-frequency components of the input signal ("low-frequency" differentiators) may be satisfactorily used for some applications.
10. Various transfer functions may be created with a general inverting gain circuit by using different combinations of impedances for input and feedback components.
11. All-pass phase lag circuits may be used to shift the phase of a signal from 0° to $-180°$ at any arbitrary frequency.
12. All-pass phase lead circuits may be used to shift the phase of a signal from 0° to $+180°$ at any arbitrary frequency.
13. An op-amp may be powered by a single power supply if the input and output dc levels can be dealt with satisfactorily.
14. Most circuits using a single power supply are used for ac signal applications, in which case capacitors are used at both input and output for blocking dc levels.

PROBLEMS

Drill Problems

5–1. The circuit of Figure P5–1 has $R_1 = R_2 = R_3 = 10$ kΩ and $R_x = 2.5$ kΩ. For $v_1 = 5$ V and $v_2 = 4.2$ V, determine
    a. v_x **b.** i_x **c.** v_r **d.** v_a **e.** v_b **f.** v_o

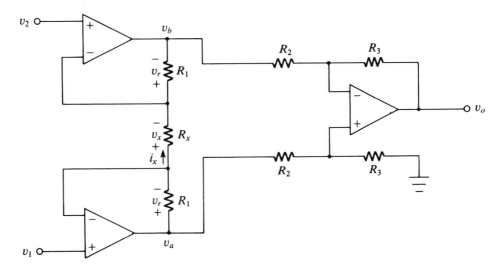

FIGURE P5–1

5–2. The circuit of Figure P5–1 has $R_1 = R_2 = 10$ kΩ, $R_3 = 20$ kΩ, and $R_x = 5$ kΩ. For $v_1 = 3$ V and $v_2 = 3.5$ V, determine

a. v_x **b.** i_x **c.** v_r **d.** v_a **e.** v_b **f.** v_o

5–3. The circuit of Figure P5–1 has $R_1 = R_2 = R_3 = 10$ kΩ. Determine the value of R_x required to establish the output as

$$v_o = 30(v_1 - v_2)$$

5–4. The circuit of Figure P5–1 has $R_1 = R_2 = 10$ kΩ and $R_3 = 20$ kΩ. Determine the value of R_x required to establish the output as

$$v_o = 100(v_1 - v_2)$$

5–5. A certain instrumentation amplifier has a gain of 20 dB and a common-mode rejection ratio of 80 dB. It is used in a noisy environment in which the signal has a level of 600 mV, and the common-mode noise level is 1 V. Determine
a. common-mode gain
b. signal output
c. noise output
d. output signal voltage-to-noise voltage ratio

5–6. A certain instrumentation amplifier has a gain of 80 and a common-mode rejection ratio of 50,000. It is used in a noisy environment in which the signal has a level of 100 mV, and the common-mode noise level is 40 mV. Determine
a. common-mode gain
b. signal output
c. noise output
d. output signal voltage-to-noise voltage ratio

5-7. For the waveform of Figure P5–7, assuming that $V_p = 80$ V and $T = 200$ ms. It is applied to a true integrator circuit whose output is

$$v_o(t) = \int_0^t v_i(t)\, dt + v_o(0)$$

Starting at $t = 0$ and with $v_o(0) = 0$, sketch the output voltage and label significant levels.

FIGURE P5–7

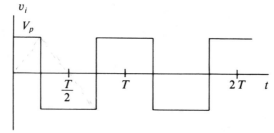

5-8. For the waveform of Figure P5–7, assume that $V_p = 60$ V and $T = 8$ ms. It is applied to an integrator circuit whose output is

$$v_o(t) = 100 \int_0^t v_i(t)\, dt + v_o(0)$$

Starting at $t = 0$ and with $v_o(0) = 0$, sketch the output voltage and label significant levels.

5-9. For the waveform of Figure P5–9, assume that $V_p = 40$ V, $t_1 = 4$ s, $t_2 = 12$ s, and $t_3 = 14$ s. It is applied to a true differentiator circuit whose output is

$$v_o(t) = \frac{dv_i(t)}{dt}$$

Sketch the output voltage and label significant levels.

FIGURE P5–9

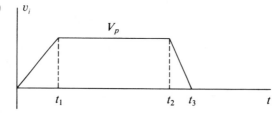

5-10. For the waveform of Figure P5–9, assume that $V_p = 16$ V, $t_1 = 20$ ms, $t_2 = 60$ ms, and $t_3 = 70$ ms. It is applied to a differentiator circuit whose output is

$$v_o(t) = 0.005 \frac{dv_i(t)}{dt}$$

Sketch the output voltage and label significant levels.

5-11. Consider the circuit of Figure P5–11(a) with $C = 0.02\ \mu$F, $R_i = 100$ kΩ, and $R_f = 470$ kΩ. The input to the circuit is the waveform of Figure P5–11(b) with $T = 0.5$ ms and $V_p = 8$ V. Under steady-state conditions, determine the peak-to-peak value of the output voltage.

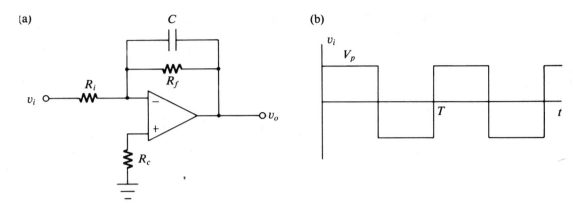

FIGURE P5–11

5–12. Consider the circuit of Figure P5–11(a) with $C = 0.1\ \mu F$, $R_i = 100\ k\Omega$, and $R_f = 1\ M\Omega$. The input to the circuit is the waveform of Figure P5–11(b) with $T = 10$ ms and $V_p = 12$ V. Under steady-state conditions, determine the peak-to-peak value of the output voltage.

5–13. For the circuit and waveform of Problem 5–11, plot the output voltage after steady-state conditions are reached.

5–14. For the circuit and waveform of Problem 5–12, plot the output voltage after steady-state conditions are reached.

5–15. Suppose that in the circuit of Problem 5–11 the input square wave were shifted slightly so that it contained a dc level of 0.5 V. At what voltage level will the output be centered?

5–16. Suppose that in the circuit of Problem 5–12 the input square wave were shifted slightly so that it contained a dc level of −0.7 V. At what voltage level will the output be centered?

5–17. Consider the circuit of Figure P5–17(a) with $R_i = 470\ \Omega$, $R_f = 20\ k\Omega$, and $C = 0.2\ \mu F$. The input to the circuit is the waveform of Figure P5–17(b) with $T = 10$ ms and $V_p = 4$ V. Under steady-state conditions, determine the peak-to-peak value of the output voltage.

FIGURE P5–17

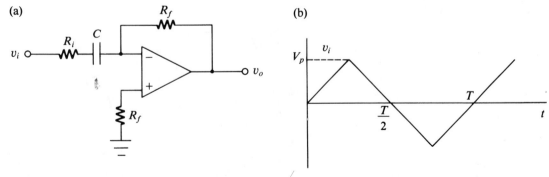

5–18. Consider the circuit of Figure P5–17(a) with $R_i = 4.7\ k\Omega$, $R_f = 100\ k\Omega$, and $C = 0.1\ \mu F$. The input to the circuit is the waveform of Figure P5–17(b) with $T = 50$ ms and $V_p = 10$ V. Under steady-state conditions, determine the peak-to-peak value of the output voltage.

5–19. For the circuit and waveform of Problem 5–17, plot the output voltage after steady-state conditions are reached.

5–20. For the circuit and waveform of Problem 5–18, plot the output voltage after steady-state conditions are reached.

5–21. Consider the circuit of Figure P5–21 with $R = 20$ kΩ, $C = 0.01$ μF, and $R_1 = 43$ kΩ. Determine the phase shift at
    a. 500 Hz   **b.** 1 kHz   **c.** 2 kHz

FIGURE P5–21

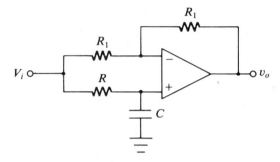

5–22. Consider the circuit of Figure P5–21 with $C = 0.02$ μF and $R_1 = 10$ kΩ. At a frequency of 1 kHz, determine the phase shift for
    a. $R = 5$ kΩ **b.** $R = 10$ kΩ **c.** $R = 20$ kΩ

5–23. Assume in the circuit of Problem 5–21 that R and C are interchanged. Repeat the calculations of Problem 5–21.

5–24. Assume in the circuit of Problem 5–22 that R and C are interchanged. Repeat the calculations of Problem 5–22.

Derivation Problems

5–25. Consider the circuit shown in Figure P5–25.
    a. Determine the transfer function $H(j\omega)$.
    b. Show that the circuit is *theoretically* capable of operation as a true integrator.
    (*Note*: This circuit has serious practical limitations due to the imperfect nature of inductors at low frequencies.)

FIGURE P5–25

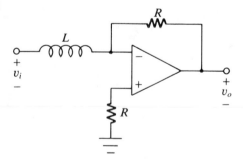

5–26. Consider the circuit shown in Figure P5–26.
    a. Determine the transfer function $H(j\omega)$.
    b. Show that the circuit is *theoretically* capable of operation as a true differentiator.
    (See note at end of Problem 5–25.)

FIGURE P5–26

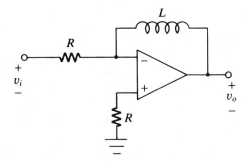

5–27. The circuit of Figure P5–27 represents an extension of the inverting gain circuit to include three-terminal circuits. The quantity Y_{21} is the *short-circuit transfer admittance*. For the input and feedback networks, the following definitions apply:

$$\overline{Y}_{21i} = \frac{\overline{I}_{2i}}{\overline{V}_i}$$

and

$$\overline{Y}_{21f} = \frac{\overline{I}_{1f}}{\overline{V}_o}$$

FIGURE P5–27

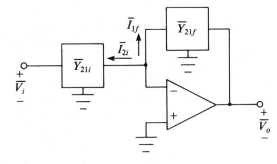

Note the defined directions of the currents. Show that

$$H(j\omega) = -\frac{\overline{Y}_{21i}}{\overline{Y}_{21f}}$$

5–28. Determine the transfer function $H(j\omega)$ of the all-pass phase lead circuit shown in Figure 5–26. The procedure follows the same form as for the all-pass lag circuit given in the text. Show that the amplitude and phase functions are given by Equations (5–106) and (5–107), respectively.

Design Problems

5–29. An ac precision integrator is desired for a particular application to perform the operation

$$v_o'(t) = -2000 \int v_i'(t) \, dt$$

where the primes indicate the ac portions of the respective functions. The lowest frequency (other than a possible dc component) of the input signal is estimated to be 2 kHz. Determine a suitable design based on the criterion $\tau_f = 10T_L$.

5–30. Suppose that in Problem 5–29 the criterion $f_b = 0.01\ f_L$ is employed. Determine a new design.

5–31. A low-frequency differentiator is desired for a particular application to perform the operation

$$v_o(t) = -0.002\ \frac{dv_i(t)}{dt}$$

Based on a periodic signal with a frequency of 500 Hz, determine a suitable design using the criterion $\tau_i = 0.01T$.

5–32. Suppose in Problem 5–31 that the criterion $f_b = 10f$ is employed. Determine a new design.

5–33. Design an all-pass phase lag circuit to produce a phase shift of $-90°$ at a frequency of 1 kHz.

5–34. Design an all-pass phase lag circuit to produce a phase shift of $-45°$ at a frequency of 1 kHz.

5–35. Design an all-pass phase lead circuit to produce a phase shift of $+90°$ at a frequency of 1 kHz.

5–36. Design an all-pass phase lead circuit to produce a phase shift of $+45°$ at a frequency of 1 kHz.

5–37. Design a phase lag network that is capable of adjusting the phase shift from $-30°$ to $-150°$ at a frequency of 1 kHz. Using a fixed capacitance, specify the range of the variable resistance.

5–38. Design a phase lag network that is capable of being adjusted to a phase shift of $-90°$ as the frequency is varied from 200 Hz to 1 kHz. Using a fixed capacitance, specify the range of the variable resistance.

5–39. In a certain all-pass phase lag circuit, it is desired to be able to adjust the phase shift from $-45°$ to $-135°$ at any frequency in the range from 500 Hz to 2 kHz. If $C = 0.01\ \mu F$, determine the range in the variable resistance that will permit the adjustments involved.

5–40. Assume in Problem 5–39 that the desired phase shift is from $+45°$ to $+135°$, but all other conditions are the same. Determine the range in the variable resistance.

5–41. An ac-coupled inverting amplifier circuit is to be designed for an application in which a single 30-V power supply (negative terminal grounded) is available. The desired signal gain is -10. To minimize loading on the source, the input impedance must be at least 20 kΩ. The output will be coupled to a circuit having a resistive impedance of 5 kΩ. As an arbitrary low-frequency design criterion, the reactances of all capacitors cannot exceed one-tenth the values of the corresponding series resistances at 25 Hz.

    a. Specify components for the design.

    b. What are the approximate linear limits of the output and input voltages?

5–42. If in the design of Problem 5–41, suppose that a noninverting gain of 25 were desired. Assuming all other conditions and specifications given in Problem 5–41, perform a new design.

LABORATORY EXERCISES

ac Integrator
LE 5–1.

1. Connect the ac integrator circuit of Example 5–6 (Figure 5–16).

2. Measure the 3-dB break frequency f_b and compare with the value calculated in the text.

3. Measure the amplitude response over the range from below the break frequency to several kilohertz.
4. Plot the amplitude response on *log-log* paper, and show that the response approaches a straight line with a negative slope well above the break frequency.
5. Apply a 1-kHz square wave with no dc component and a peak value of 10 V. Sketch the input and output waveforms, showing their time relationship.
6. Measure the positive and negative peak values of the output and compare with the results of Example 5–7 [Figure 5–17(b)].
7. Apply a 1-kHz triangular waveform and observe the output. What kind of function is this? What does it closely resemble?

Low-Frequency Differentiator
LE 5–2.

1. Connect the low-frequency differentiator of Example 5–9 (Figure 5–21).
2. Measure the 3-dB break frequency f_b and compare with the value calculated in the text.
3. Measure the amplitude response over the range from near dc to well above the break frequency.
4. Plot the amplitude response on *log-log* paper, and show that the response approaches a straight line with a positive slope well below the break frequency.
5. Apply a 1-kHz triangular wave with a peak-to-peak value of 5 V. Sketch the input and output waveforms showing their time relationship.
6. Measure the positive and negative peak values of the output and compare with the results of Example 5–10 [Figure 5–22(b)].
7. Apply a 1-kHz square wave with an arbitrary amplitude. Sketch the input and output waveforms showing their time relationship. Can you suggest an application for this signal processing operation?

All-Pass Phase Lag Circuit
LE 5–3.

1. Design a circuit that will permit the phase shift to be adjustable over the range of at least from $-45°$ to $-135°$ at a frequency of 1 kHz using a 10-kΩ potentiometer.
2. Connect the circuit and verify that the phase shift can be adjusted over the range desired at 1 kHz.
3. Take a quick check of the amplitude response to ensure that the circuit is an "all-pass" form. A few check points from 100 Hz to 100 kHz should be sufficient.
4. Based on the capacitor you selected, calculate the exact value of resistance required to produce a phase shift of $-90°$ at 1 kHz.
5. Experimentally adjust the phase shift to $-90°$ at 1 kHz. Without changing the setting of the pot, deenergize the circuit, and remove the resistor (or resistors) used to set the phase shift. Measure this total resistance with an ohmmeter. Compare this value with the calculated value.

All-Pass Phase Lead Circuit
LE 5–4.

1. Design a circuit that will permit the phase shift to be adjustable over the range of at least from $+45°$ to $+135°$ at a frequency of 1 kHz using a 10-kΩ potentiometer.
2. Connect the circuit and verify that the phase shift can be adjusted over the range desired at 1 kHz.
3. Take a quick check of the amplitude response to ensure that the circuit is an "all-pass" form.
4. Based on the capacitor you selected, calculate the exact value of resistance required to produce a phase shift of $90°$ at 1 kHz.
5. Experimentally adjust the phase shift to $+90°$ at 1 kHz. Without changing the setting of the pot, energize the circuit, and remove the resistor (or resistors) used to set the phase shift. Measure this total resistance with an ohmmeter. Compare this value with the calculated value.

Single Power Supply Operation (Inverting)
LE 5–5.

1. Design an *inverting* symmetrical-input amplifier circuit to be operated from a *single* 15-V dc power supply. The following specifications are required;

mid-frequency voltage gain $= -10$

mid-frequency input resistance $\geq 10 \text{ k}\Omega$

"nearly flat" response down to 50 Hz

As an arbitrary but reasonable approach, assume that the reactance of each capacitor must be no greater than 10% of the corresponding resistance at a frequency of 50 Hz.

2. Connect the circuit and apply a 1-kHz input signal. Sketch the form of the op-amp output and the circuit output. Increase the input level until saturation occurs, and observe the op-amp output levels at saturation.

3. With the input set to a 0.5-V peak level at 1 kHz, measure the voltage gain and compare it with the expected value.

4. Reduce the frequency to 50 Hz and measure the voltage gain. Compare it with the expected value.

Single Power Supply Operation (Noninverting)
LE 5–6.

1. Design a *noninverting* symmetrical-input amplifier circuit to be operated from a single 15-V dc power supply. The following specifications are required:

mid-frequency voltage gain $= 11$

mid-frequency input resistance $\geq 10 \text{ k}\Omega$

"nearly flat" response down to 50 Hz

As an arbitrary but reasonable approach, assume that the reactance of each capacitor must be no greater than 10% of the corresponding equivalent resistance at a frequency of 50 Hz.

2. Connect the circuit and apply a 1-kHz input signal. Sketch the form of the op-amp output and the circuit output. Increase the input level until saturation occurs, and observe the op-amp output levels at saturation.

3. With the input set to a 0.5-V peak level at 1 kHz, measure the voltage gain and compare with the expected value.

4. Reduce the frequency to 50 Hz and measure the voltage gain. Compare with the expected value.

6

COMPARATORS

OBJECTIVES

After completing this chapter, the reader should be able to:

- Explain why the differential input voltage rating for an op-amp is more important for comparator circuits than for amplifiers.
- Discuss the concept of an input-output characteristic and sketch some typical forms.
- Explain the operation of a comparator circuit.
- Explain the difference between a *saturating* circuit and a *nonsaturating* circuit and between a *noninverting comparator* and an *inverting comparator.*
- Draw the schematic diagrams and sketch the input-output characteristics of a noninverting comparator circuit and of an inverting comparator circuit and show how to change their transition levels by adding a bias voltage.
- Sketch the output waveform for a given input waveform to a comparator.
- Define *hysteresis* as it relates to a comparator.
- Define *Schmitt trigger circuit,* explain its operation, and discuss its advantages.
- Draw the schematic diagram and sketch the input-output characteristic of an inverting Schmitt trigger circuit and of a noninverting Schmitt trigger circuit.
- Sketch the output waveform for a given input waveform to a Schmitt trigger circuit.
- Discuss the operation of an LM311 comparator and the advantages of a dedicated comparator over a general-purpose op-amp.

267

6-1 INTRODUCTION

The emphasis in this chapter is on the analysis, operation, and design of comparator circuits. Comparator circuits represent the first class of circuits we have considered that are basically *nonlinear* in operation. Specifically, comparator circuits produce two or more discrete outputs, each of which is dependent on the input level. The most common comparator circuits are those that have two possible output levels.

Comparator circuits are widely employed in applications involving the selection of a finite number of possible circuit conditions. Along with purely analog-type applications, comparators are used as key elements of analog-to-digital and digital-to-analog conversion systems. Comparators are also used in oscillator and waveform generator applications.

The best performance comparator functions are achieved with special integrated circuit (IC) chips designed and optimized for the comparator function. As is true with many IC chips, however, it is difficult to understand what is actually happening in the circuit. For that reason, our major focus early in the chapter is on utilizing general-purpose operational amplifiers for comparator purposes. This approach is adequate for satisfying some of the less demanding comparator applications, but, more significantly for our purposes, it establishes the basis for the design employed in the packaged IC comparator chips.

We consider both open-loop comparators and Schmitt trigger circuits. The latter circuits employ positive feedback to enhance operational performance. We discuss using general-purpose op-amps to implement comparators, as well as the specific use of a widely employed dedicated comparator IC chip (LM311).

6-2 INPUT VOLTAGE CONSIDERATIONS

Virtually all circuits considered thus far in the text have been linear forms, and operation in the linear region has been assumed in almost all cases. The linear active region is characterized by the fact that the output voltage can vary as required to force the input differential voltage to a sufficiently small value to ensure stable feedback operation. Indeed, the assumption that $v^+ = v^-$ (or $v_d = v^+ - v^- = 0$) has been widely used in predicting circuit performance.

Comparators are examples of *nonlinear* circuits. Many nonlinear circuits are characterized by the absence of negative feedback for either all or some operating regions. Under such conditions, *the two op-amp input voltages may have completely different values*. In some circuits, there is no feedback at all for all operating conditions. In other circuits, a region of closed-loop negative feedback operation may exist for certain signal levels, in which case $v^+ = v^-$ may be assumed *for those particular conditions*. Such circuits often contain regions of operation in which the loop is broken, however, in which case equal input voltages can no longer be assumed.

In many nonlinear circuits, the output may be in one of the saturation states for certain operating conditions. Even if the circuit contains a negative feedback loop, it cannot function as such when the output is in a saturation condition.

Since the two op-amp input voltages may be different, it is necessary to consider the maximum value of the differential input voltage (as well as the individual input voltages) expected for the given circuit and to ensure that the maximum ratings for the op-amp used are sufficient. From the 741 op-amp data of Appendix C, the absolute maximum input voltages for the two inputs are given as ± 15 V for ± 15-V power supplies and the maximum differential input voltage rating is ± 30 V. For lower power supply voltages, the maximum input voltages are equal to the power supply voltages. These values for the 741 op-amp are quite liberal compared with those of some other op-amps. Some op-amps have very low values for the maximum differential input voltage rating (less than 0.5 V in some cases), so this limitation must be carefully noted in circuits where nonlinear operation occurs.

The discussion thus far is applicable to the comparator circuits of this chapter as well as to other nonlinear circuits we will consider.

6–3 INPUT-OUTPUT CHARACTERISTIC

Any circuit with a single input signal and a single output signal can be described graphically by an *input-output characteristic curve*. Such a characteristic is a graph showing the value of the output for any value of input. The linear amplifier circuits considered thus far have had such simple mathematical input-output relationships that the use of graphical forms has not been necessary. As we will see shortly, however, graphical forms are very useful in describing the input-output relationships for nonlinear circuits, and for some nonlinear circuits, it is the only way to represent the circuit operation.

To illustrate the concept of the input-output characteristic, consider the simple low-gain noninverting amplifier of Figure 6–1(a). The gain of the circuit is $A_{CL} = 1 + 10 \text{ k}\Omega/$

FIGURE 6–1

Simple noninverting amplifier and its input-output characteristic curve.

(a)

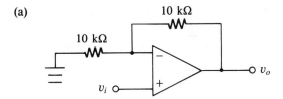

(b)

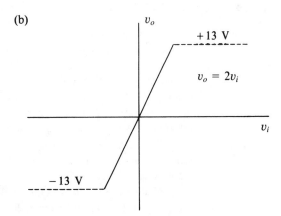

FIGURE 6–2
Example of a nonlinear input-output characteristic curve.

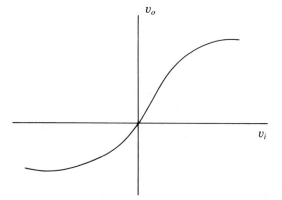

$10\ k\Omega = 1 + 1 = 2$. In the *linear range of operation,* the output voltage v_o can thus be written in terms of the input voltage v_i as

$$v_o = 2v_i \qquad\qquad (6\text{–}1)$$

A graph showing this relationship is shown in Figure 6–1(b). The voltage v_i is considered as the *independent variable,* and the voltage v_o is considered as the *dependent variable.* The slope of the curve in the linear region of operation is the gain, which is 2 in this case. If the input voltage reaches a level of about either $+6.5$ V or -6.5 V, the op-amp will saturate. At those points, the output voltage reaches a nearly constant value of about $+13$ V or -13 V for subsequent changes in the input.

Many nonlinear circuits have a number of abrupt changes in slope, or jumps, that are difficult to describe mathematically. In fact, some nonlinear circuits cannot be described at all in terms of mathematical equations.

An example of an arbitrary nonlinear input-output characteristic curve is shown in Figure 6–2.

6–4 OPEN-LOOP COMPARATORS

The simplest comparator circuits are those that operate with no feedback at all. Such comparators will be referred to as *open-loop comparators.* Circuits in which one or more conditions of operation occur at a saturation level will be referred to as **saturating circuits.** Saturating comparator circuits are relatively slow in response and thus are limited in application. By various clamping techniques, it is possible to establish reference levels well below saturation, and the switching speed is increased significantly. Such circuits will be referred to as **nonsaturating** circuits in this text.

The open-loop comparator circuits we will consider at this point are the saturating types. The two possible output levels are thus $v_o = +V_{sat}$ and $v_o = -V_{sat}$, assuming symmetrical circuit conditions.

Comparators may be classified as either *noninverting* or *inverting* according to the following: If the output assumes the *high* state when the input is *above* a certain minimum transition level, the circuit will be considered as *noninverting*. Conversely, if the output assumes the *low* state when the input is *above* a certain minimum transition level, the circuit will be considered as *inverting*.

Noninverting Comparator

The simplest open-loop comparator is the noninverting saturating form shown in Figure 6–3(a). The input signal is applied to the noninverting input, and the inverting input is grounded. If $v_i > 0$, the differential input voltage and the output voltage are both positive. In view of the very large typical open-loop gain, a positive voltage in the microvolt range will drive the output to positive saturation. For example, if $V_{sat} = 13$ V and $A_d = 200,000$, a positive voltage $v_i = 65$ μV will cause the output to reach the saturation level.

If $v_i < 0$, the differential input voltage and the output voltage are both negative. At a very small negative input voltage, the output will reach negative saturation. For the range of values in the preceding paragraph, the negative saturation level will be reached when the input voltage is $v_i = -65$ μV.

In a practical sense, the transition point for this circuit is assumed to be at 0 V. The desired mathematical operation can then be stated as

$$v_o = V_{sat} \quad \text{for } v_i > 0 \tag{6–2a}$$

$$v_o = -V_{sat} \quad \text{for } v_i < 0 \tag{6–2b}$$

where it is understood that a very small voltage is required on either side of zero to drive the output to saturation.

The input-output characteristic for the circuit is shown in Figure 6–3(b). The curve indicates that if v_i is positive by a very small amount, v_o will assume a fixed value of $+V_{sat}$. This region of operation is in the ***first quadrant*** of the coordinate system ($v_i > 0$, $v_o > 0$). Conversely, if v_i is negative by a very small amount, v_o will assume a fixed value of $-V_{sat}$. This region of operation is in the ***third quadrant*** of the coordinate system ($v_i < 0$, $v_o < 0$). The input-output characteristic curve is in obvious agreement with the mathematical description of (6–2).

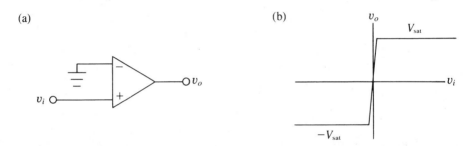

FIGURE 6–3
(a) Basic noninverting saturating comparator and (b) its input-output characteristic curve.

Inverting Comparator

The same basic circuit can be converted to an inverting form by grounding the noninverting input and applying the signal to the inverting input [see Figure 6–4(a)]. The general argument given earlier for the noninverting form applies here, except that a positive input results in a negative output, and vice versa. In this case, the mathematical operation can be expressed as

$$v_o = V_{sat} \quad \text{for } v_i < 0 \tag{6–3a}$$

$$v_o = -V_{sat} \quad \text{for } v_i > 0 \tag{6–3b}$$

The corresponding input-output characteristic is shown in Figure 6–4(b). Observe that in this case, operation is in either the **second quadrant** ($v_i < 0$, $v_o > 0$) or the **fourth quadrant** ($v_i > 0$, $v_o < 0$).

Changing the Transition Level

The two open-loop comparator circuits considered thus far have their transition points at a voltage level of zero. By applying a dc bias to either of the op-amp inputs, the transition level may be established at some arbitrary voltage level. Depending on the polarity of the bias and the op-amp terminal at which the bias is applied, several combinations are possible. Two cases will be considered here and two others will be left as exercises (Problems 6–21 and 6–22). Rather than memorize the different forms, plan to apply basic circuit analysis to each case as it arises.

   The key to the analysis is the following two basic properties: (1) when the differential voltage v_d is positive, assume that the output is in the high state, that is, $v_o = +V_{sat}$; (2) when the differential voltage v_d is negative, assume that the output is in the low state, that is, $v_o = -V_{sat}$. Understand that there is a small transition voltage region between the states that is being ignored for this purpose.

   Recall that $v_d = v^+ - v^-$. This means that when the noninverting input voltage level is higher (more positive or less negative) than the inverting input voltage level, the output of the op-amp will be in the high state. Conversely, when the noninverting input

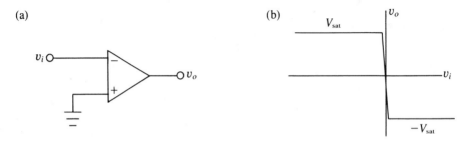

(a) (b)

FIGURE 6–4
(a) Basic inverting saturating comparator and (b) its input-output characteristic curve.

voltage level is lower (less positive or more negative), the output of the op-amp will be in the low state.

Most beginners have little difficulty determining the highest level voltage when the two voltages are both positive. When both voltages are negative, however, the voltage with the smallest magnitude has the highest algebraic level or is the most positive. Thus, a voltage level of -5 V is higher than -8 V. Further, any positive voltage is at a higher level than any negative voltage irrespective of the relative magnitudes. Thus, a voltage level of $+2$ V is higher than -5 V.

If desired, one can always resort to a mathematical formulation in which one first sets $v_d > 0$, in which case the conditions are then mathematically determined for producing the output high state. The opposite situation, that is, $v_d < 0$, can then be formulated for determining low state conditions; however, we will use the simpler rules formulated earlier.

Noninverting Comparator with Positive Bias

Consider the circuit of Figure 6–5(a), in which a positive bias level is applied to the inverting input, and the signal is applied to the noninverting input. When the signal level is lower than the positive bias level, the differential voltage is negative and the output will be at negative saturation. Conversely, when the input signal level exceeds the reference level, the differential voltage becomes positive, and the output changes to a positive saturation level.

A mathematical description of the preceding discussion can be formulated as follows:

$$v_o = -V_{\text{sat}} \quad \text{for } v_i < V_{\text{ref}} \tag{6–4a}$$

$$v_o = V_{\text{sat}} \quad \text{for } v_i > V_{\text{ref}} \tag{6–4b}$$

The corresponding input-output characteristic curve is shown in Figure 6–5(b).

(a)   (b)

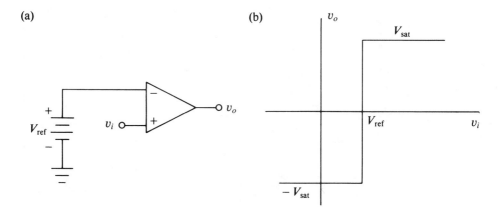

FIGURE 6–5
Basic saturating noninverting comparator with positive bias.

Inverting Comparator with Negative Bias

A circuit with a characteristic essentially opposite to the preceding one is shown in Figure 6–6(a). In this case, a negative bias voltage is applied to the noninverting input and the signal is applied to the inverting input. When the signal is negative, but with a larger magnitude than the bias voltage, the differential input is positive, and the output will be at positive saturation. Conversely, when the signal level becomes more positive than the bias level the differential voltage becomes negative, and the output will assume the negative saturation level.

A mathematical statement of the preceding is

$$v_o = V_{sat} \quad \text{for } v_i < -V_{ref} \qquad \text{(6–5a)}$$

$$v_o = -V_{sat} \quad \text{for } v_i > -V_{ref} \qquad \text{(6–5b)}$$

The corresponding input-output characteristic is shown in Figure 6–6(b).

Waveshaping

The comparator circuits previously considered, as well as others to be considered later, may be used for various waveshaping applications. For example, various periodic waveforms such as sinusoids and triangular waveforms may be converted to square waves or pulse trains.

Comparator circuits are used to generate "clean" versions of pulse data that have been disturbed by noise and degraded by signal transmission systems. As long as the zero crossings of the data are preserved, a new version of the signal, free from noise and distortion, can be constructed.

We will illustrate the process of waveshaping with several examples. The main point in each example is to use the transition point on the input-output characteristic curve as the basis for the state change. As the input voltage increases or decreases below the transition

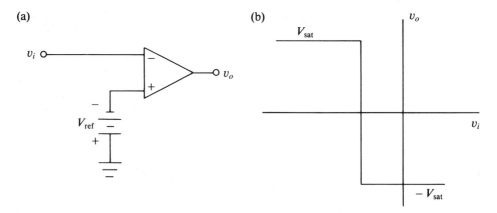

FIGURE 6–6
Basic saturating inverting comparator with negative bias.

voltage, the output will change states accordingly. The output waveform can then be sketched and shown in direct relationship to the input waveform.

As already discussed, the framework of analysis at this point is the use of general-purpose op-amps with the output changing between opposite saturation levels. Aside from the times required to go into and out of saturation, appreciable time is involved in changing between levels due to slew rate. Consequently, the square waves and pulse waves shown in the following examples are assumed to be at relatively low frequencies, so the transition times are not evident on the time scales shown. When faster transition times are required, the use of dedicated comparators, such as the LM311 (discussed in Section 6–6), are recommended.

EXAMPLE 6–1

The sinusoidal voltage of Figure 6–7(a) having a peak value of 6 V is applied as the input to the noninverting comparator circuit of Figure 6–3. Sketch the output voltage. Assume relatively low-frequency operation so that slew rate effects are not apparent, and assume $\pm V_{sat} = \pm 13$ V.

Solution

If the op-amp had an infinite gain, the transition could occur exactly at zero. In practice, however, a small voltage on either side of zero is required for the transition to occur, but this level should not be noticeable on the scale provided. For the noninverting comparator, the positive half-cycle of the output voltage corresponds to the positive half-cycle of the input voltage, and vice versa, as shown in Figure 6–7(b).

The positive and negative peak values of the output represent the saturation voltages of the op-amp and are essentially independent of the peak input voltage. Thus, the output voltage would remain nearly unchanged if the input voltage were adjusted in magnitude over a wide range. However, the magnitude of the input voltage should obviously be kept below the maximum input and differential voltage rating for the op-amp. At the other extreme, a too-small voltage level could result in a degradation in the switching point or even a failure to switch due to the finite op-amp gain and offset effects.

This circuit as given could be used to convert a sine wave (or any other oscillating waveform) to a square wave. Zener diodes could be used at the output to establish voltage levels other than zero if desired. However, our eventual objective will be to use the higher speed dedicated comparator circuits. At this point, as well as with further examples in this and the next section, the objective is to learn about basic circuit operation, so we will keep the circuits as simple as possible.

EXAMPLE 6–2

The sinusoidal voltage of Figure 6–7(a) is applied as the input to the inverting comparator of Figure 6–4. Sketch the output voltage. Assume relatively low-frequency operation so that slew rate effects are not apparent, and assume $\pm V_{sat} = \pm 13$ V.

Solution

The output of the inverting comparator has the same form as that of the noninverting comparator, except that it is reversed in phase as shown in Figure 6–7(c). The same comments as those at the end of Example 6–1 apply to this example.

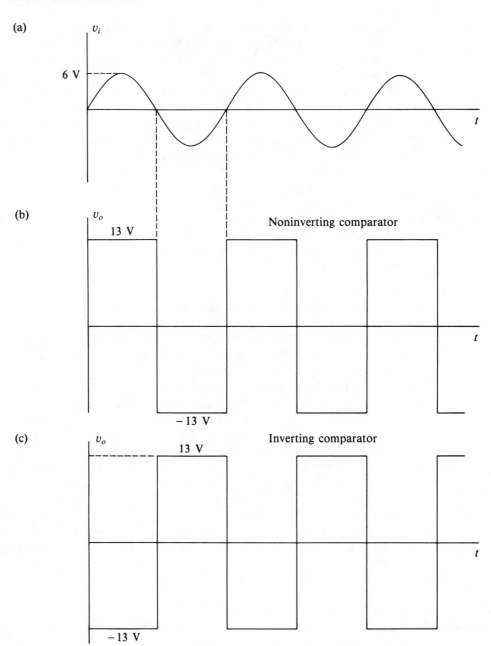

FIGURE 6–7
Waveforms of Examples 6–1 and 6–2.

EXAMPLE 6–3

Consider the circuit of Figure 6–8, and assume that the sinusoidal voltage of Figure 6–9(a) is applied as the input. Sketch the output voltage. Assume relatively low-frequency operation so that slew rate effects are not apparent, and assume $\pm V_{sat} = \pm 13$ V.

Solution
Since the signal is applied to the noninverting input, the comparator is a noninverting form. A positive voltage is applied through the voltage divider circuit to the inverting input, so the comparator is basically a saturating noninverting comparator with positive bias as first shown in Figure 6–5.

The value of the reference voltage, which establishes the transition point, is

$$V_{ref} = \frac{3 \text{ k}\Omega}{3 \text{ k}\Omega + 2 \text{ k}\Omega} \times 5 = 3 \text{ V} \qquad (6\text{–}6)$$

Since the input voltage has a peak value of 6 V, the transition point corresponds to the sine wave reaching half its peak value.

For the purpose of this analysis, we can work with basic angular units and write the input sine wave as

$$v_i = V_p \sin\theta = 6 \sin\theta \qquad (6\text{–}7)$$

where $\theta = \omega t$, if it were necessary to refer explicitly to time. Setting $v_i = 3$ V and solving for $\sin\theta$, we have

$$\sin\theta = \frac{3}{6} = 0.5 \qquad (6\text{–}8)$$

The values of θ that satisfy (6–8) in one cycle are $\theta = 30°$ ($\pi/6$ radians) and $\theta = 150°$ ($5\pi/6$ radians). The first point corresponds to an upward transition, and the second point corresponds to a downward transition.

Refer to Figure 6–9(a) and (b). For $v_i < 3$ V, the output is at negative saturation. When v_i exceeds 3 V, the output changes to positive saturation and remains there until the input drops below 3 V. The output waveform has a phase sense aligned with that of the input, but the positive portion of the cycle is shorter than the negative portion.

FIGURE 6–8
Circuit of Example 6–3.

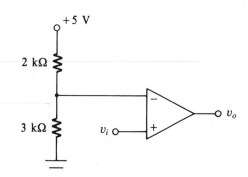

(a)

(b)

(c)

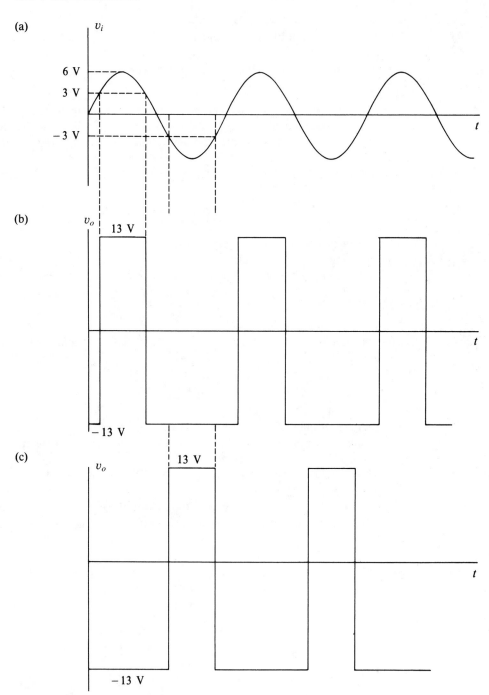

FIGURE 6–9
Waveforms of Examples 6–3 and 6–4.

In working with waveforms of this nature, it is convenient to define a *duty cycle D*. For the purpose of this example and others, the duty cycle will be defined as

$$D = \frac{\text{duration of positive interval}}{\text{period}} \times 100\% \tag{6-9}$$

Working in degrees, the positive interval is $150° - 30° = 120°$. Since the total period is $360°$, the duty cycle is

$$D = \frac{120°}{360°} \times 100\% = 33.3\% \tag{6-10}$$

Thus, the positive interval is one-third of the period.

Unlike the earlier circuits with a transition at zero, the output waveshape here depends greatly on the peak value of the input. If the input voltage were increased, the transition level would correspond to points of the sine wave closer to zero crossings, and the duty cycle would increase. Conversely, if the sine wave were reduced somewhat in amplitude, the duty cycle would decrease. In fact, if the input were reduced below 3 V, the output would simply lock in the negative saturation output level. This condition is referred to as *latchup*.

EXAMPLE 6-4

Consider the circuit of Figure 6-10, and assume that the sinusoidal voltage of Figure 6-9(a) is applied as the input. Sketch the output voltage. Assume relatively low-frequency operation so that slew rate effects are not apparent, and assume $\pm V_{\text{sat}} = \pm 13$ V.

Solution

Since the signal is applied to the inverting input, the comparator is an inverting form. A negative voltage is applied through the voltage divider circuit to the inverting input. Thus, the circuit is a saturating inverting comparator with negative bias as first shown in Figure 6-6.

The value of the reference voltage expressed as a positive value is

$$V_{\text{ref}} = \frac{3\text{ k}\Omega}{3\text{ k}\Omega + 12\text{ k}\Omega} \times 15 = 3\text{ V} \tag{6-11}$$

The transition point is, of course, at -3 V.

FIGURE 6-10
Circuit of Example 6-4.

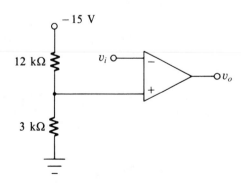

The sinusoid can be expressed again as

$$v_i = 6 \sin \theta \qquad (6\text{--}12)$$

However, the transition occurs in this case at $v_i = -3$ V. Solving for $\sin \theta$, we have

$$\sin \theta = \frac{-3}{6} = -0.5 \qquad (6\text{--}13)$$

The values of θ that satisfy (6–13) in one cycle are $\theta = 210°$ ($7\pi/6$ radians) and $\theta = 330°$ ($11\pi/6$) radians.

Refer to Figure 6–9(a) and (c). For $v_i > -3$ V, the output is at negative saturation. When v_i drops below -3 V, the output changes to positive saturation and remains there until the input increases above -3 V. The output waveform has the opposite phase sense of the input, and the positive portion of the cycle is shorter than the negative portion.

The positive interval in degrees is $330° - 210° = 120°$. The duty cycle is

$$D = \frac{120°}{360°} \times 100\% = 33.3\% \qquad (6\text{--}14)$$

which is the same as in Example 6–3.

As in Example 6–3, the duty cycle is highly dependent on the input signal level.

6–5 COMPARATORS WITH POSITIVE FEEDBACK

In this section we will discuss comparators employing *positive feedback*. Such comparators are widely known as *Schmitt trigger circuits*. The addition of positive feedback results in an effect called *hysteresis*. Hysteresis is a phenomenon in which the transition point for the input voltage is different when switching from the low state to the high state as compared with switching from the high state to the low state. Said differently, the transition process is direction-sensitive.

What is gained from this process? First, the possibility of undesirable state changes due to spurious noise pickup is minimized by employing hysteresis. Second, the switching process can be accentuated by the positive feedback in the Schmitt trigger. Finally, the hysteresis effect can be used to advantage in certain waveform generators, an example of which will be considered in Chapter 7.

Inverting Schmitt Trigger

The easiest Schmitt trigger to understand is the *inverting saturating* form, so we will consider that circuit first. The basic form of such a circuit is shown in Figure 6–11(a). The divider network consisting of R_1 and R_2 establishes a voltage at the noninverting input terminal proportional to the output voltage. The magnitude of the voltage across R_2 will be defined as the *threshold voltage,* denoted as V_T. This voltage is

$$V_T = \frac{R_2 V_{\text{sat}}}{R_1 + R_2} \qquad (6\text{--}15)$$

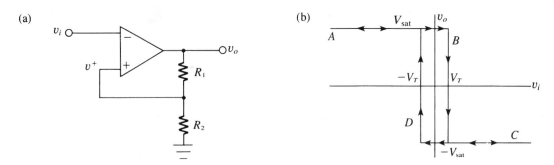

FIGURE 6–11

(a) Inverting saturating Schmitt trigger circuit and (b) its input-output characteristic curve.

Note that

$$v^+ = V_T \quad \text{when } v_o = V_{\text{sat}} \tag{6–16}$$

and

$$v^+ = -V_T \quad \text{when } v_o = -V_{\text{sat}} \tag{6–17}$$

In discussions of the principle of operation, the input-output characteristic curve of Figure 6–11(b) is very useful. For illustration, assume initially that the circuit is in a state corresponding to the left-hand portion of line A. Then $v_o = V_{\text{sat}}$, $v^+ = +V_T$, and v_i is negative. Since $v^+ = +V_T$, the input voltage v_i must slightly exceed $+V_T$ to force the op-amp differential input voltage to change signs. This pattern is indicated by the direction of arrows to the right on line A.

When v_i reaches V_T (and slightly exceeds it), the op-amp output starts to drop. The voltage at the noninverting input, which is always a constant fraction of the output voltage, likewise starts to drop, which further increases the differential input voltage and accentuates the transition process. The op-amp output thus changes from $+V_{\text{sat}}$ to $-V_{\text{sat}}$ along line B in a short time interval, limited primarily by the slew rate. Any further increase in v_i causes the input to change along line C, but the output remains in negative saturation. Since $v_o = -V_{\text{sat}}$ in this state, $v^+ = -V_T$.

To return to the previous state, v_i must assume a slightly more negative value than $-V_T$. Thus, v_i must move along C to the left to a value $-V_T$. At that point, the differential input voltage changes signs and the output returns to positive saturation along line D. Any further decrease in v_i results in movement to the left along line A, and this region is the one in which operation was initially assumed.

Observe how the switching level is a function of the direction of change. By choosing an appropriate value of V_T, one can minimize the effects of noise at the transition points. However, if V_T is too large, the accuracy of the crossover point may be degraded for certain applications, so the threshold must be selected with some care.

The rectangle on the input-output characteristic curve is called a **hysteresis loop.** Note that it is necessary to label a hysteresis loop with arrows in order to identify the proper directions that apply. It is difficult to write a simple mathematical expression to describe a Schmitt trigger, so the input-output characteristic curve containing the hysteresis loop is most useful in this case.

Noninverting Schmitt Trigger

A noninverting Schmitt trigger is shown in Figure 6–12(a). In this circuit the voltage v^+ at the noninverting input terminal is a linear combination of the input voltage v_i and the output voltage v_o. Superposition may be readily applied to the circuit by treating v_o **as if** it were an ideal voltage source in establishing an effect on v^+. (This approach is justified, since v^+ is a linear combination of both v_i and v_o, and the output impedance of the op-amp is very low.)

Let v_1^+ represent the contribution of v_i to v^+, and let v_2^+ represent the contribution of v_o to v^+. To determine v_1^+, we deenergize the "source equivalent" corresponding to v_o; that is, the right-hand side of R_f is grounded. The voltage v_1^+ is then

$$v_1^+ = \frac{R_f}{R_i + R_f} v_i \qquad (6\text{–}18)$$

To determine v_2^+, we deenergize v_i by replacing it with a short-circuit, and we have

$$v_2^+ = \frac{R_i}{R_i + R_f} v_o \qquad (6\text{–}19)$$

The net value of v^+ is then

$$v^+ = v_1^+ + v_2^+ \qquad (6\text{–}20a)$$

or

$$v^+ = \frac{R_f}{R_i + R_f} v_i + \frac{R_i}{R_i + R_f} v_o \qquad (6\text{–}20b)$$

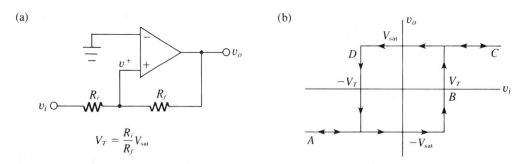

(a)

(b)

$$V_T = \frac{R_i}{R_f} V_{sat}$$

FIGURE 6–12
(a) Noninverting saturating Schmitt trigger circuit and (b) its input-output characteristic curve.

Inequalities for switching can be determined from this equation by first noting that since $v_d = v^+$, the transition point occurs for $v^+ = 0$. The requirement for the output positive state is $v^+ > 0$, and the requirement for the output negative state is $v^+ < 0$.

Refer to the input-output characteristic curve of Figure 6–12(b) for the explanation that follows. Assume initially that the circuit is in a state corresponding to the left-hand portion of line A. Thus, $v_o = -V_{sat}$, v_i is negative, and the net value of v^+ as given by (6–20b) is negative. (Both v_i and v_o are negative, so it is logical that v^+ is negative.) When $v_o = -V_{sat}$ is substituted in (6–20b), the value of v^+ becomes

$$v^+ = \frac{R_f}{R_i + R_f} v_i - \frac{R_i}{R_i + R_f} V_{sat} \qquad \text{(6–21)}$$

In order to change states, v^+ must become positive. Setting $v^+ > 0$ in (6–21), we obtain after simplification

$$v_i > \frac{R_i}{R_f} V_{sat} \qquad \text{(6–22)}$$

A threshold voltage V_T will be defined for this comparator as

$$V_T = \frac{R_i}{R_f} V_{sat} \qquad \text{(6–23)}$$

Thus, v_i must assume a positive value slightly greater than V_T before the circuit can change states, as indicated by the direction of arrows along line A in Figure 6–12(b).

Once v_i exceeds V_T by a slight amount, the output voltage changes to $+V_{sat}$ along line B. Any further increase in v_i causes the input to change along line C, but the output remains in positive saturation.

The condition for switching back is determined by first substituting $v_o = +V_{sat}$ in (6–20b), which yields

$$v^+ = \frac{R_f}{R_i + R_f} v_i + \frac{R_i}{R_i + R_f} V_{sat} \qquad \text{(6–24)}$$

To switch back, v^+ must become negative. Setting $v^+ < 0$ in (6–24), we obtain

$$v_i < -\frac{R_i}{R_f} V_{sat} \qquad \text{(6–25)}$$

where the expression on the right-hand side is recognized as $-V_T$ according to the definition of (6–23). Thus, the input voltage must assume a negative value $-V_T$ along line C before the circuit changes states again. When the necessary switching condition is established, the output drops to $-V_{sat}$ along line D. Any further decrease in v_i results in movement to the left along line A, and this region is the one in which operation was initially assumed.

Some Interesting Comparisons

Although the reader may not have observed the pattern, there is some similarity in the appearance of the circuit diagrams of the inverting Schmitt trigger and the noninverting linear amplifier. Likewise, the noninverting Schmitt trigger circuit and the inverting amplifier circuit have a similar appearance. However, there is a major difference in both cases. Linear amplifier circuits have the feedback returned to the inverting input terminal, while Schmitt trigger circuits have the feedback returned to the noninverting terminal. The first condition results in negative feedback for stable linear amplification, while the second condition results in positive feedback, a requirement for the nonlinear nature of the trigger circuits.

EXAMPLE 6–5

Consider the inverting Schmitt trigger circuit of Figure 6–13. (a) Assuming that $\pm V_{\text{sat}} = \pm 13$ V, calculate the threshold voltage. (b) Assume that the input is the triangular voltage of Figure 6–14(a). Sketch the output voltage.

Solution

(a) The threshold voltage is

$$V_T = \frac{R_2 V_{\text{sat}}}{R_1 + R_2} = \frac{33 \times 13}{33 + 10^4} = 42.8 \text{ mV} \tag{6-26}$$

(b) The transition points correspond to the input voltage becoming more positive than $+42.8$ mV in one direction and becoming more negative than -42.8 mV in the opposite direction. Because these voltages are small compared to the positive and negative peak values of the input voltage (less than 0.5%), the transition points are shifted only a small amount from the zero crossing in each case.

Refer to Figure 6–14(b). As v_i of (a) crosses zero, v_o is initially in a high state, but then it switches to the low state after v_i increases above $+V_T$. The opposite transition occurs after v_i decreases below $-V_T$. The output is thus a periodic square wave that is inverted with respect to the triangular wave, but aligned closely at opposite zero crossing points.

FIGURE 6–13
Circuit of Example 6–5

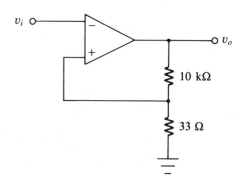

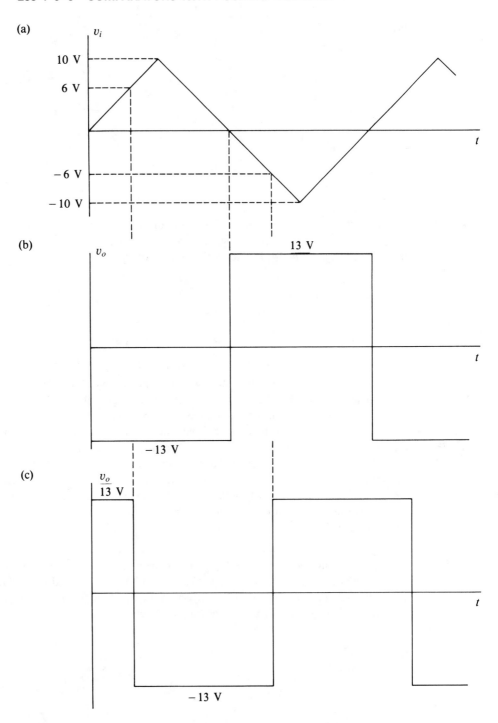

FIGURE 6–14
Waveforms of Examples 6–5 and 6–6.

FIGURE 6–15

Circuit of Example 6–6.

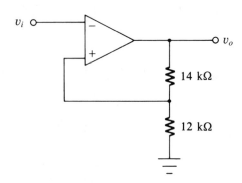

EXAMPLE 6–6

Consider the inverting Schmitt trigger circuit of Figure 6–15. (a) Assuming that $\pm V_{sat} = \pm 13$ V, calculate the threshold voltage. (b) Assume that the input is the triangular voltage of Figure 6–14(a). Sketch the output voltage.

Solution

(a) The threshold voltage is

$$V_T = \frac{R_2 V_{sat}}{R_1 + R_2} = \frac{12 \text{ k}\Omega \times 13}{12 \text{ k}\Omega + 14 \text{ k}\Omega} = 6 \text{ V} \tag{6–27}$$

(b) What makes this problem different from Example 6–5 is the much larger threshold voltage and its effect on the output waveform, as we will see shortly. Refer to Figure 6–14(c), and assume that the output is initially in the high state.

The input voltage must exceed a level of $+6$ V before the transition to the negative level occurs as noted. On the next half-cycle of the input, the voltage must drop below a level of -6 V before the opposite transition occurs.

Note that the input waveform has equal positive and negative intervals, but the effect of the large threshold is to *delay* the input waveform as compared to the case of a very small threshold. This property can be used to advantage in some applications. In fact, by using a potentiometer to form the feedback voltage, a square wave with an adjustable delay can be created.

EXAMPLE 6–7

Consider the noninverting Schmitt trigger of Figure 6–16. The input is the triangular waveform of Figure 6–17(a). Assume that $\pm V_{sat} = \pm 13$ V. It is desired to produce a square wave in which the transitions occur exactly at the peaks of the input (± 5 V). (a) Determine the value of R_f required. (b) Sketch the output waveform.

FIGURE 6–16

Circuit of Example 6–7.

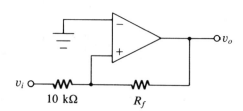

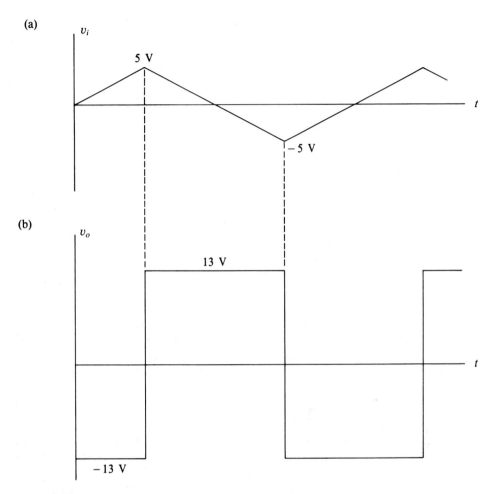

FIGURE 6-17
Waveforms of Example 6-7.

Solution

(a) From (6–23), the value of R_f is determined as

$$R_f = \frac{R_i V_{\text{sat}}}{V_T} \tag{6-28}$$

Substituting $R_i = 10$ kΩ, $V_{\text{sat}} = 13$ V, and $V_T = 5$ V, we have

$$R_f = \frac{10^4 \times 13}{5} = 26 \text{ k}\Omega \tag{6-29}$$

(b) Refer to Figure 6–17(a) and (b) in the discussion that follows. To sketch the output, note that in the first interval of time, the input is a positive voltage less than 5 V. The output must then be at -13 V. When the input reaches a level of 5 V, the output changes to 13 V. The output remains at that level until the input drops down to -5 V, at which time a negative transition for the output occurs. The pattern continues, as shown in Figure 6–17(b).

EXAMPLE 6–8

Using standard 5% resistance values, design an inverting Schmitt trigger circuit having threshold levels close to ± 50 mV based on saturation voltages of ± 13 V. The exact value of the threshold is not critical in this application.

Solution

The basic circuit diagram is that of Figure 6–11(a). From (6–15), we require that

$$50 \times 10^{-3} = \frac{R_2}{R_1 + R_2} \times 13 \tag{6–30}$$

Simplification of this relationship leads to

$$R_1 = 259R_2 \tag{6–31}$$

While there is no doubt that a number of possible values would suffice for this noncritical application, some trial and error with standard values leads to the result that if $R_1 = 39$ kΩ and $R_2 = 150$ Ω, then $R_1 = 260R_2$. The predicted values for the threshold levels with these resistors are ± 49.8 mV. Considering the leeway in design and the fact that the resistors are 5% values, this combination will be selected for the proposed design.

6–6 LM311 COMPARATOR

The various comparator circuits discussed thus far have been based on using general-purpose operational amplifiers with output switching levels at the saturation points. The primary objective has been to provide a clear understanding of circuit operation. These circuits are usable for relatively low-frequency applications where switching speed is not critical. In fact, clamping diodes or Zener reference levels may be used to establish operation well below saturation and thus provide higher speed performance. On the other hand, a point of diminishing returns is reached where general-purpose op-amps do not provide the switching speeds required for precision operations.

A class of high-speed op-amps dedicated to switching operation is available, and their use is highly recommended for applications requiring the shortest transition times. These circuits in IC form are referred to simply as *comparators*. This section will be devoted to one of the most popular of these comparators, the LM311 comparator. Actually, the LM311 is the commercial version of a set indicated as the LM111/LM211/LM311 group. The first two are the military and industrial versions, respectively. Some data sheets provide information on all in the group, but the data sheets provided here only cover the LM311. In subsequent references, we will use the shortened term *311*. A complete set of data sheets on the 311 is provided in Appendix C, and many application circuits along with specifications are given.

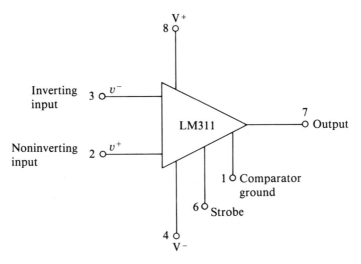

FIGURE 6–18
Diagram of LM311 with major features.

The major features of the LM311 are shown on the schematic symbol of Figure 6–18. The pin numbers correspond to those of the 8-pin DIP unit. The usual inverting and noninverting input appear on the left. On the data sheets in Appendix C, the noninverting input appears at the top, but they are reversed on Figure 6–18 in accordance with the text convention thus far.

The positive and negative power supply terminals are denoted as V^+ and V^-, respectively. It is actually possible to operate the unit with a single 5-V power supply, in which case the V^- terminal is connected to ground. However, the unit can also be operated from ± 15-V power supplies, as is customary with most op-amps.

One terminal (pin 1) will be defined for this discussion as the "comparator ground." This terminal may be externally connected to the circuit ground or it may be connected to some other voltage level.

Comparator Logic

The general operation of the LM311 comparator is based on the following logic:

1. If v^+(pin 2) $< v^-$(pin 3), the output (pin 7) is internally connected to the "comparator ground" pin (pin 1).
2. If $v^+ > v^-$, the output assumes an open collector condition. Thus, a pull-up resistor to a reference voltage is required to establish the output state.
3. If the strobe (pin 6) is left open (or alternately connected to a positive voltage), the comparator functions normally as given by (1) or (2). If, however, the strobe is connected *through a resistor* to ground, the output assumes the open collector state of (2) irrespective of the input.

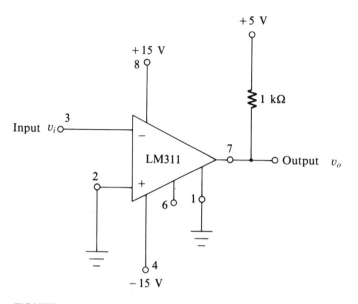

FIGURE 6-19
LM311 comparator connected as an open-loop inverting comparator.

As a general rule, any of the comparator circuits considered thus far may be adapted to the 311. However, the output levels of the 311 may be established at convenient fixed levels such as, for example, 0 and 5 V. These particular values are convenient for interfacing with transistor-transistor logic (TTL) circuits. Further, the strobe feature allows the comparator function to be switched on and off as required for some applications.

Representative 311 Connection

A representative connection of the 311 is shown in Figure 6-19. The 311 is functioning as an open-loop inverting comparator in this particular connection. The output (pin 7) is connected to a +5-V supply through a 1-kΩ pull-up resistor. The "comparator ground" (pin 1) is connected to circuit ground. These connections result in the output assuming the possible states of 0 and 5 V. The logic for the complete circuit is

$$v_o = 5\text{ V} \quad \text{for } v_i < 0 \qquad (6\text{--}32\text{a})$$
$$= 0 \quad\;\; \text{for } v_i > 0 \qquad (6\text{--}32\text{b})$$

The strobe feature is not being used in this particular circuit.

6-7   MULTISIM EXAMPLES

The examples in this section are devoted to the simulation of comparator circuits using Multisim. The 741 model is used in all cases, and transient analysis is employed.

MULTISIM EXAMPLE 6–1

Using the 741 model, develop a Multisim circuit for simulating the noninverting comparator of Example 6–1 and use it to plot the output voltage for the sinusoidal input voltage used in that example. Set the input sinusoidal frequency to 100 Hz.

Solution

The Multisim circuit schematic diagram is shown in Figure 6–20. All of the elements in the circuit have been considered in other examples, so there is nothing new to explain. To avoid having the op-amp output voltage left as a "floating" pin, a 10-kΩ resistor is connected from pin 6 to ground. The input sinusoidal voltage is set to a peak value of 6 V at a frequency of 100 Hz.

A **Transient Analysis** is performed over two cycles of the input signal with the **Maximum time step** set to 10 μs. The results are shown in Figure 6–21. The input sinusoid is represented by the dashed curve, and the resulting output square wave is represented by the solid curve. At this relatively low frequency, slew rate effects are not apparent.

FIGURE 6–20
Circuit of Multisim Example 6–1.

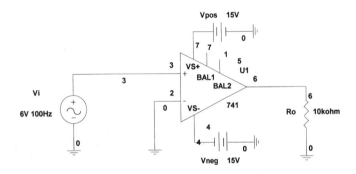

FIGURE 6–21
Waveforms of Multisim Example 6–1.

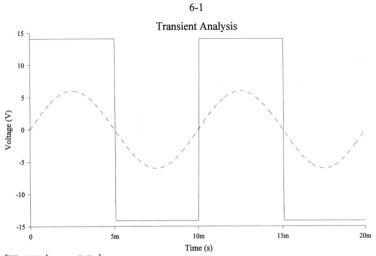

MULTISIM EXAMPLE 6–2

Modify the Multisim circuit of the preceding example to realize the circuit of Example 6–3, and use it to plot the output voltage for the sinusoidal input voltage used in that example. Set the input sinusoidal frequency in this case to 1 kHz.

Solution

The original circuit diagram was shown in Figure 6–8, and the reference voltage was obtained from a voltage divider. For simplicity, however, the Multisim realization will utilize a 3-V dc source, which achieves the same result. The modified circuit diagram is shown in Figure 6–22. As before, a 10-kΩ resistor is connected from pin 6 to ground. The input sinusoidal voltage is set to a peak value of 6 V at a frequency of 1 kHz.

A **Transient Analysis** is performed over two cycles of the input signal, and the results are shown in Figure 6–23. After some experimentation, the **Maximum time step** was set to 1 μs. The input sinusoid is represented by the dashed curve, and the resulting output square wave is represented by the solid curve. At a frequency of 1 kHz, we are beginning to see some slew rate effects.

FIGURE 6–22

Circuit of Multisim Example 6–2.

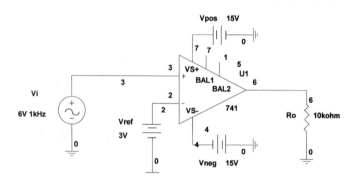

FIGURE 6–23

Waveforms of Multisim Example 6–2.

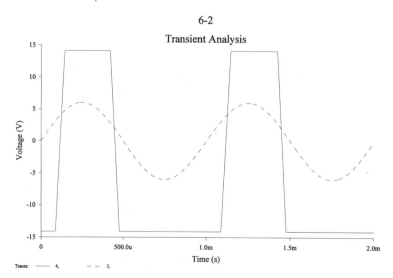

MULTISIM EXAMPLE 6–3

Using the 741 model, develop a Multisim circuit for simulating the inverting Schmitt trigger circuit of Example 6–6 and use it to plot the input and output voltages for the triangular input voltage used in that example. Set the period to 1 s.

Solution
The Multisim circuit schematic diagram is shown in Figure 6–24. The PWL voltage source is used, and the following time-voltage pairs are established for the function:

0	0
0.25	10
0.75	−10
1.25	10
0	0

These points define the triangular waveform over 1.5 cycles.

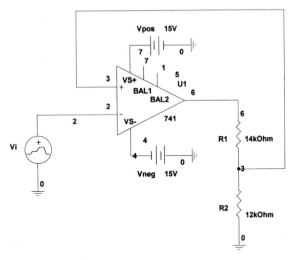

FIGURE 6–24
Circuit of Multisim Example 6–3.

A **Transient Analysis** is performed over the defined interval of the input signal, and the results are shown in Figure 6–25. The **Minimum number of time points** was set to 1000. The input triangular function is represented by the dashed curve, and the resulting output square wave is represented by the solid curve. The output is in agreement with the curve of Figure 6–14(c) except for slight differences in the saturation voltages.

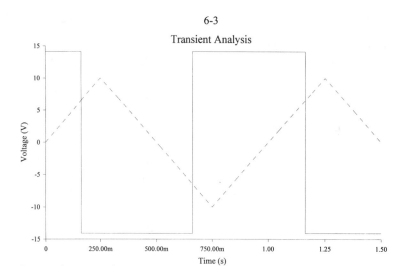

FIGURE 6–25
Waveforms of Multisim Example 6–3.

SUMMARY

1. In nonlinear circuits, such as the comparator, the two op-amp input voltages may differ widely.
2. The input-output characteristic may be used to display graphically the relationship between input and output signals.
3. Comparators are circuits used to produce a finite number of discrete outputs, each of which is a function of the input level.
4. The most common comparator circuits produce two possible output levels.
5. Comparators may be classified as either saturating or nonsaturating.
6. Comparators may be classified as either noninverting or inverting.
7. A general-purpose op-amp may be used to implement a simple noninverting or inverting open-loop saturating comparator.
8. The transition level of an open-loop comparator may be changed by the addition of a dc bias to one of the two input terminals.
9. A comparator with positive feedback is called a Schmitt trigger circuit.
10. The addition of positive feedback results in a hysteresis loop.
11. All comparator circuits can be used to change various time-varying waveforms to forms exhibiting rectangular segments.
12. Dedicated comparator chips are available, in which the switching speed is optimized.
13. Along with the basic switching functions, comparator chips often permit strobe operation.

PROBLEMS

Drill Problems
Note: The voltage waveforms of Figures 6–20 and 6–21 are to be considered as possible input signals to various comparator circuits in Problems 6–1 through 6–20 as indicated. In all circuits using

general-purpose op-amps, assume relatively low-frequency operation so that the slew rate effects are not apparent, and assume $\pm V_{sat} = \pm 13$ V.

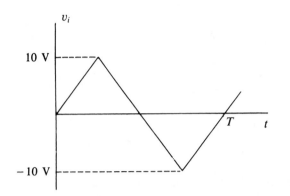

FIGURE 6–26
Waveform used in Problems 6–1 through 6–20.

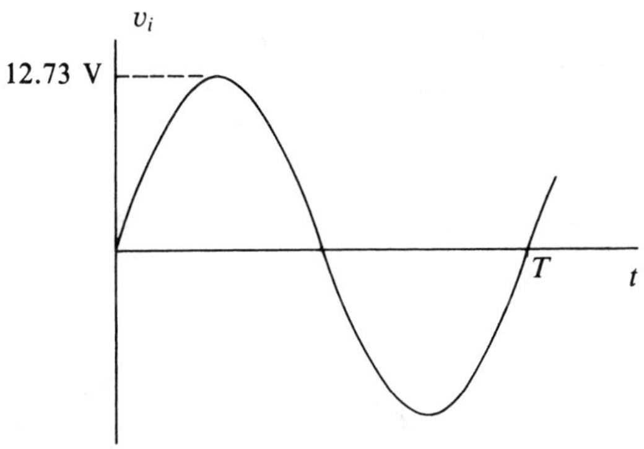

FIGURE 6–27
Waveform used in Problems 6–1 through 6–20.

6–1. **a.** Sketch the input-output characteristic curve of the circuit of Figure P6–1.
b. If the waveform of Figure 6–20 is applied as the input, sketch the output.

FIGURE P6–1

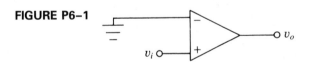

6–2. **a.** Sketch the input-output characteristic curve of the circuit of Figure P6–2.
b. If the waveform of Figure 6–20 is applied as the input, sketch the output.

FIGURE P6–2

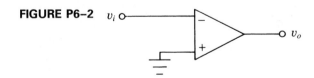

6–3. The waveform of Figure 6–21 is applied as the input to the circuit of Problem 6–1. Sketch the output.

6–4. The waveform of Figure 6–21 is applied as the input to the circuit of Problem 6–2. Sketch the output.

6–5. a. Sketch the input-output characteristic curve of the circuit of Figure P6–5.
b. If the waveform of Figure 6–20 is applied as the input, sketch the output.

FIGURE P6–5

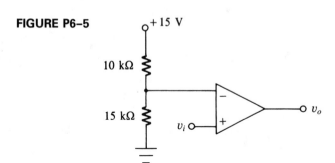

6–6. a. Sketch the input-output characteristic curve of the circuit of Figure P6-6.
b. If the waveform of Figure 6-20 is applied as the input, sketch the output.

FIGURE P6–6

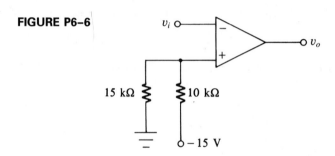

6–7. The waveform of Figure 6–21 is applied as the input to the circuit of Problem 6–5. Sketch the output.

6–8. The waveform of Figure 6–21 is applied as the input to the circuit of Problem 6–6. Sketch the output.

6–9. a. Sketch the input-output characteristic curve of the circuit of Figure P6–9.
b. If the waveform of Figure 6–20 is applied as the input, sketch the output.

FIGURE P6–9

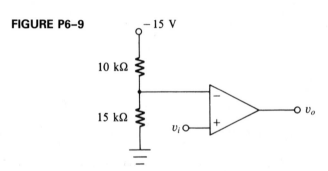

6–10. a. Sketch the input-output characteristic curve of the circuit of Figure P6–10.
  b. If the waveform of Figure 6–20 is applied as the input, sketch the output.

FIGURE P6–10

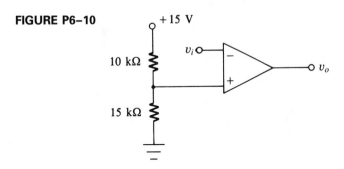

6–11. The waveform of Figure 6–21 is applied as the input to the circuit of Problem 6–9. Sketch the output.

6–12. The waveform of Figure 6–21 is applied as the input to the circuit of Problem 6–10. Sketch the output.

6–13. The waveform of Figure 6–21 is applied as the input to the 311 comparator circuit of Figure P6–13. Sketch the output.

FIGURE P6–13

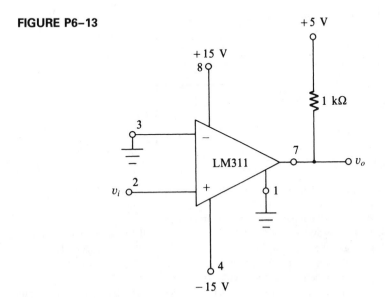

6–14. The waveform of Figure 6–21 is applied as the input to the 311 comparator circuit of Figure P6–14. Sketch the output. (Note that pin 1 is not connected to ground.)

FIGURE P6-14

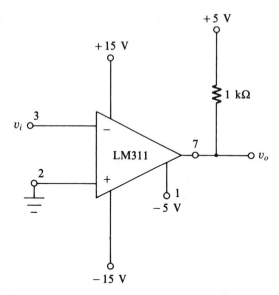

6-15. Consider the circuit of Figure P6–15 with $R_1 = 20$ kΩ and $R_2 = 100$ Ω.
  a. Calculate the threshold voltage.
  b. Sketch the input-output characteristic curve.
  c. If the waveform of Figure 6–20 is applied as the input, sketch the output.

FIGURE P6-15

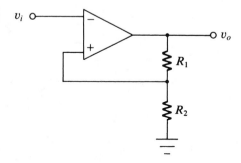

6-16. Consider the circuit of Figure P6–15 with $R_1 = 27$ kΩ and $R_2 = 68$ Ω.
  a. Calculate the threshold voltage.
  b. Sketch the input-output characteristic curve.
  c. If the waveform of Figure 6–21 is applied as the input, sketch the output.
6-17. Consider the circuit of Figure P6–15 with $R_1 = 10$ kΩ and $R_2 = 16$ kΩ.
  a. Calculate the threshold voltage.
  b. Sketch the input-output characteristic curve.
  c. If the waveform of Figure 6–20 is applied as the input, sketch the output.
6-18. Consider the circuit of Figure P6–15 with $R_1 = 10$ kΩ and $R_2 = 22.5$ kΩ.
  a. Calculate the threshold voltage.
  b. Sketch the input-output characteristic curve.
  c. If the waveform of Figure 6–21 is applied as the input, sketch the output.

6–19. Consider the circuit of Figure P6–19 with $R_i = 10 \text{ k}\Omega$ and $R_f = 26 \text{ k}\Omega$.
   a. Calculate the threshold voltage.
   b. Sketch the input-output characteristic curve.
   c. If the waveform of Figure 6–20 is applied as the input, sketch the output.

FIGURE P6–19

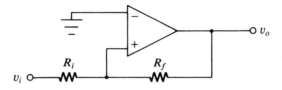

6–20. Consider the circuit of Figure P6–19 with $R_i = 30 \text{ k}\Omega$ and $R_f = 43.3 \text{ k}\Omega$.
   a. Calculate the threshold voltage.
   b. Sketch the input-output characteristic curve.
   c. If the waveform of Figure 6–21 is applied as the input, sketch the output.

Derivation Problems
6–21. Consider the circuit of Figure P6–21.
   a. Write the equations to describe the output as a function of the input.
   b. Sketch the input-output characteristic curve.

FIGURE P6–21

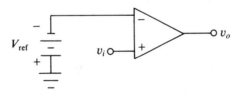

6–22. Consider the circuit of Figure P6–22.
   a. Write the equations to describe the output as a function of the input.
   b. Sketch the input-output characteristic curve.

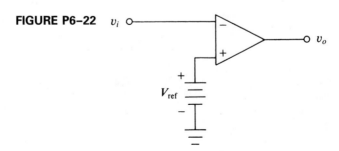

FIGURE P6–22

6–23. The circuit of Figure P6–23 is one form of a *window comparator*. Assume that $\pm V_{sat} = \pm 13$ V and that forward-biased diode drops $= 0.7$ V.

    a. Write expressions for the input-output characteristic with arbitrary values of V_{ref1} and V_{ref2}.

    b. Express the result of (a) for $V_{ref1} = V_{ref2} = 6$ V.

    c. Sketch the result of (b).

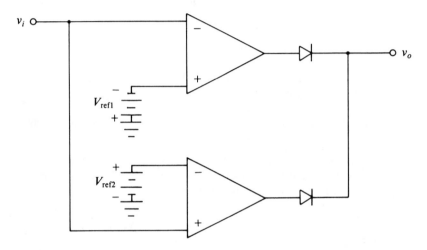

FIGURE P6–23

6–24. The circuit of Figure P6–24(a) is one form of a *pulse-width modulator*. The quantity v_i is an input time-varying signal which, for the purpose of this development, can be assumed to be essentially constant during the width of a sampling period T. The voltage v_s is a periodic sawtooth waveform as shown in (b). The signal is adjusted, through the possible addition of a bias level and scaling, so that its amplitude range is the same as the sawtooth wave; that is, $0 < v_i < V_p$.

    a. Show that during an interval of one period T, the output v_o is a pulse whose width t_p is given by

$$t_p = \frac{T}{V_p} v_i = K v_i$$

    where $K = T/V_p$ is a fixed constant. Thus, the pulse width in each interval is directly proportional to the input signal v_i.

    b. Assume that $V_p = 5$ V. Sketch the output waveforms corresponding to $v_o = 1$ V and $v_o = 3$ V.

(a)

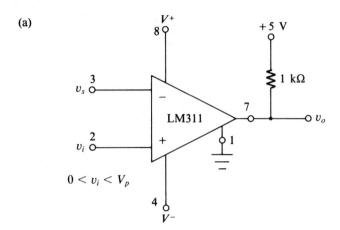

(b)

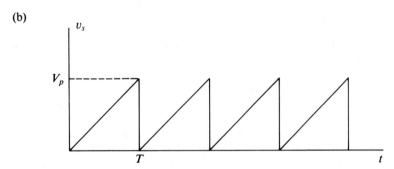

FIGURE P6–24

Design Problems

6–25. Using a general-purpose op-amp, design an open-loop zero crossing detector based on the following logic:

$$v_o \simeq 13 \text{ V} \quad \text{for } v_i > 0$$
$$\simeq -13 \text{ V} \quad \text{for } v_i < 0$$

6–26. Using a general-purpose op-amp, design an open-loop zero crossing detector based on the following logic:

$$v_o \simeq 13 \text{ V} \quad \text{for } v_i < 0$$
$$\simeq -13 \text{ V} \quad \text{for } v_i > 0$$

6–27. Using a 311 comparator, design an open-loop zero crossing detector based on the following logic:

$$v_o = 5 \text{ V} \quad \text{for } v_i > 0$$
$$= 0 \quad \text{for } v_i < 0$$

A dual ±15-V and a 5-V power supply are available.

6–28. Using a 311 comparator, design an open-loop zero crossing detector based on the following logic:

$$v_o = 5 \text{ V} \quad \text{for } v_i < 0$$
$$= 0 \quad \text{for } v_i > 0$$

6–29. An open-loop comparator is desired to provide the following logic:

$$v_o \simeq 13 \text{ V} \quad \text{for } v_i > 7.5 \text{ V}$$
$$\simeq -13 \text{ V} \quad \text{for } v_i < 7.5 \text{ V}$$

Design the circuit using a general-purpose op-amp and 5% resistors as required. Dual ±15-V power supplies are available.

6–30. An open-loop comparator is desired to provide the following logic:

$$v_o \simeq 13 \text{ V} \quad \text{for } v_i < 5 \text{ V}$$
$$\simeq -13 \text{ V} \quad \text{for } v_i > 5 \text{ V}$$

Design the circuit using a general-purpose op-amp and 5% resistors as required. Dual ±15-V power supplies are available.

6–31. An inverting Schmitt trigger circuit is to be designed for use as a zero crossing detector. Based on saturation voltages of ±13 V, the hysteresis transition voltages desired are about ±25 mV. Determine an acceptable design using 5% resistors.

6–32. Repeat the design of Problem 6–31 if the hysteresis transition voltages desired are about ±60 mV.

6–33. A noninverting Schmitt trigger circuit is to be designed for use in a certain function generator circuit. Based on saturation voltages of ±13 V, the desired transition voltages are ±2 V. Determine an acceptable design using 5% resistors.

6–34. Repeat the design of Problem 6–33 if the desired transition voltages are about ±6 V.

6–35. A circuit is to be designed to convert a 6-V sine wave to a square wave oscillating between −13 V and +13 V. A control is to be provided for possible delay adjustment as follows:
  a. At one extreme setting, the output square wave will simply be 180° out of phase with the sine wave.
  b. At the other extreme setting, the output square wave will undergo its transitions at positive and negative peaks of the input sine wave.
  Specify the value of any fixed resistance and the range of any variable resistance used.

6–36. A circuit is to be designed to convert a 6-V peak sine wave to a square wave oscillating between −13 V and +13 V. A control is to be provided for possible delay adjustment as follows:
  a. At one extreme setting, the output square wave will be delayed 30° with respect to the input.
  b. At the other extreme setting, the output square wave will undergo its transitions at positive and negative peaks of the input sine wave.
  Specify the value of any fixed resistance and the range of any variable resistance used.

LABORATORY EXERCISES

Noninverting Open-Loop Comparator
LE 6–1.

1. Connect the open-loop comparator of Figure 6–3.
2. Apply a 1-kHz sine wave input with a peak value of several volts. Sketch the forms of v_i and v_o showing their time relationships. Measure appropriate levels.

3. Investigate the effect on the output of varying the level of the input voltage.
4. Increase the frequency to 10 kHz and observe the output.
5. Increase the frequency to 100 kHz and observe the output. Explain the severe degradation in the waveform.

Inverting Open-Loop Comparator
LE 6–2.

Perform the exercises outlined in LE 6–1 for the circuit of Figure 6–4.

Comparator with Bias
LE 6–3.

1. Connect the circuit of Figure 6–4 using a variable low-voltage dc source for V_{ref}. (A fixed dc voltage and a potentiometer may be used if a variable voltage is not available.) Initially set V_{ref} to a small value.
2. Apply a 1-kHz sine wave with a peak value of several volts for v_i. With v_i fixed, investigate the effect on the output of varying the level of V_{ref}. Sketch the input and output waveforms and their time relationships for at least two values of V_{ref}.
3. Set V_{ref} to some convenient value in the range of several volts. With V_{ref} fixed and the sine wave input of step (2), investigate the effect on the output of varying the level of v_i. Sketch the input and output waveforms and their time relationships for at least two values of V_{ref}.

Inverting Schmitt Trigger (Small Threshold)
LE 6–4.

1. Connect the circuit of Figure 6–11(a). Initially set $R_1 = 100$ kΩ and $R_2 = 1$ kΩ.
2. Apply a 1-kHz sine wave input with a peak value of several volts. Sketch the forms of v_i and v_o showing their time relationships. Measure appropriate levels.
3. Investigate the effect on the output of varying the level of the input voltage.

Inverting Schmitt Trigger (Large Threshold)
LE 6–5.

Perform the exercises outlined in LE 6–4 for the circuit of Figure 6–4.

LM311 Comparator
LE 6–6.

1. Using an LM311 comparator, connect the circuit of Figure 6–19.
2. Apply a 1-kHz sine wave to the input. Sketch the forms of the input and output waveforms and show their time relationships. Measure appropriate levels.
3. Increase the frequency to 100 kHz and observe the output. How does this output compare with the expected output of a 741 comparator operating at the same frequency?

7

OSCILLATORS
AND WAVEFORM
GENERATORS

OBJECTIVES

After completing this chapter, the reader should be able to:

- Discuss two ways of classifying oscillator circuits according to output waveform and method of generating the waveform.
- Explain the differences among the *bistable, monostable*, and *astable* multivibrator circuits.
- Draw the schematic diagram for and analyze the operation of an astable multivibrator implemented with an op-amp and design one to produce a specified output frequency.
- Show how a 555 timer is connected for astable operation and design the circuit required for 555 astable operation at a specified frequency.
- Show how a 555 timer is connected for monostable operation and design the circuit required to achieve a specified pulse width.
- Draw the schematic diagram for and analyze the operation of a square wave/triangular wave function generator and design one to produce a specified output frequency.
- State the Barkhausen criterion for oscillation in both rectangular and polar form.
- Draw the schematic diagram for and analyze the operation of a Wien bridge oscillator and design the circuit for producing a specified output frequency.
- Discuss how additional negative feedback can be applied to a Wien bridge oscillator to stabilize the loop.
- Draw the schematic diagram for and analyze the operation of a phase shift oscillator and design the circuit for producing a specified output frequency.
- Discuss the general operation of an 8038 waveform generator and its capabilities.

7-1 INTRODUCTION

Most electronic equipment, both analog and digital, requires one or more periodic waveforms to perform various timing, control, or modulation functions. Such waveforms can be sinusoids, square waves, triangular waves, or other forms. Consequently, it is necessary to generate these various waveforms using appropriate electronic components.

The classical term for a circuit used to generate a waveform is *oscillator*, and this term is still used. More recent terms include *function generator* and *waveform generator*. Other terms are also used in specific cases (e.g., *multivibrator*), as will be seen.

A wide variety of oscillator circuit forms have been developed. We consider some of the most widely used forms suitable for implementation with op-amps and related IC technology in this chapter. As a general rule, these circuit forms employ one or more op-amps or related active devices and an *RC* circuit.

The forms best suited for communications applications at higher frequencies usually employ an *LC* circuit with the active device. However, the topic of *LC* oscillators is not within the intended scope of this text, so the interested reader should consult communications circuit texts.

Within the chapter, we consider both sinusoidal and nonsinusoidal oscillators. Among the nonsinusoidal oscillators we discuss are an op-amp multivibrator, the 555 IC timer, and a square wave/triangular wave generator. Sinusoidal oscillators include the Wien bridge oscillator and the phase shift oscillator. We also provide an introduction to the 8038 IC waveform generator, which produces several waveforms simultaneously.

7-2 CLASSIFICATION SCHEME

A common characteristic of all oscillators is that some form of positive feedback is present. The effect of the positive feedback is to provide enough signal level to maintain oscillations of the proper form. Oscillations are obviously very undesirable in linear amplifier circuits, so the corresponding design strategies are based on minimizing the possibility of oscillation. In contrast, oscillators are designed specifically to produce a controlled and predictable oscillation, so the design strategy is quite different. Indeed, most oscillators do not have an external input because an oscillator circuit is usually an internal feedback loop having one or more points at which the output signal may be extracted. Oscillator inputs, when present, are normally used to control the frequency or to synchronize the oscillations with an external reference.

Because of the large number of different types of oscillator circuits and design techniques, it is very difficult to classify them in any simple manner. The classification schemes that will be used here are somewhat oversimplified, but they should be sufficient to categorize many of the modern op-amp oscillator circuits and the linear integrated circuit oscillator modules.

First, oscillators may be classified as either ***sinusoidal*** or ***nonsinusoidal***. By the terms, it follows that sinusoidal oscillators produce sinusoidal waveforms, and all other oscillators can be classified as nonsinusoidal.

Strictly speaking, very few oscillators produce a pure sinusoid, since there is usually some harmonic distortion present. ***Harmonics*** are frequencies that are integer multiples of the primary output frequency. Thus, a 1-kHz "sinusoidal" oscillator may have a small output at 2 kHz, 3 kHz, and so on. On an oscilloscope, the waveform may appear to be a perfectly good sinusoid, since it is difficult visually to determine small levels of harmonic distortion. A spectrum analyzer or frequency-selective voltmeter may be used to determine the distortion level.

A major point of the preceding discussion is that some attention should be paid to the harmonic distortion level of a "sinusoidal" oscillator. Depending on the application, this level could affect the choice of a circuit. For classification purposes, we will designate any oscillator whose objective is to produce a sinusoid as a sinusoidal oscillator even though some distortion may be present.

A second classification scheme for oscillators is based on the technique used for generating the basic waveforms. The two primary methods considered in this text will be denoted as (1) the *timing method* and (2) the *unity-loop-gain method*.

The *timing method* employs *RC* timing circuits and components that change states (for example, comparators) at certain critical levels. The primary waveforms generated by such circuits are usually nonsinusoidal. However, by appropriate waveshaping circuits, such functions may be converted to approximate sinusoids. Many of the currently available waveform generator IC chips utilize the timing method. Many of the earlier oscillator circuits using the timing method were referred to as *relaxation* oscillators.

The *unity-loop-gain method* employs the concept that steady-state oscillations will be generated in any feedback circuit in which the loop gain can be maintained at unity. This concept was the earliest approach used in oscillator design, and many types of oscillators covering a wide frequency range have evolved from this approach. The basis for the unity-loop-gain method will be developed in Section 7–6.

7–3 OPERATIONAL AMPLIFIER ASTABLE MULTIVIBRATOR

The first oscillator that will be considered is a multivibrator circuit. In its basic form, this circuit requires only an op-amp, three resistors, and a capacitor. However, the approach employed in this circuit is representative of oscillators utilizing the timing principle.

Multivibrator Classification

Before we discuss the details of this circuit, we will review the classification of multivibrator circuits. Historically, multivibrator circuits date back to the early days of vacuum tubes. In general, there are three types of multivibrators: (1) *bistable* (or *flip-flop*), (2) *monostable* (or *one-shot*), and (3) *astable* (or *free-running*).

The *bistable multivibrator* (or *flip-flop*) has two stable states and will remain in either state until a proper trigger is received. While it is possible to implement a bistable multivibrator with op-amps, there is very little reason to emphasize such an approach, since digital flip-flop circuits are superior and are readily available.

The *monostable multivibrator* (or *one-shot*) has one stable state. The circuit will remain in the stable state until a trigger pulse (or level) is received. The circuit then changes states for a specified period, but then it returns to the original stable state. Both digital and linear monostable IC chips are available. The use of the 555 chip as a mono-stable multivibrator will be discussed in Section 7–4.

The *astable* (or *free-running*) multivibrator has no stable states. Consequently, it continually changes back and forth between two states at a predictable rate. The remainder of this section will be devoted to the analysis and design of an op-amp astable multivibrator circuit.

Analysis of Circuit

Refer to the circuit diagram of Figure 7–1. Observe that the circuit consists of an inverting Schmitt trigger plus an RC timing circuit. In the discussion that follows, references will be made to both this circuit and the waveforms of Figure 7–2 as the need arises. It is very difficult to analyze the buildup of oscillations in a circuit such as this. Instead, one must assume a certain steady-state starting point within a cycle, develop the response for the remainder of the cycle, and show that the conditions that follow for the beginning of the next cycle are the same as those assumed at the beginning of the preceding cycle.

The reference time $t = 0$ will be used as the starting point for the analysis even though steady-state conditions are assumed to have already been established. A transition in the output state is assumed at $t = 0$. To avoid ambiguity in the transition interval, we will use the terminology $t = 0^+$ to refer to the time immediately after the transition occurs. However, $t = 0^+$ is treated like $t = 0$ in continuous mathematical expressions.

Assume that the output v_o is at positive saturation immediately after the transition, as shown in Figure 7–2(a); that is, $v_o(0^+) = V_{sat}$. Observe from the circuit diagram of Figure 7–1 that a fraction of v_o is fed back to the noninverting input. The form of this voltage $v^+(t)$ is illustrated in Figure 7–2(b). For convenience, the magnitude V_1 of the feedback voltage is defined as

$$V_1 = \frac{R_2}{R_1 + R_2} V_{sat} \tag{7–1}$$

Thus, $v^+(0^+) = V_1$.

FIGURE 7–1

Astable multivibrator circuit implemented with an operational amplifier.

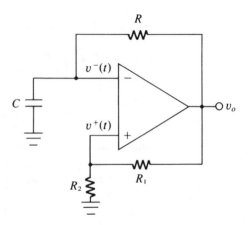

(a)

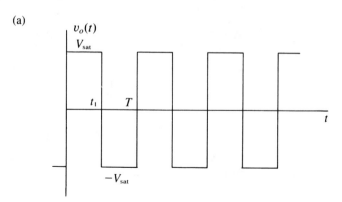

(b)

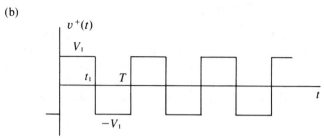

(c)

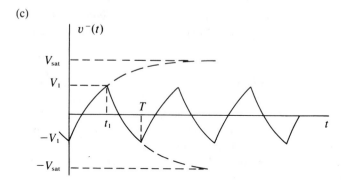

FIGURE 7–2
Waveforms in op-amp astable multivibrator circuit.

Next, refer to the voltage $v^-(t)$ at the inverting input as shown in Figure 7–2(c). For reasons that may be puzzling at this point, assume that the initial value of the inverting voltage is $v^-(0^+) = -V_1$. Observe that the voltage at the noninverting terminal is clearly more positive than the voltage at the inverting terminal, which is compatible with the assumption of positive saturation.

Consider the timing portion of the op-amp circuit consisting of C, R, and the output of the op-amp. Since the op-amp output voltage is momentarily a constant value V_{sat}, the ouput is acting as if it were a dc voltage source with voltage V_{sat}. Consequently, this part

FIGURE 7–3
Equivalent circuit model for astable timing circuit when output is in positive saturation.

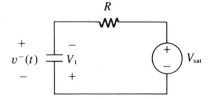

of the circuit may be redrawn as shown in Figure 7–3, which is a first-order circuit containing an initial capacitor voltage and a "dc equivalent source."

First-Order Timing Circuits

First-order circuits are those that can be simplified through various network theorems to one containing a single capacitance and a single resistance *or* a single inductance and a single resistance. When such circuits are excited by a combination of dc sources and/or initial conditions, the response can be written by inspection. It is shown in various network analysis texts that the voltage or current in such a circuit may be expressed in general form as

$$y(t) = y(\infty) + [y(0^+) - y(\infty)]e^{-t/\tau} \qquad (7\text{–}2)$$

where the symbol y is used to represent either voltage or current. In this equation, $y(0^+)$ is the initial value, $y(\infty)$ is the final value, and τ is the time constant. For circuits containing an equivalent capacitance C and an equivalent resistance R, the time constant is

$$\tau = RC \qquad (7\text{–}3)$$

We will now apply this concept to the circuit of Figure 7–3 to determine $v^-(t)$. The initial value of this voltage is $v^-(0^+) = -V_1$ as already assumed. However, the capacitor is connected through the resistance R to a constant voltage V_{sat}, which is acting as if it were a dc voltage. Eventually, if the circuit were allowed to settle into a dc condition (no oscillations), the voltage across the capacitor would reach V_{sat}. Therefore, $v^-(\infty) = V_{\text{sat}}$ is used in (7–2). Actually, v^- will never reach V_{sat} because a switching action will occur before it gets there. However, the capacitor charges toward V_{sat} *as if* it would reach that value. (The capacitor "does not know" that switching will occur.) The time constant is simply $\tau = RC$.

Substituting the preceding values in (7–2), we can express the form of $v^-(t)$ for some interval of time as

$$v^-(t) = V_{\text{sat}} - (V_1 + V_{\text{sat}})e^{-t/\tau} \qquad (7\text{–}4)$$

This function is shown as a combination of a solid curve and a dashed curve in Figure 7–2(c). The break will be explained shortly.

As the voltage $v^-(t)$ moves toward V_{sat}, it eventually reaches a point where it slightly exceeds the voltage at the noninverting input. When this happens, the differential voltage $v^+ - v^-$ becomes negative, and the op-amp output switches to the negative saturation level $-V_{\text{sat}}$, as shown in Figure 7–2(a). Simultaneously, the voltage at the noninverting terminal switches to $-V_1$ as shown in Figure 7–2(b). Now that the output voltage has switched to a negative level, the capacitor ceases to charge in a positive direction. Instead, it now starts to charge toward $-V_{sat}$, as shown in Figure 7–2(c). When the capacitor voltage reaches $-V_1$ and drops slightly below this level, the differential voltage becomes positive again, causing the output to change back toward positive saturation. This condition is the one assumed at the beginning of the analysis, and a full cycle has been completed.

Period and Frequency

A fundamental question of interest is the relationship between the period (or frequency) and the circuit parameters. Assume for this purpose a symmetrical situation with equal magnitudes for the saturation voltages, as depicted in Figure 7–2. Let t_1 represent the time for one half-period. From (7–4), the voltage must reach V_1 when $t = t_1$; that is, $v^-(t_1) = V_1$. Substitution of these values results in the equation

$$V_1 = V_{sat} - (V_1 + V_{sat})e^{-t_1/\tau} \tag{7–5}$$

Some simplification yields

$$e^{t_1/\tau} = \frac{V_{sat} + V_1}{V_{sat} - V_1} \tag{7–6}$$

The quantity t_1/τ may be determined by taking the natural logarithms of both sides. From this operation and multiplication of both sides by τ, we have

$$t_1 = \tau \ln\left(\frac{V_{sat} + V_1}{V_{sat} - V_1}\right) \tag{7–7}$$

The total period T is

$$T = 2t_1 = 2\tau \ln\left(\frac{V_{sat} + V_1}{V_{sat} - V_1}\right) \tag{7–8}$$

The voltages in the argument of the logarithmic function of (7–8) may be eliminated by first substituting V_1 from (7–1) expressed in terms of V_{sat}. With this substitution, V_{sat} appears as a common factor for both numerator and denominator and may be canceled. After some subsequent algebraic simplification, the period becomes

$$T = 2\tau \ln\left(\frac{R_1 + 2R_2}{R_1}\right) \tag{7–9a}$$

or

$$T = 2RC \ln\left(1 + \frac{2R_2}{R_1}\right) \tag{7–9b}$$

The frequency f is, of course, $f = 1/T$. However, the expression for T is easier to work with and will be used in much of the work that follows.

From (7–9a) and (7–9b), the period is directly proportional to the time constant $\tau = RC$. Thus, changing R or C always changes the period in the same ratio. The most convenient way to implement a continuously tunable oscillator is to employ a variable resistor for R.

The variation of period with the voltage levels V_1 and V_{sat}, or equivalently with R_1 and R_2, is more complex because of the logarithmic relationship. The period increases with an increase in R_2, and the period decreases with an increase in R_1. However, the resulting change in either case is not linear.

Design Considerations

In the design of a multivibrator circuit, there are usually more component values to be selected than there are constraints, so one or more components can usually be arbitrarily selected. Because of the logarithmic relationship concerning the resistances R_1 and R_2, it is often better to make arbitrary choices for R_1 and R_2 as a starting point.

While different design objectives and operating frequency ranges could influence such a choice, one particular constraint deserves mention. If the resistances are selected such that $R_2 = 0.86R_1$, the logarithmic function in (7–9b) becomes unity, and the expression for period becomes

$$T = 2\tau \quad \text{for } R_2 = 0.86R_1 \tag{7–10}$$

Verification of this point will be left as an exercise for the reader in Problem 7–17. Resistance ratios that are very high or very low should be avoided.

Two precautions concerning the circuit deserve some attention. First, the voltage across the capacitor assumes both positive and negative values. Thus, an electrolytic (or any type of polarized) capacitor should **not** be used. Second, the op-amp used must have a maximum differential input voltage specification sufficient for the range involved. From Figure 7–2(b) and (c), it is noted that the peak values of this voltage slightly exceed $\pm 2V_1$. This factor could also be a design consideration in the selection of the voltage level V_1, which is a function of the resistances R_1 and R_2.

EXAMPLE 7–1

For the circuit of Figure 7–4, determine the (a) feedback voltage, (b) period, and (c) frequency. Assume $\pm V_{sat} = \pm 13$ V.

Solution

Comparing the circuit with the basic form of Figure 7–1, it is noted that $R = 12$ kΩ, $C = 0.02$ μF, $R_1 = 30$ kΩ, and $R_2 = 20$ kΩ.

(a) The feedback voltage V_1 is

$$V_1 = \frac{R_2 V_{sat}}{R_1 + R_2} = \frac{20 \text{ k}\Omega}{30 \text{ k}\Omega + 20 \text{ k}\Omega} \times 13 = \frac{2}{5} \times 13 = 5.2 \text{ V} \tag{7–11}$$

(b) The period can be calculated from (7–8) or either form of (7–9). Arbitrarily choosing (7–9b), we have

$$T = 2RC \ln\left(1 + \frac{2R_2}{R_1}\right) = 2 \times 12 \times 10^3 \times 0.02 \times 10^{-6} \ln\left(1 + \frac{2 \times 20 \text{ k}\Omega}{30 \text{ k}\Omega}\right)$$
$$= 0.00048 \ln 2.3333 \tag{7–12}$$
$$= 0.00048 \times 0.84730 = 0.4067 \text{ ms}$$

FIGURE 7–4

Circuit of Example 7–1.

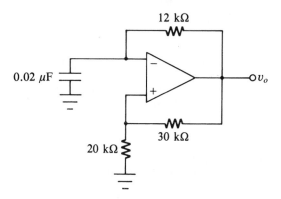

(c) The frequency is

$$f = \frac{1}{T} = \frac{1}{0.4067 \times 10^{-3}} = 2459 \text{ Hz} \qquad (7\text{–}13)$$

EXAMPLE 7–2

It is desired to design an astable multivibrator of the form shown in Figure 7–1. The following approach is somewhat arbitrarily assumed: The feedback resistances are selected as $R_1 = 150 \text{ k}\Omega$ and $R_2 = 100 \text{ k}\Omega$. (a) Determine the time constant τ required to produce a 1-kHz output. (b) For $C = 0.01 \ \mu\text{F}$, determine the required value of R. (c) Determine the maximum value of the differential input voltage if $\pm V_{\text{sat}} = \pm 14 \text{ V}$.

Solution

(a) With the choices of R_1 and R_2 given in the problem statement, the expression for the period in (7–9b) reduces to

$$T = 2\tau \ln\left(1 + \frac{2 \times 10^5}{1.5 \times 10^5}\right) = 1.695\tau \qquad (7\text{–}14)$$

The frequency $f = 1$ kHz corresponds to $T = 1/10^3 = 10^{-3}$ s. This value requires that $1.695\tau = 10^{-3}$. The time constant is thus

$$\tau = 0.5900 \times 10^{-3} \text{ s} \qquad (7\text{–}15)$$

(b) With $C = 0.01 \ \mu\text{F}$ and $\tau = RC$, the value of R is determined as

$$R = \frac{0.5900 \times 10^{-3}}{0.01 \times 10^{-6}} = 59 \text{ k}\Omega \qquad (7\text{–}16)$$

It is of interest to note that the parallel combination of R_1 and R_2 is 60 kΩ, which is very close to the required value of R. This condition is very nearly the optimum situation for minimizing the effects of input bias currents. Adjustment of the resistance levels of R_1 and R_2 (but not their ratio) can sometimes be used to achieve or at least approach this condition.

(c) From (7–1), the value of V_1 is

$$V_1 = \frac{100 \times 10^3}{100 \times 10^3 + 150 \times 10^3} \times 14 = 5.6 \text{ V} \qquad \textbf{(7–17)}$$

The peak values for the differential input voltage are slightly greater than $\pm 2 \times 5.6 = \pm 11.2$ V.

7–4 THE 555 TIMER

The 555 integrated circuit timer was first introduced by Signetics, Inc., but other manufacturers have since provided versions of the circuit. The widespread popularity of this chip is shown by the fact that several books devoted to the 555 timer and its numerous applications have been published. A general overview of the operation of the timer and its basic configurations is given in this section; additional data and a detailed circuit diagram are provided in Appendix C in the specification sheets. A more realistic starting point for discussion is the operational block diagram shown in Figure 7–5.

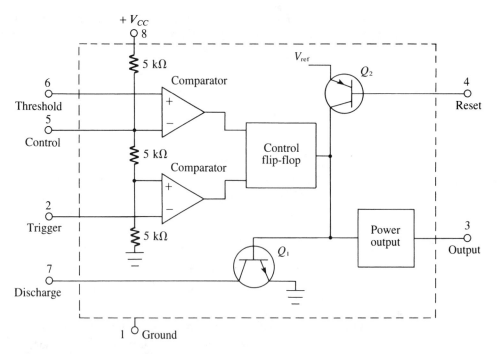

FIGURE 7–5
Block diagram of major operations contained on 555 timer chip.

General Description

The 555 timer will operate with a dc supply voltage from +5 V to +18 V. Because of this flexibility, it can be made to be compatible with common logic levels as well as with op-amp voltage levels. The positive terminal of the dc power supply is connected to pin 8 ($+V_{CC}$), and the negative terminal is connected to pin 1 (ground). The ground terminal serves as the common point from which all subsequent voltages are referred.

The *output* (pin 3) of the 555 timer can assume one of two possible levels. The "high" level is slightly less than V_{CC} (typically 0.5 V less than V_{CC}), and the "low" level is approximately 0.1 V. In subsequent discussions and on schematics, the output voltage in the high state will be denoted as V_H, and the output voltage in the low state will be denoted as V_L.

Observe that there are two comparators in the circuit. These will be referred to as the upper and lower comparators, respectively. The inverting terminal of the upper comparator is connected to a point whose dc potential is readily observed to be $2V_{CC}/3$, assuming that pin 5 is not connected externally to a dc circuit. The noninverting input of the upper comparator is called the *threshold terminal* (pin 6).

The noninverting terminal of the lower comparator is connected to a point whose dc potential is $V_{CC}/3$, assuming again that pin 5 is not connected to a dc voltage. The inverting input of the lower comparator is called the *trigger terminal* (terminal 2).

Through operation of the two comparators, the connections to the control flip-flop, and the power output state, the logic for the 555 timer is summarized in Table 7–1. The voltages $V_{CC}/3$ and $2V_{CC}/3$ are the transition levels of the comparators in the basic form, as previously discussed. When an external dc circuit is connected to the control input (pin 5), the transition levels will be modified. Note that the condition of the *discharge terminal* (pin 7) is a function of which state exists at the output. When the output is high, the discharge transistor is reverse biased. In this case, the discharge terminal appears essentially as an open circuit. When the output is low, the discharge transistor is forward biased. The discharge terminal then appears nearly as a short to ground.

The *reset terminal* (pin 4) allows the 555 timer to be disabled. If this terminal is grounded or the potential is reduced below 0.4 V, the output assumes the low state irrespective of the operation underway. If this feature is not needed in a given application, the reset terminal is connected to V_{CC}.

TABLE 7–1

Four possible states for 555 timer

Condition	Trigger* (pin 2)	Threshold* (pin 6)	Output State (pin 3)	Discharge State (pin 7)
A	Below $V_{CC}/3$	Below $2V_{CC}/3$	High	Open
	Below $V_{CC}/3$	Above $2V_{CC}/3$	Remembers last state	Remembers last state
C	Above $V_{CC}/3$	Below $2V_{CC}/3$	Remembers last state	Remembers last state
D	Above $V_{CC}/3$	Above $2V_{CC}/3$	Low	Ground

* Transition levels are based on no signal at control input.

The *control input* (pin 5) is used for special-purpose functions where the reference levels of the comparators are to be varied. When this function is not required, pin 5 should be bypassed to ground with a 0.01-μF capacitor.

The 555 timer can be operated in either the *astable* mode or the *monostable* mode. Each of these configurations will be discussed in some detail.

Astable Operation

The basic circuit for astable (free-running) operation of the 555 timer is shown in Figure 7–6. The timing circuit consists of R_A, R_B, and C. The trigger (2) and threshold (6) terminals are connected to the timing capacitor, and the discharge terminal is connected to the junction point between the resistors. For the simplest form of astable operation, the reset and control functions are not used, so pin 4 is connected to V_{CC}, and pin 5 is bypassed to ground with a 0.01-μF capacitor.

The timing portion of the circuit is delineated in Figure 7–7, and the associated waveforms are shown in Figure 7–8. In the discussion that follows, reference will be made to both figures as required.

As in the case of the op-amp multivibrator discussed in the preceding section, we will assume a convenient point of reference at the beginning of a cycle, develop the successive steps within a cycle, and show that the conditions at the end of a full cycle are the same as those assumed at the beginning. Steady-state oscillatory conditions will be assumed, but the convenient reference time $t = 0$ will be used for the beginning.

The waveforms for the output voltage $v_o(t)$ and the capacitor voltage $v_C(t)$ are shown in Figure 7–8. Assume that a transition from a low to a high state occurs at $t = 0$. At $t = 0^+$,

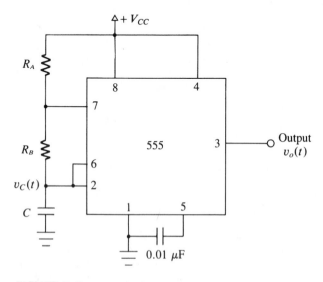

FIGURE 7–6
Astable multivibrator using 555 timer.

(a) Output high state

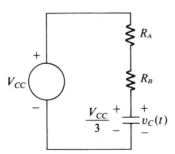

(b) Output low state

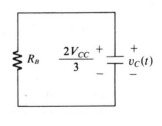

FIGURE 7–7
Equivalent circuits of timing circuit portion of 555 astable circuit.

the output voltage is V_H, and the discharge terminal is open. Assume that the capacitor voltage is $V_{CC}/3$ at $t = 0^+$. Assume also that the trigger and threshold conditions of A of Table 7–1 applied just before the transition.

The model for the timing circuit during the output high state is shown in Figure 7–7(a). This circuit is a first-order circuit with a dc source and an initial capacitor voltage. Thus, the capacitor voltage $v_C(t)$ may be expressed in the form of (7–2). The initial value of $v_C(0^+) = V_{CC}/3$. The capacitor charges toward V_{CC} (although switching action prevents it from reaching that level), so $v_C(\infty) = V_{CC}$. The time constant τ_H during the high

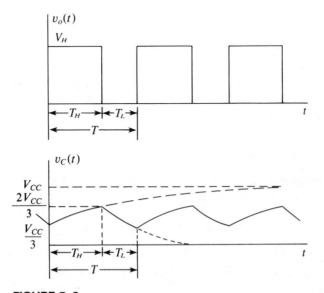

FIGURE 7–8
Waveforms in 555 astable circuit.

interval is

$$\tau_H = (R_A + R_B)C \qquad (7\text{–}18)$$

Substituting these values in (7–2), we have

$$v_C(t) = V_{CC} - \frac{2}{3} V_{CC} e^{-t/\tau_H} \qquad (7\text{–}19)$$

for the capacitor voltage during the output high interval.

As $v_C(t)$ rises above $V_{CC}/3$, condition C of Table 7–1 applies as long as this voltage is less than $2V_{CC}/3$, and the output remains in the high state. However, when $v_C(t)$ slightly exceeds $2V_{CC}/3$, the upper comparator initiates a change, and condition D of Table 7–1 applies. The output assumes the low state, and the discharge terminal becomes grounded.

The equivalent circuit during the output low state is shown in Figure 7–7(b). The action of the discharge terminal has removed the effects of R_A and V_{CC} from the timing circuit. The timing circuit now consists of a capacitor initially charged to $2V_{CC}/3$ in parallel with a resistance R_B. The capacitor voltage will now start to decrease exponentially as shown in Figure 7–8(b).

A mathematical expression for the capacitor discharge voltage can be readily determined. It is convenient to define a new time scale t' that starts at the beginning of the discharge phase. The initial value is $2V_{CC}/3$, and the final value would be $v_C(\infty) = 0$, *if* the capacitor were allowed to fully discharge. The time constant τ_L during this low phase is

$$\tau_L = R_B C \qquad (7\text{–}20)$$

Substituting these values in (7–2), we have

$$v_C(t') = \frac{2V_{CC}}{3} e^{-t'/\tau_L} \qquad (7\text{–}21)$$

for the capacitor voltage during the output low interval.

As $v_C(t')$ drops below $2V_{CC}/3$, condition C applies as long as this voltage is greater than $V_{CC}/3$. However, when $v_C(t)$ drops slightly below $V_{CC}/3$, the lower comparator initiates a change, and condition A of Table 7–1 applies. The output then assumes the high state, and the discharge terminal is opened. This condition is the one assumed at the beginning, and one full cycle has been completed.

Period and Frequency

Mathematical expressions for the period and frequency will now be developed. Let T_H represent the time during which the output assumes the high state, and let T_L represent the corresponding time in the low state. The time T_H is determined by setting $t = T_H$ and $v_C(T_H) = 2V_{CC}/3$ in (7–19). Solution of this equation yields

$$T_H = \tau_H \ln 2 = 0.693\tau_H = 0.693(R_A + R_B)C \qquad (7\text{–}22)$$

The time T_L is determined by setting $t' = T_L$ and $v_C(T_L) = V_{CC}/3$ in (7–21). Solution of this equation yields

$$T_L = \tau_L \ln 2 = 0.693\tau_L = 0.693 R_B C \qquad (7\text{–}23)$$

The total period T is

$$T = T_H + T_L = 0.693(R_A + 2R_B)C \qquad (7\text{–}24)$$

Finally, the frequency $f = 1/T$ is

$$f = \frac{1.443}{(R_A + 2R_B)C} \qquad (7\text{–}25)$$

It can be observed from the expressions for T_H and T_L as well as from Figure 7–8 that the high interval is longer than the low interval in the basic circuit configuration. The different lengths may be completely unimportant in some applications, while they may pose some problems in others. They are a result of the fact that the resistance during the charging interval is $R_A + R_B$, while the resistance during the discharge interval is R_B. It might seem that a symmetrical duty cycle could be approached by choosing R_B to be much greater than R_A. However, one must be very cautious with that approach for the following reason: During the output low interval, the discharge terminal is grounded. Thus, R_A is placed directly across the power supply, and the resulting power dissipation may be excessive. For example, if R_A were 100 Ω and V_{CC} were 15 V, the power dissipated in R_A would be 2.25 W! A power calculation based on V_{CC} and R_A should always be made in a circuit design with a relatively small value of R_A because of this potential difficulty. A further restriction is that current in the discharge transistor is limited to 0.2 A. During the low state, this current is approximately V_{CC}/R_A, so a check on this value should also be made when R_A is relatively small.

Duty Cycle

The standard definition of duty cycle D as used for most pulse generator circuits is

$$D = \frac{\text{high interval}}{\text{period}} \times 100\% = \frac{T_H}{T} \times 100\% \qquad (7\text{–}26)$$

Using this standard definition and substituting (7–22) and (7–24) in (7–26), we obtain

$$D = \frac{R_A + R_B}{R_A + 2R_B} \times 100\% \qquad (7\text{–}27)$$

Since $T_H > T_L$, the duty cycle for the basic circuit is always greater than 50%; that is, $D > 50\%$. Some specific references on the 555 utilize a reverse definition, replacing T_H in (7–26) with T_L. The reason for reversing the definition is to have as the resulting duty cycle a value less than 50%. We have conformed to the more widely employed definition of duty cycle.

Fifty Percent Duty Cycle

A number of methods may be used for establishing exact or nearly exact time intervals for T_H and T_L in applications requiring such symmetry. One relatively simple concept is illustrated in Figure 7–9. Assume momentarily that the diode D_1 is ideal. During the high inter-

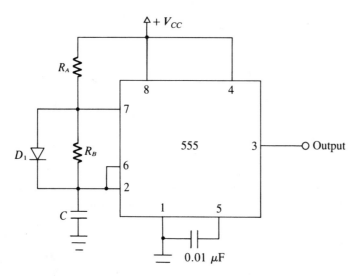

FIGURE 7–9
Means for achieving approximately equal high and low intervals in 555 astable mode using a diode.

val, the diode will be forward biased, and the time constant is $\tau_H = R_A C$. When the discharge transistor turns on, the upper terminal of the diode is grounded, and the diode will thus be reversed biased during the low interval. The time constant during the low interval is $\tau = R_B C$, which is the same as in earlier developments. Thus, if R_A were equal to R_B, the two time constants would be the same if the diode were ideal. In practice, some difference in the intervals can occur with this method because the diode is not ideal. The difference is less when the resistances R_A and R_B are both large compared with the diode forward resistance and when the diode forward voltage drop is small compared with V_{CC}.

One method for achieving essentially perfect symmetry is illustrated in Figure 7–10. In this method, a nonsymmetrical square wave having exactly twice the desired frequency is generated with the standard 555 timer circuit discussed in detail earlier. The 555 output is then applied to a divide-by-2 logic counter circuit, in which each change of state is initiated by either a negative transition or a positive transition (but not both), depending on the type of circuit. (A simple toggle flip-flop will suffice.) The output of this circuit will

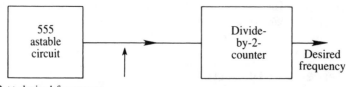

FIGURE 7–10
Means for achieving exactly equal high and low intervals using a counter.

have exactly half the frequency of the 555 timer, and the intervals T_H and T_L of the output waveform will be exactly the same.

Other methods for achieving symmetry can be found in the books and application notes pertaining to the 555 timer. One particular method not previously observed by this author in the literature is considered in Problem 7–20.

Monostable Operation

The basic circuit for monostable (one-shot) operation of the 555 timer is shown in Figure 7–11. The timing circuit consists of R and C. The threshold (6) and discharge (7) terminals are connected to the timing capacitor. For the simplest form of monostable operation, the reset and control functions are not used, so pin 4 is connected to V_{CC} and pin 5 is bypassed to ground with a 0.01-μF capacitor.

Monostable operation requires an input trigger pulse to initiate the output pulse. The trigger input (pin 2) is the point at which this signal is applied. The concept for triggering is based on condition A of Table 7–1; that is, if the trigger input drops below the reference level of the lower threshold, the output assumes the high state for a specified interval. While it is possible to apply the trigger pulse directly to pin 2, a triggering circuit, such as shown on the left of Figure 7–11, is usually better. The capacitor "differentiates" the input pulses so that only narrow trigger pulses are applied to the 555 timer. The diode limits the positive peak level of the pulses. The time constant R_iC_i should be chosen to be small compared with the monostable pulse width T_p, which will be discussed shortly. The use of this triggering circuit minimizes the possibility of false triggering or "misses."

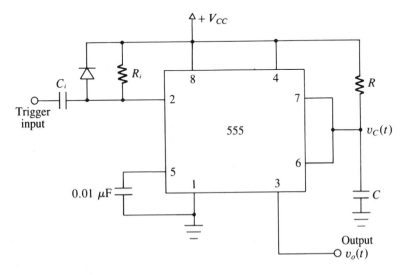

FIGURE 7–11
Monostable multivibrator using 555 timer.

FIGURE 7–12

Timing circuit portion of 555 monostable circuit.

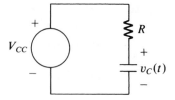

The timing portion of the circuit is delineated in Figure 7–12, and the associated waveforms are shown in Figure 7–13. In the discussion that follows, references will be made to both figures as required.

Assume that a negative trigger pulse is received at $t = 0$. The output then assumes the high state as shown in Figure 7–13(a). At $t = 0^+$, the output voltage is V_H, and the discharge terminal is open. The equivalent circuit based on the preceding assumptions is shown in Figure 7–12, which is a first-order circuit with a dc source. The initial value of the capacitor voltage is $v_C(0^+) = 0$. The capacitor charges toward V_{CC} (although switching action prevents it from reaching that level), so $v_C(\infty) = V_{CC}$. The time constant τ is

$$\tau = RC \tag{7-28}$$

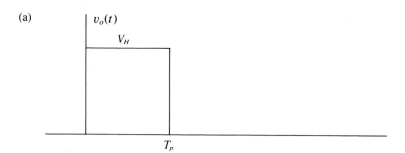

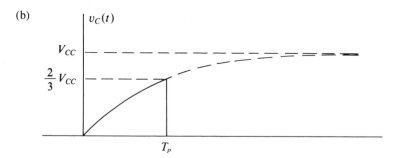

FIGURE 7–13

Waveforms in 555 monostable circuit following excitation with a single negative trigger pulse.

Substituting these values in (7–2), we have

$$v_C(t) = V_{CC} - V_{CC}e^{-t/\tau} \tag{7-29}$$

which is illustrated in Figure 7–13(b).

When $v_C(t)$ slightly exceeds $2V_{CC}/3$, the upper comparator initiates a change, and condition D of Table 7–1 applies. The output assumes the low state, and the discharge terminal becomes grounded. The capacitor is very quickly discharged, and both $v_o(t)$ and $v_C(t)$ will remain low until a new trigger pulse is received.

Pulse Width

Let T_p represent the pulse width produced by the 555 monostable timer as illustrated in Figure 7–13(a). The time T_p is determined by setting $t = T_p$ and $v_c(T_p) = 2V_{CC}/3$ in (7–29). Solution of this equation yields

$$T_p = \tau \ln 3 = RC \ln 3 = 1.1RC \tag{7-30}$$

EXAMPLE 7–3 Consider the 555 astable circuit of Figure 7–14. Determine the (a) high state-time interval, (b) low state-time interval, (c) period, (d) frequency, and (e) duty cycle.

Solution
Comparing the specific circuit of Figure 7–14 with the general circuit form of Figure 7–6, we note that $R_A = 1$ kΩ, $R_B = 3$ kΩ, and $C = 0.02$ μF.

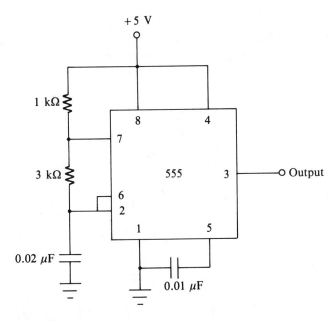

FIGURE 7–14
Circuit of Example 7–3.

(a) The high state-time interval is

$$T_H = 0.693(R_A + R_B)C = 0.693(1 \times 10^3 + 3 \times 10^3)(0.02 \times 10^{-6}) = 55.44 \ \mu s \quad \textbf{(7–31)}$$

(b) The low state-time interval is

$$T_L = 0.693R_BC = 0.693 \times 3 \times 10^3 \times 0.02 \times 10^{-6} = 41.58 \ \mu s \qquad \textbf{(7–32)}$$

(c) The period is

$$T = T_H + T_L = 55.44 + 41.58 = 97.02 \ \mu s \qquad \textbf{(7–33)}$$

(d) The frequency is

$$f = \frac{1}{T} = \frac{1}{97.02 \times 10^{-6}} = 10.31 \text{ kHz} \qquad \textbf{(7–34)}$$

(e) The duty cycle is

$$D = \frac{T_H}{T} \times 100\% = \frac{55.44 \ \mu s}{97.02 \ \mu s} \times 100\% = 57.14\% \qquad \textbf{(7–35)}$$

From the preceding work and the circuit diagram, it is deduced that the circuit produces a 10.31-kHz square wave in which the levels are about 0 and 5 V, and with the 5-V level occupying about 57% of the period.

EXAMPLE 7–4

Design an astable 555 timer circuit of the form shown in Figure 7–6 to product a 1-kHz square wave. For this application, the symmetry of the square wave is not critical, so the simplified choice $R_A = R_B$ will be made. Determine appropriate values for the design.

Solution
The frequency $f = 1$ kHz corresponds to $T = 1$ ms. From (7–24) or (7–25), a constraint between R_A, R_B, and C to produce the required period or frequency can be determined. In this case, we also set $R_B = R_A$, and the constraint is

$$R_AC = 4.81 \times 10^{-4} \qquad \textbf{(7–36)}$$

There are many combinations of R_A and C that would satisfy the given product. However, in the range of reasonable choices, we will select after some trial and error

$$C = 0.01 \ \mu F$$
$$R_A = R_B = 48.1 \text{ k}\Omega$$

It should be stressed that the simplified choice of $R_A = R_B$ leads to a noticeably nonsymmetrical waveform. In fact, $T_H = 0.667$ ms, $T_L = 0.333$ ms, so the high interval is twice the width of the low interval. The duty cycle is thus $D = 66.7\%$. This situation is not tolerable in some applications.

**EXAMPLE
7–5**

Consider the astable 555 timer design of Example 7–4, but assume that the period of 1 ms is divided as follows: $T_H = 0.55$ ms and $T_L = 0.45$ ms. Selecting $C = 0.01\ \mu F$ as in Example 7–4, determine R_A and R_B.

Solution
The expression for T_H is given by (7–22), and the expression for T_L is given by (7–23). Since T_H involves both R_A and R_B, while T_L involves only R_B, it is usually easier to work with T_L first. Setting the expression for T_L equal to 0.45 ms, we obtain

$$R_B C = 6.494 \times 10^{-4} \tag{7–37}$$

Selecting $C = 0.01\ \mu F$ again, we obtain $R_B = 64.94\ k\Omega$.
    Next, the expression for T_H is set equal to 0.55 ms. The constraint obtained is

$$(R_A + R_B)C = 7.937 \times 10^{-4} \tag{7–38}$$

Substituting the values of R_B and C previously determined, we obtain $R_A = 14.43\ k\Omega$. The three component values are summarized as follows:

$$C = 0.01\ \mu F$$
$$R_A = 14.43\ k\Omega$$
$$R_B = 64.94\ k\Omega$$

**EXAMPLE
7–6**

For the 555 monostable circuit of Figure 7–15, determine the pulse width.

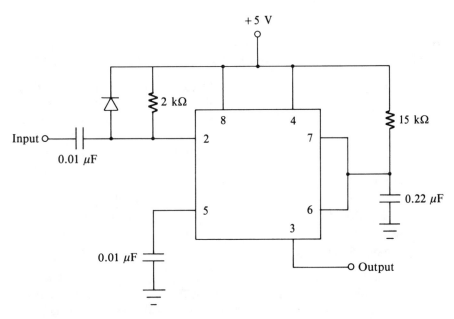

FIGURE 7–15
Circuit of Example 7–6.

Solution

Comparing the specific circuit of Figure 7–15 with the general circuit of Figure 7–11, we note that $R = 15$ kΩ and $C = 0.22$ μF. The period is

$$T_p = 1.1RC = 1.1 \times 15 \times 10^3 \times 0.22 \times 10^{-6} = 0.00363 \text{ s} = 3.63 \text{ ms} \quad \textbf{(7–39)}$$

EXAMPLE 7–7

Design a monostable 555 timer circuit of the form shown in Figure 7–11 to produce an output pulse 1 ms wide.

Solution

Since $T_p = 1$ ms, (7–30) can be used to establish the required RC product. We have

$$RC = 9.1 \times 10^{-4} \qquad \textbf{(7–40)}$$

There are many choices of R and C that would satisfy the constraint. In the range of reasonable choices, we will select

$$C = 0.01 \ \mu\text{F}$$
$$R = 91 \text{ k}\Omega$$

If the trigger circuit shown on the left-hand side of Figure 7–11 is used, values of R_i and C_i should be selected. These values are not critical, but the product R_iC_i should typically be much smaller than T_p. Reasonable choices might be $C = 0.001$ μF and $R = 10$ kΩ.

7–5 SQUARE/TRIANGULAR WAVE FUNCTION GENERATOR

A useful circuit, which simultaneously generates both a square wave and a triangular wave, and which can be implemented with two op-amps and several passive components, is shown in Figure 7–16. It will be denoted in this text as a *square/triangular wave function generator*.

Note that the circuit consists of a noninverting comparator with hysteresis (Schmitt trigger circuit) and an integrator. The output of the comparator is applied as the input to the integrator, and its output is simultaneously applied as the input to the comparator. This connection results in a complete feedback loop. We will refer to the circuit of Figure 7–16 and the waveforms of Figure 7–17 in the development that follows.

Loop Analysis

Assume that the saturation voltages of the comparator op-amp are $\pm V_{\text{sat}}$. The saturation voltages of the integrator do not enter into the analysis, but it is necessary to assume that they are comparable to those of the comparator for reasons that will be clear shortly.

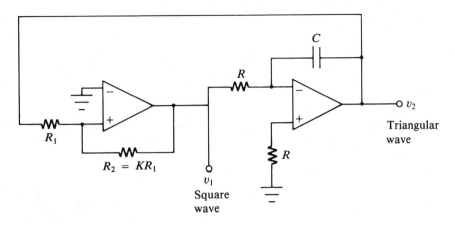

FIGURE 7-16
Square/triangular wave function generator.

FIGURE 7-17
Waveforms for square/triangular wave function generator.

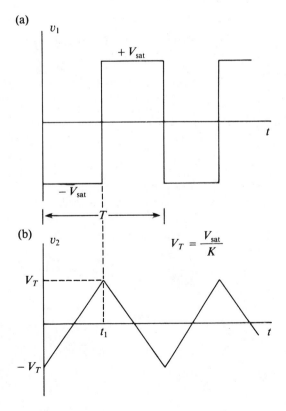

The threshold or transition voltages for the comparator are $\pm V_T$ where V_T is given by

$$V_T = \frac{R_1}{R_2} V_{sat} = \frac{R_1}{KR_1} V_{sat} = \frac{V_{sat}}{K} \tag{7-41}$$

Referring to Figure 7–17, assume at $t = 0$ that the comparator output v_1 undergoes a negative transition to a level $-V_{sat}$, as shown in part (a). This means that the output v_2 of the integrator, which is also the input to the comparator, had just reached a level of $-V_T$, as shown in part (b). (Actually, v_2 would have been slightly more negative than $-V_T$ to cause the transition.) It is thus reasonable to assume that $v_2(0^+) = -V_T$.

The output $v_2(t)$ of the integrator following the transition can be expressed as

$$v_2(t) = -\frac{1}{RC} \int_o^t v_1(t)\, dt + v_2(0^+) \tag{7-42}$$

Substitution of $v_1(t) = -V_{sat}$ and $v_2(0^+) = -V_T$ in (7–42) results in

$$v_2(t) = \frac{-1}{RC} \int_o^t (-V_{sat})\, dt - V_T = \frac{V_{sat}}{RC} t - V_T \tag{7-43}$$

From this equation, it is seen that $v_2(t)$ increases linearly from a level of $-V_T$, as shown in Figure 7–17(b).

Period and Frequency

The comparator remains in the low state until v_2 reaches the level $+V_T$. Let t_1 represent the time at which this condition is reached. Substituting $v_2(t) = V_T$ and $t = t_1$ in (7–43), and subsequent solution for t_1, yields

$$t_1 = \frac{2RCV_T}{V_{sat}} = \frac{2RC}{K} \tag{7-44}$$

where the relationship of (7–41) was employed.

After the transition occurs, the input voltage to the integrator now causes the output to *decrease* linearly, as shown in Figure 7–17(b). A second equation to describe v_2 during that interval could be written, but it is not necessary for our purposes. When $v_2(t)$ reaches the level $-V_T$, the comparator output drops to $-V_{sat}$, and a full cycle is completed.

Assuming symmetrical values for saturation voltages, the period T can be expressed as

$$T = 2t_1 = \frac{4RC}{K} \tag{7-45}$$

The corresponding frequency f is

$$f = \frac{K}{4RC} \tag{7-46}$$

Note that it is necessary for the integrator to remain in the linear range of operation. Assuming comparable saturation voltages for the two op-amps, this condition is assured by setting $V_T < V_{sat}$ or $K > 1$. Note that the peak value of the triangular wave is V_T.

As for the op-amp multivibrator, a nonpolarized capacitor must be employed in the integrator, since the voltage across the capacitor assumes both polarities.

EXAMPLE 7–8

Consider the circuit of Figure 7–18 and assume that $\pm V_{sat} = \pm 13$ V for both op-amps. Determine the (a) period, (b) frequency, (c) peak value of square wave, and (d) peak value of triangular wave.

Solution

Comparing the specific circuit of Figure 7–18 with the general circuit of Figure 7–16, we note that $R = 82$ kΩ, $C = 0.01$ μF, $R_1 = 10$ kΩ, and $R_2 = 20$ kΩ. The value of K is

$$K = \frac{R_2}{R_1} = \frac{20 \text{ k}\Omega}{10 \text{ k}\Omega} = 2 \qquad (7\text{–}47)$$

(a) The period is

$$T = \frac{4RC}{K} = \frac{4 \times 82 \times 10^3 \times 0.01 \times 10^{-6}}{2} = 1.64 \text{ ms} \qquad (7\text{–}48)$$

(b) The frequency is

$$f = \frac{1}{T} = \frac{1}{1.64 \times 10^{-3}} = 610 \text{ Hz} \qquad (7\text{–}49)$$

(c) The peak value of the square wave is simply the saturation voltage, that is, 13 V. The square wave thus oscillates between $+13$ V and -13 V.

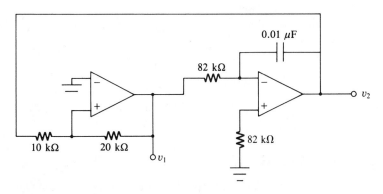

FIGURE 7–18
Circuit of Example 7–8.

(d) The peak value of the triangular wave is the threshold voltage V_T, which is

$$V_T = \frac{V_{\text{sat}}}{K} = \frac{13}{2} = 6.5 \text{ V} \tag{7–50}$$

The triangular wave thus oscillates between 6.5 V and -6.5 V.

EXAMPLE 7–9

Using op-amps with $\pm V_{\text{sat}} = \pm 13$ V, design a circuit to generate a 1-kHz triangular wave with a peak-to-peak output voltage of 8 V.

Solution

The square/triangular wave function generator can be used with the output taken as v_2 in Figure 7–16. A peak-to-peak output voltage of 8 V corresponds to a peak value of 4 V. Since this value must represent the threshold voltage of the comparator, we determine K as

$$K = \frac{V_{\text{sat}}}{V_T} = \frac{13}{4} = 3.25 \tag{7–51}$$

The values of R_1 and R_2 must be selected to satisfy

$$\frac{R_2}{R_1} = K = 3.25 \tag{7–52}$$

After some trial and error, it turns out that the standard values $R_1 = 12$ kΩ and $R_2 = 39$ kΩ satisfy this relationship exactly.

From the relationship for frequency in (7–46), the RC product is determined as

$$RC = \frac{K}{4f} = \frac{3.25}{4 \times 10^3} = 0.8125 \times 10^{-3} \text{ s} \tag{7–53}$$

At this point, either R or C must be selected somewhat arbitrarily. After some trial and error, it is determined that if C is selected as 0.012 μF, the value of R is calculated as $R = 67.7$ kΩ. This value of R is very close to the standard value of 68 kΩ, which should suffice for noncritical applications.

To summarize, the circuit will be of the form shown in Figure 7–16 with

$$R_1 = 12 \text{ k}\Omega$$
$$R_2 = 39 \text{ k}\Omega$$
$$R = 68 \text{ k}\Omega$$
$$C = 0.012 \text{ }\mu\text{F}$$

7–6 BARKHAUSEN CRITERION

This section concerns the concept of producing oscillations by maintaining unity loop gain. Consider the block diagram of a feedback system as given in Figure 7–19. This block diagram differs somewhat from the corresponding block diagram for negative feedback as

FIGURE 7–19
Block diagram of feedback loop for which oscillations are desired.

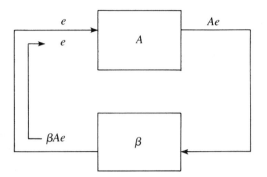

given in Chapter 3. First, there is no need for an input summing block for the basic oscillator circuit since there is no external signal input. Second, the basic negative feedback system of Chapter 3 assumed that the feedback signal was **subtracted** from the input signal, while the feedback signal in Figure 7–19 is applied directly (without inversion) to the input. Thus, an extra inversion appeared in the loop-gain expression for the feedback amplifier that does not appear for the present case. However, this is somewhat arbitrary since some references assume an adding circuit for all feedback circuitry (as was done in Figure 3–8 for the inverting amplifier). The reader should take care in using various references on feedback amplifiers and oscillators because of some possible sign ambiguities resulting from these differences in conventions.

Loop Gain

The *loop gain* G_L will be defined as the product of all gain blocks around a closed loop. Thus, for the system block diagram of Figure 7–19, the loop gain is determined as

$$G_L = A\beta \qquad (7\text{–}54)$$

When G_L is a function of steady-state frequency, it will be denoted as $G_L(j\omega)$. Similarly, A and β can be represented as $A(j\omega)$ and $\beta(j\omega)$ if required.

Conditions for Oscillation

The Barkhausen criterion will now be developed intuitively. Assume in Figure 7–19 that a small "error" signal e exists at the input. (It could even be noise.) The output of the block A is Ae. That signal is applied as the input to the β block, and its output is βAe. To sustain oscillations, this quantity must equal the assumed input; that is, $\beta Ae = e$. Cancelling the common factors and noting that $G_L = \beta A$, we have

$$G_L = 1 \qquad (7\text{–}55)$$

Thus, the loop gain must be maintained at unity level in order to sustain an oscillation. This is the basic concept of the Barkhausen criterion, but some further clarification is needed.

In general, G_L is a function of frequency, so for the complex equation $G_L(j\omega) = 1$, two possible forms can be used: (1) **polar form** and (2) **rectangular form.** Form (1) involves both magnitude and phase, while form (2) involves both real and imaginary parts. Thus, each form requires two separate constraints to be satisfied.

If the **polar form** is used, $G_L(j\omega)$ is expressed as

$$G_L(j\omega) = |G_L(j\omega)| \: \underline{/\text{arg}[G_L(j\omega)]} \qquad (7\text{–}56)$$

where $|G_L(j\omega)|$ is the magnitude of the loop gain and $\text{arg}[G_L(j\omega)]$ is the "argument" or angle. The value unity can be expressed as $1 \underline{/0°}$ in complex polar form. The Barkhausen criterion can be expressed in polar form as

$$|G_L(j\omega)| = 1 \qquad (7\text{–}57)$$
$$\text{arg}[G_L(j\omega)] = k360° \qquad (7\text{–}58)$$

where k is an integer.

Stated in words, the magnitude of the product of all gains around the loop must be unity, and the net sum of all phase shifts must be an integer multiple of $360°$. This, of course, includes $0°$.

If the **rectangular form** is used, $G_L(j\omega)$ is expressed as

$$G_L(j\omega) = \text{Re}[G_L(j\omega)] + j\,\text{Im}[G_L(j\omega)] \qquad (7\text{–}59)$$

where Re[ ] indicates the real part of the quantity in brackets and Im[ ] indicates the corresponding imaginary part. The value unity can be expressed as $1 + j0$ in complex rectangular form. The Barkhausen criterion can be expressed in rectangular form as

$$\text{Re}[G_L(j\omega)] = 1 \qquad (7\text{–}60)$$
$$\text{Im}[G_L(j\omega)] = 0 \qquad (7\text{–}61)$$

The two separate approaches yield identical results. Depending on the circuit, the polar form is easier to apply in some cases, while the rectangular form is easier to apply in others.

EXAMPLE 7–10

A hypothetical feedback loop is shown in Figure 7–20. The loop contains one amplifier with an inverting gain, a resistive attenuation network, an all-pass phase shift network, and an amplifier having unspecified gain and phase shift. At some reference frequency, the polar forms of the various gains and transfer functions are shown. What gain magnitude A and phase shift θ for amplifier 2 would result in steady-state sinusoidal oscillations at the reference frequency?

Solution

This system is not intended to represent any particular useful application, but it was fabricated to illustrate the Barkhausen criterion. Note that the transfer function of the inverting amplifier is represented as a positive magnitude with an angle of $180°$. Note also that the transfer function of the attenuation network is represented as a "gain" of 0.02 at an angle of $0°$. Finally, note that the all-pass phase shift network has a magnitude response of unity.

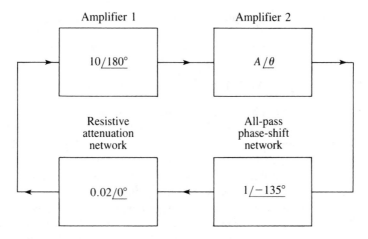

FIGURE 7–20
Feedback loop of Example 7–10.

The loop gain G_L can be expressed as

$$G_L = (10 \underline{/180°})(A \underline{/\theta})(1 \underline{/-135°})(0.02 \underline{/0°}) \qquad \textbf{(7–62a)}$$

or

$$G_L = 0.2A \underline{/\theta + 45°} \qquad \textbf{(7–62b)}$$

Applying the polar form of the Barkhausen criterion as given by (7–57) and (7–58), we have

$$0.2A = 1 \qquad \textbf{(7–63)}$$

or

$$A = 5 \qquad \textbf{(7–64)}$$

and

$$\theta + 45° = k360° \qquad \textbf{(7–65)}$$

The simplest choice for k is $k = 0$, which leads to

$$\theta = -45° \qquad \textbf{(7–66)}$$

However, in a sense it does not matter which integer value is selected. To illustrate this point, suppose that $k = 1$ is selected. In this case, (7–65) would lead to $\theta = 315°$. But $\theta = 315°$ corresponds to the same location in the complex plane as $-45°$, so the two results are the same mathematically. With some inspection, the integer value of k that will result in the simplest form of the angle can be selected.

To summarize this problem, if amplifier 2 has a gain magnitude of 5 and a phase shift of $-45°$, the loop will theoretically produce oscillations at the reference frequency. One extension to this problem is that any closed-loop system having sufficient gain and phase shift possesses the potential for oscillation, whether intended or not. Even if negative feedback is intended, there is a certain combination of gain and phase shift around the loop that could create oscillations. This fact is frequently painfully rediscovered by anyone who does linear circuit design!

7-7 WIEN BRIDGE OSCILLATOR

While numerous oscillator circuits utilizing the unity-loop-gain oscillation principle have been developed over the years, the Wien bridge oscillator is among the most widely used for the generation of sinusoids in the frequency range below 1 MHz or so. The circuit utilizes relatively few components, is straightforward in design, and is easily tunable. The last feature is particularly useful for oscillators in which the frequency must be adjustable. We will focus on this oscillator as a means of demonstrating the Barkhausen criterion and discuss some of the design techniques for circuits utilizing the Wien bridge.

The basic configuration for the Wien bridge oscillator is shown in Figure 7–21. The block with $+A$ is any type of noninverting amplifier with gain A. (The earliest Wien bridge oscillators utilized vacuum tubes.) It is assumed that the amplifier has infinite input impedance (no loading) and zero output impedance.

Loop Analysis

The loop is momentarily broken at the amplifier input, and the transfer function relating the feedback voltage to the output voltage is determined. This is the β portion of the block diagram of Figure 7–19. Let $\overline{V}_i$ and $\overline{V}_o$ represent the steady-state phasor forms for v_i and v_o, respectively. In the steady state, the capacitances C_1 and C_2 are replaced by impedances $1/j\omega C_1$ and $1/j\omega C_2$, respectively.

Let $\overline{Z}_1$ represent the parallel combination of R_1 and $1/j\omega C_1$. We have

$$\overline{Z}_1 = \frac{R_1 \times (1/j\omega C_1)}{R_1 + 1/j\omega C_1} = \frac{R_1}{1 + j\omega R_1 C_1} \tag{7-67}$$

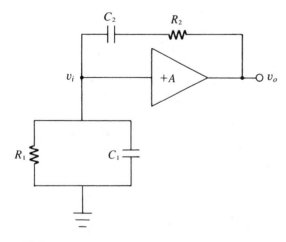

FIGURE 7-21
Basic form of the Wien bridge oscillator circuit.

Let $\bar{Z}_2$ represent the series combination of R_2 and $1/j\omega C_2$. We have

$$\bar{Z}_2 = R_2 + \frac{1}{j\omega C_2} = \frac{1 + j\omega R_2 C_2}{j\omega C_2} \tag{7-68}$$

The feedback transfer function $\beta(j\omega)$ can be expressed as

$$\beta(j\omega) = \frac{\bar{V}_i}{\bar{V}_o} = \frac{\bar{Z}_1}{\bar{Z}_1 + \bar{Z}_2} \tag{7-69}$$

Substitution of $\bar{Z}_1$ and $\bar{Z}_2$ from (7-67) and (7-68) in (7-69) and simplification lead to

$$\beta(j\omega) = \frac{j\omega R_1 C_2}{1 - \omega^2 R_1 R_2 C_1 C_2 + j\omega(R_1 C_1 + R_2 C_2 + R_1 C_2)} \tag{7-70}$$

The gain of the forward block is A, so the loop gain $G_L(j\omega)$ is

$$G_L(j\omega) = \frac{j\omega A R_1 C_2}{1 - \omega^2 R_1 R_2 C_1 C_2 + j\omega(R_1 C_1 + R_2 C_2 + R_1 C_2)} \tag{7-71}$$

While it is possible to complete the analysis with this general formula, virtually all Wien bridge oscillators use equal resistances and equal capacitances. Thus, we will set $R_1 = R_2 = R$ and $C_1 = C_2 = C$. With these substitutions, (7-71) reduces to

$$G_L(j\omega) = \frac{j\omega A R C}{1 - \omega^2 R^2 C^2 + j3\omega R C} \tag{7-72}$$

Conditions for Oscillation

From the Barkhausen criterion, we require that $G_L(j\omega) = 1$. In this case, it is easier to work with the rectangular form than with the polar form. Setting the right-hand side of (7-72) equal to unity and simplifying, we have

$$1 - \omega^2 R^2 C^2 + j3\omega R C = j\omega A R C \tag{7-73}$$

This equation has both real and imaginary parts, so it is equivalent to two different equations. For the equality to hold, both the real and the imaginary parts must be separately satisfied. The real-part requirement is

$$1 - \omega^2 R^2 C^2 = 0 \tag{7-74}$$

The imaginary-part requirement is

$$3\omega R C = \omega A R C \tag{7-75}$$

These two equations allow two important values to be determined: (1) frequency of oscillation and (2) gain requirement to sustain oscillation. The solution of (7-74) yields the frequency of oscillation, and the solution of (7-75) yields the requirement for oscillation.

Let ω_o represent the radian frequency of oscillation, and let $f_o = \omega_o/2\pi$ represent the corresponding frequency in hertz. From (7–74), this frequency is

$$f_o = \frac{1}{2\pi RC} \tag{7–76}$$

The gain requirement as determined from (7–75) is

$$A = 3 \tag{7–77}$$

Thus, the Wien bridge circuit will theoretically produce sinusoidal oscillations at a frequency f_o if a noninverting gain of 3 is provided by the active device. The gain of 3 compensates exactly for the fact that the transfer function of the feedback network is exactly $\frac{1}{3}$ at a frequency f_o (see Problem 7–18).

The oscillation frequency is inversely proportional to the RC product, assuming that both resistances and both capacitances are equal. Theoretically, the frequency can be varied by varying either R or C. In practice, it is often easier to vary R on a continuous basis and vary C by discrete steps. A common design strategy in tunable oscillators is to switch fixed capacitors for different frequency ranges. Two identical capacitors are switched into the circuit at each frequency. Two identical variable resistors, which are referred to as *ganged* potentiometers, are mounted on the same shaft and are used to vary the frequency on a continuous basis in each frequency range.

Gain Considerations

A gain of 3 can be readily achieved with an op-amp noninverting amplifier by choosing $R_f = 2R_i$ in the basic noninverting amplifier. An implementation using this approach is shown in Figure 7–22. However, there is a basic problem with this oscillator circuit as well as others utilizing the unity-loop-gain principle. The problem is that of maintaining the loop gain **exactly** at the required value. If the loop gain drops below unity, oscillations will stop. Conversely, if the loop gain increases above unity, the output amplitude will gradually increase, and operation will eventually shift toward saturation. In this case, the output will likely deviate considerably from a sine wave, and a large level of distortion will exist.

FIGURE 7–22
Basic form of Wien bridge oscillator implemented with op-amp.

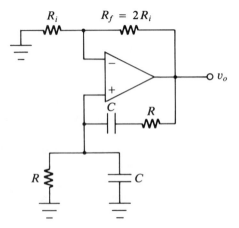

If the amplifier gain is variable, it may be adjusted experimentally in the laboratory to produce an acceptable waveform at a given time, but slight drifts or changes in circuit components will soon shift operation either below or above the critical level.

Good oscillator design practice requires a means of stabilizing the loop gain so that it can be maintained exactly at the required level. This stability is most often achieved through the use of a nonlinear element in the amplifier feedback network. The nonlinear element is connected in a manner such that if the output amplitude starts to increase above a certain level, the gain is reduced. Conversely, if the output amplitude starts to decrease, the gain is increased.

The closed-loop gain of the basic noninverting amplifier is

$$A_{CL} = 1 + \frac{R_f}{R_i} \tag{7-78}$$

A nonlinear element whose resistance increases with increasing signal level could be connected as all or a portion of R_i. Conversely, a nonlinear element whose resistance decreases with increasing signal level could be connected as all or part of R_f. Among the components that have been used to stabilize the amplitude of Wien bridge oscillators are field-effect transistors, incandescent bulbs, diodes, and thermistors.

Practical Circuit

A practical form of a typical Wien bridge oscillator using a representative stabilizing method is shown in Figure 7–23. The stabilizing element in this case is an N-channel junction field-effect transistor (JFET). The output amplitude of the circuit is monitored through a diode negative peak detector. The resulting dc voltage at the gate of the JFET starts to

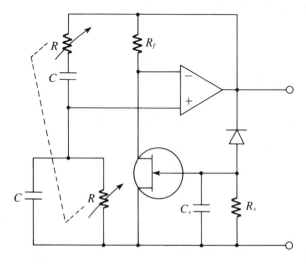

FIGURE 7–23
A typical practical form of a Wien bridge oscillator circuit using an FET for amplitude stabilization.

FIGURE 7–24
Circuit of Example 7–11.

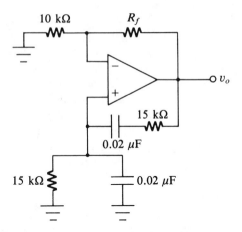

become more negative when the amplitude of the oscillation starts to increase. As a result, the JFET is biased further into the negative gate region, and its effective resistance from drain to source increases. Thus, the gain is decreased according to (7–78) and the amplitude is brought back to a stable level. If the output peak starts to decrease, the opposite pattern occurs. The R_sC_s product should be chosen sufficiently large to hold the peak with negligible change for a very long interval compared with a period at the longest oscillation period. However, if there were no discharge path for C_s, the gate voltage would not be able to decrease if the output amplitude dropped, and oscillations would then cease. Thus, some realistic upper limit for the time constant is necessary.

EXAMPLE 7–11

For the Wien bridge oscillator circuit of Figure 7–24, determine the (a) oscillation frequency and (b) value of R_f required.

Solution
Comparing the specific circuit of Figure 7–24 with the general form of Figure 7–22, we see that $R = 15$ kΩ, $C = 0.02$ μF, and $R_i = 10$ kΩ.

(a) The oscillation frequency f_o is

$$f_o = \frac{1}{2\pi RC} = \frac{1}{2 \times 15 \times 10^3 \times 0.02 \times 10^{-6}} = 530.5 \text{ Hz} \qquad \text{(7–79)}$$

(b) To achieve a gain of 3, the value of R_f must be

$$R_f = 2R_1 = 2 \times 10^4 = 20 \text{ k}\Omega \qquad \text{(7–80)}$$

EXAMPLE 7–12

A variable Wien bridge oscillator of a form similar to that of Figure 7–23 is to be designed to produce an output sinusoid that can be adjusted from 100 Hz to 1 kHz. The two capacitor values are selected to be $C = 0.01$ μF. Determine the required range in the resistances.

Solution

From the expression for frequency in (7–76), it is observed that the frequency is inversely proportional to R, and vice versa. Solving for R, we obtain

$$R = \frac{1}{2\pi f_o C} \tag{7–81}$$

The largest value of f_o requires the smallest value of R, and the smallest value of f_o requires the largest value of R. Let R_{min} and R_{max} represent the minimum and maximum values of R. We have

$$R_{min} = \frac{1}{2\pi \times 10^3 \times 0.01 \times 10^{-6}} = 15.915\text{ k}\Omega \tag{7–82}$$

$$R_{max} = \frac{1}{2\pi \times 10^2 \times 0.01 \times 10^{-6}} = 159.15\text{ k}\Omega \tag{7–83}$$

Thus, the resistances must be adjustable from 15.915 kΩ to 159.15 kΩ in order to tune the oscillator over the required range. It is assumed that resistances are aligned so that their values are equal.

The variation of resistance from a low value to a high value can be achieved by a series connection of a fixed resistance and a potentiometer for each section. No attention was paid in this example to the possibility of a standard potentiometer size, but it is possible to design around such a constraint. See Problem 7–41 for further details.

7–8 PHASE SHIFT OSCILLATOR

Another oscillator that can be readily implemented with an op-amp and an RC network is the phase shift oscillator. Like the Wien bridge oscillator, the basic form of the phase shift oscillator dates back to the vacuum tube era.

The classical form of the phase shift oscillator that will be initially assumed for the basic circuit analysis is shown in Figure 7–25. In contrast to the noninverting amplifier requirement for the Wien bridge oscillator, the phase shift oscillator requires an *inverting amplifier*. We could leave this requirement unspecified and let the mathematics lead to that conclusion, but to maintain consistency on the schematic diagrams, we will assume an inverting amplifier at the outset and label the circuit accordingly. The amplifier is assumed to have infinite input impedance and zero output impedance.

Loop Analysis

The steady-state form of the circuit suitable for loop analysis is shown in Figure 7–25(b). All node voltages have been labeled on the circuit. The loop is momentarily broken at the amplifier input, and the transfer function relating the feedback voltage to the output voltage is determined. The result is the β portion of the block diagram.

(a)

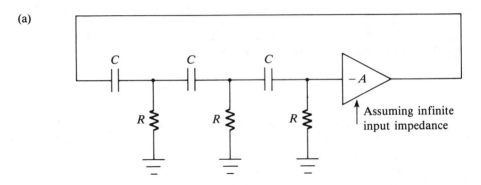

(b)

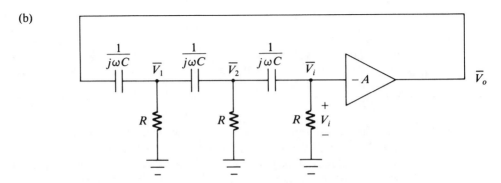

FIGURE 7–25

Classical form of the phase shift oscillator and the steady-state form suitable for analysis.

Node voltage analysis is probably the best procedure to determine the β transfer function. The various steps involved are somewhat messy and will not be given in the text, but will be left as a guided derivation exercise [Problem 7–19(a)]. The result is

$$\beta(j\omega) = \frac{\overline{V}_i}{\overline{V}_o} = \frac{-j\omega^3 R^3 C^3}{1 - 6\omega^2 R^2 C^2 + j(5\omega RC - \omega^3 R^3 C^3)} \qquad (7\text{–}84)$$

The loop gain $G_L(j\omega)$ is

$$G_L(j\omega) = \frac{jA\omega^3 R^3 C^3}{1 - 6\omega^2 R^2 C^2 + j(5\omega RC - \omega^3 R^3 C^3)} \qquad (7\text{–}85)$$

Conditions for Oscillation

From the Barkhausen criterion, we require that $G_L(j\omega) = 1$. Again, the details will be left as a guided derivation exercise [Problem 7–19(b)]. The two results obtained are as follows:

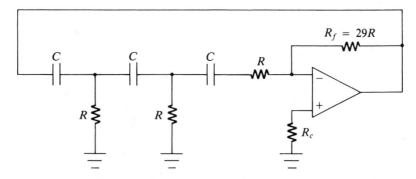

FIGURE 7–26
Phase shift oscillator form adapted for op-amp implementation.

$$f_o = \frac{1}{2\pi\sqrt{6}RC} \tag{7–86}$$

and

$$A = 29 \tag{7–87}$$

Op-Amp Implementation

The development thus far has assumed that the input impedance of the inverting amplifier is infinite. However, the impedance of the op-amp inverting amplifier is R_i. While it might be possible to select R_i to be considerably larger than the impedance level of the phase shift circuit, this approach is not too satisfactory.

The best approach is to absorb the last resistor in the phase shift circuit as a part of the actual amplifier design as shown in Figure 7–26. Since the op-amp inverting input assumes a virtual ground, the transfer function between the op-amp output and the node to the left of the op-amp input resistance is the same as in the earlier development. Further, the op-amp output voltage is $-R_f/R_i$ times this voltage. Thus, oscillation is assured by setting $R_f = 29R_i$.

EXAMPLE 7–13

For the phase shift oscillator circuit of Figure 7–27, determine (a) the oscillation frequency and (b) value of R_f required.

Solution

(a) The oscillation frequency is

$$f_o = \frac{1}{2\pi\sqrt{6}RC} = \frac{1}{2\pi \times \sqrt{6} \times 12 \times 10^3 \times 0.01 \times 10^{-6}} = 541.5 \text{ Hz} \tag{7–88}$$

(b) To produce oscillations, the value of R_f must be

$$R_f = 29R = 29 \times 12 \times 10^3 = 348 \text{ k}\Omega \tag{7–89}$$

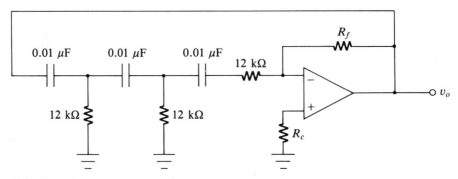

FIGURE 7–27
Circuit of Example 7–13.

EXAMPLE 7–14

Design a phase shift oscillator to produce oscillations at 1 kHz. The capacitor values are selected as $C = 0.01 \ \mu\text{F}$.

Solution

From the expression for frequency in (7–86), the value of R can be determined in terms of f_o and C. The latter values can be substituted in that expression, and we have

$$R = \frac{1}{2\pi\sqrt{6}f_o C} = \frac{1}{2\pi\sqrt{6} \times 10^3 \times 0.01 \times 10^{-6}} = 6497 \ \Omega \qquad (7\text{–}90)$$

The value of R_f is then determined as

$$R_f = 29R = 29 \times 6497 = 188.4 \ \text{k}\Omega \qquad (7\text{–}91)$$

Neither of the resistance values are standard, so some adjustments will be required.

7–9 THE 8038 WAVEFORM GENERATOR

Linear integrated circuit technology has developed to the point where complete function generators having a variety of waveforms are available on a single chip. There is still a definite place for oscillators implemented using the unity-loop-gain concept as discussed in the last two sections, but sophisticated IC generator chips are capable of satisfying many of the design requirements. One of the most popular function generator chips, namely, the 8038, is an example of the level of capability available in linear integrated circuits. Specifications and ratings on the 8038 function generator appear in Appendix C. This unit was introduced by Intersil, Inc., but it is now also available from other manufacturers.

One particular connection for the 8038 function generator is shown in Figure 7–28. (Variations on this connection are given with the specifications sheets.) This connection has provisions for minimizing the distortion level with the adjustable resistances on the lower right. For less stringent requirements, a simpler arrangement without variable resistances is possible.

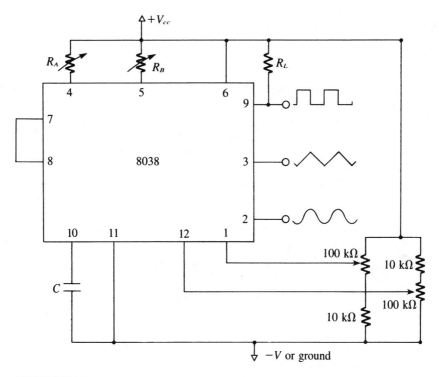

FIGURE 7-28
Possible connection for 8038 function generator with provision for adjusting distortion to minimum level.

The resistances R_A and R_B are shown as adjustable in Figure 7-28. However, if perfectly symmetrical waveforms are desired, the two values should theoretically be the same; that is, $R_A = R_B = R$.

The 8038 generator simultaneously produces a sine wave, a square wave, and a triangular wave. By varying the duty cycle on the square wave, one can also generate a pulse waveform. The frequency range can be adjusted from 0.001 Hz to 1 MHz. The frequency of operation is established by R_A, R_B, and the timing capacitor C. For the case where $R_A = R_B = R$, the frequency f is

$$f = \frac{0.3}{RC} \qquad (7\text{-}92)$$

Data concerning other cases are given in the specifications sheets.

7-10 MULTISIM EXAMPLES

Multisim will now be employed to simulate the operation of several oscillator circuits. In addition to op-amp oscillator circuits, the 555 library model will be introduced and used to create an astable multivibrator circuit.

MULTISIM EXAMPLE 7–1 Using the 741 model, develop a Multisim circuit for simulating the op-amp astable multivibrator design of Example 7–2 and plot the output voltage over several cycles.

Solution

The circuit diagram created with Multisim is shown in Figure 7–29. All component types have appeared in numerous earlier circuits so no additional explanations are required for the circuit construction.

A **Transient Analysis** is performed from $t = 0$ to $t = 5$ ms. To ensure high accuracy, the **Maximum time step** was set to 1 μs, and the results are shown in Figure 7–30. Several comments need to be made about the results.

First, there is an initial *startup* interval extending to nearly 1 ms, and this time interval should not be used in measuring the period. Second, the two cursors have been utilized for measuring the period. The cursors may be activated by left-clicking on the eighth button from the right on the row above the graphs. Each cursor may be moved by left-clicking on it and "dragging" it while holding the left button down. A window opens that provides a significant amount of information about the coordinates of the cursor and other mathematical properties.

From the cursor results, the period is determined as

$$T = 2.4356 - 1.3376 = 1.098 \text{ ms} \tag{7–93}$$

This corresponds to a frequency of $1/(1.098 \times 10^{-3}) = 910.7$ Hz, which is about 9% below the design goal. The reason for this discrepancy is the significant amount of transition time required at the beginning and end of each cycle as a result of the slew rate limitation. This transition time is very noticeable on the response curve.

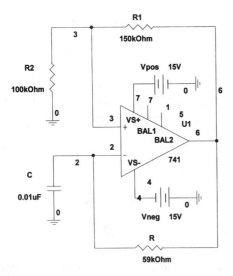

FIGURE 7–29
Circuit of Multisim Example 7–1

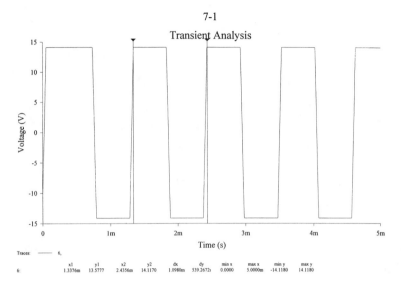

7-1

Transient Analysis

Traces: —— 6,

	x1	y1	x2	y2	dx	dy	min x	max x	min y	max y
6:	1.3376m	13.5777	2.4356m	14.1170	1.0980m	539.2672i	0.0000	5.0000m	-14.1180	14.1180

FIGURE 7–30
Waveform of Multisim Example 7–1.

MULTISIM EXAMPLE 7–2

Using the 741 model, develop a Multisim circuit for simulating the square/triangular wave function generator design of Example 7–9 and plot the two op-amp output voltages over several cycles.

Solution

The circuit diagram created with Multisim is shown in Figure 7–31. Since two op-amps are involved, in order to use node numbers that correspond as closely as possible to the op-amp pin terminal, a more elaborate system for node numbering has been used here. All node numbers for the op-amp on the left have been followed with A, and those for the op-amp on the right have been followed with B. For example, the node number corresponding to the noninverting input for the left-hand op-amp is 3A. It is not necessary to be so elaborate, but this helps to keep track of the relationship between node numbers, op-amps, and op-amp pin numbers.

A **Transient Analysis** is performed from $t = 0$ to $t = 5$ ms. To ensure high accuracy, the **Maximum time step** was set to 1 μs, and the results are shown in Figure 7–32. The voltages involved are those of node 6A and 6B.

As in the preceding example, there is an initial *startup* interval before steady-state conditions are reached, and this time interval should not be used in measuring the period. From the two cursors, the value of the period is calculated as

$$T = 3.1873 - 2.0503 = 1.137 \text{ ms} \tag{7–94}$$

As in the preceding example, the period is longer than expected, and the effect is primarily a result of slew rate limitations.

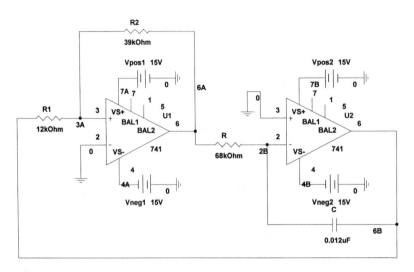

FIGURE 7–31
Circuit of Multisim Example 7–2.

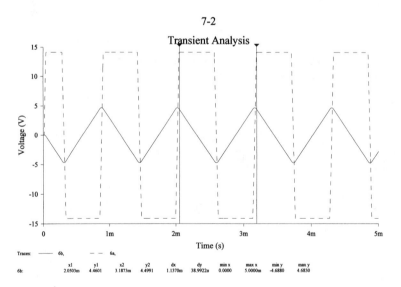

FIGURE 7–32
Waveforms of Multisim Example 7–2.

MULTISIM EXAMPLE 7–3

Using the 555 model, develop a Multisim circuit for simulating the circuit of Example 7–3 and plot the output voltage over several cycles.

Solution

The original circuit was shown in Figure 7–14, and the corresponding Multisim version is shown in Figure 7–33. The **555** model is obtained from the **Mixed** parts bin near

FIGURE 7–33
Circuit of Multisim Example 7–3

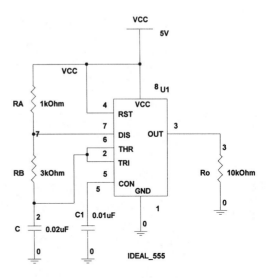

the bottom of the **Components Toolbars.** Note also that the **VCC** voltage source obtained from the **Sources** bin has been used. This source has a value of 5 V with respect to ground, and the ground connection is automatically made.

A **Transient Analysis** is performed from $t = 0$ to $t = 400$ μs, and the results are shown in Figure 7–34. To ensure high accuracy, the maximum step size was set to 0.1 μs. The output voltage is that of node 3. Note that a resistance has been connected from the output node to ground to avoid a "floating node."

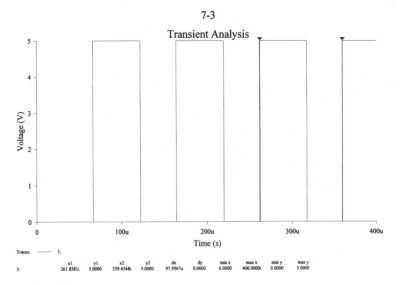

FIGURE 7–34
Waveform of Multisim Example 7–3.

Following an initial startup interval, the period is measured from the cursors as

$$T = 359.4548 - 261.8581 = 97.5967 \ \mu s \tag{7-95}$$

This compares quite favorably with the calculated period in the text of 97.02 μs.

MULTISIM EXAMPLE 7–4

Using the 741 model, develop a Multisim circuit for simulating the Wien bridge oscillator design of Example 7–12 at a frequency of 100 Hz. Plot the output voltage over an interval of 20 ms.

Solution

The Multisim schematic diagram is shown in Figure 7–35. The resistance values correspond to operation at 100 Hz.

Anyone who has built a Wien bridge oscillator circuit without an amplitude-stabilizing element will know how sensitive such circuits are to slight changes in element values. The reason is that the exact point at which oscillations can be sustained without unwanted growth is a delicate one, and it is difficult to predict accurately the amplitude of the oscillations. This same phenomenon arises in a software simulation. By assuming different initial conditions for various circuit voltages, a range of possible amplitude conditions can be created. Thus, this simulation must be interpreted with some caution.

For this example, node 3 was set to an initial voltage of 1 V. To accomplish this, first double left-click on the wire corresponding to node 3. When the **Node** window opens, left-click on **Use IC for Transient Analysis.** Then set the value in the corresponding space to **1 V.** Next, proceed to perform a **Transient Analysis** in the normal fashion, but when the **Transient Analysis** window opens, set the **Initial conditions** option to **User-defined.** From that point on, the transient analysis follows previous procedures.

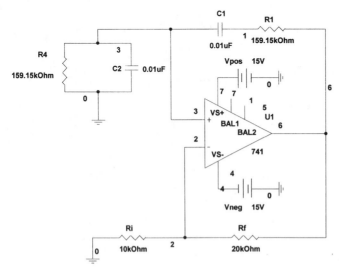

FIGURE 7–35
Circuit of Multisim Example 7–4.

7-4

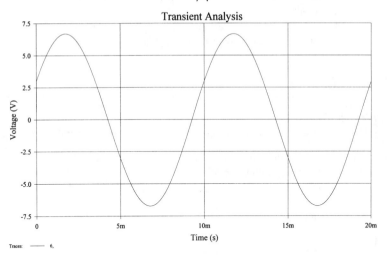

FIGURE 7–36
Waveform of Multisim Example 7–4.

The op-amp output based on this particular situation over an interval of 20 ms with a **Maximum time step** of 10 μs is shown in Figure 7–36. This corresponds to about two cycles at the design frequency of 100 Hz.

SUMMARY

1. Oscillators may be classified according to their output as either sinusoidal or nonsinusoidal.
2. Two methods considered in this chapter for generating oscillations are the timing method and the unity-loop-gain method.
3. The three types of multivibrator circuits are the bistable, monostable, and astable types.
4. An astable multivibrator may be implemented with an op-amp inverting Schmitt trigger circuit and an RC timing circuit.
5. The 555 integrated-circuit timer is a widely used timing circuit capable of operation as both an astable and a monostable multivibrator.
6. A waveform generator that simultaneously produces both a square wave and a triangular wave may be implemented with two op-amps and several passive components.
7. The Barkhausen criterion is a mathematical statement describing the conditions required to generate oscillations.
8. To generate oscillations, the magnitude of the loop gain must be unity and the angle must be either 0° or an integer multiple of 360°.
9. The Wien bridge oscillator is a widely employed sinusoidal generator employing a noninverting amplifier and an RC network.
10. One advantage of the Wien bridge oscillator is its ease of tuning.
11. Along with the required positive feedback, most practical Wien bridge oscillators employ some negative feedback to provide stabilization of the loop at the critical gain level.

12. The phase shift oscillator is a sinusoidal generator employing an inverting amplifier and an *RC* phase shift circuit.

13. One of the most popular IC waveform generator chips is the 8038, which simultaneously generates a sine wave, a square wave, and a triangular wave.

PROBLEMS

Drill Problems

Note: Unless otherwise indicated, assume $\pm V_{sat} = \pm 13$ V in all general-purpose op-amps.

7–1. Consider the circuit of Figure P7–1 with $R_1 = 10$ kΩ, $R_2 = 3$ kΩ, $R = 18$ kΩ, and $C = 0.068$ μF. Determine the following:
  a. feedback voltage
  b. period
  c. frequency

FIGURE P7–1

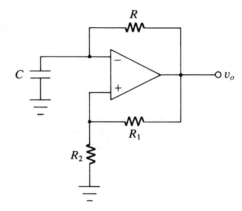

7–2. Consider the circuit of Figure P7–1 with $R_1 = 24$ kΩ, $R_2 = 12$ kΩ, $R = 47$ kΩ, and $C = 0.082$ μF. Determine the following:
  a. feedback voltage
  b. period
  c. frequency

7–3. Consider the circuit of Figure P7–1 with $R = 22$ kΩ, $C = 0.047$ μF, and $R_1 = 10$ kΩ.
  a. Determine the value of R_2 required to produce oscillations at 1 kHz.
  b. Determine the value of the feedback voltage when the value of (a) is used.

7–4. Consider the circuit of Figure P7–1 with $R = 15$ kΩ, $C = 0.033$ μF, and $R_2 = 10$ kΩ.
  a. Determine the value of R_1 required to produce oscillations at 1 kHz.
  b. Determine the value of the feedback voltage when the value of (a) is used.

7–5. Consider the 555 astable circuit of Figure P7–5 with $R_A = 10$ kΩ, $R_B = 200$ kΩ, and $C = 0.1$ μF. Determine the
  a. high state interval
  b. low state interval
  c. period
  d. frequency
  e. duty cycle

FIGURE P7–5

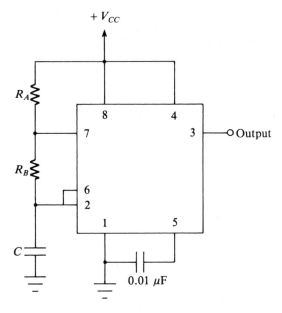

7–6. Consider the 555 astable circuit of Figure P7–5 with $R_A = 2$ kΩ, $R_B = 6.8$ kΩ, and $C = 0.022$ μF. Determine the

   a. high state interval   **b.** low state interval   **c.** period

   d. frequency   **e.** duty cycle

7–7. Consider the 555 monostable circuit of Figure P7–7 with $R = 200$ kΩ and $C = 2$ μF. Determine the pulse width.

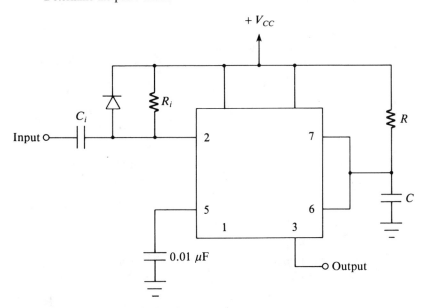

FIGURE P7–7

7–8. Consider the 555 monostable circuit of Figure P7–7 with $R = 3$ kΩ and $C = 0.0068$ μF. Determine the pulse width.

7–9. Consider the circuit of Figure P7–9 with $R_1 = 12$ kΩ, $R_2 = 36$ kΩ, $R = 20$ kΩ, and $C = 0.02$ μF. Determine the following:

a. period
b. frequency
c. peak value of square wave
d. peak value of triangular wave

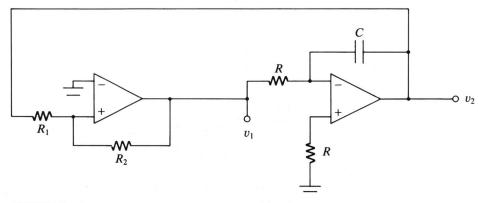

FIGURE P7–9

7–10. Consider the circuit of Figure P7–9 with $R_1 = 6.8$ kΩ, $R_2 = 30$ kΩ, $R = 100$ kΩ, and $C = 0.1$ μF. Determine the following:

a. period
b. frequency
c. peak value of square wave
d. peak value of triangular wave

7–11. For the loop of Figure P7–11, determine the values of A and θ that will produce steady-state sinusoidal oscillations.

FIGURE P7–11

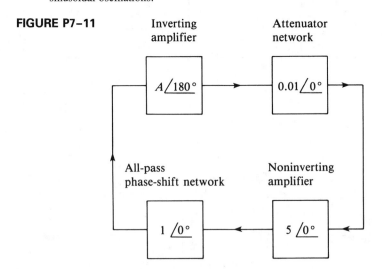

7-12. For the loop of Figure P7–12, determine the values of A and θ that will produce steady-state sinusoidal oscillations.

FIGURE P7–12

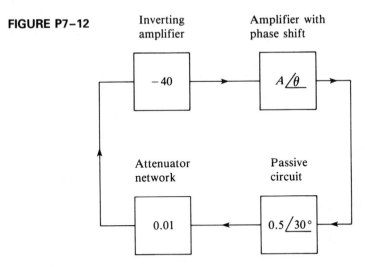

Inverting
amplifier

Amplifier with
phase shift

-40

$A\underline{/\theta}$

Attenuator
network

Passive
circuit

0.01

$0.5\underline{/30°}$

7-13. Consider the circuit of Figure P7–13 with $R = 22$ kΩ, $C = 0.047$ μF, and $R_i = 12$ kΩ. Determine the following:
a. oscillation frequency
b. value of R_f required

FIGURE P7–13

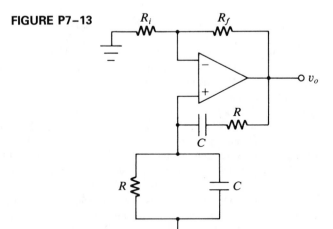

R_i

R_f

v_o

R

C

R

C

7-14. Consider the circuit of Figure P7–13 with $R = 3$ kΩ, $C = 0.005$ μF, and $R_f = 10$ kΩ. Determine the following:
a. oscillation frequency
b. value of R_i required
7-15. Consider the circuit of Figure P7–15 with $R = 100$ kΩ and $C = 0.1$ μF. Determine the following:
a. oscillation frequency
b. value of R_f required

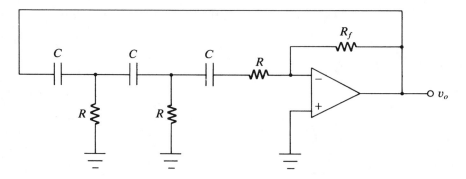

FIGURE P7–15

7–16. Consider the circuit of Figure P7–15 with $R = 1 \text{ k}\Omega$ and $C = 0.005 \ \mu\text{F}$. Determine the following:

   a. oscillation frequency

   b. value of R_f required

Derivation Problems

7–17. For the op-amp astable multivibrator circuit of Figure 7–1, show that if the resistances are selected such that $R_2 = 0.86R_1$, the period is $T = 2\tau$ as given by equation (7–10).

7–18. For the Wien bridge oscillator feedback circuit with equal resistances and equal capacitances, verify that the transfer function at a frequency $f_o = \omega_o/2\pi$ has a magnitude response of $^1/_3$ and an angle of $0°$; that is,

$$\beta(j\omega_o) = \frac{1}{3} \ \underline{/0°}$$

7–19. a. Consider the phasor form of the phase shift oscillator as given in Figure 7–25(b). Momentarily, break the loop at the amplifier input and consider $\overline{V_i}$ as the output of the phase shift network and $\overline{V_o}$ as its input. Write node voltage equations at nodes 1 and 2 as well as the $\overline{V_i}$ node. Eliminate $\overline{V_1}$ and $\overline{V_2}$ from the three equations and show that the transfer function relating $\overline{V_i}$ to $\overline{V_o}$ is the function given by (7–84).

   b. Using the results of (a) and forming the loop-gain function of (7–85), apply the Barkhausen criterion to obtain the results of (7–86) and (7–87).

*7–20. One method for achieving equal high and low time intervals for the 555 timer in astable operation is illustrated in Figure P7–20. First, a resistance is connected between the control terminal (pin 5) and ground (pin 1) to lower the threshold levels for the two comparators. (Refer to Figure 7–5.) A convenient choice is $10 \text{ k}\Omega$ as shown. The actual downward voltage shift in the upper threshold level is greater than that in the lower threshold level. During the portion of the cycle that $v_C(t)$ is exponentially increasing toward V_{CC}, the lowered threshold level for triggering can offset the effect of the longer time constant if the value of R_A is set correctly.

   a. Show that the choice of the 10-kΩ resistance from pin 5 to ground results in threshold levels of $V_{CC}/2$ and $V_{CC}/4$ for the upper and lower comparators, respectively.

   b. Show that the high interval T_H is

$$T_H = \tau_H \ln 1.5 = 0.405\tau_H$$

where $T_H = (R_A + R_B)C$.

---

*This method was shown to the author by Professor W. H. Thornton of Old Dominion University.

c. Show that the low interval T_L is

$$T_L = \tau_L \ln 2 = 0.693\tau_L$$

where $T_L = R_B C$.

d. Show that the two intervals are equal if

$$R_A = 0.71 T_B$$

e. Show that for the symmetrical conditions of part (d), the period is

$$T = 1.39 R_B C$$

In practice, a variable resistance can be used to adjust the symmetry.

FIGURE P7–20

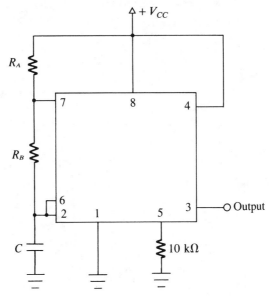

7–21. It is possible to produce either pulse-width modulation or frequency modulation with the 555 timer in the astable mode by connecting an external signal to the control terminal (pin 5). Assume in the astable schematic diagram of Figure 7–6 that the bypass capacitor is removed and that a signal control current source i is connected to pin 5. (The current source could be implemented with a voltage source in series with a large resistance if desired.) Refer back to Figure 7–5 as necessary.

a. Let V_{UT} and V_{LT} represent the upper and lower threshold voltages, respectively. Show that these values are

$$V_{UT} = \frac{2}{3}V_{CC} + 3333i$$

$$V_{LT} = \frac{1}{3}V_{CC} + 1667i$$

b. Show that the high state-time interval is

$$T_H = \tau_H \ln \frac{2V_{CC} - 5000i}{V_{CC} - 10,000i}$$

c. Show that the low state-time interval is the same as without modulation; that is,

$$T_L = \tau_H \ln 2$$

d. Assume that the current i is sufficiently small that $10,000t \ll V_{CC}$. With this approximation and the property that $\ln ab = \ln a + \ln b$, show that

$$T_H \simeq \tau_H \ln 2 + \tau_H \ln\left(1 + \frac{7500i}{V_{CC}}\right)$$

e. An approximation for the logarithmic function is that $\ln(1 + x) \simeq x$ for $x \ll 1$. Using this approximation, show that

$$T_H \simeq \tau_H \ln 2 + \frac{7500\tau_H i}{V_{CC}}$$

The first term on the right is the value of T_H based on $i = 0$ as developed in the text. The second term is a variation in the high state interval which, to the validity of the approximation, is directly proportional to the signal current i.

7–22. It is possible to produce pulse width modulation with the 555 timer in the monostable mode by connecting an external signal to the control terminal (pin 5). Assume in the monostable schematic diagram of Figure 7–11 that the bypass capacitor is removed and that a signal control current source i is connected to pin 5. (The current source could be implemented with a voltage source in series with a large resistance, if desired.) Refer back to Figure 7–5 as necessary.

a. Let V_{UT} represent the upper threshold voltage. Show that it is

$$V_{UT} = \frac{2}{3}V_{CC} + 3333i$$

b. Show that the pulse width T_p is

$$T_p = \tau \ln\left(\frac{3V_{CC}}{V_{CC} - 10,000i}\right)$$

c. Use the property that $\ln(a/b) = \ln a - \ln b$ and show that

$$T_p = \tau \ln 3 - \tau \ln\left(1 - \frac{10,000i}{V_{CC}}\right)$$

d. Assume that $10,000i \ll V_{CC}$, and use the logarithmic function approximation given in Problem 7–21 to show that

$$T_p \simeq \tau \ln 3 + \frac{10,000i}{V_{CC}}$$

The first term on the right is the value of T_p based on $i = 0$ as developed in the text. The second term is a variation in the pulse width which, to the validity of the approximation, is directly proportional to the signal current i.

Design Problems

7–23. Consider the op-amp astable multivibrator design of Example 7–2, but assume that the values of R_1 and R_2 are selected as $R_1 = 220$ kΩ and $R_2 = 22$ kΩ. Repeat the design and analysis of Example 7–2 if all other given conditions are unchanged.

7–24. Consider the op-amp astable multivibrator design of Example 7–2, but assume that the values of R_1 and R_2 are selected as $R_1 = 10$ kΩ and $R_2 = 47$ kΩ. Repeat the design and analysis of Example 7–2 if all other conditions are unchanged.

7–25. Consider the op-amp astable multivibrator design of Example 7–2, but assume that the values of R_1 and R_2 are selected in accordance with the simplifying constraint of (7–10). For

$R_1 = 10$ kΩ, complete the design and analysis of Example 7–2 if all other conditions are unchanged.

7–26. An astable op-amp multivibrator is to be designed to produce a 2-kHz square wave. The following somewhat arbitrary assumptions are made: feedback voltage = 2 V, $R_1 = 10$ kΩ, $C = 0.01$ μF. Complete the design of the circuit by specifying the values of R and R_2.

7–27. Design a square/triangular wave function generator to operate at a frequency of 2 kHz with the triangular wave having a peak value of 3 V.

7–28. Design a square/triangular wave function generator to operate at a frequency of 100 Hz with the triangular wave having a peak-to-peak value of 10 V.

7–29. Design an astable 555 timer circuit to produce a 2-kHz square wave with a duty cycle of 60%. Select $C = 0.01$ μF.

7–30. Repeat the design of Problem 7–29 if the duty cycle is to be 52%.

7–31. Design an astable 555 timer circuit to produce a 100-Hz square wave with a duty cycle of 70%. Select $C = 0.1$ μF.

7–32. Repeat the design of Problem 7–31 if the duty cycle is to be 51%.

7–33. Design a monostable 555 timer circuit to produce an output pulse 10 μs wide.

7–34. Design a monostable 555 timer circuit to produce an output pulse 2 ms wide.

7–35. Design a monostable 555 timer circuit to produce an output pulse 400 ms wide.

7–36. Design a monostable 555 timer circuit to produce an output pulse one minute long.

7–37. Design a Wien bridge oscillator circuit to produce a sinusoid at 5 kHz.

7–38. Design a Wien bridge oscillator circuit to produce a sinusoid at 100 Hz.

7–39. A variable Wien bridge oscillator is to be designed to produce an output sinusoid that can be adjusted from 25 Hz to 1 kHz. The two capacitor values are selected to be $C = 0.02$ μF. Determine the required range in the resistances.

7–40. Repeat Problem 7–39 if the frequency range is to be from 300 Hz to 5 kHz and the capacitor values are selected to be $C = 0.005$ μF.

7–41. The circuit of Figure P7–41 shows a method for achieving nearly exact limits on the tuning range of a Wien bridge oscillator while using a stock dual potentiometer, which in this case has sections that vary from zero to 10 kΩ. Note the presence of the series resistance R_1. As the potentiometer is varied from one limit to the other, it is desired that the frequency vary from 100 Hz to 1 kHz. Determine the values of R_1 and C required. (*Note*: Possible amplitude stabilization has not been shown here.)

FIGURE P7–41

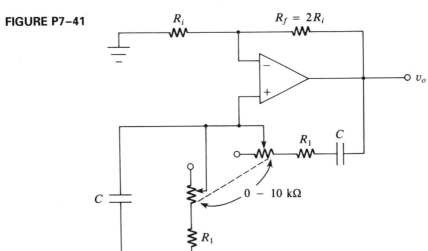

7–42. Repeat the design of Problem 7–41 if the desired tuning range is from 1 kHz to 10 kHz.

7–43. A Wien bridge oscillator of the form shown in Figure P7–13 is to employ a nonlinear resistor for R_i as a means for stabilizing the loop. The value of the resistance (in ohms) can be approximated as

$$R_i = 4000 + 1000V_i$$

where V_i is the peak voltage (in volts) across R_i. If the desired peak value of the *output voltage* is 9 V, determine the value of R_f required.

7–44. A Wien bridge oscillator of the form shown in Figure P7–13 is to employ a nonlinear resistor for R_f as a means for stabilizing the loop. The value of the resistance (in ohms) can be approximated as

$$R_f = \frac{30{,}000}{2 + V_f}$$

where V_f is the peak voltage (in volts) across R_f. If the desired peak value of the *output voltage* is 6 V, determine the value of R_i required.

7–45. Design a phase shift oscillator to produce oscillations at 8 kHz.

7–46. Design a phase shift oscillator to produce oscillations at 200 Hz.

LABORATORY EXERCISES

Op-Amp Astable Multivibrator
LE 7–1.

1. Connect the circuit of Example 7–2.
2. Record and measure the appropriate levels of the waveforms at the two op-amp inputs and the output showing their time relationships.
3. Measure the frequency and compare with the expected value.

555 Astable Timer
LE 7–2.

1. Using a 555 IC chip and a 5-V dc supply, connect the circuit of Example 7–4.
2. Record and measure the two levels of the output waveform (pin 3).
3. Measure the high and low state-time intervals (T_H and T_L) and determine the frequency. Compare the measured frequency with the expected value.

555 Astable Timer
LE 7–3.

Perform the exercises outlined in LE 7–2 for the circuit of Example 7–5.

555 Monostable Timer
LE 7–4.

1. Using a 555 IC chip and a 5-V dc supply, connect the circuit of Example 7–7.
2. In order to observe the output without using a storage scope, it is necessary to provide a periodic pulse train to the input so that the monostable is successively triggered on a periodic basis. To accomplish this, apply a periodic square wave having a peak-to-peak amplitude slightly less

than 5 V to the input. The period of the square wave should be somewhat longer than 1 ms. Observe the output waveform and measure the two levels.

3. Observe the effect of varying the frequency of the input square wave.
4. Measure the output pulse width and compare with the expected value.

Square/Triangular Wave Function Generator
LE 7–5.

1. Connect the circuit of Example 7–9.
2. Record and measure the appropriate levels of the waveforms at the two op-amp outputs and show their relationships.
3. Measure the frequency and compare with the expected value.

Wien Bridge Oscillator
LE 7–6.

1. Connect the circuit of Example 7–11 (Figure 7–24). Provide a means for adjusting R_f around the value of 20 kΩ by using a potentiometer for all or a portion of the resistance.
2. Carefully adjust R_f until oscillations having the appearance of a sine wave are obtained.
3. Measure the frequency and compare with the expected value.

8

ACTIVE
FILTERS

OBJECTIVES

After completing this chapter, the reader should be able to:

- Define a filter and discuss its general characteristics.
- Explain the differences between active and passive filters and the advantages and disadvantages of each type.
- Sketch the amplitude response of an ideal low-pass, a high-pass, a band-pass, and a band-rejection filter.
- Discuss the differences among a Butterworth, a Chebyshev, and a maximally flat time-delay filter.
- Apply to a filter the procedures for frequency scaling and for impedance scaling.
- Determine the number of poles required in a low-pass or high-pass Butterworth filter to meet specified filtering requirements.
- Design a low-pass and a high-pass Butterworth active filter using unity-gain sections.
- Design a low-pass and a high-pass Butterworth active filter using equal-component sections.
- Apply curves to determine the response of a two-pole band-pass filter as a function of center frequency and bandwidth or Q.
- Design a two-pole band-pass active filter using a multiple-feedback circuit and a single op-amp.
- Discuss the concept of a state-variable filter, show its basic circuit form, and design one to achieve either a low-pass, a high-pass, a band-pass or a band-rejection circuit.

8–1 INTRODUCTION

Before we discuss the specific principles of active filters, it is instructive to consider the role that active filters play within the general framework of filter theory and design. Electrical filters have been an important part of the evolution of the electrical field from the beginning. Indeed, it would have been impossible to achieve many outstanding technological accomplishments without the use of electrical filters.

Because of the importance of filters, much research and development has been performed in the areas of filter theory, design, and implementation. Many books and thousands of articles have been written on the subject. Therefore, one chapter in a book must necessarily assume a very modest profile in dealing with a subject so vast. The treatment here will consider some key aspects of filter usage and workable design material. While much of the underlying theoretical basis will necessarily be omitted, useful design procedures will be developed within the chapter. After completing this chapter, the reader should have acquired both an appreciation of how filters work in general as well as proficiency in actually designing certain common types of specific filters to meet required specifications.

Section 8–2 is a general discussion of the various classifications of filters on a somewhat broader level than the remainder of the chapter, to show the varied types of filters and their qualitative characteristics.

We will focus primarily on low-pass and high-pass filters of the widely employed Butterworth types. Means for predicting their performance and for determining the complexity required to meet specific filtering requirements are covered. Two common circuit forms for designing these filters will be given; the forms are denoted as *unity-gain designs* and *equal-component designs,* respectively.

We discuss two-pole band-pass response functions and their relationships to basic *RLC* resonance curves, then cover the design of such circuits using a single op-amp with multiple feedback.

We also discuss the state-variable filter and its evolution. Finally we present techniques for designing a state-variable filter to achieve a low-pass, a high-pass, a band-pass, or a band-rejection filter.

8–2 FILTER CLASSIFICATIONS

For our purposes, a ***filter*** is defined as any circuit that produces a prescribed frequency response characteristic, of which the most common objective is to pass certain frequencies while rejecting others. Some circuits classified as filters have objectives other than frequency response criteria, but our considerations here will be limited to the common amplitude and phase filter interpretations.

Filters may be classified in a number of different ways, and like most other classification schemes, filter classes often contain ambiguities and contradictions. The classifications to be given here are simplified as much as possible to convey some of the major categories within the space limitations.

Passive Filters Versus Active Filters

The first classification to be considered is that of *passive filters* versus *active filters.* Passive filters consist of combinations of resistance, capacitance, and inductance. These *RLC* passive filters were the major types used for many years, and such filter technology still represents the dominant form above the audio frequency range. Passive *RLC* structures are capable of achieving relatively good filter characteristics in applications ranging from the audio frequency range to the upper limit of the lumped parameter range.

A problem occurs with passive *RLC* filters at the lower end of the audio frequency range. Inductance values increase as the required frequency decreases, creating several problems. First, inductors are somewhat imperfect devices due to internal losses, but these losses increase markedly in the very large inductance range required at lower frequencies. These heavy losses degrade significantly the quality factor (Q) for each coil, and the associated filter responses have large deviations from their desired forms. Second, the actual physical sizes of the large inductance values limit their usefulness. Third, the costs of such inductances are certainly not trivial considerations.

Active filters consist of combinations of resistance, capacitance, and one or more active devices (such as op-amps) employing feedback. Such filters are theoretically capable of achieving the same responses as passive *RLC* filters. Since inductances are not required, the difficulties associated with them at low frequencies are thus eliminated. Indeed, active *RC* filters operating at very low frequencies can be readily implemented. Active-filter frequency response characteristics can be made to approach the ideal forms very closely, and their costs and physical sizes are very reasonable.

While active filters are capable of circumventing most of the low-frequency limitations of passive filters, they introduce a few disadvantages and problems of their own. Since they are active, power is required to make them operate properly. A closely related fact is that active components in general are less reliable than passive circuits. Finally, active filters employ feedback, and there is always a possibility of instability.

In spite of these limitations, the use of active filters has grown, and they now play a very prominent role in filter technology. Active filters are superior to passive filters in the majority of low- and very low-frequency applications. There is a range of frequencies in which arguments could be made for either passive or active filters, depending on the various design constraints. However, as the region of application extends into the so-called radio frequency (RF) range, passive *RLC* filters take the lead. As a general rule with some probable exceptions, the practical limit of most *RC* active filters is perhaps 100 kHz or so, and most active filters are used well below that frequency.

From the title of this chapter, it is obvious that our interest is directed toward active *RC* filters because such devices fit within the objectives of this book. Passive *RLC* filters are considered in various books devoted to network synthesis and filter theory and design. However, many of the concepts that will be developed here can be related to passive filters. The next several classification schemes, for example, apply equally well to both passive and active filters.

Band Classifications

From the point of view of the amplitude response, most filters can be classified as *low-pass, high-pass, band-pass,* or *band-rejection.* It is helpful to define some ideal block characteristics to represent these different filter types, and such models are shown in Figure 8–1.

The amplitude response for the low-pass ideal block characteristic form is shown in Figure 8–1(a). The quality f_c represents the ***cutoff frequency.*** In this ideal case, the amplitude response for $f < f_c$ is unity, so frequencies in this range are passed by the filter. However, for $f > f_c$, the amplitude response is zero, so frequencies in this range are completely eliminated by the filter.

The high-pass ideal block characteristic is shown in Figure 8–1(b). Observe that its nature is inverted with respect to the low-pass filter; that is, frequencies above the cutoff frequency f_c are passed by the filter, and lower frequencies are rejected.

The band-pass ideal block characteristic is shown in Figure 8–1(c). The band-pass amplitude response is characterized by a center frequency f_o and a bandwidth B.

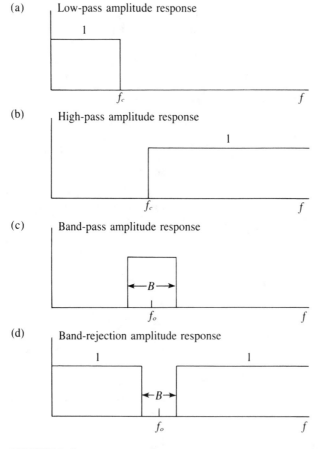

FIGURE 8–1
Ideal block form of the amplitude response for (a) low-pass, (b) high-pass, (c) band-pass, and (d) band-rejection filters.

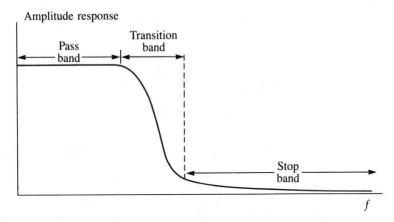

FIGURE 8–2
Representative amplitude response of realistic low-pass filter.

Frequencies falling within the band-pass region are passed by the filter, while components either below the lower band edge or above the upper band edge are rejected.

The band-rejection ideal block characteristic is shown in Figure 8–1(d). This filter passes all frequencies except those within a certain band-rejection region. Parameters f_o and B are used for this filter, but in this case they refer to the center and width, respectively, of the rejection band.

For each of the ideal filter forms, we can define a **pass band** and a **stop band.** The **pass band** is the range of frequencies that is transmitted through the filter, and the **stop band** is the range of frequencies that is rejected.

Actual filters do not possess the ideal block characteristics. Using a low-pass filter for reference, the form of the amplitude response for a realistic filter is illustrated in Figure 8–2. For such nonideal characteristics, it is useful to define three regions: (1) **pass band,** (2) **stop band,** and (3) **transition band.** The exact boundaries between these three regions are somewhat arbitrary. Typically, the amplitude response is specified to be within a certain range from unity in the pass band, and the response is specified to be below a certain level in the stop band. The connecting region between these levels then identifies the transition band.

As a general rule, the flatter the pass band and the lower the stop band, the more complex the filter. This basic trade-off has led to much research and development in filter theory through the years. The ideal block filter characteristic can never be reached, but as it is approached, the complexity of the associated filter designs increases markedly. Fortunately, the vast majority of filter applications can be met with realistic filter characteristics. The typical approach is for the systems designer to determine the minimum level of filtering required, and a filter meeting or exceeding the specifications is then determined.

Attenuation

In specifications of filtering requirements, the concept of **attenuation** is often used. Assume that the maximum pass-band amplitude response for a certain filter is M_o, and the amplitude response at some arbitrary frequency ω is $M(\omega)$. The relative attenuation $\alpha_{dB}(\omega)$ measured in decibels (dB) is defined as

$$\alpha_{dB}(\omega) = 20 \log_{10}\left[\frac{M_o}{M(\omega)}\right] \tag{8-1a}$$

$$\alpha_{dB}(\omega) = -20 \log_{10}\left[\frac{M(\omega)}{M_o}\right] \tag{8-1b}$$

Note that this attenuation is *relative* to the maximum level of transmission in the pass band. One can also use the total attenuation as a parameter, but it may be misleading for filters that have a fixed attenuation in the pass band. For example, suppose that a certain filter has 6-dB attenuation at the point of maximum output in the pass band, and an attenuation of 40 dB at some point of interest in the stop band. The unqualified value of 40 dB could lead one to believe that the filter is actually better than it really is. A more meaningful way to characterize the response is to say that the *relative* attenuation is 34 dB, which sheds a different light on the response.

In view of the nonideal nature of filter responses, one method of classifying filters is according to the type of approximation to the block characteristic employed. A number of useful approximations have been developed. The amplitude response forms for a few of the major types using low-pass characteristics are surveyed in Figures 8–3, 8–4, and 8–5. A very brief introduction to each follows.

Butterworth Response

The form of the **Butterworth amplitude characteristic** is illustrated in Figure 8–3. The Butterworth amplitude response is also referred to as a **maximally flat amplitude response** because of the mathematical structure of its development. The response always decreases as the frequency increases. The Butterworth function will be one of the primary forms developed in depth, so a detailed discussion will be delayed to Section 8–4.

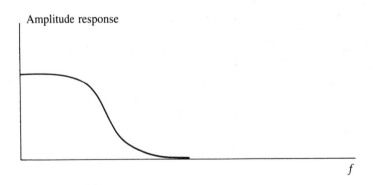

FIGURE 8–3
General form of amplitude response of Butterworth filter.

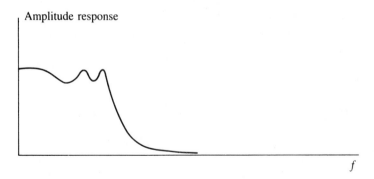

FIGURE 8–4
Form of amplitude response of a particular Chebyshev filter.

Chebyshev Response

The form of one particular ***Chebyshev amplitude characteristic*** is illustrated in Figure 8–4. The Chebyshev response is referred to as an ***equiripple*** response because the pass band is characterized by a series of ripples that have equal maximum levels and equal minimum levels. The number of ripples is a function of the number of reactive elements in the design. The response shown is one particular example within the Chebyshev class. Chebyshev filters have a sharper slope than Butterworth filters and are thus capable of achieving more attenuation in the stop band for a given number of reactive elements. However, their time delay and phase characteristics are less ideal than those of the Butterworth filter, and they tend to exhibit a ringing effect with transient signals.

Maximally Flat Time-Delay Response

A different approach to the approximation problem is that of the ***maximally flat time-delay (MFTD)*** filter. With the MFTD filter, the phase response is optimized so that all frequency components have nearly constant time delay through the filter. The general form of the associated amplitude response for a low-pass case is illustrated in Figure 8–5. At first

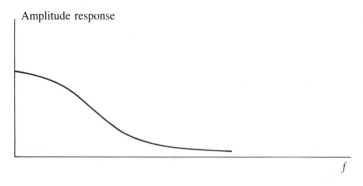

FIGURE 8–5
General form of amplitude response of maximally flat time-delay filter.

glance, this response resembles the Butterworth characteristic in that the response always decreases as the frequency increases. Compared with the Butterworth response, however, the MFTD amplitude response is not as constant in the pass band, and the attenuation is not as high in the stop band. MFTD filters are used in phase-sensitive applications, where constant time delay is very important but the attenuation requirements are moderate.

Comparison

There are various other approximations, but these three types are among the most widely employed. Within the group of these three filter characteristics, the Chebyshev amplitude response has the sharpest rate of attenuation increase above cutoff, but its phase and time-delay characteristics are the poorest. In contrast, the MFTD filter has the most ideal time-delay and phase characteristics, but its amplitude response is the poorest. The Butterworth filter is a reasonable compromise between these extremes, and as a result, it is a very popular choice.

The major emphasis on low-pass and high-pass filter design in this text will utilize Butterworth filter characteristics. While such forms cannot solve all filtering problems, they are certainly capable of solving many. Their ease of design and implementation makes them most suitable for the level intended for this book. Readers requiring more demanding filter requirements than can be met with these filters are advised to consult a specialist!

While Butterworth filters are capable of achieving any basic form of filtering, the most common applications are in low-pass and high-pass forms. For band-pass and band-rejection functions, we will consider a class of active *RC* circuits that have operating characteristics equivalent to those of passive *RLC* resonant circuits.

8–3 FREQUENCY AND IMPEDANCE SCALING

Before considering specific filter designs, we will discuss a widely employed method for dealing with general filter design procedures. Actual filter design requirements vary over a considerable range of frequency and impedance levels, so it would be impossible to tabulate sufficient data for all possible cases. When specific design formulas are used to determine element values, they are often cumbersome and unwieldy.

The most widely used method for tabulating filter design data is based on providing element values for a ***normalized circuit.*** The normalized circuit is developed on the basis of very simplified cutoff or center frequencies and convenient impedance levels. Typically, most of the element values in a normalized design will be in the neighborhood of 1 Ω, 1 F, and so on. Obviously, such element values are quite unrealistic, but they serve as the normalized prototype from which a workable design can be developed. The techniques for converting a normalized design to an actual design are straightforward and can be readily mastered. In fact, with some practice, the conversion can be done almost intuitively.

Conversion of a normalized design to a realistic design can be achieved through the process of two scaling operations: (1) ***frequency scaling*** and (2) ***impedance scaling.*** Frequency scaling is used to change the frequency of the normalized design to the required frequency of the actual design. Impedance scaling is used to change element values to more realistic or workable values. Each of the procedures will now be discussed in detail.

Frequency Scaling

Frequency scaling in an *RC* active filter may be most easily achieved by changing all filter capacitance values in the same proportion *or* by changing all filter resistance values in the same proportion (but not both). The basic rule to remember is that the frequency changes in *inverse* proportion to the change in either *C* or *R*. For example, if all capacitor values in an *RC* active low-pass filter are halved, the cutoff frequency is doubled. In turn, if all resistance values are halved, the cutoff frequency is again doubled. Thus, the process of changing *either* the capacitance values *or* the resistance values may be used to shift the frequency range of an *RC* filter.

When the capacitance values are changed, the impedance level in the modified frequency range is the same as the impedance level in the original circuit in the corresponding frequency range. However, when resistance values are changed, the impedance level is modified. If the resistances are increased, the impedance level in the new frequency range increases in direct proportion to the resistance changes. The change in impedance level may cause difficulty in some applications, while it may result in an advantage in others.

Note that *all filter* capacitances or *all filter* resistances must be changed in the same proportion. The modifier "filter" is used here because active *RC* circuits will typically contain some resistances used only for establishing gains and for bias compensation. Such resistances need not necessarily be scaled. The criterion is whether a resistance is truly part of the filter network itself or simply a resistance value used to establish an amplifier gain or to compensate for bias.

Impedance Scaling

Impedance scaling is the process of changing the relative impedance levels of all components in the filter by a specified amount. This process is performed for the purpose of acquiring realistic and workable element values for the circuit. In the most basic form, to be discussed now, impedance scaling does not alter the frequency scale. (When the resistance values are changed to perform frequency scaling, the impedance level also changes as a secondary effect, but the primary process of impedance scaling leaves the frequency response unchanged.)

Impedance scaling is performed by changing all filter resistance values in a common proportion *and* by changing all filter capacitance values in the opposite proportion. The impedance level change is directly proportional to the resistance change, or it is inversely proportional to the capacitance change. For example, if all filter resistances in an *RC* active filter are doubled and all filter capacitances are halved, the new filter will have exactly the same frequency response as the old filter, but the impedance level at a given frequency will be twice as great as before. Note that since resistance and capacitance values are changed in opposite directions, the frequency change that would have resulted from changing only one parameter is effectively canceled by the change in the other parameter.

Impedance level scaling is required to ensure that element values fall within practical limits for implementation. In fact, with some care, it is often possible to select an impedance scaling factor that will force one or more component values to assume readily available standard values, as will be demonstrated later.

The earlier discussion about resistance values associated only with op-amp operation for frequency scaling holds equally true for impedance scaling. Thus, resistance values used only for establishing closed-loop op-amp gains and in bias compensating circuits can be adjusted independently of the filter resistances.

Having stated the impedance scaling rule in its general form, we will give one modification. Active *RC* filters consisting of several isolated stages have the advantage that the transfer function of one stage does not affect the transfer function of other stages. This permits the impedance level of one stage to be adjusted independently of other stages. Thus, it is possible to employ one impedance scaling factor for one stage and a different factor for a second stage. Such a procedure could permit a better set of component values in certain cases. Bear in mind that if different impedance scaling levels are selected, ***all** pertinent components in a given stage must be modified by the impedance scaling constant for that particular stage.* One can, of course, use the same impedance scaling factor for the entire filter, but the possibility of different constants offers a degree of additional flexibility.

Most normalized filter design data are tabulated on the basis of radian frequencies. The reason for such tabulation is that radian frequencies are related more easily to the Laplace transform techniques from which most of these filter designs are obtained. However, most "real-life" design specifications are given in terms of cyclic frequency in hertz, so one must be careful about the 2π factor. This author has observed experienced designers implementing unworkable designs because they momentarily overlooked this basic concept.

Procedure for Scaling

Assume that the normalized design has some specific reference radian frequency $\overline{\omega}_r$ of importance. This frequency is typically the cutoff frequency of a low-pass or a high-pass filter, or it might be the center frequency of a band-pass filter. Often, the normalized design will be based on a reference radian frequency of 1 rad/s. However, the notation and approach here will provide some generality to the concept.

Assume that there is a cyclic reference frequency f_r in the actual filter that is to correspond to $\overline{\omega}_r$ in the normalized design. Let $\omega_r = 2\pi f_r$ represent the radian frequency corresponding to f_r. A frequency scaling constant K_f is defined as

$$K_f = \frac{\omega_r}{\overline{\omega}_r} = \frac{2\pi f_r}{\overline{\omega}_r} \tag{8–2}$$

Note that the normalized frequency is specified from the beginning in radian form, while the final design is initially specified in cyclic form because these forms are commonly encountered in practice. Irrespective of how they arise, however, the frequency scaling constant must have the same units in both numerator and denominator.

After the frequency scaling constant is determined, it is helpful to note in which direction the component values must change. For example, if the actual design is to be higher in frequency than the normalized design, either capacitance or resistance values will be reduced. By this intuitive approach, it is easier to identify what is happening with K_f rather than relying on a rigid formula.

We have seen that frequency scaling can be performed by changing either capacitance or resistance values. In practice, changing the capacitances is used more often for the process of converting to a higher frequency (the most common case), since the resulting capacitance values are smaller. Resistance changes are most often used to create tunable filter types, since it is usually much easier to vary resistances than capacitances. The procedures to be discussed will emphasize changing the capacitance for frequency scaling, but the possibility of varying the resistance should also be recognized.

An impedance scaling constant K_r will be used in the scaling operation. This constant will be defined as

$$K_r = \frac{\text{impedance level of final circuit}}{\text{impedance level of normalized circuit}} \qquad (8-3)$$

When one is starting with normalized designs, K_r will almost always be larger than unity. However, one may have occasion to take a workable circuit and convert it to a lower impedance level, in which case K_r could be less than unity. The intuitive concept to remember is that the circuit having the highest impedance level will have the largest values of resistances and the smallest values of capacitances, and vice versa.

To convert a normalized active filter design to a realistic design, use the following recommended sequence of steps:

1. Perform the frequency scaling operation first by dividing all filter capacitances by K_f. (Alternately, all resistance values could be divided by K_f in the event that resistance changes are to be used.) Remember the intuitive point that a higher frequency corresponds to smaller element values and vice versa.
2. Select an impedance scaling constant K_r. There is no single way that is best for this procedure, as it will depend on the various constraints. In some cases, a value of K_r may be selected such that one or more final element values turn out to be readily available in stock components. From various design examples considered later, some general guidelines will evolve.
3. Multiply all resistance values by K_r and divide all capacitance values by K_r. Remember the intuitive point: A higher impedance level corresponds to larger resistance values and smaller capacitance values, and vice versa. Note that the frequency scaling process is completely specified while the impedance scaling process leaves some choice in the matter. For this reason, frequency scaling should normally be performed first.

The following two steps are not always necessary, but they provide additional flexibility to ease the overall design process:

4. For filters with more than one section, different impedance scaling constants may be selected for the different sections if desired.
5. For any active filter, resistances that are used only to set gain or attenuation may be scaled completely independently of any other impedance scaling for either that or other sections. However, a common scale factor must necessarily be used for all resistances in a particular network.

EXAMPLE 8–1

Consider the normalized design of a simple passive RC low-pass filter shown in Figure 8–6(a). The 3-dB normalized radian break frequency $\overline{\omega}_b$ is 1 rad/s.

(a) Perform a frequency scaling operation on the circuit by changing the capacitance so that the 3-dB cyclic break frequency f_b is 1 kHz. (b) It is desired to implement the filter to achieve the desired break at 1 kHz using a 0.001-μF capacitor. Determine the required value of resistance. (c) After the preceding design is completed, assume that it is desired to replace the fixed resistance by a variable resistance that would permit the 3-dB frequency to be continuously adjusted from 500 Hz to 5 kHz. Determine the required range of the variable resistance.

Solution

(a) The actual filter at 1 kHz must have the same response as the normalized design at 1 rad/s. The frequency scaling constant K_f is thus defined as

$$K_f = \frac{2\pi \times 10^3 \text{ rad/s}}{1 \text{ rad/s}} = 6283.185 \tag{8–4}$$

where the units have been expressed in both numerator and denominator to emphasize that both must be converted to the same form.

To scale the frequency range, we divide the 1-F capacitance in the normalized design by K_f, so the modified capacitance is $1/6283.185 = 159.155\ \mu$F, as shown in Figure 8–6(b). The resistance value is not changed. The new circuit has the same impedance level as the old circuit, but the 3-dB break frequency is now 1 kHz. Note the intuitive idea that the capacitance is reduced to shift to a higher frequency.

(b) The circuit shown in Figure 8–6(b) is still not very good because of the large capacitance and the small resistance. An impedance scaling operation should be performed to create more desirable element values. While a number of different strategies could be used, the specific method to be used here is to force the final capacitance to be the standard value of 0.001 μF. Thus, an impedance scaling constant K_r is to be selected such that 159.155 μF converts to 0.001 μF. This forces the following constraint:

$$\frac{159.155 \times 10^{-6}}{K_r} = 0.001 \times 10^{-6} \tag{8–5}$$

The constant is then determined to be $K_r = 159{,}155$. Since the capacitance is reduced by this factor, the 1-Ω resistance must be increased by the same factor. The final circuit is shown in Figure 8–6(c).

(c) The stated objective in this step illustrates an important case where it is desired to change the frequency scale by changing the resistance. Rather than scaling the frequency on paper, however, the actual resistance is to be varied in the circuit.

Using an intuitive approach, we have established that a resistance of 159.154 kΩ is required for a break frequency of 1 kHz. To change the break frequency to 500 Hz, we must double the resistance, so the maximum resistance is 2×159.154 k$\Omega = 318.310$ kΩ. To change the break frequency to 5 kHz, we must divide the resistance by 5, so the minimum resistance is $159.154 \times 10^3/5 = 31.83$ kΩ. Thus, in order to allow adjustment of the break frequency from 500 Hz to 5 kHz, the resistance must be adjustable from 318.31 kΩ down to 31.83 kΩ.

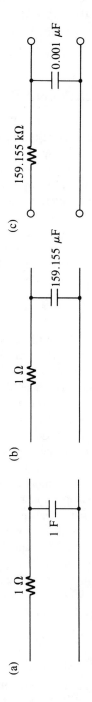

FIGURE 8-6
Simple passive RC filter used to illustrate frequency scaling and impedance scaling.

Incidentally, this circuit could have been designed more easily with the one-pole roll-off model concept developed in Chapter 1, and the reader may wish to verify the equivalence of the results. However, the powerful approaches of frequency and impedance scaling were illustrated with this simple circuit.

8-4 BUTTERWORTH FILTER RESPONSES

During the brief introduction to a number of the most common filter approximations in Section 8–2, it was indicated that the Butterworth form was one of the most widely employed filter types. In this section, the Butterworth form will be discussed in some detail, and quantitative data concerning the relative performance will be presented.

The emphasis in this section will be on relating the actual filter performance to the complexity or order of the approximation. In this manner, it will be possible to determine if given specifications can be achieved with a Butterworth filter and to determine the particular filter complexity required. The actual design of such filters will be considered in later sections. It should be noted at the outset that the performance specifications apply equally well to both passive and active Butterworth filters, so the material presented in this section has broader scope than just active filters.

Number of Poles

A common parameter used in specifying filters is the *order* of the filter. In Laplace transform terminology, the term *order* is synonymous with the **number of poles.** The actual mathematical significance of these designations is related to the transfer function of the filter, and since both designations are widely used in filter specifications, they will both be used here. For our purposes, the order or number of poles is the number of nonredundant reactive elements in the circuit. Both inductors and capacitors are reactive elements, but since active filters contain no inductors, the order or number of poles will be the number of nonredundant capacitors. None of the circuit diagrams considered later will contain redundant capacitors, so the order or number of poles will be the number of capacitors in the circuit. (This does not include capacitors external to the filter network, such as power supply bypass capacitors and op-amp compensating capacitors.)

For a given filter type, the performance generally becomes closer to the ideal block characteristic as the number of poles increases. Thus, a higher-order filter will have a flatter pass-band response and a lower stop-band response (more attenuation) than one of lower order. However, the ideal block characteristic can never be fully attained.

Butterworth functions for low-pass and high-pass designs will be considered. Band-pass and band-rejection forms will be developed in later sections using a different approach. Next we will consider the mathematical forms for the low-pass and high-pass Butterworth functions.

Low-Pass Butterworth

The Butterworth functions utilize a convenient reference frequency at which the amplitude response drops to $1/\sqrt{2}$ of its maximum pass-band level. This corresponds to the response being down 3.01 dB, and this value is usually rounded to 3 dB. This frequency will be referred to as the ***cutoff frequency,*** but it should be understood that it is not an abrupt cutoff. Let f_c represent the cutoff frequency, and let n represent the order (number of poles) of the approximation. The amplitude response $M(\omega)$ of the Butterworth low-pass function is given by

$$M(\omega) = \frac{1}{\sqrt{1 + (f/f_c)^{2n}}} \qquad \qquad \textbf{(8–6)}$$

The maximum value of $M(\omega)$ occurs at $f = 0$, and it has been established at unity for convenience; that is, $M(0) = 1$. With some passive filters, the maximum value is less than unity, and the filter is said to have a ***flat loss.*** Conversely, some active filters have a maximum value greater than unity. One can readily incorporate these differences into (8–6) by putting a constant in the numerator, but since it is the relative response that is of primary interest, the simpler form will be assumed here.

Let $M_{\text{dB}}(\omega)$ represent the decibel form of the response relative to the maximum level. Since the maximum level of (8–6) is unity, the decibel form can be expressed as

$$M_{\text{dB}}(\omega) = 20 \log_{10} \left[\frac{1}{\sqrt{1 + (f/f_c)^{2n}}} \right] \qquad \qquad \textbf{(8–7a)}$$

or

$$M_{\text{dB}}(\omega) = -10 \log_{10} \left[1 + (f/f_c)^{2n} \right] \qquad \qquad \textbf{(8–7b)}$$

where some basic properties of the logarithmic function were used in converting from (8–7a) to (8–7b).

High-Pass Butterworth

The cutoff frequency f_c for high-pass Butterworth filters has essentially the same meaning as for low-pass filters except, of course, it is at the low end of the pass band. The response is down by 3.01 dB at this point. The amplitude response $M(\omega)$ of the Butterworth high-pass function is given by

$$M(\omega) = \frac{1}{\sqrt{1 + (f_c/f)^{2n}}} \qquad \qquad \textbf{(8–8)}$$

The decibel form $M_{\text{dB}}(\omega)$ of the high-pass response can be expressed as

$$M_{\text{dB}}(\omega) = 20 \log_{10} \frac{1}{\sqrt{1 + (f_c/f)^{2n}}} \qquad \qquad \textbf{(8–9a)}$$

or

$$M_{dB}(\omega) = -10 \log_{10}\left[1 + \left(\frac{f_c}{f}\right)^{2n}\right]$$ (8–9b)

Comparing (8–9a) and (8–9b) with (8–7a) and (8–7b), we note that the high-pass form is derived from the low-pass form by simply replacing f/f_c with f_c/f. This transformation "inverts" the low-pass form and converts it to the high-pass form.

Amplitude Response Curves

Plots of the stop-band amplitude response for Butterworth filters with orders 2 through 7 are given in Figure 8–7. All the curves start at the -3 dB level, which corresponds to $f = f_c$. The response curves are shown as *relative* decibel output based on a reference dc gain of unity as given by (8–7a) and (8–7b). In the event that the filter has a flat loss or gain, the curves still apply, but a constant decibel level could be either added to or subtracted from the curves given. Alternately, the curves could simply be interpreted as the decibel response *relative* to the maximum response, which is usually the most meaningful approach.

Attenuation in a positive sense corresponds to gain in a negative sense. Thus, if a relative attenuation of 20 dB is specified, one could look for a level of -20 dB on the given curves.

The abscissa, labeled "Normalized frequency," is interpreted differently for low-pass and high-pass filters. For low-pass filters, the abscissa is interpreted as f/f_c. For example, *normalized frequency* = 1 corresponds to $f = f_c$, and *normalized frequency* = 10 corresponds to $f = 10f_c$. For high-pass filters, the abscissa is interpreted as f_c/f. In this case, *normalized frequency* = 10 corresponds to $f = 0.1f_c$. While the curves are shown only for a range of one decade above cutoff, this is the region where most specifications are directed, and the curves are most useful in evaluating relative filter performance.

In some applications, it is desired to inspect more closely the flatness of the response within the pass band. For example, it may be necessary to maintain a response to be flat within a fraction of 1 dB over a specified frequency range. The preceding equations may be readily used to compute the amplitude response within the pass band for a given filter order, and the results could be plotted. However, the various curves are so close to each other over various parts of the pass-band range that the graphical results are judged to be less useful in the pass band than in the stop band. Instead, some useful tabulated data will be provided.

Table 8–1 provides a list of certain normalized frequencies at which the relative amplitude response is down by a specified amount. The relative amplitude response definition is the same as for Figure 8–7, and data for two-pole through seven-pole filters are provided.

To illustrate these data, assume that it is desired to determine how close to the standard cutoff frequency the response is when down by 1 dB from the dc level. For $n = 2$, the low-pass frequency is $0.713f_c$, and the high-pass frequency is $f_c/(0.713) = 1.403f_c$. Moving horizontally across the table, note that at the other extreme, for $n = 7$, the low-pass frequency is $0.908f_c$, and the high-pass frequency is $f_c/(0.908) = 1.101f_c$.

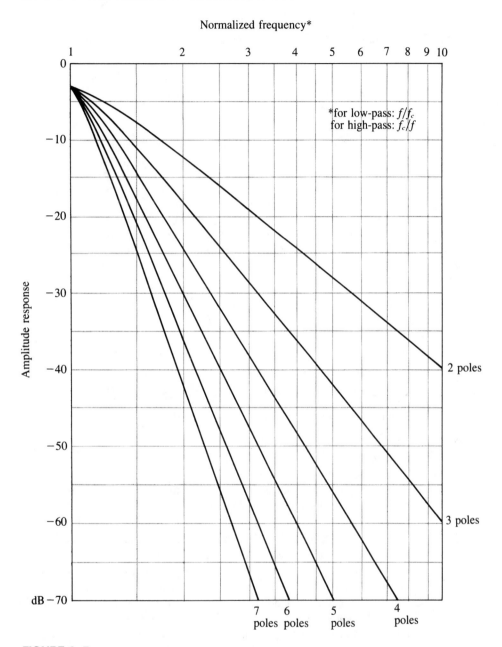

FIGURE 8-7
Stop-band amplitude response of Butterworth filters.

TABLE 8–1
Data for determining pass-band relative amplitude response for
Butterworth filters

Amplitude Response (dB)	f/f_c for Low-Pass f_c/f for High-Pass					
	$n = 2$	$n = 3$	$n = 4$	$n = 5$	$n = 6$	$n = 7$
−0.01	0.219	0.363	0.468	0.545	0.603	0.648
−0.02	0.261	0.408	0.511	0.584	0.639	0.681
−0.05	0.328	0.476	0.573	0.640	0.690	0.727
−0.1	0.391	0.534	0.625	0.687	0.731	0.764
−0.2	0.466	0.601	0.683	0.737	0.775	0.804
−0.5	0.591	0.704	0.769	0.810	0.839	0.860
−1	0.713	0.798	0.845	0.874	0.894	0.908
−2	0.875	0.914	0.935	0.948	0.956	0.962
−3.01	1	1	1	1	1	1

EXAMPLE 8–2

A low-pass filter is desired for a given application. The specifications are as follows:

1. relative attenuation ≤ 3 dB for $f \leq 800$ Hz
2. relative attenuation ≥ 23 dB for $f \geq 2$ kHz

Specify the minimum number of poles for a Butterworth filter that will satisfy the requirements.

Solution
These specifications are given in a typical manner. The requirement of (1) specifies a "pass band" in which the response drops no more than 3 dB. The requirement of (2) specifies a "stop band" in which the response is required to be at or below a certain level. The region between (1) and (2) can then be interpreted as a "transition band."

The specifications as given are not very demanding, since a relatively wide transition band is provided. The curves of Figure 8–7 may be readily used. We interpret the cutoff frequency to be $f_c = 800$ Hz. The actual frequency $f = 2$ kHz corresponds to a normalized frequency $f/f_c = 2000/800 = 2.5$. At an abscissa of 2.5 on the normalized frequency scale, we drop down to determine the response curves that will achieve the required attenuation. The response of a two-pole function is down by only about 16 dB at this frequency, so it is inadequate. However, a three-pole response is down by about 24 dB at the normalized frequency of 2.5, so it more than meets the specifications. The minimum number of poles required in the Butterworth filter is ***three.***

Before leaving this problem, we should note that the relative amplitude response curve for any three-pole Butterworth filter will meet specifications as given for any frequencies having the same ratios as those given. For example, if the frequency of specification (1) had been 5 kHz and the frequency of specification (2) had been 12.5 kHz, a three-pole Butterworth filter would again be the correct solution, since 12.5 kHz/5 kHz = 2.5 is the same normalized frequency as determined in the problem. The utility of the normalized frequency concept is that the various results apply to any frequency combinations having the same relative ratio.

EXAMPLE 8-3

For a three-pole Butterworth filter with a cutoff frequency of 800 Hz as determined in Example 8-2, determine the relative amplitude response at the following frequencies: dc, 400 Hz, 800 Hz, 2 kHz, 4 kHz, and 8 kHz.

Solution

Since the 3-dB frequency was established as 800 Hz, the normalized frequency is $f/f_c = f/800$. For all but dc and 400 Hz, the normalized frequency is calculated, and the response is read from Figure 8-7. For dc, the relative amplitude response is always 0 dB. For 400 Hz, the function of (8-7b) can be used to compute the response. In fact, this equation could be used to calculate the response at any frequency, and this approach is more accurate than the graphical approach. However, the graphical approach is usually sufficient for routine analysis and design. The results are summarized as follows:

f	$f/800$	Relative Amplitude Response (dB)
dc	0	0
400 Hz	0.5	−0.067
800 Hz	1	−3
2 kHz	2.5	−24
4 kHz	5	−42
8 kHz	10.5	−60

EXAMPLE 8-4

A high-pass filter is desired for a given application. The specifications are as follows:

1. relative attenuation ≤ 3 dB for $f ≥ 500$ Hz
2. relative attenuation ≥ 46 dB for $f ≤ 125$ Hz

Specify the minimum number of poles for a Butterworth filter that will satisfy the requirements.

Solution

The manner in which the specifications are given suggests that the 3-dB cutoff frequency can be established at 500 Hz. For a high-pass filter, the frequency of 125 Hz corresponds to an inverted normalized frequency of $f_c/f = 500/125 = 4$. At a normalized frequency of 4, the smallest number of poles satisfying the specification is *four.* The attenuation at this point is slightly greater than 48 dB, so the specification is met with some reserve.

---

*EXAMPLE 8–5

A low-pass filter is desired for a given application. The specifications are as follows:

1. relative attenuation ≤ 0.5 dB for $f \leq 1$ kHz
2. relative attenuation ≥ 20 dB for $f \geq 2$ kHz

Specify the minimum number of poles for a Butterworth filter that will satisfy the requirements.

Solution

A significant point to note at the outset is that the "pass-band" upper frequency limit of 1 kHz is specified as the "0.5 dB down frequency." Presumably, the application here requires a more rigid pass-band bound than in the preceding examples.

Let $f_c' = 1$ kHz represent the modified cutoff frequency corresponding to the -0.5-dB point. The frequency referred to f_c' at which the response must be down 24 dB is 2 kHz/1 kHz = 2. However, the curves of Figure 8–7 are based on f/f_c where f_c is the 3-dB frequency. It can be readily deduced that since f_c is larger than f_c', f/f_c will be smaller than f/f_c'. Table 8–1 provides data on the normalized frequencies at which the response is down by 0.5 dB. The frequency values can be interpreted as f_c'/f_c for our purposes.

To use the curves of Figure 8–7, we first compute f/f_c by the operation

$$\frac{f}{f_c} = \left(\frac{f}{f_c'}\right) \times \left(\frac{f_c'}{f_c}\right) \qquad \textbf{(8–10)}$$

where the ratio f_c'/f_c is determined from the table. For $f/f_c' = 2$, the values of f_c'/f_c are read from Table 8–1 in the row corresponding to -0.5 dB and are multiplied by $f/f_c' = 2$ to determine the values of f/f_c for Figure 8–7. For each order, a new value of f/f_c is computed, and the attenuation is determined from the curve. The lowest value of n satisfying the requirement is the solution.

Some trial and error may be involved in the process. For example, for $n = 3$, $f_c'/f_c = 0.704$ and $f/f_c = 2 \times 0.704 = 1.408$. The attenuation at this frequency is between 9 and 10 dB. However, for $n = 5$, $f_c'/f_c = 0.810$, $f/f_c = 2 \times 0.810 = 1.62$, and the response is down by about 21 dB. Thus, $n = 5$ is the smallest number of poles that will meet the specifications for a Butterworth filter as given. The required 3-dB frequency for the Butterworth filter is $f_c = 1$ kHz/0.810 = 1235 Hz. Thus, establishing the 3-dB frequency of a five-pole filter at 1235 Hz will result in a pass-band flat to within 0.5 dB up to 1 kHz.

Incidentally, the manner in which the specifications are given suggests the possibility of a Chebyshev filter, in which the pass-band variation is controlled to a specified

---

* This example represents a different approach than normally used with Butterworth filters. It may be bypassed without loss of continuity.

ripple level (0.5 dB in this case). In fact, for filter specifications reasonably more demanding than those appearing in this chapter, it is recommended that some of the specialized references on filter design be consulted. While the types of filters considered in this chapter are very important, they will not cover all the possible requirements by any means.

8–5 LOW-PASS AND HIGH-PASS UNITY-GAIN DESIGNS

Design data for certain low-pass and high-pass Butterworth active filters will be presented in this section. These active circuits utilize what are referred to as *finite gain realizations;* that is, the gain of the active device (op-amp) is established at a relatively low finite level by feedback. In particular, the gain in the circuits of this section will be set at unity by using voltage followers for the active stages.

All of the designs to be presented have a normalized 3-dB radian cutoff frequency of $\omega_c = 1$ rad/s. This value represents the defined upper edge of the pass band for low-pass designs and the lower edge of the pass band for high-pass designs.

The unity-gain designs are based on combinations of two-pole and three-pole sections. All even-numbered realizations employ only two-pole sections, while odd-numbered realizations employ one three-pole section and as many two-pole sections as required to realize the required numbers of poles.

Low-Pass Designs

The forms of the normalized two-pole and three-pole unity-gain sections for the low-pass designs are shown in Figure 8–8. The filter resistances have been established as 1 Ω in the normalized designs. However, as noted, the bias compensating resistances are set at their optimum values, but they do not affect the filter characteristics. It is common practice not to display units on schematics of normalized designs, but all component values are assumed to be in their basic units, that is, ohms and farads. Note that the op-amps are functioning as voltage followers.

The values of the capacitances are given in Table 8–2 for different low-pass filter orders (number of poles). The units for these normalized capacitances are *farads.* Rows containing only C_1 and C_2 represent two-pole sections, while rows containing C_1, C_2, and C_3 represent *three-pole* sections. Do not make the mistake of using the form of a three-pole section when only C_1 and C_2 are given, since the filter will not work properly.

The various sections are connected in cascade to implement the composite filter. Since each section contains a voltage follower at the output, theoretically the sections could be connected in any order. In practice, however, some sections are more peaked than others, and more easily overloaded. Such sections are best located closer to the output. The data are arranged in the order for achieving the best implementation; that is, the section in the top row for a given number of poles should be connected at the input, and at the other extreme, the bottom row should be connected at the output section.

It should be strongly emphasized that *a higher-order Butterworth filter is not a cascade of lower-order Butterworth filters.* For example, consider a four-pole Butterworth

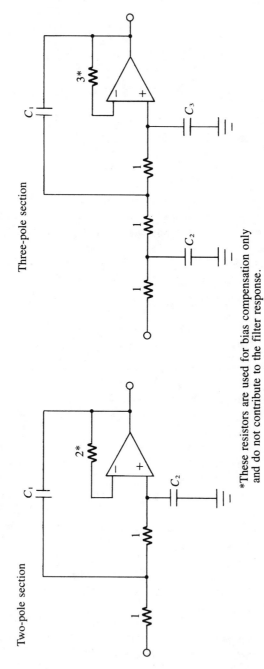

FIGURE 8-8

Normalized unity-gain low-pass Butterworth active filter designs. (See Table 8–2 for capacitance values.)

TABLE 8–2
Capacitance values for low-pass
Butterworth active filter designs

Poles	C_1	C_2	C_3
2	1.414	0.7071	
3	3.546	1.392	0.2024
4	1.082	0.9241	
	2.613	0.3825	
5	1.753	1.354	0.4214
	3.235	0.3089	
6	1.035	0.9660	
	1.414	0.7071	
	3.863	0.2588	
7	1.531	1.336	0.4885
	1.604	0.6235	
	4.493	0.2225	

filter which consists of two two-pole sections. Observe from Table 8–2 that the capacitors in the two sections are all different, and none are equal to the values for a two-pole Butterworth filter. (Because of the symmetry of the mathematics involved, it does turn out that one of the three sections for a six-pole filter is the same as a two-pole Butterworth filter, but this is a "mathematical coincidence.")

High-Pass Designs

The two-pole and three-pole sections for the high-pass filters are shown in Figure 8–9. For the high-pass filters, the capacitors are all normalized to a level of 1 F. The resistance values for the circuits are tabulated in Table 8–3. The various procedures for determining the data and implementing the circuits discussed earlier for the low-pass filter designs apply here as well.

The extent to which the final active design matches the theoretical form depends the closeness of the element values to their ideal values as well as the performance of the op-amp in the frequency range involved. Polystyrene capacitors are recommended when possible, but high-quality mica and Mylar® capacitors may also be used. Do not use electrolytic or other polarized capacitors.

The closed-loop bandwidth of the op-amp and the slew rate should be sufficient for the frequency range involved. In a sense, the high-pass filter is really a sort of band-pass filter, since the finite bandwidth and slew rate of the op-amp will cause a roll-off at high frequencies. This factor should be considered in the overall design for any particular application.

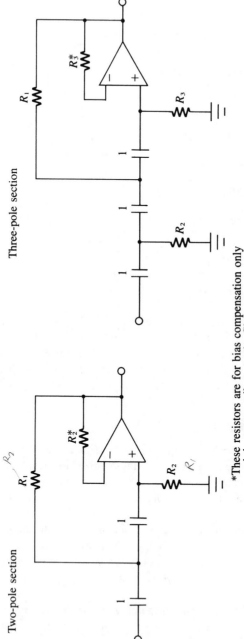

FIGURE 8–9

Normalized unity-gain high-pass Butterworth active filter designs. (See Table 8–3 for resistance values.)

TABLE 8–3
Resistance values for high-pass
Butterworth active filter designs

Poles	R_1	R_2	R_3
2	0.7072	1.414	
3	0.2820	0.7184	4.941
4	0.9242	1.082	
	0.3827	2.614	
5	0.5705	0.7386	2.373
	0.3091	3.237	
6	0.9662	1.035	
	0.7072	1.414	
	0.2589	3.864	
7	0.6532	0.7485	2.047
	0.6234	1.6038	
	0.2226	4.494	

The use of the design data for low-pass and high-pass filters will now be illustrated with several examples.

EXAMPLE 8–6

Design a low-pass active filter to meet the specifications of Example 8–2.

Solution
In Example 8–2, the requirements called for a 3-dB cutoff frequency of 800 Hz, and it was determined that the specifications could be met with a three-pole Butterworth filter. The normalized data are obtained from Table 8–2, and one three-pole section of the form given in Figure 8–8 is used. The normalized circuit diagram with element values given is shown in Figure 8–10(a).

Conversion to the proper frequency range and a realistic impedance level is achieved in a two-step process. The frequency scaling constant K_f is determined such that 1 rad/s converts to 800 Hz. We have

$$K_f = \frac{2\pi \times 800 \text{ rad/s}}{1 \text{ rad/s}} = 5026.548 \qquad \textbf{(8–11)}$$

For the first step, the circuit is scaled to the proper frequency range by dividing all capacitance values by K_f, and the corresponding circuit is shown in Figure 8–10(b).

The next step in the process is to perform an impedance scaling that will yield realistic values of both resistances and capacitances. There is no single impedance factor K_r

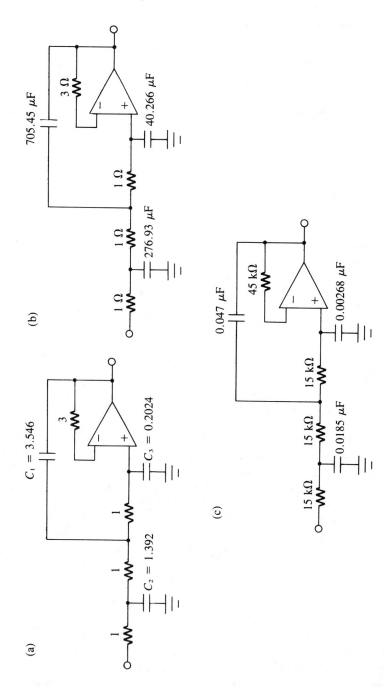

FIGURE 8-10

Circuit design of Example 8-6.

that is best for this purpose. For example, one might try to achieve available standard values whenever possible. However, some of the values required may not always correspond to standard values, and some "tweaking" or combining of several component values may be necessary. Some trial and error is often required, and a keen intuition is useful. The latter capability develops with experience.

After some trial and error, a constant $K_r = 15,000$ was selected. All resistance values are multiplied by K_r, and capacitance values are divided by K_r. The circuit obtained is shown in Figure 8–10(c). The advantage of this design is that the filter resistance values are standard, and the three calculated capacitance values are all very close to standard values. For noncritical applications, one could start with the standard values of 0.047 μF, 0.018 μF, and 0.0027 μF for C_1, C_2, and C_3, respectively. Additional "tweaking" could be performed if necessary.

EXAMPLE 8–7
Design a five-pole, low-pass, active Butterworth filter with a 3-dB cutoff frequency of 2 kHz.

Solution
The normalized data are obtained from Table 8–2, and the forms of the two sections are given in Figure 8–8. Observe that one two-pole section and one three-pole section are required. The complete normalized schematic diagram is shown in Figure 8–11(a).

The frequency scaling constant K_f is determined such that 1 rad/s converts to 2 kHz. We have

$$K_f = \frac{2\pi \times 2 \times 10^3 \text{ rad/s}}{1 \text{ rad/s}} = 12,566.37 \qquad \textbf{(8–12)}$$

All capacitor values are first divided by K_f to convert the basic circuit to the proper frequency range, and the resulting circuit is shown in Figure 8–11(b).

As indicated earlier in the chapter, different impedance scaling constants may be used for different sections of an active filter provided that all element values in a given section are modified by the same constant. After some trial and error, an impedance scaling constant $K_r = 27,000$ was selected for the first section and a constant $K_r = 22,000$ was selected for the second section. Resistance values are multiplied and capacitance values are divided by their respective constants, and the final circuit is shown in Figure 8–11(c).

Observe that the capacitance values in the first stage are quite close to the standard values of 0.0051 μF, 0.0012 μF, and 0.0039 μF, and the capacitance values in the second stage are close to the standard values of 0.012 μF and 0.0011 μF.

EXAMPLE 8–8
Design a two-pole, high-pass, active Butterworth filter with a 3-dB cutoff frequency of 1 kHz.

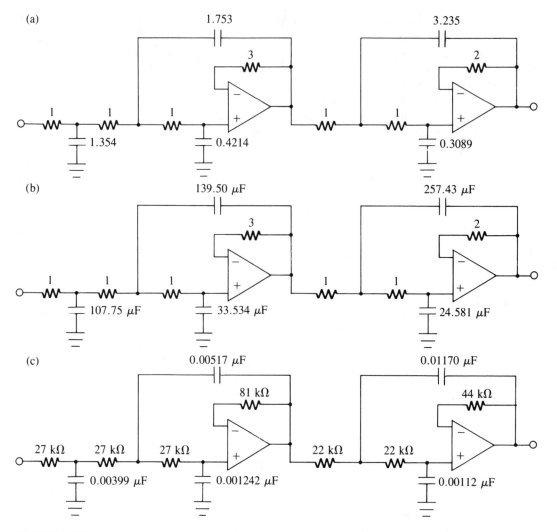

FIGURE 8–11
Circuit design of Example 8–7.

Solution

The normalized data are obtained from Table 8–3, and one two-pole section of the form given in Figure 8–9 is used. The normalized circuit diagram with element values given is shown in Figure 8–12(a).

The frequency scaling constant K_f is

$$K_f = 2\pi \times 10^3 = 6283.185 \qquad (8\text{–}13)$$

Conversion to the proper frequency range is obtained by dividing the two capacitances by K_f, and the result is shown in Figure 8–12(b).

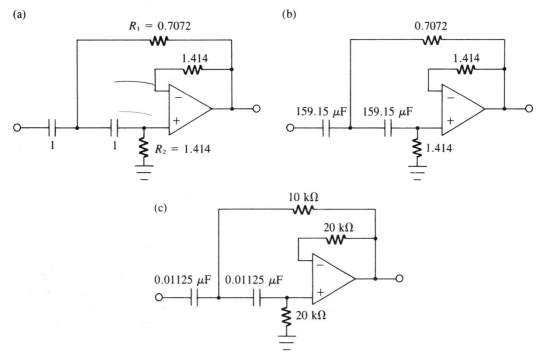

FIGURE 8–12
Circuit design of Example 8–8.

For this case, one of the two resistance values is twice the other value. This suggests the possibility of choosing a scaling constant such that the two final resistance values are 10 kΩ and 20 kΩ, which are two of the 5% values that are in the right proportion. We thus require that $1.414 \times K_r = 20,000$, which yields $K_r = 14,144$. Dividing the two equal capacitance values by K_r results in required capacitance values of 0.01125 μF, which is close to 0.011 μF. The resulting circuit is shown in Figure 8–12(c).

8–6   LOW-PASS AND HIGH-PASS EQUAL-COMPONENT DESIGNS

The design of low-pass and high-pass filters using unity-gain (voltage follower) amplifiers was covered in the last section; now we will examine a different approach for implementing finite gain filters. The approach is based on the fact that if the gain of the amplifier in a two-pole section is increased above the unity level, the values of the two capacitances in a low-pass section and the values of the two resistances in a high-pass section can be made to be equal. For each case, there is a critical value of gain that achieves the exact value required. The result is that each two-pole section will have two equal resistances and two equal capacitances. The resulting circuits will be denoted, for obvious reasons, as *equal-component designs*.

TABLE 8–4
Comparison of the number of stages required with unity-gain designs and equal-component designs

Number of Poles	Number of Stages	
	Unity-Gain Design	Equal-Component Design
2	1	1
3	1	2
4	2	2
5	2	3
6	3	3
7	3	4

With equal-component designs, the composite filter is normally implemented with combinations of two-pole and one-pole sections. For an even number of poles, only two-pole sections are used. For an odd number of poles, all sections but one are two-pole types and a single one-pole section is required. Thus, for certain combinations, the number of stages (and active devices) is greater for equal-component designs than for unity gain designs. A comparison is given in Table 8–4.

Low-Pass Designs

The form of the two-pole equal-component section for the low-pass designs is shown in Figure 8–13. Note that the two filter resistances each have a value of 1 Ω, and the two

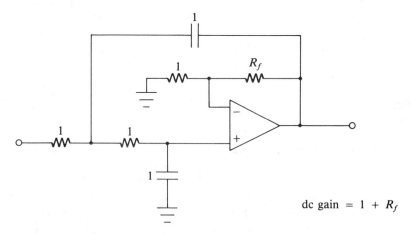

$$\text{dc gain} = 1 + R_f$$

FIGURE 8–13
Normalized equal-component low-pass two-pole section.

(a)

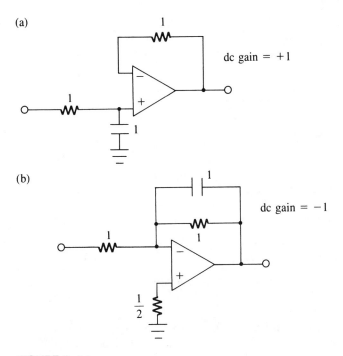

(b)

FIGURE 8–14
Two forms of normalized one-pole low-pass filter sections.

capacitances each have a value of 1 F. The gain-setting resistances have values of 1 Ω and R_f. As we will see shortly, the value of R_f is critical in establishing the response required for a given stage. Note that the gain of the amplifier is $A_{CL} = 1 + R_f/1 = 1 + R_f$ on a normalized basis.

A single one-pole section is required when the order of the filter is odd. One form of a low-pass one-pole input circuit is shown in Figure 8–14(a). This circuit is basically a simple RC one-pole passive filter accompanied by a voltage follower, which isolates the next section from the two-pole section. An alternate version of a one-pole low-pass circuit is shown in Figure 8–14(b). Although the input impedance of this circuit is less than that of the first circuit, it has the advantage that it is a constant resistive impedance, which could prove to be an advantage in some applications. The dc gain of the first circuit is +1, while the dc gain of the latter circuit is −1; so the choice could also depend on which value is desired in a given application.

The actual design data required are given in Table 8–5. The data are arranged in the order for achieving the best implementation, that is, the first section on the left should be connected at the input, etc. Note that one-pole sections are required for three-pole, five-pole, and seven-pole designs. All designs require at least one two-pole section, and as the order increases, more sections are required. The normalized value of R_f required for each section is given in the table.

TABLE 8–5
Feedback resistance values for low-pass and high-pass
equal-component value Butterworth designs

Poles	One-Pole Section	First Two-Pole Section Rf	Second Two-Pole Section Rf	Third Two-Pole Section Rf	Net* dc Gain
2		0.5858			1.5858 (4.00 dB)
3	Required	1			2 (6.02 dB)
4		0.1523	1.2346		2.5749 (8.22 dB)
5	Required	0.3819	1.3820		3.2917 (10.35 dB)
6		0.06838	0.5858	1.4824	4.2058 (12.48 dB)
7	Required	0.1979	0.7530	1.5550	5.3653 (14.59 dB)

* May be inverting if inverting form of one-pole section is used.

With the exception of the one-pole stage when required, all stages have dc and low-frequency gains exceeding unity. The net dc gain is thus the product of all individual stage gains. The net dc gain of the filter for each case has been calculated and is tabulated in the last column of Table 8–5. Both the linear magnitude and the corresponding decibel values are listed. Note that the magnitude of the gain is listed, but a possible inversion arising from using a one-pole inverting stage is not considered.

The presence of a dc gain other than unity may be an advantage in some cases, a disadvantage in some cases, and of no consequence in other cases. However, one must be aware of this property to properly deal with the filter output in subsequent circuits. To apply the curves of Figure 8–7, one must interpret these results as the response *relative* to the maximum or dc response. For example, suppose one is using a four-pole filter in which the dc gain is 8.22 dB. At the cutoff frequency, the response will *not* be -3 dB, as was the case for the unity-gain designs. Instead, the response will be $8.22 - 3 = 5.22$ dB, which is still a gain. However, the *relative* response is down 3 dB, and the shape of the response curve is the same.

High-Pass Designs

The form of the two-pole equal-component section for the high-pass designs is shown in Figure 8–15. As in the case of the low-pass designs, there are two possible forms for the one-pole section, and these are shown in Figure 8–16. Neither of these circuits offers a constant input resistance in this case, but other considerations may favor one circuit over the other.

The results of Table 8–5 may be applied directly for the high-pass designs in the same manner as for the low-pass designs except, of course, that the sections to which the

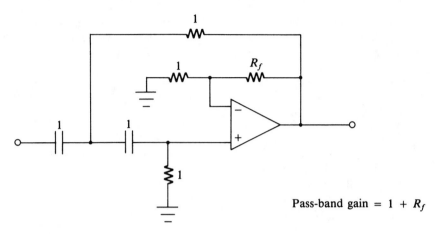

Pass-band gain $= 1 + R_f$

FIGURE 8–15
Normalized equal-component high-pass two-pole section.

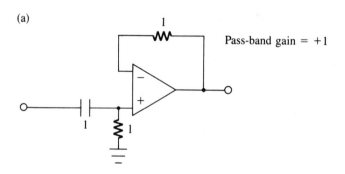

(a)

Pass-band gain $= +1$

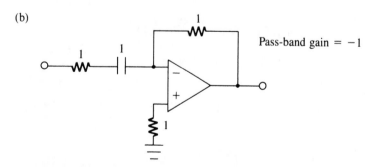

(b)

Pass-band gain $= -1$

FIGURE 8–16
Two forms of normalized one-pole high-pass filter sections.

data apply have different forms. The concept of the pass-band gain exceeding unity applies equally well to high-pass design, except it is the *pass-band* flat gain at which the values are applicable.

EXAMPLE
8–9

Perform a new design of the low-pass filter of Example 8–6 using the equal-component approach.

Solution

To achieve three poles with an equal-component design approach, we use a one-pole section followed by a single two-pole section. No specific preference is given for the input stage, so the noninverting form will be arbitrarily selected. Utilizing the circuit diagrams of Figures 8–13 and 8–14(a) and the data of Table 8–5, the normalized design is shown in Figure 8–17(a).

As determined in Example 8–16, the frequency scaling constant is $K_f = 5026.548$. For the first step, the circuit is scaled to the proper frequency range by dividing the three capacitances by K_f, and the corresponding circuit is shown in Figure 8–17(b).

A number of possible values of K_r could produce satisfactory results. The ideal choice is a value that produces stock values for the capacitors and the resistors. After a little trial and error, it was determined that a reasonable approach is to force the capacitors to assume the final value of 0.018 μF. This process is achieved by the following constraint:

$$\frac{198.94 \times 10^{-6}}{K_r} = 0.018 \times 10^{-6} \qquad \textbf{(8–14)}$$

Solving for K_r, we have

$$K_r = 11,052 \qquad \textbf{(8–15)}$$

Using this value, the ideal value of all the resistances would be 11,052 Ω. However, this value differs from the standard value of 11 kΩ by less than 0.5%. For noncritical applications, the value of 11 kΩ should suffice. The circuit diagram with these rounded values is shown in Figure 8–17(c).

Note that the bias compensation in the second stage is not optimum. At the relatively low impedance level of 11 kΩ, bias currents are usually not too troublesome; however, if it were necessary to optimize this condition, the two gain-setting resistances of the second stage could be increased in the same proportion. The standard values of 43 kΩ for *both* of the two gain-setting resistances would result in nearly optimum bias compensation for the second stage.

EXAMPLE
8–10

Perform a new design of the high-pass filter of Example 8–8 using the equal-component approach.

Solution

The design objective is to implement a two-pole high-pass filter with a cutoff frequency of 1 kHz. Thus, a single two-pole section of the form given in Figure 8–15 is used. The value of R_f is determined from Table 8–5 to be 0.5858, and the normalized circuit is shown in Figure 8–18(a).

(a)

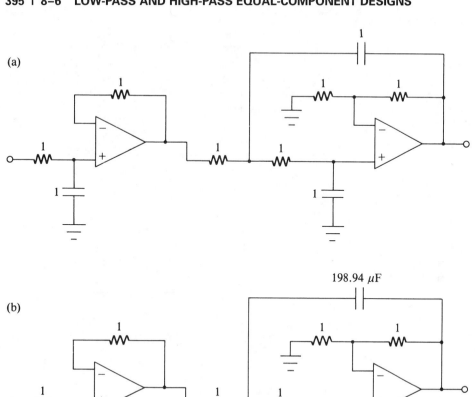

(b)

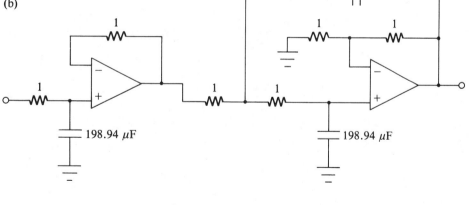

(c)

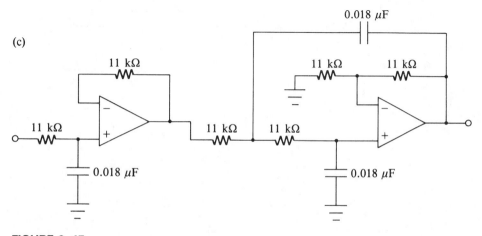

FIGURE 8–17
Design of Example 8–9.

(a)

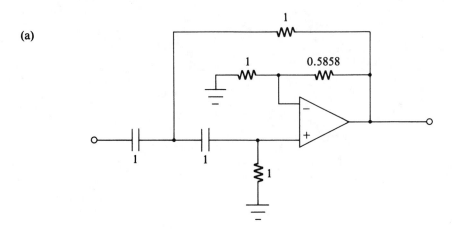

(b)

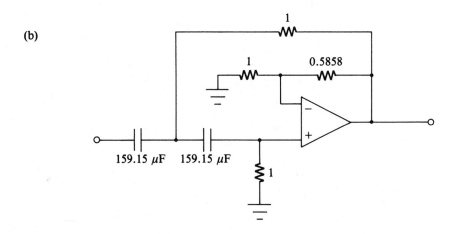

(c)

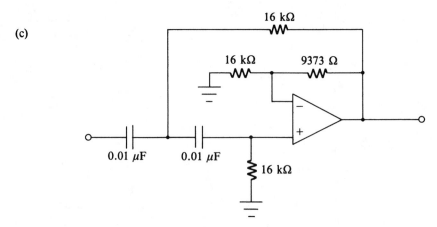

FIGURE 8–18
Design of Example 8–10.

As determined in Example 8–8, the frequency scaling constant is $K_f = 6283.185$. For the first step, the circuit is scaled to the proper frequency range by dividing the two capacitances by K_f, and the corresponding circuit is shown in Figure 8–18(b).

The value of K_r was determined, after a little trial and error, by forcing the two capacitors to assume the final value of 0.01 μF. This process is achieved as follows:

$$\frac{159.15 \times 10^{-6}}{K_r} = 0.01 \times 10^{-6} \tag{8-16}$$

Solving for K_r, we obtain

$$K_r = 15{,}915 \tag{8-17}$$

Using this value, the ideal value of three of the resistances would be 15,915. However, this value differs from the standard value of 16 kΩ by less than 0.6%. For non-critical applications, the value of 16 kΩ should suffice. The final circuit diagram of Figure 8–18(c) shows this rounded value for three of the four resistances.

The value of R_f is set at 0.5858 times the rounded value of 16 kΩ in order to establish the required gain of 1.5858. This yields a value for R_f of 9373 Ω, which is not a standard value. Some adjustment would be required to achieve this value. For example, the series combination of a 9100-Ω and a 270-Ω resistor should suffice.

8-7 TWO-POLE BAND-PASS RESPONSE

Before considering any specific active band-pass filter designs, it is desirable to investigate the value of the amplitude response forms that can be obtained. In general, band-pass filters are among the most complex of all filter types, and much effort has been directed toward developing a myriad of band-pass filters for various system requirements. Since the bulk of band-pass filter requirements are in frequency ranges in which op-amps are not well suited, band-pass active filters are more limited in their ability to compete with passive designs. Nevertheless, there are some low-frequency applications for band-pass filters in which active designs are significantly superior to passive forms.

The most common band-pass active filters are two-pole forms. While limited compared to passive higher-order designs such as those used in high-frequency communications systems, they are capable of achieving good results in less demanding applications.

The two-pole forms we will consider are mathematically equivalent to certain configurations involving series or parallel RLC resonant circuits. This is a good example of how active RC circuits can perform operations equivalent to passive RLC circuits.

Transfer Function

The steady-state transfer function $H(j\omega)$ of any two-pole band-pass filter can be expressed as

$$H(j\omega) = \frac{M_o}{1 + jQ\left(\dfrac{f}{f_o} - \dfrac{f_o}{f}\right)} \tag{8-18}$$

where M_o is the maximum gain within the band, and f_o is called the **geometric center frequency.** (In passive *RLC* circuits, f_o is called the **resonant frequency.**) The parameter Q is a measure of the selectivity or sharpness of the filter, and its quantitative meaning will be discussed next.

The amplitude response $M(\omega)$ corresponding to (8–18) can be determined as

$$M(\omega) = \frac{M_o}{\sqrt{1 + Q^2\left(\dfrac{f}{f_o} - \dfrac{f_o}{f}\right)^2}} \qquad (8\text{–}19)$$

As was done with other forms earlier in the chapter, it is convenient to divide by M_o and work with the relative amplitude response. The relative decibel response $M_{dB}(\omega)$ corresponding to (8–19) can be expressed as

$$M_{dB}(\omega) = 20 \log_{10}\frac{M(\omega)}{M_o} = 20 \log_{10}\left[\frac{1}{\sqrt{1 + Q^2\left(\dfrac{f}{f_o} - \dfrac{f_o}{f}\right)^2}}\right] \qquad (8\text{–}20a)$$

or

$$M_{dB}(\omega) = -10 \log\left[1 + Q^2\left(\frac{f}{f_o} - \frac{f_o}{f}\right)^2\right] \qquad (8\text{–}20b)$$

The general form of $M(\omega)$ in (8–19) is illustrated in Figure 8–19 as it would appear on a **linear** scale. Observe that the roll-off on the high-frequency side occurs at a slower rate than on the low-frequency side. In other words, a point at which the response is down from the maximum by a specified amount is farther away from the center frequency on the high side than on the low side.

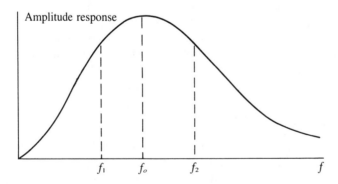

FIGURE 8–19
Illustration of band-pass filter response on linear scales.

Bandwidth and Q

Let f_1 and f_2 represent frequencies on the low and high sides, respectively, at which the response is $1/\sqrt{2}$ times the peak response (-3.01 dB down). The bandwidth B is defined as

$$B = f_2 - f_1 \qquad (8\text{–}21)$$

The frequencies f_1 and f_2 have **geometric** symmetry about the center frequency f_o. This property means that the following relationship is satisfied:

$$f_o = \sqrt{f_1 f_2} \qquad (8\text{–}22)$$

The parameter Q is related to the center frequency and bandwidth by

$$Q = \frac{f_o}{B} \qquad (8\text{–}23)$$

As Q is increased, the filter becomes more selective; that is, the 3-dB bandwidth is smaller for a given center frequency. At higher values of Q, the frequencies f_1 and f_2 are approximately the same distance on either side of f_o, and this narrow-band response approximates **arithmetic** symmetry. However, the true correct condition is that of geometric symmetry as given by (8–22).

Band-Pass Response Curves

A family of curves for the amplitude response of the two-pole band-pass response is given in Figure 8–20. Observe that the horizontal scale is the normalized frequency f/f_o, and the scale is **logarithmic** in form. The response curves are symmetrical on this logarithmic scale. However, on a linear scale, the curves are somewhat skewed as observed qualitatively in Figure 8–19. Note that the relative amplitude response is shown over a frequency range from $0.1f_o$ to $10f_o$.

Observe the marked effect of the parameter Q on the amplitude response. At the lowest value of Q shown ($Q = 1$), the response drops off very slowly on either side of f_o. At the other extreme, for the highest value of Q shown ($Q = 100$), the response drops off very sharply on either side of f_o.

The geometric center frequency f_o is a very convenient parameter to use in analyzing and designing two-pole band-pass filters. However, since it is at the geometric center rather than the arithmetic center, confusion sometimes arises when it is desired to locate the two 3-dB band edges precisely. It can be shown that the arithmetic center between the two 3-dB frequencies is at a frequency $f_o\sqrt{1 + (1/2Q)^2}$, which is larger than f_o. The two 3-dB frequencies f_1 and f_2 are then given by

$$f_1 = f_o\sqrt{1 + \left(\frac{1}{2Q}\right)^2} - \frac{B}{2} \qquad (8\text{–}24)$$

$$f_2 = f_o\sqrt{1 + \left(\frac{1}{2Q}\right)^2} + \frac{B}{2} \qquad (8\text{–}25)$$

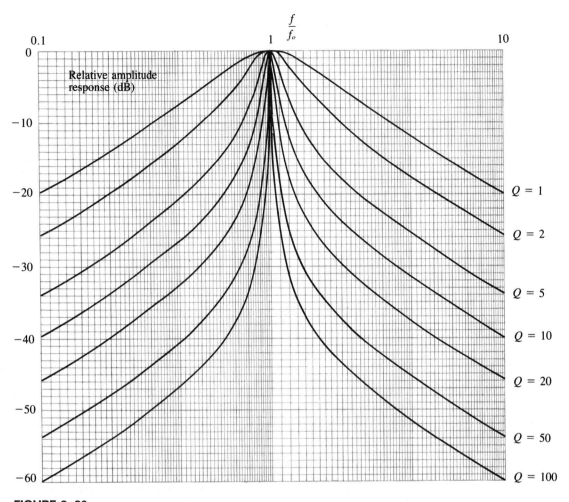

FIGURE 8–20
Two-pole band-pass amplitude response curves.

EXAMPLE 8–11

A certain two-pole band-pass filter response is desired with a geometric center frequency of 2 kHz and a 3-dB bandwidth of 400 Hz. (a) Calculate Q, f_1, and f_2. (b) Determine the relative decibel amplitude response at each of the following frequencies: 200 Hz, 400 Hz, 1 kHz, 1.5 kHz, 2 kHz, 3 kHz, 4 kHz, 10 kHz, and 20 kHz.

Solution
(a) We readily calculate Q as

$$Q = \frac{f_o}{B} = \frac{2000 \text{ Hz}}{400 \text{ Hz}} = 5 \qquad \textbf{(8–26)}$$

From (8–24) and (8–25), the two band-edge frequencies f_1 and f_2 are calculated as

$$f_1 = 2000\sqrt{1 + \left(\frac{1}{2 \times 5}\right)^2} - \frac{400}{2} = 1810 \text{ Hz} \qquad \textbf{(8–27)}$$

$$f_2 = 2000\sqrt{1 + \left(\frac{1}{2 \times 5}\right)^2} + \frac{400}{2} = 2210 \text{ Hz} \qquad \textbf{(8–28)}$$

Observe that the arithmetic center between the two 3-dB frequencies is 2010 Hz, which is close to 2 kHz. The difference between the geometric and arithmetic center frequencies becomes quite small as Q increases.

(b) The relative decibel response can be computed with (8–20b), or the curves can be used if Q and the frequency range permit. Since $Q = 5$ is one of the curves provided, and all frequencies are in the range from $0.1f_o$ to $10f_o$, this approach will be used. A convenient tabulation of the data follows:

f	f/f_o	Relative Amplitude Response (dB)
200 Hz	0.1	−33.9
400 Hz	0.2	−27.6
1 kHz	0.5	−17.6
1.5 kHz	0.75	−9.8
2 kHz	1	0
3 kHz	1.5	−12.6
4 kHz	2	−17.6
10 kHz	5	−27.6
20 kHz	10	−33.9

8–8 MULTIPLE-FEEDBACK BAND-PASS FILTER

In this section, a two-pole band-pass filter utilizing one operational amplifier will be presented. Based on the theoretical development leading to this circuit, it is commonly referred to as a *multiple-feedback band-pass filter*. The multiple-feedback filter is capable in theory of achieving band-pass filter designs of the form discussed in Section 8–7, and is thus operationally equivalent to an *RLC* resonant circuit. As we will see, however, this circuit is best suited to designs utilizing relatively low values of Q.

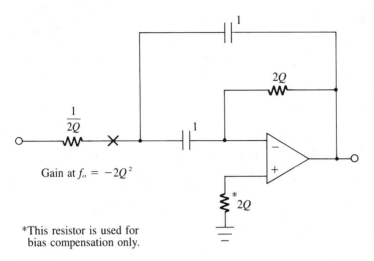

FIGURE 8–21
Normalized form of band-pass filter using single operational amplifier.

The basic form of the normalized circuit is shown in Figure 8–21. The circuit is normalized to have a geometric center frequency $\omega_o = 1$ rad/s. Observe that the normalized resistance values are functions of the desired circuit Q. The Q parameter is a ratio of two frequency values, and both frequencies are changed by the same ratio during a scaling operation. Thus, *the desired Q is established directly in the normalized design, and it remains unchanged during the scaling process.*

From the data provided on the figure, it should be observed that the gain at the geometric center frequency is inverting and its magnitude increases with the square of Q. This marked increase of gain as Q increases may pose serious problems at moderate to high values of Q because of possible overload of the op-amp as well as bandwidth and slew rate limitations.

Achieving a Gain of One

A partial solution to the problem is to place an attenuator network ahead of the filter to reduce the overall gain to a nominal level. The normalized form of a possible circuit is shown in Figure 8–22. The actual derivation of the normalized design will be left as an exercise in Problem 8–15, but two criteria have been employed in the circuit:

1. The open-circuit voltage is $1/2Q^2$ times the input voltage on the left. This factor exactly cancels the magnitude of the gain of the filter at f_o, so the **overall** magnitude of the gain at f_o is unity.
2. The net Thevenin resistance looking back to the left in the circuit of Figure 8–22 must equal the value of the left-hand resistance (normalized value $1/2Q$) in Figure 8–21. This requirement results in the same resonant frequency and Q as for the basic design of Figure 8–21.

FIGURE 8–22

Normalized attenuator network used in band-pass filter of Figure 8–21 to establish unity gain at f_o. (This circuit replaces the first resistor.)

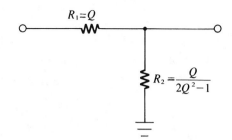

The attenuator network replaces the first resistance in the circuit of Figure 8–21. The complete normalized design is shown in Figure 8–23. Since the attenuator network becomes an integral part of the complete design, it must be scaled by the same factor as all other resistances in the scaling process.

Although the use of this circuit improves the design, the spread of element values and the associated sensitivity of this circuit still make it difficult to use for high values of Q. As an approximate rule of thumb, this circuit should be limited to Q values less than about 20 or so. Where higher values of Q are required, the state-variable filter of Section 8–9 is recommended.

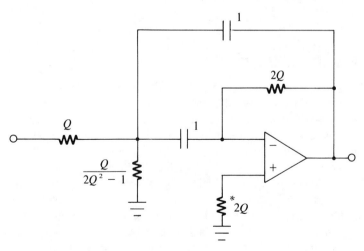

*This resistor is used for bias compensation only.

FIGURE 8–23

Normalized band-pass filter using single op-amp with attenuator network used to establish unity gain at f_o.

EXAMPLE 8-12

Design a band-pass filter using a single op-amp to meet the specifications of Example 8–11. Include an attenuator network to achieve unity gain at f_o.

Solution

In Example 8–11, the requirements called for a geometric center frequency $f_o = 2000$ Hz and $Q = 5$. The basic form of the circuit is obtained from Figure 8–23, and as applied to the Q values specified here, the normalized circuit is given in Figure 8–24(a).

First, the frequency scaling operation will be performed. The response of the normalized design at 1 rad/s must correspond to the response of the actual design at 2 kHz. The constant K_f is thus determined as

$$K_f = \frac{2\pi \times 2 \times 10^3 \text{ rad/s}}{1 \text{ rad/s}} = 12{,}566.37 \tag{8-29}$$

The capacitor values are then divided by K_f, and the corresponding circuit is shown in Figure 8–24(b).

While a number of possible solutions could be investigated for an appropriate impedance scaling constant, a brief trial-and-error approach led to the selection of the standard value of 0.01 μF for the capacitors. The impedance scaling constant K_r must be determined such that $79.577 \times 10^{-6}/K_f = 0.01 \times 10^{-6}$ F, which leads to $K_r = 7957.7$. The resistance values are multiplied by K_r, and the capacitance values are changed to 0.01 μF, as shown in Figure 8–24(c). As a starting point, the standard resistance values of 39 kΩ, 820 Ω, and 82 kΩ could be used. The directions of these values will contribute to increased gain, so an additional fixed resistance could be added in series with the 39-kΩ resistance if necessary.

8-9 STATE-VARIABLE FILTER

The *state-variable filter* is one of the most versatile of all the active filter types. The same structure can be used for implementing a low-pass, a high-pass, or a band-pass filter; and with the addition of a summing circuit, it can even be used for a band-rejection filter.

The evolution of the state-variable filter is rather interesting and should be mentioned. The earliest applications of operational amplifiers were in analog computers, which have been used for many years to solve differential equations and to simulate physical systems. A technique evolved for "programming" a differential equation by a certain systematic layout and interconnection of various integrators and summers. As operational amplifiers became available in integrated-circuit form, the concept of realizing a filter by "programming" the corresponding differential equation was conceived. Thus, the state-variable filter represents essentially the same implementation strategy as that of simulating the applicable differential equation on an analog computer.

The term *state-variable* is related to a form of system analysis called *state-variable theory,* which provides a systematic means of formulating the differential equations of large systems. As part of such a formulation, a state diagram showing the interconnection of all the mathematical operations may be constructed. The state diagram is, in reality, a

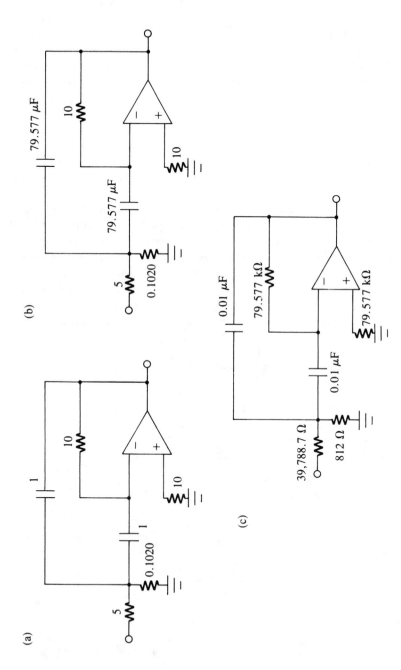

FIGURE 8–24

Circuit design of Example 8–12.

mathematical form of an analog computer simulation. The reader, of course, need not understand all this background to utilize state-variable designs properly.

While it is theoretically possible to form state-variable sections of any order, most designs are based on two-pole sections. The two-pole section is less sensitive to parameter variations than those of higher order. Further, it is possible to create sections of any order by cascading sections of lower order. Incidentally, there is no need to form a one-pole section since any first-order function can be created with a simple passive RC circuit.

Two-Pole Circuit Form

There are several variations of the two-pole state-variable circuit, all of which are functionally equivalent. The form that will be emphasized here is shown in Figure 8–25 in normalized form. Observe that the normalized values of all passive components except R_Q are unity.

Several element values have been labeled for purposes of clarity and explanation. The two resistances denoted as R_t and the two capacitances denoted as C_t will be collectively referred to as the ***four tuning elements.*** Note that the two tuning capacitances have identical values, and the two tuning resistances have identical values. When frequency scaling, impedance scaling, or tuning is performed, the two capacitors *or* the two resistors must be varied together. One common approach to implementing frequency-adjustable units is to switch in or out fixed capacitance values (two at a time) and to employ ganged variable resistors (two on a shaft) for continuous frequency adjustment over a range.

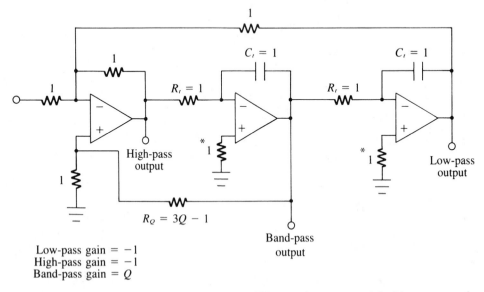

Low-pass gain $= -1$
High-pass gain $= -1$
Band-pass gain $= Q$

*These resistors are used for bias compensation only and do not contribute to the filter response.

FIGURE 8–25
State-variable normalized filter.

The value of R_Q establishes the required Q for the circuit. The meaning of Q for band-pass and band-rejection circuits has been discussed at length, but the interpretation for low-pass and high-pass forms will be given later.

The impedance level of the voltage divider network containing R_Q and the 1-Ω resistance on the left may be adjusted independently of all other resistances in the circuit provided that these two resistances remain in exactly the same proportion. After the initial form of a circuit is designed, one may desire to alter the impedance level here so that the optimum bias compensating resistance is seen from both terminals.

The impedance level of the three 1-Ω resistances connected to the inverting input of the left-hand summing networks may be adjusted independently of other parts of the circuit. Again, it is necessary that these three components be scaled in exactly the same proportion.

The reader should understand that it may not be necessary to use different impedance scaling constants at different points, and the simplest approach "on paper" is to use the same constant. However, flexibility in the choice of element values as well as the realization of optimum bias resistance levels may result from exploiting the possibility of different levels for different parts of the circuit. Each of the different circuit forms will now be discussed.

Band-Pass Form

The input signal is applied on the extreme left, and the output is taken to the right of the first integrator as noted. The circuit can be assumed to be normalized to a geometric center frequency $\omega_o = 1$ rad/s. The desired band-pass Q is established by setting $R_Q = 3Q - 1$. Note that the gain at ω_o is Q, so care must be taken not to overload any amplifier at high values of Q.

Low-Pass and High-Pass Forms

For Butterworth two-pole low-pass and high-pass filters, the normalized circuit has a 3-dB frequency $\omega_o = 1$ rad/s. Since the only types of low-pass and high-pass characteristics being considered in this book are Butterworth forms, the 1 rad/s value identifies the upper 3-dB frequency for the low-pass case and the lower 3-dB frequency for the high-pass case. However, it should be emphasized that for other types of filter characteristics (for example, Chebyshev, maximally flat time-delay, and so on), the cutoff frequency of the normalized section may not be 1 rad/s. Such a discussion is not within the intended objective, but the caution is given for anyone wanting to extend the material in this book to other filter forms.

While the Q parameter has a direct physical meaning for band-pass filters and, as we will see later, for band-rejection filters, its meaning for low-pass and high-pass filters is more subtle and not as easily associated with common physical characteristics. The value of Q for such cases relates to the relative damping of the transfer function. The Q value or values must be used in the two-pole sections as specified, but there is no simple, observable relationship as for band-pass and band-rejection functions.

For either two-pole low-pass or high-pass Butterworth filters, the value of Q is $Q = 1/\sqrt{2} = 0.7071$, so the corresponding value of the resistance is $R_Q = 3Q - 1 = 3(0.7071) - 1 = 1.1213$. Thus, the same value of R_Q establishes the two-pole section as either a low-pass or a high-pass Butterworth filter. If a low-pass form is desired, the output is taken on the extreme right, as shown in Figure 8–25, and if a high-pass form is desired, the output is taken to the right of the summing circuit on the left. Observe from the information associated with the figure that the maximum pass-band gain of both forms is -1.

When low-pass and high-pass filters of order higher than two are desired, it is necessary to cascade two or more sections. Consider first the case where n is even (4, 6 and so on). In this case, two or more two-pole state-variable sections are cascaded to create the proper order. As previously noted, the 1 rad/s normalized cutoff frequency applies to all sections. However, *each section requires a different value of Q*. Recall, as stated earlier, that a higher-order Butterworth filter is not a cascade of lower-order Butterworth filters. Rather, each section must be set to a unique value of Q such that the combined effect of all the sections produces the required overall effect.

A tabulation of Q values for different orders of Butterworth filters is given in Table 8–6. The form of the extra first-order section for odd n will be discussed later. The sections are connected in cascade with the output of the first section connected as the input to the next section, and so on. The values are listed in the best order of implementation; that is, the value on the left in the table is the Q of the section located nearest the input, and so on.

When n is odd (3, 5, and so on), it is necessary to add an additional one-pole section to the cascade of two-pole sections. This one-pole section should be connected at the input as indicated by the relative order in Table 8–6 for each of the odd cases. While the various state-variable sections may be used for either low-pass or high-pass functions by changing the output point, the one-pole input circuits must be different for the two cases. The form of the extra first-order section is the same as those used in the equal-component designs considered in Section 8–6, that is, Figure 8–14 for low-pass and Figure 8–16 for high-pass.

TABLE 8–6

Values of Q for two-pole sections of filters used to implement low-pass and high-pass Butterworth filters

Poles	One-Pole Section	Q of Two-Pole Section	Q of Two-Pole Section	Q of Two-Pole Section
2		0.7071		
3	Required	1.0000		
4		0.5412	1.3065	
5	Required	0.6180	1.6181	
6		0.5177	0.7071	1.9320
7	Required	0.5549	0.8019	2.2472

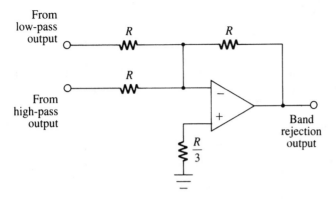

FIGURE 8-26
Summing circuit used to create band-rejection response in state-variable filter.

Band-Rejection Form

The band-rejection form is obtained by summing the output of the low-pass section to the output of the high-pass section. Thus, an additional op-amp is required as shown in Figure 8–26. Since the gains of the low-pass and high-pass sections are both −1, the additional inversion for both inputs of the summing circuit results in a net gain of +1 for the band-rejection circuit.

The geometric center frequency f_o is now the geometric center of the rejection band, and f_1 and f_2 now represent the points on either side of the center at which the response is 3 dB below the levels of the high- and low-frequency flat bands. These concepts are illustrated in Figure 8–27. The Q parameter is still defined as

$$Q = \frac{f_o}{B}$$

(8–30)

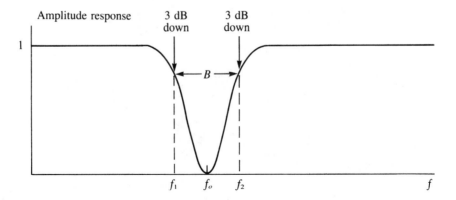

FIGURE 8-27
General form of amplitude response of band-rejection filter.

However, the frequency values f_o and B now refer to a rejection band rather than a pass band.

EXAMPLE 8–13 Design a state-variable filter that can serve as either a low-pass or a high-pass Butterworth two-pole filter if the desired 3-dB cutoff frequency is 1 kHz.

Solution
From Table 8–6, $Q = 0.7071$, and the value of R_Q in Figure 8–25 is $R_Q = 1.1213$. All other normalized values are the same as in Figure 8–25. The frequency of 1 rad/s in the normalized circuit must convert to 1 kHz in the final circuit, so the constant K_f is

$$K_f = \frac{2\pi \times 1000 \text{ rad/s}}{1 \text{ rad/s}} = 6283.185 \qquad \textbf{(8–31)}$$

The two 1-F capacitors in Figure 8–25 are then divided by K_f, and the resulting values are each 159.2 μF.

Selection of the impedance scaling constant is somewhat arbitrary, but after some trial and error, a choice was made to set $K_f = 10^4$. All the 1-Ω resistances then convert to 10 kΩ, the value of R_Q becomes 11.213 kΩ, and the two capacitors assume values of 0.01592 μF. The final circuit diagram is shown in Figure 8–28. Standard capacitance values of 0.016 μF should suffice for the capacitors, and a standard value of 11 kΩ could be used for R_Q.

The circuit as designed could be used as either a low-pass filter or a high-pass filter in accordance with where the output is taken. As a band-pass filter, however, this design would be quite limited, since $Q = 0.7071$, and the selectivity would be far too low for most applications. This example illustrates that while the state-variable filter can be used for several types of filter functions, the Q value required for one type may be quite different from that required for other types.

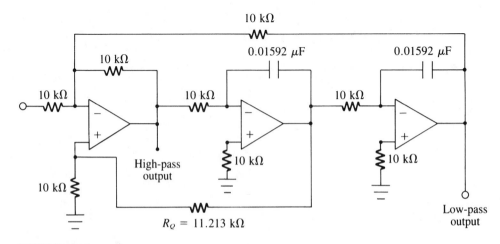

FIGURE 8–28
Circuit design of Example 8–13.

**EXAMPLE
8–14**

Design a state-variable, two-pole, band-pass filter with a geometric center frequency of 500 Hz and a 3-dB bandwidth of 10 Hz.

Solution
The required value of Q is

$$Q = \frac{f_o}{B} = \frac{500}{10} = 50 \qquad \text{(8–32)}$$

The normalized design of Figure 8–25 is used as the starting point, with $R_Q = 3Q - 1 = 3 \times 50 - 1 = 149$. The normalized frequency of 1 rad/s must convert to 500 Hz, so the frequency scaling constant is

$$K_f = \frac{2\pi \times 500 \text{ rad/s}}{1 \text{ rad/s}} = 3141.593 \qquad \text{(8–33)}$$

The capacitance values are divided by K_f, and the new value of each is 318.31 μF. After some trial and error, the resistance scaling constant was selected as $K_r = 2 \times 10^4$. All resistance values except R_Q then become 20 kΩ, and the two values of C become 0.01592 μF. The value of the Q adjustment resistance is $R_Q = 2.98$ MΩ. The design is shown in Figure 8–29. The 20-kΩ resistances are standard values, and a value of 0.016 μF should suffice for C. A 3-MΩ resistance could be used as a starting point for R_Q, but some delicate adjustments might be needed in view of the large value of Q. Depending on the acceptable tolerance, the next smaller standard value of resistance in series with a variable resistance could be used if necessary.

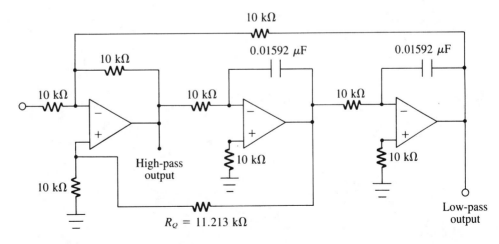

FIGURE 8–29
State-variable band-pass filter design of Example 8–14.

The gain of this circuit at 500 Hz is 50, so care must be taken not to overload any stages. In high-Q circuits such as this, it may be necessary to attenuate the input signal if it is too high. If this is necessary, an isolation stage between the attenuator and the first stage may be required to avoid the undesirable interaction between the attenuator and the gain constants of the summing circuit.

EXAMPLE 8–15

An undesired interfering component at 500 Hz is to be eliminated in a system, and a band-rejection filter is desired. Design a state-variable two-pole filter to accomplish the purpose if the width between 3-dB points in the rejection band is to be set at 10 Hz.

Solution
The required value of Q is

$$Q = \frac{500}{10} = 50 \qquad (8\text{–}34)$$

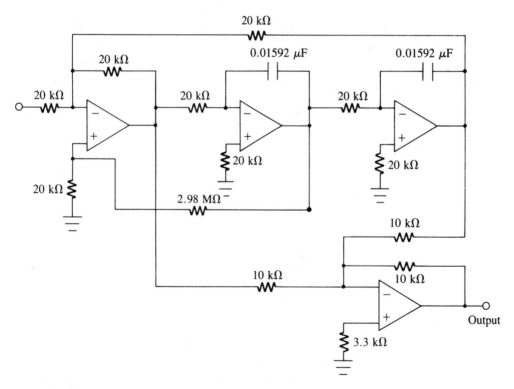

FIGURE 8–30
State-variable band-rejection filter design of Example 8–15.

Since the center frequency and the bandwidth are the same as in Example 8–14, the design of that example can be used as the basis for this band-rejection filter. All that is required is to sum the low-pass and high-pass outputs, and the complete circuit is shown in Figure 8–30. The same point about the high gain made in Example 8–14 applies here.

In a notch filter such as this, the actual level of attenuation at the notch frequency (500 Hz) will be a critical function of the parameter tolerances. Consequently, some trimming or adjustments of the element values may be required in order to obtain optimum rejection at 500 Hz.

8–10 MULTISIM EXAMPLES

The frequency response of an active filter may be readily displayed with Multisim. The examples that follow illustrate this process with several of the designs of this chapter.

MULTISIM EXAMPLE 8–1

Using the 741 model, develop a Multisim circuit for simulating the low-pass active filter design of Example 8–6 and perform a frequency response analysis.

Solution
The original circuit was shown in Figure 8–10(c), and the corresponding Multisim circuit is shown in Figure 8–31. The bias compensating resistance has been eliminated in the circuit since it does not contribute to the frequency response.

An **AC Analysis** is performed on the circuit with the frequency swept from 100 Hz to 10 kHz, and the results are shown in Figure 8–32. One of the cursors is used to determine the 3-dB down frequency. To the resolution possible with the cursor, the response is down about 3 dB at 800 Hz as expected.

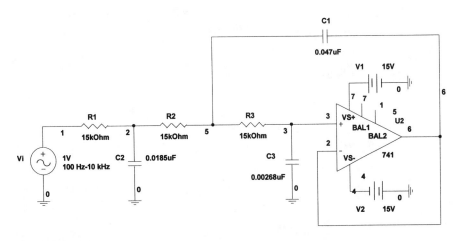

FIGURE 8–31
Circuit of Multisim Example 8–1.

FIGURE 8–32

Frequency response of Multisim Example 8–1.

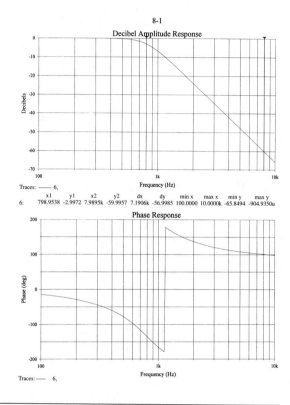

MULTISIM EXAMPLE 8–2

Using the 741 model, develop a Multisim circuit for simulating the band-pass active filter design of Example 8–12 and perform a frequency response analysis.

Solution

The original circuit was shown in Figure 8–24(c), and the corresponding Multisim circuit is shown in Figure 8–33. An **AC Analysis** is performed on the circuit with the frequency

FIGURE 8–33

Circuit of Multisim Example 8–2.

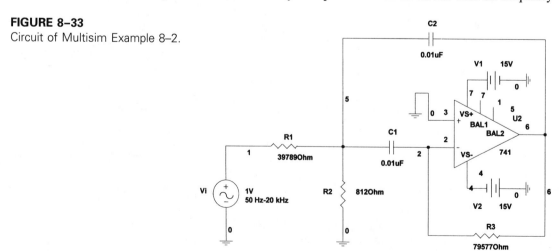

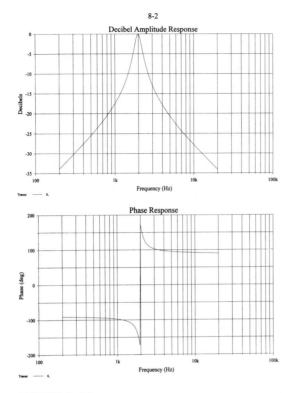

FIGURE 8–34
Frequency response of Multisim Example 8–2.

swept from 200 Hz to 20 kHz, and the results are shown in Figure 8–34. The center frequency is at 2 kHz as expected.

MULTISIM EXAMPLE 8-3

Using the 741 model for each stage, develop a Multisim circuit for simulating the low-pass state variable active filter design of Example 8–13 and perform a frequency response analysis.

Solution

The original circuit was shown in Figure 8–28, and the corresponding Multisim circuit is shown in Figure 8–35. A decision was made in the example to flip the op-amps so that the inverting inputs would be at the top. This is achieved by a left-click on **Edit** followed by a left-click on **Flip Vertical.** Note that the positive power supply connection is now on the bottom and the negative power supply connection is now on the top.

An **AC Analysis** is performed on the circuit with the frequency swept from 100 Hz to 10 kHz, and the results are shown in Figure 8–36. The results are as expected.

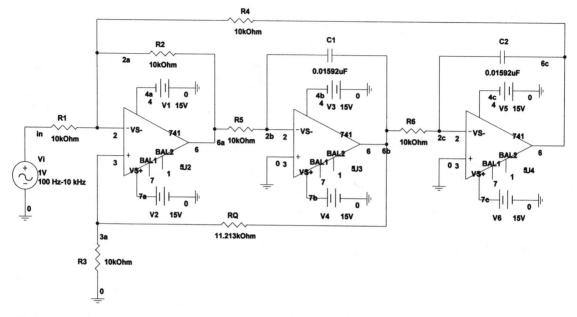

FIGURE 8–35
Circuit of Multisim Example 8–3.

FIGURE 8–36
Frequency response of Multisim Example 8–3.

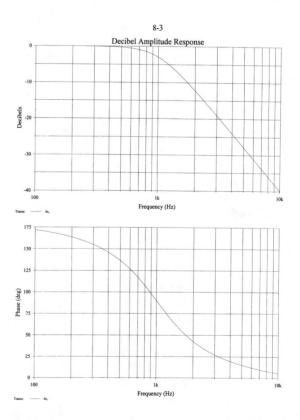

MULTISIM EXAMPLE 8–4

Using the 741 model for each stage, develop a Multisim circuit for simulating the band-pass state variable active filter design of Example 8–14 and perform a frequency response analysis.

Solution

The original circuit was shown in Figure 8–29, and the corresponding Multisim circuit is shown in Figure 8–37. As in the previous example, the op-amps are "flipped." Actually, this circuit was readily adapted from the previous circuit by changing some of the element values and taking the output from a different op-amp.

Recall that the gain of this circuit at the center frequency is Q, which is 50 in this case. When a frequency response is performed, Multisim automatically sets the input voltage to 1 V. This makes the output at center frequency to appear to be 50 V, which would be impossible in the actual circuit! A more desirable input voltage for testing this circuit in a laboratory would be 20 mV, which would result in 1 V at center frequency and would provide a convenient reference. In the **AC Analysis** mode, Multisim develops a linear model at the operating point and this makes the output appear to reach the impossible level of 50 V.

Because of the relatively narrow frequency range, a linear sweep was used in this case. This is achieved by selecting the **Sweep type** as **Linear.** The frequency response over the range from 400 Hz to 600 Hz is shown in Figure 8–38. The center frequency value of 50 V corresponds to about 34 dB above a level of 1 V and this is close to the value shown on the graph.

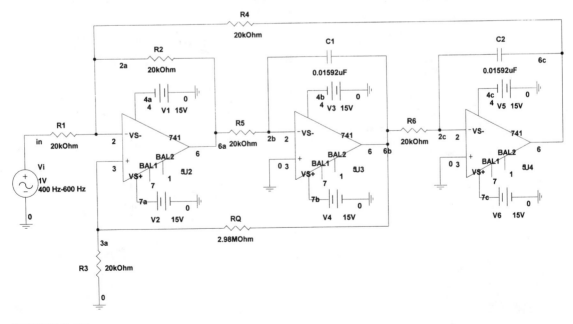

FIGURE 8–37
Circuit of Multisim Example 8–4.

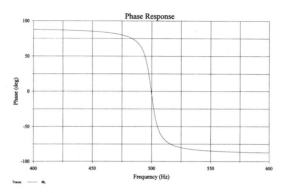

8-4

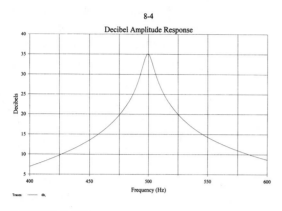

FIGURE 8–38

Frequency response of Multisim Example 8–4.

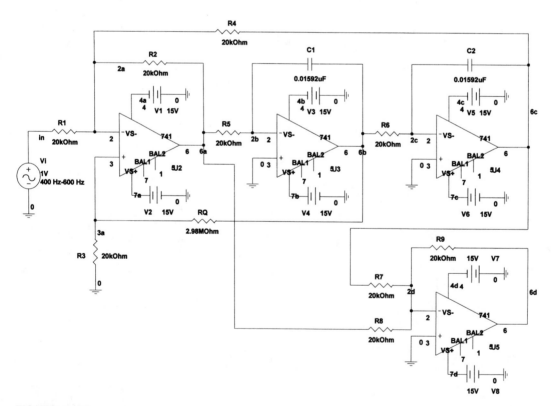

FIGURE 8–39

Circuit of Multisim Example 8–5.

FIGURE 8–40
Frequency response of Multisim Example 8–5.

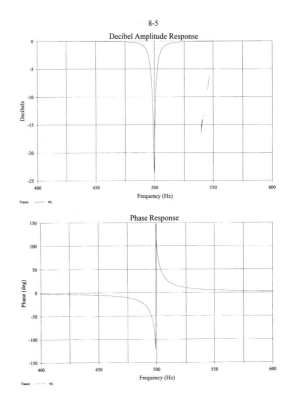

MULTISIM EXAMPLE 8–5

Using the 741 model for each stage, develop a Multisim circuit for simulating the band-rejection state variable active filter design of Example 8–15 and perform a frequency response analysis.

Solution
The original circuit was shown in Figure 8–30, and the corresponding Multisim circuit is shown in Figure 8–39. A fourth op-amp has been added in this case, and all are "flipped."

The frequency response over the range from 400 Hz to 600 Hz is shown in Figure 8–40. The notch is quite evident in the response.

SUMMARY

1. Filters may be classified as either passive or active according to how they are implemented.
2. Filters may be classified as either low-pass, high-pass, band-pass, or band-rejection according to the location of the pass band.
3. There are a variety of different filter characteristics, but three of the most common types are the Butterworth, the Chebyshev, and the maximally flat time-delay.

4. Frequency scaling may be applied to a filter circuit to change the frequency range while preserving the relative shape of the response.
5. Impedance scaling may be applied to a filter circuit to change the sizes of the component values while maintaining the exact response and frequency scale.
6. As the number of filter poles increases, the pass-band shape becomes flatter, and the response in the stop band drops at a faster rate.
7. Low-pass and high-pass active filters may be implemented with a cascade arrangement of unity-gain stages, each of which contains a voltage follower and an appropriate RC circuit.
8. Low-pass and high-pass active filters may also be implemented with a cascade arrangement of stages having equal-component values, but in which the gain of each stage must be set at a specified value.
9. A two-pole band-pass active filter may be implemented, for low or moderate Q, with a single op-amp utilizing multiple feedback.
10. A state-variable filter is a special circuit form that has evolved from earlier analog computer circuits.
11. A state-variable filter can be used to implement a low-pass, a high-pass, a band-pass, or a band-rejection filter.

PROBLEMS

Drill Problems

8–1. The circuit of Figure P8–1 is a simple passive RC high-pass filter with a 3-dB break frequency $\omega_b = 1$ rad/s.

FIGURE P8–1

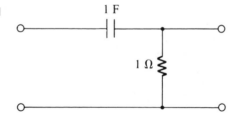

1 F

1 Ω

    a. Perform a frequency scaling operation on the circuit by changing the capacitance so that the actual 3-dB cyclic break frequency is $f_b = 1$ kHz.
    b. It is desired to implement the filter to achieve the desired break at 1 kHz using a 0.001-μF capacitor. Determine the required value of resistance.
    c. After the preceding design is completed, assume that it is desired to replace the fixed resistance by a variable resistance that would permit the 3-dB frequency to be adjusted continuously from 500 Hz to 5 kHz. Determine the required range of the variable resistance.
    d. Compare the results here with those of Example 8–1. What conclusion can be drawn about the relationship between simple passive one-pole RC low-pass and high-pass filters?
8–2. a. For the circuit of Example 8–1 [Figure 8–6(a)], perform a frequency scaling operation on the circuit by changing the capacitance so that the 3-dB break frequency is 120 kHz.
    b. It is desired to implement the filter using a 200-pF capacitor. Determine the required value of resistance.
    c. Using the value of the capacitance of (b), determine the range of resistance required to vary the 3-dB frequency from 60 kHz to 240 kHz.

8–3. A 6-pole low-pass Butterworth filter has a 3-dB cutoff frequency of 2 kHz.
   a. Use Figure 8–7 to determine the decibel response at the following frequencies: 3 kHz, 4 kHz, 5 kHz, 6 kHz, 7 kHz.
   b. Use an appropriate equation in Section 8–4 to compute the decibel response at each of the preceding frequencies.

8–4. A 7-pole low-pass Butterworth filter has a 3-dB cutoff frequency of 8 kHz.
   a. Use Figure 8–7 to determine the decibel response at the following frequencies: 12 kHz, 16 kHz, 24 kHz, 32 kHz.
   b. Use an appropriate equation in Section 8–4 to compute the decibel response at each of the preceding frequencies.

8–5. For the filter of Problem 8–3, use Table 8–1 to determine the frequencies within the pass band at which the response is −2 dB, −1 dB, and −0.5 dB.

8–6. For the filter of Problem 8–4, use Table 8–1 to determine the frequencies within the pass band at which the response is −2 dB, −1 dB, and −0.5 dB.

8–7. A 4-pole high-pass Butterworth filter has a 3-dB cutoff frequency of 600 Hz.
   a. Use Figure 8–7 to determine the decibel response at the following frequencies: 300 Hz, 150 Hz, 100 Hz, 80 Hz.
   b. Use an appropriate equation in Section 8–4 to compute the decibel response at each of the preceding frequencies.

8–8. A 3-pole high-pass Butterworth filter has a 3-dB cutoff frequency of 1.5 kHz.
   a. Use Figure 8–7 to determine the decibel response at the following frequencies: 1 kHz, 500 Hz, 300 Hz, 150 Hz.
   b. Use an appropriate equation in Section 8–4 to compute the decibel response at each of the preceding frequencies.

8–9. For the filter of Problem 8–7, use Table 8–1 to determine the frequencies within the pass band at which the response is −2 dB, −1 dB, and −0.5 dB.

8–10. For the filter of Problem 8–8, use Table 8–1 to determine the frequencies within the pass band at which the response is −2 dB, −1 dB, and −0.5 dB.

8–11. A two-pole band-pass filter has a center frequency of 1.5 kHz and a 3-dB bandwidth of 300 Hz. Determine:
   a. the Q
   b. the exact 3-dB frequencies f_1 and f_2
   c. Use Figure 8–20 to determine the decibel response at 150 Hz and 15 kHz.
   d. Use an appropriate equation from Section 8–7 to compute the decibel response at the preceding two frequencies.

8–12. A two-pole band-pass filter has a center frequency of 800 Hz and a Q of 20. Determine:
   a. the 3-dB bandwidth
   b. the exact 3-dB frequencies f_1 and f_2
   c. Use Figure 8–20 to determine the decibel response at 80 Hz and 8 kHz.
   d. Use an appropriate equation from Section 8–7 to compute the decibel response at the preceding two frequencies.

Derivation Problems

8–13. Show that the amplitude response of low-pass Butterworth filters well above cutoff (that is, $f \gg f_c$) decreases by about $6n$ dB per octave. (An octave is a doubling of the frequency.)

8–14. Show that the amplitude response of low-pass Butterworth filters well above cutoff (that is, $f \gg f_c$) decreases by $20n$ dB per decade. (A decade corresponds to multiplying the frequency by a factor of 10.)

8–15. Consider the resistive voltage divider circuit shown in Figure 8–22. This circuit replaces the input resistance of value $1/2Q$ in the multiple-feedback band-pass filter of Figure 8–21 in order to establish unity overall gain. The following requirements are imposed:

     1. The open-circuit voltage must be $1/2Q^2$ times the input voltage.
     2. The Thevenin equivalent resistance looking back must equal the normalized value $1/2Q$.

Derive expressions for R_1 and R_2 and verify the results given in Figures 8–22 and 8–23.

8–16. Show that the transfer function of the circuit of Figure P8–16 is

$$H(j\omega) = \frac{\overline{V}_o}{\overline{V}_i} = \frac{K\overline{Y}_1\overline{Y}_2}{\overline{Y}_1\overline{Y}_2 + \overline{Y}_1\overline{Y}_4 + \overline{Y}_2\overline{Y}_4 + \overline{Y}_3\overline{Y}_4 + \overline{Y}_2\overline{Y}_3(1 - K)}$$

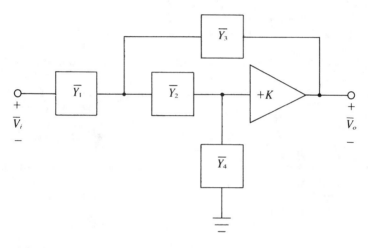

FIGURE P8–16

Design Problems

8–17. Design a two-pole low-pass Butterworth active filter using a unity-gain section to achieve a 3-dB frequency of 1 kHz. Select the two filter resistances as 10 kΩ each.

8–18. Design a two-pole low-pass Butterworth active filter using a unity-gain section to achieve a 3-dB frequency of 200 Hz. Select the two filter resistances as 30 kΩ each.

8–19. Repeat the unity-gain design of Problem 8–17 if the largest capacitor is selected as 0.01 μF.

8–20. Repeat the unity-gain design of Problem 8–18 if the largest capacitor is selected as 0.02 μF.

8–21. Perform a new design of the filter of Problem 8–17 based on an equal-component section with the two capacitances selected as 0.01 μF each.

8–22. Perform a new design of the filter of Problem 8–18 based on an equal-component section with the two capacitances selected as 0.039 μF each.

8–23. A low-pass filter is desired for a given application. The specifications are as follows:

     1. relative attenuation $\leq$ 3 dB for $f \leq$ 1 kHz
     2. relative attenuation $\geq$ 35 dB for $f \geq$ 4 kHz

Specify the minimum number of poles for a Butterworth filter that will satisfy the requirements.

8–24. A low-pass filter is desired for a given application. The specifications are as follows:

1. relative attenuation ≤ 3 dB for $f \leq 1$ kHz
2. relative attenuation ≥ 23 dB for $f \geq 2$ kHz

Specify the minimum number of poles for a Butterworth filter that will satisfy the requirements.

8–25. A high-pass filter is desired for a given application. The specifications are as follows:

1. relative attenuation ≤ 3 dB for $f \geq 1$ kHz
2. relative attenuation ≥ 40 dB for $f \leq 200$ Hz

Specify the minimum number of poles for a Butterworth filter that will satisfy the requirements.

8–26. A high-pass filter is desired for a given application. The specifications are as follows:

1. relative attenuation ≤ 3 dB for $f \geq 1$ kHz
2. relative attenuation ≥ 30 dB for $f \leq 400$ Hz

Specify the minimum number of poles for a Butterworth filter that will satisfy the requirements.

8–27. Design a unity-gain filter to meet the specifications of Problem 8–23. Select the filter resistances as 10 kΩ each.

8–28. Design a unity-gain filter to meet the specifications of Problem 8–24. Select the filter resistances as 10 kΩ each.

8–29. Perform a new unity-gain design to meet the specifications of Problem 8–23 if the largest capacitance is selected as 0.01 μF.

8–30. Perform a new unity-gain design to meet the specifications of Problem 8–24 if the largest capacitance in *each section* is selected as 0.01 μF.

8–31. Perform a new design using equal-component sections to meet the specifications of Problem 8–23. All capacitances are to be selected as 0.01 μF each.

8–32. Perform a new design using equal-component sections to meet the specifications of Problem 8–24. All capacitances are to be selected as 0.01 μF each.

8–33. Design a unity-gain filter to meet the specifications of Problem 8–25. Select the two capacitances as 0.01 μF each.

8–34. Design a unity-gain filter to meet the specifications of Problem 8–26. Select the two capacitances *for each section* as 0.01 μF each.

8–35. Design a band-pass filter using a single op-amp to achieve a center frequency of 1 kHz and a 3-dB bandwidth of 100 Hz. The capacitances are to be selected as 0.01 μF each.

8–36. Design a band-pass filter using a single op-amp to achieve a center frequency of 500 Hz and a 3-dB bandwidth of 25 Hz. The capacitances are to be selected as 0.02 μF each.

8–37. Design a state-variable Butterworth active filter that can be used as either a low-pass or a high-pass filter with a 3-dB frequency of 2 kHz. Select the filter-tuning resistances at 10 kΩ each. Show the output connections for low-pass or high-pass.

8–38. Repeat the design of Problem 8–37 if the capacitances are selected as 0.01 μF each.

8–39. Design a state-variable two-pole band-pass filter with a geometric center frequency of 2 kHz and a 3-dB bandwidth of 20 Hz. Select the filter-tuning resistances to be 10 kΩ each.

8–40. Using the result of Problem 8–39, design a band-rejection filter with the geometric center frequency of rejection at 2 kHz and a rejection bandwidth of 20 Hz.

LABORATORY EXERCISES

General Discussion

In each of the exercises that follow, you will be asked to measure the amplitude response of an active filter. This will require the use of a variable frequency sine-wave input. It will be assumed that (1) the oscillator output amplitude does not vary with frequency over the range involved, and (2) the output impedance of the oscillator is small compared with the input impedance of the filter. The second condition is adequately met with a 50-Ω output impedance, which is typical of many function generators.

The most convenient way to measure amplitude response is with an ac voltmeter having a decibel scale. Most analog multimeters and some digital multimeters have such a scale. By adjusting the signal amplitude, a reference filter output level of 0 dB is established on a convenient voltage scale in the pass band. For low-pass filters, a frequency of about $0.1f_c$ is usually adequate, and for band-pass filters, the frequency of maximum output is usually employed. For high-pass filters, use a frequency $10f_c$ unless either finite bandwidth or slew rate effects limit the response. For band-rejection filters, use a frequency $0.1f_o$. With the decibel scale, the relative level in decibels can be read at different frequencies. When it is necessary to change scales, most analog meters are designed so that -10 dB is added to the reading for each scale change in a more sensitive direction.

If a decibel is not available, a convenient filter output voltage level V_{ref} may be established in the pass band. Let V_m represent the measured voltage at any arbitrary frequency. The relative decibel response MRdB is then computed as

$$M_{\text{dB}}^R = 20 \log_{10} \frac{V_m}{V_{\text{ref}}}$$

The quantity M_{dB}^R differs from the amplitude response M_{dB} only by a possible constant pass-band gain or loss.

The voltage level $V_{\text{ref}} = $ V is highly recommended when practicable.

Low-Pass Unity-Gain Active Filter
LE 8–1.

1. Connect the circuit of Example 8–6 [Figure 8–10(c)].
2. Measure the amplitude response of the filter over a frequency range from about 80 Hz to about 8 kHz. (See discussion at beginning of section.)
3. Carefully measure the 3-dB cutoff frequency and compare with the expected value.
4. Plot the amplitude response on semilog graph paper with decibel response on the linear scale and frequency on the logarithmic scale. Compare a few points with the corresponding points on Figure 8–7.

Low-Pass Equal-Component Active Filter
LE 8–2.

Perform the steps outlined in LE 8–1 for the circuit of Example 8–9 [Figure 8–17(c)]. Remember that the gain of this circuit is greater than unity at low frequencies.

High-Pass Equal-Component Active Filter
LE 8–3.

1. Connect the circuit of Example 8–8 [Figure 8–12(c)].
2. Measure the amplitude response of the filter over a frequency range from about 100 Hz to about 10 kHz. (See discussion at beginning of section.)

3. Carefully measure the 3-dB cutoff frequency and compare with the expected value.

4. Plot the amplitude response on semilog graph paper with decibel response on the linear scale and frequency on the logarithmic scale. Compare a few points with the corresponding points on Figure 8–7.

High-Pass Equal-Component Active Filter
LE 8–4.

Perform the steps outlined in LE 8–3 for the circuit of Example 8–10 [Figure 8–18(c)]. Remember that the gain of this circuit is greater than unity in the pass band.

Multiple-Feedback Band-Pass Filter
LE 8–5.

1. Connect the circuit of Example 8–12 [Figure 8–24(c)].

2. Measure the amplitude response of the filter over a frequency range from about 200 Hz to about 20 kHz. (See discussion at beginning of section.)

3. Carefully measure the center frequency and the 3-dB bandwidth. Compare with the expected values.

4. Plot the amplitude response on semilog graph paper with decibel response on the linear scale and frequency on the logarithmic scale. Compare a few points with the corresponding points on Figure 8–20.

State-Variable Low-Pass/High-Pass Filter
LE 8–6.

1. Connect the circuit of Example 8–13 (Figure 8–28).

2. Verify that the amplitude response based on the "low-pass output" has the correct form by checking the response at a few frequencies.

3. Carefully measure the low-pass 3-dB cutoff frequency and compare with the expected value.

4. Verify that the amplitude response based on the "high-pass output" has the correct form by checking the response at a few frequencies.

5. Carefully measure the high-pass 3-dB cutoff frequency and compare with the expected value.

State-Variable Band-Pass Filter
LE 8–9.

1. Connect the circuit of Example 8–14 (Figure 8–29).

2. Verify that the amplitude response has the correct form. (*Caution:* The gain of this circuit at the center frequency is very large, so the input signal level must be kept small to avoid overloading the op-amp.)

3. Carefully measure the center frequency and the 3-dB bandwidth. Compare with the expected values. (*Note:* Unless an accurate frequency counter is employed, don't expect to be able to determine the bandwidth to any high accuracy because of the high Q.)

State-Variable Band-Rejection Filter
LE 8–10.

1. Connect the circuit of Example 8–15 (Figure 8–30).

2. Verify that the amplitude response has the correct form.

3. Carefully measure the center frequency of rejection and the 3-dB rejection bandwidth. Compare with the expected values. (See note in LE 8–9).

9

RECTIFIER, DIODE, AND POWER CIRCUITS

OBJECTIVES

After completing this chapter, the reader should be able to:

- Compare passive and active signal-processing rectifier circuits and indicate the major advantage of the latter types.
- Draw the schematic diagrams for and analyze the operation of various signal-processing rectifier circuits.
- Compare the desired objectives of the peak detector, the envelope detector, and the sample-and-hold circuit, draw their schematic diagrams, and analyze their operations.
- Analyze the discharge of a capacitor into a resistor to predict the change in voltage occuring in a specified time interval.
- Indicate the objective of an ideal clamping circuit, and draw the schematic diagrams for and analyze the operation of various clamping circuits.
- Indicate the objective of an ideal limiting circuit, and draw the schematic diagrams for and analyze the operation of various limiting circuits.
- Draw the schematic diagram for and discuss the operation of a typical power supply circuit.
- Predict the ripple level associated with a single-capacitor filter in a power supply.
- Draw the schematic diagram for and analyze the operation of a regulator circuit employing an op-amp.
- Discuss the general operation of a switching regulator and indicate its primary advantage over a conventional regulator.
- Discuss the application of IC regulators such as the LM117/LM317 units to achieve voltage regulation.

9–1 INTRODUCTION

An alternate title to this chapter could have been "Circuits with Diodes and Op-Amps." While this title would probably not have been quite as appropriate as the chosen title, it does delineate the fact that most of the circuits we will discuss contain both op-amps and diodes to perform their desired functions. Most of these circuits are either nonlinear or have nonlinear regions of operations.

A major topic of this chapter is *rectification,* which is the process of eliminating either the negative portion or the positive portion of a signal (half-wave rectifier), or the process of inverting all portions on one side of zero and thus producing a signal with one polarity only (full wave). We consider both signal-processing rectifier circuits (small-signals) and power supply rectifier circuits (large voltage and power levels).

We then introduce a class of circuits containing diodes, op-amps, and capacitors referred to collectively as *holding circuits.* The presence of capacitors allows the circuit to perform a memory function. Included in this category are peak detectors, envelope detectors, and sample-and-hold circuits. Other circuits containing op-amps and diodes include clamping circuits and limiters. Clamping circuits perform a shift on an input signal level, while limiters limit the attainable level of a signal.

Also, we consider the operation of voltage regulation for power supplies. The evolution of the concept is developed using an op-amp, with emphasis on the use of dedicated IC regulators. Finally, we discuss the operation of switching regulators.

9–2 SIGNAL-PROCESSING RECTIFIER CIRCUITS

This group of circuits is referred to collectively as ***precision signal-processing rectifier circuits.*** The modifier *signal-processing* delineates the types of situations in which these circuits are intended. In signal-processing applications, the voltage, current, and power levels are usually quite low, and these circuits may be used for a variety of purposes in such cases. However, the circuits we will discuss here are generally not suited for power supply rectifier applications, where the voltage, current, and power levels are usually much higher.

Series Half-Wave Rectifier

The justification for active feedback rectifier circuits can be established by first reviewing two of the classical signal-processing, passive half-wave rectifier circuits. Consider first the *series* half-wave rectifier circuit shown in Figure 9–1(a). If the diode were ideal, the output voltage v_o would be the same as the input voltage v_i for v_i positive, and the output voltage would be zero for v_i negative. However, junction diodes exhibit an exponential voltage-current relationship. For many applications, this exponential relationship may be closely approximated by a constant voltage drop over a wide range of operation. For small signal levels, the voltage drop is in the range of 0.6 V to 0.7 V for silicon diodes and 0.2 V to 0.3 V for germanium diodes. Since the majority of diodes are silicon types, and since the inaccuracies are greater at larger voltage drops, we will assume the larger 0.7-V value for subsequent calculations dealing with small signal diode circuits. The reader should understand, however, that this value has been "standardized" for convenience, and the actual value could be somewhat different.

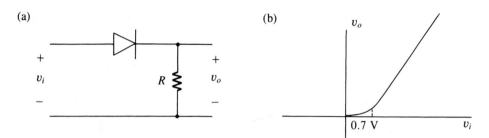

FIGURE 9–1
(a) Passive series half-wave rectifier circuit and (b) its input-output characteristic curve.

✴   The realistic diode behavior in the circuit of Figure 9–1(a) results in very little current flow until the input voltage approaches 0.7 V. Further, once conduction is established, the output can be approximated as

$$v_o \simeq v_i - 0.7 \tag{9–1}$$

The form of the input-output characteristic is illustrated in Figure 9–1(b). For input voltages well in excess of the "cut-in" value of 0.7 V, the curve exhibits a reasonably linear relationship between the input and output voltages. However, for small positive voltages, the curve is quite nonlinear. For $v_i < 0$, the curve indicates no output voltage. Diodes exhibit a small reverse current, but it is assumed to be negligible in this example. Further, the reverse-biased Zener or avalanche breakdown phenomenon is not shown here.

For less demanding applications, the nonideal nature of the input-output characteristic of the passive circuit may suffice. However, for applications where the exact level of the rectified signal is important, this circuit may introduce too much error. In particular, at very low positive signal levels, this circuit would be hopelessly inadequate for precision applications.

Shunt Half-Wave Rectifier

Before introducing precision circuits, we will briefly consider one other passive circuit. A **shunt** half-wave rectifier circuit is shown in Figure 9–2(a), and its input-output characteristic is shown in Figure 9–2(b). When the input voltage is positive, the diode is reverse biased, and the output voltage with no load connected is equal to the input voltage. (The

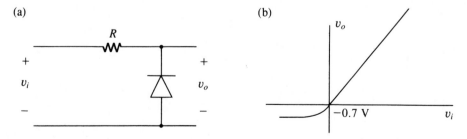

FIGURE 9–2
(a) Passive shunt half-wave rectifier circuit and (b) its input-output characteristic curve.

presence of a load resistance will change the characteristic.) When the input voltage is negative, the diode becomes forward biased, and it appears directly across the line. If the diode were ideal, it would be a short circuit under these conditions, and the output voltage would be zero. However, the nearly constant voltage drop results in $v_o \approx -0.7$ V over a wide range of negative input voltages. A further restriction is that the positive range of the input voltage must be below the avalanche or Zener voltage, at which point the linear relationship would no longer be valid. Thus, while the forward characteristic of the shunt circuit is more linear than that of the series circuit, it suffers from other limitations.

One final comment on the passive circuits concerns the polarities of the input and output variables. Both circuits of Figure 9–1 and 9–2 were designed to provide an output in the first quadrant. If the diode in either circuit were reversed, the major output would shift to the third quadrant. (See Problems 9–37 and 9–38.)

Saturating Half-Wave Rectifier

The first precision rectifier that will be considered is the half-wave, single-diode, noninverting form shown in Figure 9–3(a). Most nonlinear op-amp circuits are a bit tricky to analyze, and some practice is required to develop the required insight. The main difficulty is in predicting when a given diode is forward biased and when it is reverse biased. As a general rule, if in doubt, make an assumption and see if the results so obtained justify the assumption. Thus, if forward bias is assumed, the predicted current through the diode should be in the direction of the arrow, and if reverse bias is assumed, the predicted voltage across the open circuit should be in the direction to justify the assumption.

In the circuit of Figure 9–3, if v_i is positive, the input differential voltage $(v^+ - v^-)$ becomes positive, causing the op-amp output to become positive. Because of the feedback, however, the differential voltage, while positive, actually is very small and approaches zero in the limit to maintain stable, closed-loop conditions. This forces the circuit output v_o, for all practical purposes, to be the same as v_i as in a voltage follower circuit. As for the diode drop, the op-amp output v'_o assumes a value such that $v_o = v_i$ as required by the loop. This means that v'_o is forced to be

$$v'_o \approx v_o + 0.7 \text{ V} \tag{9-2}$$

so that $v_o = v_i$ as required.

(a)   (b)

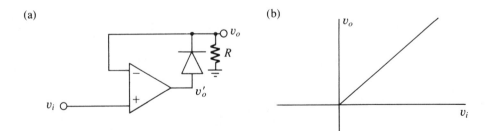

FIGURE 9–3
(a) Active saturating half-wave precision rectifier circuit and (b) its input-output characteristic curve.

A different way of looking at this circuit is that the nonlinear diode is placed in the forward block of the feedback circuit. The net gain, however, is virtually independent of this nonlinear forward characteristic because of the large value of loop gain, and the resulting closed-loop characteristic is quite linear in this region. The process of placing the nonlinear diode characteristic in the forward block of a feedback circuit having large loop gain will occur frequently in this chapter, so the reader should carefully note the concept.

When $v_i < 0$, the differential voltage becomes negative. The voltage v_o' also becomes negative, and the diode is reverse biased. The loop is then broken, and the op-amp swings down to negative saturation. However, the output terminal is now separated from both the op-amp output and the input signal v_i, so $v_o = 0$.

The input-output characteristic is shown in Figure 9–3(b). Excellent linearity results in this case, since the op-amp compensates by feedback for the nonlinear diode characteristic. Linearity is maintained down to a very low input signal voltage level. The mathematical form of the input-output characteristic curve can be expressed to a very high degree of accuracy as

$$v_o = v_i \quad \text{for } v_i \geq 0 \tag{9–3a}$$

$$v_o = 0 \quad \text{for } v_i \leq 0 \tag{9–3b}$$

This circuit represents an example of a nonlinear circuit in which essentially linear operation is achieved over one region ($v_i > 0$) and nonlinear op-amp operation (specifically, saturation) results in the other region ($v_i < 0$). In the first region $v^+ = v^-$ can be assumed, as previously noted. However, in the latter region, the voltages at the two inputs will be quite different, so care must be taken to ensure operation below the maximum rated value of the differential input voltage.

Since the output swings into negative saturation for a negative input, the circuit is a saturating form. Thus, the frequency response is somewhat limited.

Before leaving this particular circuit, we will note that reversing the diode will shift the linear region of operation from the first quadrant ($v_i > 0$ and $v_o > 0$) to the third quadrant ($v_i < 0$ and $v_o < 0$). (See Problem 9–39.)

Nonsaturating Half-Wave Rectifier, Second Quadrant

A *nonsaturating* half-wave precision rectifier circuit is shown in Figure 9–4(a). When $v_i > 0$, the voltage at the inverting input tends to become positive, forcing the op-amp output to go negative. However, this results in forward bias for $D1$, and the op-amp output can drop only by about 0.7 V below the inverting input potential. Diode $D2$ will be reverse biased in this case. The stable feedback arrangement forces the inverting input potential to be essentially zero. (This is not a contradiction of the earlier statement that it is positive. Recall that a very small differential voltage is required to maintain the output at a nonzero value.) The result is that the output v_o is zero when the input is positive.

When $v_i < 0$, the voltage at the inverting input tends to become negative, forcing the op-amp output to go positive. Forward bias for $D2$ and reverse bias for $D1$ result. The circuit is now functioning like an inverting amplifier circuit with a nonlinear diode in the forward block. However, the feedback forces v_o to be $-(R_f/R_i)$ times the input voltage, so v_o' assumes the positive value required to achieve this condition.

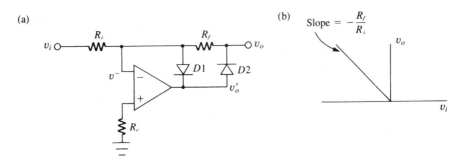

FIGURE 9–4
(a) Active nonsaturating half-wave precision rectifier circuit with positive output for negative input and (b) the input-output characteristic curve.

A mathematical representation for the circuit operation can be expressed as

$$v_o = 0 \qquad \text{for } v_i > 0 \tag{9–4a}$$

$$v_0 = -\frac{R_f v_i}{R_i} \quad \text{for } v_i < 0 \tag{9–4b}$$

The voltage v_o' at the op-amp output is

$$v_o' \simeq -0.7 \text{ V} \qquad \text{for } v_i > 0 \tag{9–5a}$$

$$v_o' \simeq \frac{R_f}{R_i} v_i + 0.7 \text{ V} \quad \text{for } v_i < 0 \tag{9–5b}$$

While v_o is the actual output variable of interest, detailed circuit design requires consideration of the behavior of v_o'. Observe that there is an abrupt change in the value of v_o' when v_i changes signs.

The input-output characteristic curve is shown in Figure 9–4(b). Because of the inverting nature of the circuit, the region of nonzero output shifts to the second quadrant ($v_i < 0, v_o > 0$). Thus, the negative portion of an input signal is multiplied by a negative gain, so the output is actually positive.

This circuit has the advantages that it is a precision rectifier and it is nonsaturating. The inverting characteristic may pose a minor annoyance in some applications, but it is readily circumvented by the use of additional inversion.

Nonsaturating Half-Wave Rectifier, Fourth Quadrant

Operation can be shifted to the fourth quadrant by reversing the two diodes. The corresponding circuit is shown in Figure 9–5(a), and its input-output characteristic is shown in Figure 9–5(b). A mathematical description in this case can be expressed as

$$v_o = -\frac{R_f}{R_i} v_i \quad \text{for } v_i > 0 \tag{9–6a}$$

$$v_0 = 0 \qquad \text{for } v_i < 0 \tag{9–6b}$$

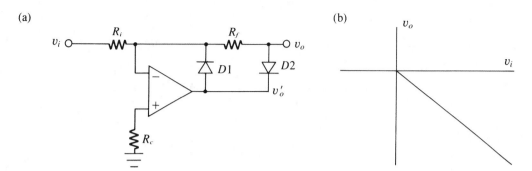

FIGURE 9–5
(a) Active nonsaturating half-wave precision rectifier circuit with negative output for positive input and
(b) the input-output characteristic curve.

The voltage v_o' the op-amp output in this case is

$$v_o' \simeq -\frac{R_f}{R_i}v_i - 0.7 \text{ V} \quad \text{for } v_i > 0 \tag{9–7a}$$

$$v_o' \simeq -0.7 \text{ V} \qquad\qquad \text{for } v_i < 0 \tag{9–7b}$$

Nonsaturating Full-Wave Rectifier

All of the rectifier circuits considered thus far have been **half-wave** circuits; that is, the output is nonzero only for one specific polarity of the input signal. The last precision rectifier that will be considered is an example of a **full-wave** circuit. This type of circuit is also referred to as an **absolute-value** circuit.

The full-wave rectifier circuit of interest is shown in Figure 9–6(a). The left-hand portion of the total circuit is a half-wave rectifier circuit of the same form previously considered in Figure 9–5(a). The right-hand portion of the total circuit is an inverting summing circuit.

Assume first that $v_i < 0$. From the characteristic of Figure 9–5(b) or from Equations (9–6a) and (9–6b), the output of the left-hand rectifier circuit is $v_a = 0$. Thus one input to the summing circuit has a value of zero. However, v_i is also applied as an input to the summer. The gain for this input is -1, and the output is $v_o = -v_i$. Since $v_i < 0$, v_o will be inverted and will thus be positive, corresponding to the second quadrant of the input-output characteristic curve shown in Figure 9–6(b).

Assume next that $v_i > 0$. From the characteristic of Figure 9–5(b) or from Equations (9–6a) and (9–6b), the output of the left-hand rectifier circuit is $v_a = -v_i$. (Note that $R_f = R_i = R$.) The voltage v_a appears as one input to the summing circuit, and the gain for that input is -2. As before, v_i also appears directly as an input to the summer. The net output is then $v_o = -v_i - 2v_a = -v_i - 2(-v_i) = v_i$. Since $v_i > 0$, v_o will be positive, corresponding to the first quadrant of the input-output characteristic of Figure 9–6(b).

(a)

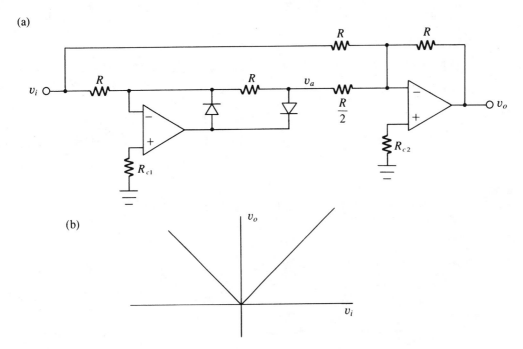

(b)

FIGURE 9–6
(a) Active nonsaturating full-wave precision rectifier (absolute-value) circuit and (b) its input-output characteristic curve.

Along with the graphical form, a mathematical representation may be summarized as

$$v_o = -v_i \quad \text{for } v_i < 0 \qquad \text{(9–8a)}$$
$$v_o = v_i \quad \text{for } v_i > 0 \qquad \text{(9–8b)}$$

The two expressions of (9-8a) and (9-8b) can be combined in a single equation as

$$v_o = |v_i| \qquad \text{(9–9)}$$

where the vertical lines represent the magnitude or absolute value of the input.

Like the preceding two half-wave precision circuits, this circuit is a nonsaturating form.

EXAMPLE 9–1

Consider the precision half-waver rectifier circuit of Figure 9–4(a) with $R_i = 10$ kΩ and $R_f = 20$ kΩ. Determine v_o, v^- and v_o' for each of the following values of input voltage: (a) $v_o' = 5$ V and (b) $v_i = -5$ V. Assume that $\pm V_{\text{sat}} = \pm 13$ V.

Solution
(a) For $v_i = 5$ V, $D1$ is forward biased and $D2$ is reverse biased. The loop is closed through $D1$, and the voltages are

$$v_o = 0 \qquad \text{(9–10)}$$

$$v^- = 0 \qquad\qquad (9\text{–}11)$$

$$v'_o \approx -0.7 \text{ V} \qquad\qquad (9\text{–}12)$$

(b) For $v_i = -5$ V, $D1$ is reverse biased and $D2$ is forward biased. The loop is now closed through $D2$ and the voltages are

$$v_o = \frac{-R_f}{R_i}v_i = \frac{-20 \times 10^3}{10 \times 10^3}v_i = -2 \times (-5) = 10 \text{ V} \qquad\qquad (9\text{–}13)$$

$$v^- = 0 \qquad\qquad (9\text{–}14)$$

$$v'_o = v_o + 0.7 = 10 + 0.7 = 10.7 \text{ V} \qquad\qquad (9\text{–}15)$$

The last result indicates that v'_o is closer to saturation than v_o, and this point should be considered in a circuit in which the full limits of the available dynamic range are to be utilized. Thus, the maximum achievable value for v_o is about 12.3 V for a saturation voltage of 13 V.

EXAMPLE 9–2   The sinusoidal waveform of Figure 9–7 (a) is applied as the input to the circuit of Example 9–1. Sketch the output voltage.

(a)

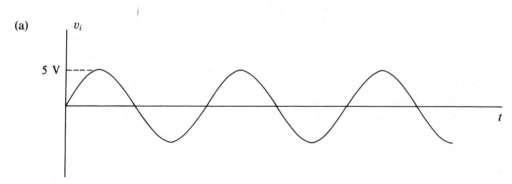

(b)

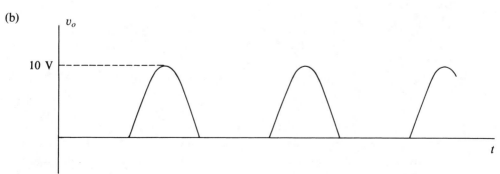

FIGURE 9–7
Waveforms of Example 9–2.

Solution

Referring back to the input-output characteristic of Figure 9–4(b) and the work of Example 9–1, the output is zero during the positive half-cycles of the input signal. However, the negative half-cycles of the input signal are multiplied by -2, that is, they are inverted and increased in amplitude by a factor of 2. The resulting output waveform is shown in Figure 9–7(b).

9–3 HOLDING CIRCUITS

This section emphasizes circuits that store or "remember" a particular level of a signal for some interval of time. Depending on the focus of the design and the particular application, this category includes such circuits as the *peak detector,* the *envelope detector,* and the *sample-and-hold* circuit.

Ideal Peak Detector

An ideal peak detector is a circuit that produces a voltage equal to the positive or negative peak value of the input voltage. A positive peak detector produces an output equal to the positive peak value of the input, and a negative peak detector produces an output equal to the negative peak value of the input.

The operation of an ideal peak detector is illustrated in Figure 9–8 for a positive peak detector. When the input v_i of (a) is turned on, the output v_o of (b) quickly reaches a level equal to the peak value of the input. As long as the peak value of the input voltages remains the same, the output remains at a constant dc level. If at some later time, the peak value of the input voltage increases to a higher level, the output voltage quickly increases to the new level. If, however, the peak value of the input voltage decreases at a later time, the output voltage of a practical peak detector may not immediately respond to the new level. This situation will be discussed in more detail later.

Passive Peak Detector

First consider the classical passive positive peak detector circuit shown in Figure 9–9. Initially, the diode will conduct when the input signal is sufficiently positive, and the capacitor will charge to a voltage level given by the peak positive level minus the diode forward voltage drop. Letting $V_i(peak)$ represent the peak value of the input, we have

$$v_o \simeq V_i(peak) - 0.7 \qquad \text{(9–16)}$$

After the capacitor is charged, the positive voltage across the capacitor tends to keep the diode reverse biased most of the time. However, any load, whether intended to or not, will discharge the capacitor and reduce the output voltage. If the input is an alternating voltage, conduction will occur for a short portion of each cycle to restore the capacitor voltage to the peak level. Conduction begins to occur at the point where the input voltage exceeds the voltage existing on the capacitor.

(a)

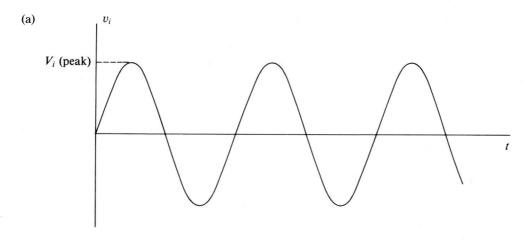

(b)

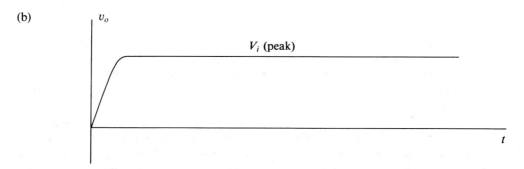

FIGURE 9–8
Waveforms of ideal peak detector.

This passive circuit is widely used where the peak level attained across the capacitor is not critical. However, the circuit suffers from the fact that the forward diode voltage drop causes an error in the output peak level. Further, there is very little isolation between the source, the capacitor, and the load, and the time constant for the circuit may be difficult to adjust properly.

FIGURE 9–9
Passive peak detector circuit (can also be used as an envelope detector).

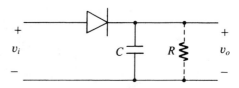

Active Saturating Peak Detector

The first active peak detector to be considered is shown in Figure 9–10. Initially, when $v_i > 0$, the op-amp output will be positive, and the diode will be forward biased. Operation in the linear active region requires that $v^- = v^+$, forcing the circuit output voltage v_o to reach a value equal to the peak value of the input:

$$v_o = V_i(\text{peak}) \qquad (9\text{–}17)$$

The capacitor charges quickly to this value through the low-resistance path consisting of the op-amp output resistance and the diode forward resistance. Feedback forces the op-amp output v_o' at this point in the cycle to be

$$v_o' = v_o + 0.7 \qquad (9\text{–}18)$$

After the capacitor is charged to the peak input voltage and the input voltage drops below this value, the differential input voltage $v^+ - v^-$ becomes negative, and the op-amp output voltage in turn becomes negative. The diode then becomes reverse biased, and the loop is broken. The op-amp output voltage then swings down to negative saturation. The voltage across the capacitor remains close to the peak value of the input voltage, although the effect of any load will result in some depletion during the remainder of the input cycle. During the next cycle, the op-amp will move back into the linear region for the interval in which the input voltage exceeds the capacitor voltage. During that interval the capacitor is charged back to the peak input voltage.

This circuit is a "precision" circuit in that the error introduced by the diode drop has been virtually eliminated. However, as the input signal changes polarity, the op-amp swings back and forth from saturation to linear operation, so the circuit is a saturating form. Consequently, its application is generally limited to relatively low-frequency operation.

Because the capacitor holds a positive voltage at the inverting terminal when the input voltage at the noninverting terminal decreases, the peak value of the differential input voltage may be rather large, particularly if the input voltage becomes negative. This fact should be carefully considered in selection of the maximum differential input voltage rating for an op-amp to be used in such a circuit. As an additional point of caution, the 10-

FIGURE 9–10

Active precision saturating peak detector circuit (can also be used as an envelope detector).

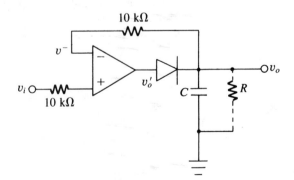

kΩ resistance (a typical value) between the capacitor and the inverting input is used to protect the op-amp input from an excessive capacitor discharge current that can arise when a power supply is turned off. The other 10-kΩ resistor is the corresponding bias compensating resistance.

For both the passive circuit of Figure 9-9 and the active circuit of Figure 9-10, the effect of the discharge time constant must be carefully considered. Assume that R in each of the figures represents the net effective load resistance as viewed from the capacitor terminals. This resistance could represent the actual input resistance to the next stage, an external resistance, or a combination of both. Let $\tau = RC$ represent the discharge time constant.

The optimum time constant depends on the application. For peak detector circuits where it is desired to hold a steady dc value for a relatively long time, it is usually desirable for τ to be relatively large. Even for peak detectors, however, if a short-term peak value is desired and the detector must be able to readjust the level over a period of time, some discharge path may be desired. If some means of discharge were not present, a peak detector could theoretically never readjust itself to a smaller peak value once the capacitor were charged. Thus, while a peak detector generally necessitates a relatively long time constant, this consideration must be tempered with the practical requirement for possible readjustment of the level in some cases. If T represents the period of the signal, it is necessary that $\tau \gg T$ in order to hold the peak value with negligible degradation during a cycle.

Ideal Envelope Detector

An ideal **envelope detector** is a circuit that extracts a slow variation in the envelope of a high-frequency signal. A positive envelope detector extracts the positive envelope, and a negative envelope detector extracts the negative envelope.

Envelope detectors are primarily used in radio receivers for the detection of *amplitude-modulated* (AM) waveforms. The operation of an ideal positive envelope detector is illustrated in Figure 9-11. A simplified amplitude-modulated signal is shown in Figure 9-11(a). In a practical situation, the high-frequency variation is, typically, 100 times or more as great as the modulating waveform. For ease of illustration here, the high frequency is only a few times as large as the modulating frequency, but the concept is equally valid. An ideal circuit should produce, for the first step of the detection process, the waveform shown in Figure 9-11(b). In most circuits, the dc level is also removed, but that process is not shown here.

Basically, any circuit that may be used as a peak detector may also be used as an envelope detector, and the circuit diagrams appear essentially the same. The major difference is in the selection of the discharge time constant τ for the circuit. The time constant should be small enough to allow the capacitor voltage to follow the envelope and large enough to prevent this voltage from following the high-frequency carrier variations.

Let T_m represent the shortest modulation period (worst-case condition), and let T_c represent the carrier period. Good envelope detector response requires that the following two inequalities be satisfied:

$$\tau \ll T_m \qquad\qquad (9\text{-}19)$$

and

$$\tau \gg T_c \qquad\qquad (9\text{--}20)$$

It is usually possible to satisfy both inequalities, since the typical carrier frequency may be several orders of magnitude greater than the highest modulation frequency.

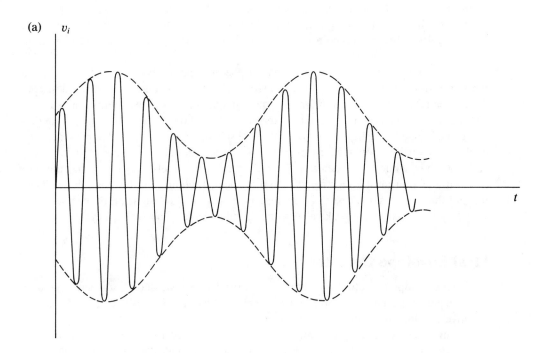

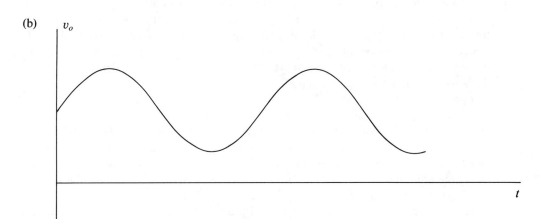

FIGURE 9–11
Illustration of the operation of an ideal envelope detector for an amplitude-modulated (AM) signal.

Discharge Voltage

For any peak or envelope detector circuit in which the discharge is a result of an equivalent single resistance R in parallel with a single capacitance C, the voltage level resulting from the discharge interval may be determined from the properties of the basic exponential function. Let $v_C(t)$ represent the voltage across the capacitor as a function of time, and let $V_C(\text{peak})$ represent the peak voltage. [In the circuit of Figure 9–10, $v_C(t) = v_o(t)$ and $V_C(\text{peak}) = V_i(\text{peak})$. However, the different symbols given here allow the result to be generalized somewhat.]

The exponential form of the capacitor voltage during the discharge interval is illustrated in Figure 9–12, and it may be expressed as

$$v_C(t) = V_C(\text{peak})e^{-t/\tau} \tag{9–21}$$

where $\tau = RC$ and t is measured from the beginning of the discharge interval; that is, $t = 0$ is the point at which the capacitor starts to discharge. The result of (9–21) can be used to determine the capacitor voltage after a certain time t_1, or a time t_1 can be determined at which the voltage has reached a certain level. The latter form as determined from (9–21) is

$$t_1 = \tau \ln \frac{V_C(\text{peak})}{v_C(t_1)} \tag{9–22}$$

In many applications involving peak detectors, τ will be much greater than the time interval over which the capacitor is allowed to discharge. In this case, a very good approximation for the *magnitude of the change* in voltage $|\Delta v_C|$ that occurs over a time interval Δt is

$$|\Delta v_C| \simeq \frac{V_C(\text{peak})\Delta t}{\tau} = \frac{V_C(\text{peak})\Delta t}{RC} \quad \text{for } \Delta t \ll \tau \tag{9–23}$$

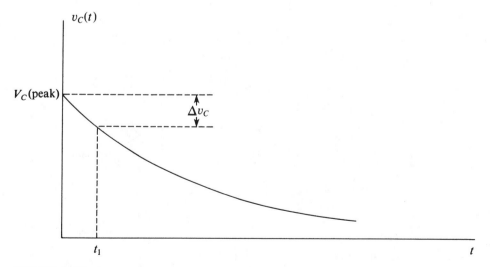

FIGURE 9–12
Exponential discharge of the voltage across a capacitor.

where the change represents a *reduction* of the capacitor voltage. This approximation is quite good when $|\Delta v_C| \ll V_C(\text{peak})$.

Peak Detector with Isolation

Where it is desirable to eliminate virtually all loading on the capacitor, isolation may be achieved with the circuit of Figure 9–13. The right-hand op-amp should be selected to have as high an input impedance as possible. A good choice would be one with FET input circuitry. The diode D2 should be one with very low reverse leakage current. Initially, when $v_i > 0$, the left-hand op-amp output is positive, D1 is reverse biased, D2 is forward biased, and the capacitor is being charged. The connection with the 10-kΩ resistor (a typical value) between the two inverting terminals and the feedback forces the capacitor to charge to the peak input voltage. The left-hand op-amp output v_a at the point of the peak input is

$$v_a \simeq V_i(\text{peak}) + 0.7 \tag{9–24}$$

When v_i begins to drop, D2 becomes reverse biased, leaving the capacitor charged with virtually no loading. The output of the right-hand op-amp remains at the peak input voltage, and since the differential input voltage of the left-hand op-amp is now negative, its output starts to drop. However, diode D1 becomes forward biased and the left-hand op-amp then functions as a voltage follower. The voltage $v_a(t)$ during this phase is

$$v_a(t) \approx v_i(t) - 0.7 \tag{9–25}$$

This constraint forces the op-amp to remain in the linear region, and thus the circuit is a *nonsaturating* form.

Either the peak or the envelope detector circuits that have been considered thus far could be easily modified to work for negative peaks or envelopes by mere reversal of all the diodes. All the arguments discussed apply except that all polarities and references are reversed in sign.

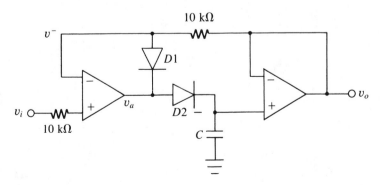

FIGURE 9–13
Active precision nonsaturating peak detector circuit with isolation of capacitor.

Ideal Sample-and-Hold Circuit

An ideal sample-and-hold circuit is one that samples the value of an analog signal at a certain point and holds that value for a specified interval, when it repeats the process based on the new value of the signal. Sample-and-hold circuits are used in conjunction with analog-to-digital conversion, a major topic of Chapter 10.

The operation of an ideal sample-and-hold circuit is illustrated in Figure 9–14. At the beginning of each time interval of duration T, the output voltage v_o quickly assumes the level of the input voltage v_i. It then holds that value until the beginning of the next interval.

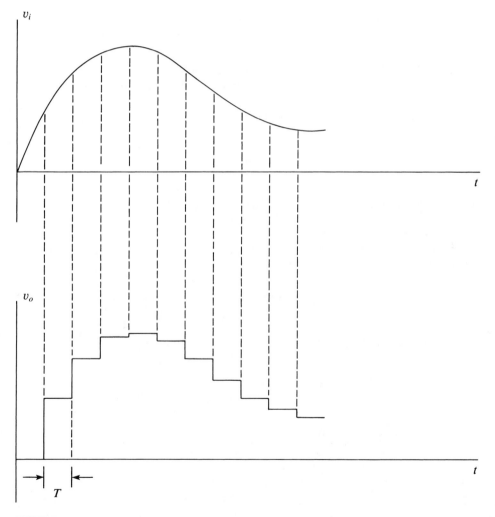

FIGURE 9–14
Illustration of the operation of a sample-and-hold circuit.

Typical Sample-and-Hold Circuit

The sample-and-hold circuit of Figure 9–15 is similar in certain ways to the peak detector of Figure 9–13. However, a significant difference is that the diode D2 in the peak detector circuit is replaced by a switch, the details of which are not shown here. In practice, analog FET switches are commonly used. Such switches are capable of being switched by convenient voltages such as TTL levels. The switch is turned on for a short interval of time, and the capacitor quickly charges (or discharges) to the value of the analog signal at that time. The switch is then opened, and the capacitor holds the voltage until the next charge interval occurs. Likewise, the voltage v_o remains constant at a voltage level equal to the analog signal at the preceding sample point. Thus, instead of retaining the peak value (as does the peak detector), the sample-and-hold circuit retains the preceding actual level of the input signal.

Because the analog signal may be either greater than or less than the preceding capacitor voltage, it is necessary to employ two parallel diodes with opposite directions if saturation is to be avoided in the left-hand op-amp. Depending on the direction of the inequality between the input and the output, one of the two diodes will clamp the op-amp output and prevent saturation during the hold interval. The left-hand op-amp output $v_a(t)$ will then be either one of the following:

$$v_a(t) \approx v_i(t) - 0.7 \quad \text{if } v_i < v_o \tag{9-26}$$

$$v_a(t) \approx v_i(t) + 0.7 \quad \text{if } v_i > v_o \tag{9-27}$$

With all the holding circuits, the rate of change of the voltage across the capacitor is limited by the peak current rating $i(\text{max})$ of the op-amp. In a short time Δt, the maximum voltage change $\Delta v(\text{max})$ that can occur is approximately equal to $i(\text{max})\,\Delta t/C$.

EXAMPLE 9–3

A peak detector circuit of the form shown in Figure 9–10 is to be designed for an application in which a short-term memory of the peak level of an input signal is desired. In order that the level may readjust to long-term changes, however, a discharge path will be pro-

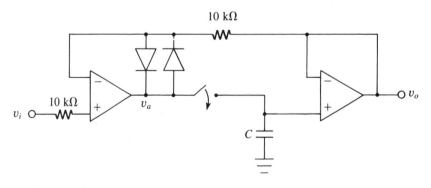

FIGURE 9–15
Sample-and-hold circuit.

vided through a resistance R. A design criterion is somewhat arbitrarily established that the capacitor voltage will drop to 95% of its peak level in a time interval of 10 ms when it is not recharged. Determine suitable values for R and C.

Solution
Given that the specification of the discharge level is somewhat arbitrary, the approximation of (9–23) should suffice. The magnitude of the change is 5% of the peak value and $|\Delta v_c| = 0.05\ V_C(\text{peak})$. Substituting this value and $\Delta t = 10$ ms $= 0.01$ s in (9–23), we have

$$0.05V_C(\text{peak}) = \frac{V_C(\text{peak})\ \Delta t}{\tau} = \frac{V_C(\text{peak}) \times 0.01}{\tau} \qquad (9\text{–}28)$$

$$\tau = \frac{0.01}{0.05} = 0.2\ \text{s} = RC \qquad (9\text{–}29)$$

Values of R and C must be selected such that their product is 0.2. Reasonable choices are $C = 1\ \mu\text{F}$ and $R = 200\ \text{k}\Omega$.

If an exact solution employing (9–22) is utilized, the reader is invited to show that the time constant determined is $\tau = 0.195$ s.

EXAMPLE 9–4

Consider the isolated peak detector circuit of Figure 9–13, which is repeated in Figure 9–16(a) with various quantities of interest shown. Assume that the input is a random signal that varies from -10 V to $+10$ V. (a) Draw an equivalent circuit at the point where $v_i = 10$ V with the capacitor fully charged, and determine the values of all variables shown. (b) Draw an equivalent circuit at the point where $v_i = -10$ V *after the peak of (a) has occurred*, and determine the values of all variables shown.

Solution
(a) For $v_i = 10$ V, the equivalent circuit is shown in Figure 9–16(b). In effect, a loop is established around both op-amps. Diode $D1$ is reverse biased and is shown with a dashed line. However, diode $D2$ is forward biased as the left hand op-amp charges the capacitor to the value of 10 V.

The output stage is acting as a voltage follower, and this voltage is fed back to the inverting input of the first stage. Since $D1$ is reverse biased, no current can flow through the 10-kΩ resistance, and the voltage at the inverting input of the first stage must therefore equal the output voltage. The values of several variables can then be summarized as

$$v_o = v_C = v^- = v_i = 10\ \text{V} \qquad (9\text{–}30)$$
$$i_R = 0 \qquad (9\text{–}31)$$

The output of the first stage is

$$v_a = 10 + 0.7 = 10.7\ \text{V} \qquad (9\text{–}32)$$

(b) For the analysis intended, it is necessary to assume that the positive peak of Solution (a) has already occurred. For that condition and for $v_i = -10$ V, the equivalent

(a)

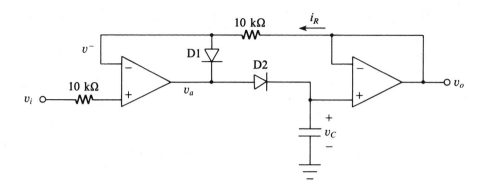

(b) At positive peak

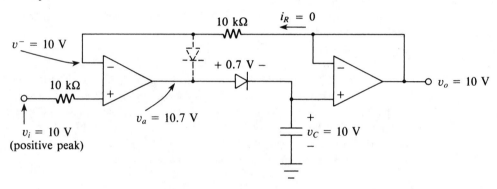

(c) At negative peak

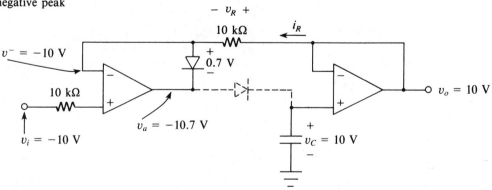

FIGURE 9–16
Circuit analysis of Example 9–4.

circuit is shown in Figure 9–16(c). Diode $D2$ is now reverse biased, so the capacitor holds the voltage of 10 V. However, $D1$ is now forward biased, and the circuit acts like two voltage followers. For the right-hand voltage follower, we have

$$v_o = v_C = 10 \text{ V} \tag{9–33}$$

For the left-hand voltage follower, we have

$$v^- = v_i = -10 \text{ V} \tag{9–34}$$

and

$$v_a = -10 - 0.7 = -10.7 \text{ V} \tag{9–35}$$

For this condition, v_R appears across the upper 10-kΩ resistance with value $v_R = v_o - v^- = 10 - (-10) = 20$ V. The value of i_R is then

$$i_R = \frac{20 \text{ V}}{10^4 \ \Omega} = 2 \text{ mA} \tag{9–36}$$

The left-hand op-amp sinks this current.

9–4   CLAMPING CIRCUITS

Clamping circuits are used to establish a certain fixed voltage level for the maximum or minimum value of a given signal. Ideally, the shape of the signal is not changed in the process, but rather the signal is either "lifted up" or "pushed down," depending on the sign and direction of the desired clamp. One of the most widely used clamping circuits is the **dc restorer,** which is employed in television circuits to reestablish a dc level for the video signal after it has lost the dc component following amplification by ac-coupled stages.

In the discussion of clamping circuits that follows, input and output waveforms illustrating the phenomena will be shown. These waveforms are assumed to apply **after** the initial clamping process has been established. To provide some generality, different positive and negative peak values for the input waveform will be assumed. The positive peak value will be denoted as V_a, and the negative peak value will be denoted as $-V_b$.

Passive Positive Clamping Circuit

The first circuit to be considered is the classical passive version shown in the middle of Figure 9–17. On the first negative half-cycle of the input voltage, the diode becomes forward biased. The capacitor charges through the low-resistance series path to a voltage V_C, the approximate value of which is

$$V_C \simeq V_b - 0.7 \tag{9–37}$$

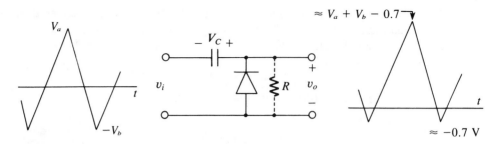

FIGURE 9–17
Passive clamping circuit and examples of steady-state input and output waveforms.

The positive terminal of the capacitor voltage will be on the right. For nearly a full cycle after the capacitor is charged, the diode is reverse biased, and no current flows. The output voltage $v_o(t)$ is the input voltage plus this capacitor voltage:

$$v_o(t) = v_i(t) + V_C \simeq v_i(t) + V_b - 0.7 \qquad \textbf{(9–38)}$$

The output voltage is thus "lifted" by a value equal to the magnitude of the negative peak voltage minus one diode forward drop. Conduction through the diode occurs for only a small portion of each subsequent cycle, and during this interval, any charge that may have been depleted through the external load is restored. The waveforms for $v_i(t)$ and $v_o(t)$ shown in Figure 9–17 display the steady-state pattern after the initial charging phase is established.

Observe that the circuit is "clamped" at a voltage level approximately equal to -0.7 V. If the diode were ideal, the clamp would be at zero. However, the nonzero diode drop results in a clamping level slightly below zero, even though the intent of the circuit is to clamp at zero. In spite of the error, this passive circuit is widely used in noncritical applications.

Active Positive Clamping Circuit

An active precision version of the clamping circuit is shown in Figure 9–18. On the first negative half-cycle of the input voltage, the differential input voltage is positive, and the op-amp output voltage in turn is positive. The diode becomes forward biased, and the

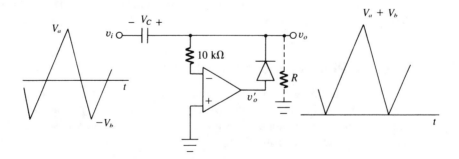

FIGURE 9–18
Active clamping circuit and examples of steady-state input and output waveforms.

capacitor is quickly charged to the value required to maintain virtual ground at the inverting terminal. The peak voltage V_C across the capacitor is reached at the negative peak point for the input and is

$$V_C = V_b \tag{9-39}$$

with the positive terminal on the right. The output voltage $v_o(t)$ after the clamp is established is

$$v_o(t) = v_i(t) + V_C = v_i(t) + V_b \tag{9-40}$$

The voltage added by the capacitor is just sufficient to "lift" the signal up to the point where the minimum value is zero. Once again, we see how the op-amp can be used to overcome the diode voltage drop through feedback.

For nearly a full cycle after the capacitor is charged, the differential input voltage becomes negative, the op-amp output drops to negative saturation and the diode becomes reverse biased. The feedback loop is broken during that time, and the only change in capacitor voltage is a result of discharge through R. Conduction will occur for a short portion of each cycle to restore the voltage across C to the correct value. During the point when conduction occurs and the capacitor is being recharged, the op-amp output assumes the value

$$v_o' \simeq 0.7 \text{ V} \tag{9-41}$$

Both the passive and the active circuits were intended to clamp the signal at a lower level of zero. Reversing the diode in either circuit will "push down" the signal below a desired upper clamping level of zero. Of course, the clamping level for the passive circuit is approximately 0.7 V above ground.

Changing the Clamping Level

We will next consider the process of clamping at a reference voltage level other than zero. While passive circuits can be devised for this purpose, we will restrict the consideration to the active version. An example of such a circuit is shown in Figure 9–19. The process of establishing the clamp is very similar to that encountered with the clamp at ground, but

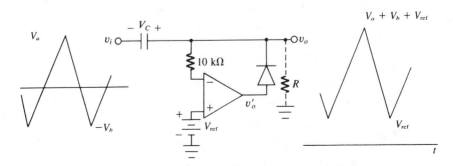

FIGURE 9-19
Active clamping circuit with nonzero reference, and examples of steady-state input and output waveforms.

there is one difference. The capacitor voltage now charges to a value such that $v^- = V_{ref}$. To achieve this, the capacitor charges to a value V_C such that

$$V_C = V_b + V_{ref} \qquad (9\text{--}42)$$

with the positive terminal on the right. The output voltage $v_o(t)$ after the clamp has been established is

$$v_o(t) = v_i(t) + V_C = v_i(t) + V_b + V_{ref} \qquad (9\text{--}43)$$

During the interval when the capacitor is being recharged, the op-amp voltage v_o' assumes the value

$$v_o' \simeq V_{ref} + 0.7 \qquad (9\text{--}44)$$

Clamping below the level of V_{ref} occurs when the diode is reversed. The sign of V_{ref} determines whether the clamp is positive or negative, but the *sense* of the clamp (that is, above or below) is a function of the diode direction.

In all of the clamping circuits considered, a discharge resistance R is shown. This resistance may represent an actual added resistance, or any output load, or a combination of the two. The time constant $\tau = RC$ determines the rate at which the voltage established across the capacitor will discharge. For some applications, it may be desirable to hold the clamp for very long periods of time, while in other cases, some means for adjusting to different levels must be provided. In particular, if the negative peak value decreases and no provision for discharge has been provided, the circuit will have difficulty reestablishing the proper clamping level. Some of the discussion in Section 9–3 concerning the time constant for peak detector circuits applies to clamping circuits.

As in previous nonlinear op-amp circuits, care must be taken to ensure that the differential input voltage magnitude does not exceed the specification for the particular op-amp. Specifically, the reader is invited to verify that the maximum value of the differential input voltage magnitude is equal to the peak-to-peak value of the input voltage, and this maximum occurs at the point in the cycle when the output voltage is at the value farthest removed from the clamping level.

EXAMPLE 9–5

Consider the clamping circuit of Figure 9–19 with $V_{ref} = 1.5$ V. Assume that the input voltage varies from -3 V to $+5$ V. Determine (a) the voltage across the capacitor V_C, (b) the peak value of the circuit output voltage, (c) the approximate op-amp output voltage during the charging interval, and (d) the maximum value of the op-amp differential input voltage.

Solution

(a) Using the notation established in Figure 9–19, we have $V_a = 5$ V and $V_b = 3$ V ($-V_b = -3$ V). From (9–42), the capacitor voltage is

$$V_C = 3 + 1.5 = 4.5 \text{ V} \qquad (9\text{--}45)$$

(b) The input signal is shifted upward by 4.5 V. The minimum level is $-3 + 4.5 = 1.5$ V as expected. The peak value $V_o(peak)$ of the circuit output is

$$V_o(peak) = 9.5 \text{ V} \qquad (9\text{--}46)$$

(c) The op-amp output voltage during the charging interval is

$$v'_o \simeq 1.5 + 0.7 = 2.2 \text{ V} \tag{9-47}$$

(d) The maximum differential input voltage magnitude occurs when $v^- = V_o(\text{peak}) = 9.5$ V. The differential voltage at this point is

$$v_d = v^+ - v^- = 1.5 - 9.5 = -8 \text{ V} \tag{9-48}$$

Thus, the maximum magnitude of the differential input voltage is the peak-to-peak value of the input signal voltage, as expected.

9-5 LIMITING CIRCUITS

Limiting circuits are used to confine the level of a signal to a specified range of operation. Most limiters generally have one or more linear ranges of operation in which the signal is unaltered in shape. However, if the signal level increases above or drops below a certain level, a nonlinear operation, designed to limit the signal level, comes into effect. In many cases, the output may simply be a constant for further increases or decreases in the input. In other cases, a more complex gain adjustment may take place.

Op-Amp/Zener Limiter

The first limiting circuit to be considered is shown in Figure 9–20(a). The back-to-back Zener diodes are assumed to be identical, and the Zener breakdown voltage of each is V_Z. However, when a given diode is operating in its Zener region, the other diode is forward biased, and the net voltage across the series combination is about $V_Z + 0.7$ V. It is assumed that $V_Z + 0.7 < V_{\text{sat}}$ for the op-amp circuit.

(a)   (b)

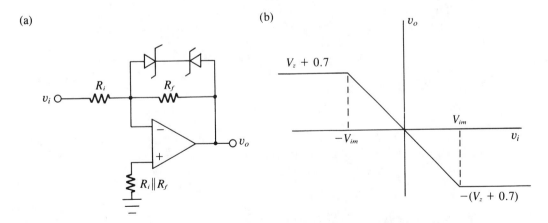

FIGURE 9–20
(a) Symmetrical limiting circuit used to limit peak value of output signal and (b) the input-output characteristic curve.

First, assume that the input signal level is sufficiently small so that reverse Zener breakdown does not occur at any point in the input cycle. In this case, one or the other of the two diodes will be reverse biased, and the series combination will act as an open circuit. Under this condition, the circuit will be functioning as an inverting amplifier with gain $-R_f/R_i$, and operation will be in the linear portion of the input-output characteristic shown in Figure 9–20(b).

Zener breakdown will occur when the output voltage reaches a level V_{om} approximately equal to

$$V_{om} = V_Z + 0.7 \qquad\qquad (9\text{–}49)$$

(Recall that the output voltage appears across R_f in the inverting amplifier.) The magnitude level of input voltage V_{im} at which breakdown occurs must satisfy the equation

$$|A_{CL}|V_{im} = \frac{R_f}{R_i}V_{im} = V_Z + 0.7 \qquad\qquad (9\text{–}50)$$

Solving for V_{im}, we obtain

$$V_{im} = \frac{V_{om}}{|A_{CL}|} = \frac{R_i}{R_f}(V_Z + 0.7) \qquad\qquad (9\text{–}51)$$

The quantity V_{im} represents the approximate level of the input voltage at which limiting starts to occur. Since the assumed forward-biased diode drop of 0.7 V is reached gradually, and this value is not precise, some gradual limiting begins to occur below the level of V_{im} predicted by (9–51). Thus, if no distortion of the signal is acceptable, some margin between the peak signal level and V_{im} should be provided. The level at which limiting occurs can be adjusted within limits by the selection of V_Z or the values of R_i and R_f.

Once breakdown occurs, the magnitude of the output voltage remains at approximately $V_Z + 0.7$ for further increases in the magnitude of the input voltage. The output magnitude is thus "limited" by this process.

If the peak value of the input signal continually increases beyond the limiting value, the resulting output signal is "clipped." For that reason, limiting circuits are also referred to as *clipping circuits* in some applications.

As is true with any design involving Zener diodes, some attention should be paid to the maximum diode ratings. Once Zener breakdown has been established, the diode must absorb any increase in input current, since the current through R_f will remain essentially constant for further increases in the input voltage. An analysis of the circuit under worst-case conditions should be performed.

It should be noted that when the output of a basic inverting or noninverting amplifier circuit is driven into the saturation region, the circuit automatically assumes the role of a limiting circuit, whether intended to or not. In this case, the limiting occurs as a natural part of the internal circuitry, which has only a finite dynamic range.

Limiting Below a Reference Level

A different form of limiting circuit is shown in Figure 9–21(a). For the circuit to operate exactly as described here, it must be assumed that negligible loading occurs at the output. For $v_i < V_{ref}$, the op-amp differential input voltage and the op-amp output voltage are

(a)

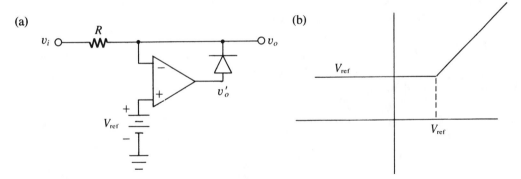

(b)

FIGURE 9–21
(a) Example of limiting circuit used to restrict linear operation above a reference level and (b) the input-output characteristic curve.

both positive. The diode will then be forward biased, and the op-amp will be acting as a voltage follower with an input V_{ref}, forcing the output to be $v_o = V_{ref}$.

When $v_i > V_{ref}$, the op-amp differential input voltage becomes negative. The op-amp output drops down to negative saturation, and the feedback loop is broken. The input v_i is then coupled directly to the output, so $v_o = v_i$ for this condition. The corresponding input-output characteristic (without loading) is shown in Figure 9–21(b). By reversing the diode and/or the reference polarity, one can achieve a number of possible limiting forms, some of which are considered in the problems at the end of the chapter.

EXAMPLE 9–6

Consider the circuit of Figure 9–22, and assume that the two diodes are identical and that they each have Zener voltages of 5.1 V. (a) Determine the output limiting voltage. (b) Determine the input limiting voltage. (c) For $v_i = 0.5$ V, determine v_o, i_i, i_R, i_Z, and i_o. (d) For $v_i = 15$ V, repeat the calculations of (c).

Solution
(a) The output limiting voltage V_{om} is

$$V_{om} = V_Z + 0.7 = 5.1 + 0.7 = 5.8 \text{ V} \qquad (9\text{–}52)$$

FIGURE 9–22
Circuit of Example 9–6.

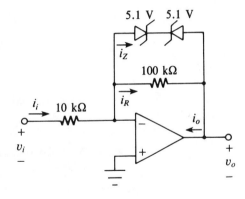

(b) The magnitude of the closed-loop gain is

$$|A_{CL}| = \frac{R_f}{R_i} = \frac{100 \text{ k}\Omega}{10 \text{ k}\Omega} = 10 \tag{9-53}$$

The input limiting voltage is

$$V_{im} = \frac{V_{om}}{|A_{CL}|} = \frac{5.8}{10} = 0.58 \text{ V} \tag{9-54}$$

(c) $v_i = 0.5$ V

Since v_i is less than the input limiting voltage, the diodes may be considered to have no significant effect, and the amplifier operates with a linear gain of -10. The output voltage is

$$v_o = -10 \times 0.5 = -5 \text{ V} \tag{9-55}$$

The input current is

$$i_i = \frac{v_i}{10^4} = \frac{0.5 \text{ V}}{10^4 \ \Omega} = 50 \ \mu\text{A} \tag{9-56}$$

All of this current flows through the 100-kΩ resistor, since the diodes are open circuits. Thus,

$$i_R = i_o = 50 \ \mu\text{A} \tag{9-57}$$

and

$$i_Z = 0 \tag{9-58}$$

(d) $v_i = 15$ V

The input voltage is considerably greater than the input limiting level and

$$v_o = -5.8 \text{ V} \tag{9-59}$$

The input current is

$$i_i = \frac{v_i}{10^4} = \frac{15 \text{ V}}{10^4 \ \Omega} = 1.5 \text{ mA} \tag{9-60}$$

The voltage across the 100-kΩ resistance must be 5.8 V with the positive terminal on the left. Thus, i_R must be

$$i_R = \frac{5.8 \text{ V}}{10^5} = 58 \ \mu\text{A} \tag{9-61}$$

The diodes must absorb the difference between i_i and i_R, that is,

$$i_Z = i_i - i_R = 1.5 \text{ mA} - 58 \ \mu\text{A} = 1.442 \text{ mA} \tag{9-62}$$

The currents i_R and i_Z come together on the right, and so i_o is

$$i_o = i_i = 1.5 \text{ mA} \tag{9-63}$$

9–6 POWER SUPPLY REVIEW

A major application of op-amps and linear integrated circuits is that of voltage regulation in power supplies. The most common type of power supply is one that converts the commercially available ac sine wave to a prescribed dc voltage.

Before we discuss voltage regulation, it is necessary to understand the general operation of a power supply. Most basic electronic texts devote one or more chapters to the detailed analysis and operation of power supplies. Since most readers will have previously studied from such a text, the treatment here will be abbreviated. The intent is to provide sufficient background for understanding the operation and application of IC regulators we will consider in Section 9–7.

Refer to the diagram of Figure 9–23 in the discussion that follows. The form of a typical power supply and the nature of the pertinent voltage waveforms are shown. The input is connected to the 120-V ac commercial power system. Both a switch and a fuse are on the transformer input side.

Transformer

The input side of the transformer is referred to as the *primary,* and the output side is called the *secondary*. The transformer is used to either increase ("step-up") or decrease ("step-down") the voltage level, depending on the requirements of the power supply. The voltage level is decreased in most IC circuit applications. Let N_1 represent the number of transformer primary turns, and let N_2 represent the corresponding number of secondary turns. A transformer is described by its *turns ratio,* which can be specified as either N_1/N_2 or as N_2/N_1.

Let v_1 and i_1 represent the instantaneous primary voltage and current, and let v_2 and i_2 represent the instantaneous secondary voltage and current. The following relationships apply to the transformer:

$$\frac{v_2}{v_1} = \frac{N_2}{N_1} \tag{9–64}$$

and

$$\frac{i_2}{i_1} = \frac{N_1}{N_2} \tag{9–65}$$

Although stated in terms of instantaneous variables, the preceding relationships apply equally well to *root-mean-square* (rms) quantities. In fact, most transformer voltage and current ratings are specified in rms units.

Rectifier

The next step in the ac-to-dc conversion process is that of *rectification,* which we considered earlier as it related to small signal applications. The rectifier in a power supply converts the sinusoidal voltage having both polarities to a voltage of one polarity, as illustrated by the middle waveform in Figure 9–23. Most power supply rectifiers use a *full-wave rectifier*. This voltage is referred to as *pulsating dc*.

The particular rectifier shown in Figure 9–23 consists of four diodes and is referred to as a *bridge* rectifier, which is one of the most common types of rectifier circuits.

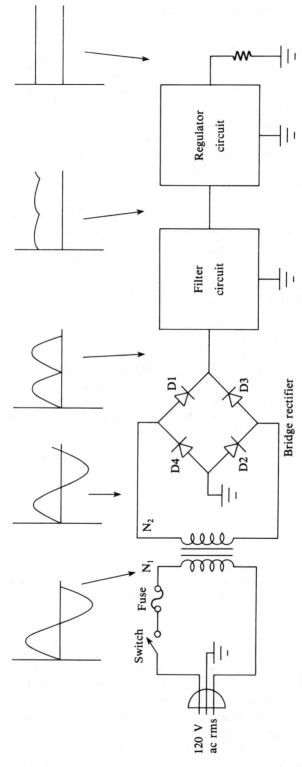

FIGURE 9–23
Diagram of a typical power supply using a bridge rectifier.

Bridge Rectifier

When the input voltage is positive on the top line with respect to the bottom line, $D1$ and $D2$ are forward biased, and $D3$ and $D4$ are reverse biased. Thus, the positive level of the top line is coupled to the output through $D1$, and the negative level of the bottom line is coupled through $D2$ to ground.

When the input voltage is positive on the bottom line with respect to the top line, $D3$ and $D4$ are forward biased, and $D1$ and $D2$ are reverse biased. The positive level of the bottom line is now coupled to the output through $D3$, and the negative level of the top line is coupled through $D4$ to ground. This latter step "flips over" the negative half-cycle, and the output is thus positive for a full cycle.

Note that for either half-cycle, there are *two* forward diode drops in the conduction paths. Further, at the current levels of most power supplies, each diode drop is somewhat greater than the value of 0.7 V assumed earlier for small signal values. Finally, all trans-formers exhibit losses. *As a result of these different factors, it is difficult to predict the exact value of the output voltage,* although reasonable estimates can be made.

Filter Circuit

The filter circuit provides a considerable amount of smoothing on the pulsating dc wave-form and removes much of the ripple. Before the advent of electronic regulation, the filter circuit provided all the smoothing for the output, and many earlier filter circuits consisted of several stages, usually with both inductance and capacitance. Circuits of this type are still used in certain specialized applications and are found in older equipment.

The development of electronic regulation, particularly the introduction of the pack-age regulator chip, had a major impact on the design of power supplies. Assuming that some minimum level of filtering is provided by the filter circuit, the regulator is capable of providing additional smoothing and removal of ripple. The regulator also compensates for variations in the input ac line voltage and maintains a nearly constant output voltage as input conditions change widely.

Single-Capacitor Filter

Because of the additional smoothing offered by the regulator, many power supply circuits used in consort with a regulator utilize the simple *single-capacitor filter,* whose form is shown in Figure 9–24. On the first half-cycle of the pulsating dc waveform, the capacitor charges to the peak value of the input voltage as shown in Figure 9–25. When the input

FIGURE 9–24
Single-capacitor filter.

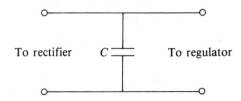

To rectifier C To regulator

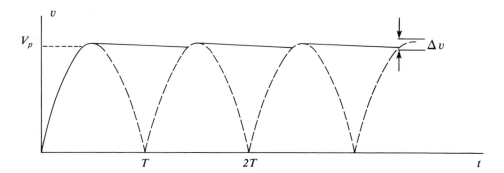

FIGURE 9–25
Voltage waveform resulting from a single-capacitor filter.

voltage decreases below this value, the two diodes in the bridge rectifier that would be forward biased into a resistive load *now become reverse biased* because of the larger capacitor voltage. If there were no load and no loss in the capacitor, the voltage would theoretically remain at the peak value. However, the presence of a load results in a slow decrease of the voltage as shown (assuming that the capacitance is sufficiently large).

At some point just before the next peak, the input voltage to the rectifier exceeds the capacitor voltage, and two diodes in the bridge rectifier become forward biased again. The capacitor then quickly charges back up to the peak value. Because of the short charging time, the peak charging current can be rather large and is a factor to consider in selecting rectifier diodes.

Ripple Voltage

An estimate for the ripple voltage will now be developed. Refer to Figure 9–25 and assume that the change in voltage Δv is small in comparison to the total voltage level. Assume also that the load current I_L remains constant. Strictly speaking, this assumption is valid only if a regulator is used, but it is a reasonable approximation even without the regulator. The voltage change in a time Δt can be shown from basic circuit theory to be

$$\Delta v = \frac{1}{C} I_L \Delta t \approx \frac{I_L T}{C} \qquad (9\text{–}66)$$

where it has been assumed that $\Delta t \approx T$ in Figure 9–25.

Let f represent the frequency of the rectified voltage. For 60-Hz voltage and a *full-wave* rectifier, the frequency is $f = 120$ Hz since there are *two* rectified pulses per cycle of the input voltage. Since the period T of the pulses is $T = 1/f$, (9–66) becomes

$$\Delta v = \frac{I_L}{fC} = \frac{I_L}{120C} \qquad (9\text{–}67)$$

for 60-Hz commercial ac and a *full-wave rectifier*.

Equation (9–67) provides an estimate of the *peak-to-peak* variation of the ripple on the voltage. Note that it is directly proportional to the load current and inversely proportional to the capacitance.

EXAMPLE 9–7

A certain transformer is specified from primary to secondary as 120 V/24 V, with rms units implied. Determine the (a) turns ratio, (b) primary rms current when secondary rms current is 2 A, and (c) peak value of secondary voltage.

Solution

(a) From the voltages specified, the turns ratio may be expressed as

$$\frac{N_1}{N_2} = \frac{120}{24} = 5 \qquad (9\text{–}68)$$

(b) Let $I_{1(\text{rms})}$ and $I_{2(\text{rms})}$ represent the primary and secondary rms currents, respectively. From the basic transformer equations,

$$\frac{I_{1(\text{rms})}}{I_{2(\text{rms})}} = \frac{N_2}{N_1} = \frac{1}{5} \qquad (9\text{–}69)$$

Solving for $I_{1(\text{rms})}$, we have

$$I_{1(\text{rms})} = \frac{1}{5} \times I_{2(\text{rms})} = \frac{1}{5} \times 2 = 0.4 \text{ A} \qquad (9\text{–}70)$$

The primary side has the highest voltage and the lowest current.

(c) Let $V_{2(\text{rms})}$ represent the rms value of the secondary, and let V_{2p} represent the peak value. For a sinusoidal voltage, the relationship between them is

$$V_{2p} = \sqrt{2}V_{2(\text{rms})} \qquad (9\text{–}71)$$

Substitution of $V_{2(\text{rms})} = 24$ V in (9–71) yields

$$V_{2p} = 33.94 \text{ V} \approx 34 \text{ V} \qquad (9\text{–}72)$$

If this transformer were used with a bridge rectifier circuit, the actual peak voltage would be somewhat less than the value given by (9–72) because of the diode drops and possible transformer losses. At small current levels, the peak output voltage could be estimated as about $34 - 2 \times 0.7 = 32.6$ V. At large current levels, the peak voltage would be expected to be somewhat lower.

EXAMPLE 9–8

A certain 60-Hz full-wave rectifier circuit employs a single-capacitor filter. The dc load current to the regulator is 0.4 A, and a 1000-μF capacitor is used. Estimate the peak-to-peak ripple voltage.

Solution

From (9–67), the ripple voltage is estimated as

$$\Delta v = \frac{I_L}{120C} = \frac{0.4}{120 \times 1000 \times 10^{-6}} = 3.33 \text{ V} \tag{9–73}$$

Note that this is a considerable level of ripple and would be intolerable for most applications if the voltage at this point were connected directly to the load. A regulator is assumed to be employed here, however, and it will provide the additional smoothing required.

9–7 VOLTAGE REGULATORS

A voltage regulator circuit is a circuit that maintains a constant voltage across a load even though the load current requirement may change. Most applications of voltage regulators are in power supply circuits where the overall current and power requirements are typically much higher than those encountered in conventional signal-processing voltage and current amplifier stages. Consequently, special design and construction techniques are required to ensure that all circuit components are operated within their rated specifications. The question of heat transfer must be considered, and heat sinks must usually be provided.

Basic Concept

We will develop the concept of the voltage regulator through several steps by first considering a very low-power circuit. Assume that a certain voltage reference of value V_{ref} is available (for example, a battery). Assume that a certain load voltage V_L is to be established, and let $V_L \geq V_{\text{ref}}$. It is desired that no current actually be drawn from the reference. As long as the load current is quite small (for example, a few milliamperes), the noninverting amplifier of Figure 9–26 may be used for this purpose. The reference voltage is applied to the noninverting input, at which point the input impedance is very high,

FIGURE 9–26
Low-power circuit used to illustrate concept of voltage regulation.

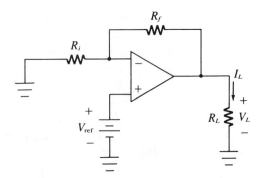

so the current drawn from the reference is nearly zero. Resistors R_i and R_f are selected such that

$$V_L = \left(1 + \frac{R_f}{R_i}\right) V_{\text{ref}} \tag{9-74}$$

For all practical purposes, the voltage V_L is independent of the load current as long as the maximum op-amp output current is not exceeded. Thus, a constant output voltage has been established. The load current is furnished by the op-amp, and the reference voltage, which is used to establish the load voltage, is actually isolated from the load. Observe that if the load and reference voltages were to be equal, the voltage follower form of the noninverting amplifier would be used.

Pass Transistor

While this circuit illustrates the concept of voltage regulation, it is limited in its basic form to a very small level of power. However, the power level may be increased significantly by the use of a *pass transistor,* provided that an unregulated source of power is also available. For reasons that will be clearer later, the magnitude of the voltage level of the unregulated voltage must be somewhat larger at all times (typically at least two volts higher) than the desired regulated voltage. Further, the unregulated source must be capable of supplying the load power required (plus some extra power). The pass transistor is a power transistor having sufficient power ratings as well as an adequate heat sink configuration.

The circuit with the addition of the pass transistor and the unregulated voltage supply are shown in Figure 9–27. A positive unregulated voltage V_u and a positive regulated load voltage V_L have been assumed, so an NPN pass transistor and a positive reference voltage are used. For negative voltages, a PNP transistor and a negative reference voltage may be used.

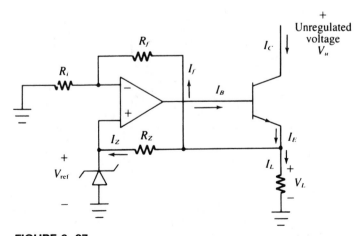

FIGURE 9-27
Addition of pass transistor and unregulated voltage to basic regulator circuit.

Observe from Figure 9–27 that the base-emitter voltage drop of the pass transistor is part of the forward block of the closed-loop feedback system, and its effect is virtually eliminated as far as the load voltage is concerned. In other words, negative feedback forces the op-amp output voltage to be sufficiently large to overcome the base-emitter voltage drop so that the constraint of (9–74) is satisfied.

The base-emitter voltage drop of a power transistor is typically larger than the small signal assumed value of 0.7 V. In addition, the collector-base junction must be reverse biased for linear operation, and thus the collector voltage must be more positive than the base for an NPN transistor. These two properties both contribute to the requirement that the unregulated voltage must be somewhat larger in magnitude than the regulated voltage.

Reference Voltage

The next step in the evolution of the regulator circuit is the use of the output voltage to generate its own reference. This is achieved by means of a Zener diode as shown in Figure 9–28. Assume that the desired Zener current is I_Z. It is readily shown that the resistance R_Z is given by

$$R_Z = \frac{V_L - V_{ref}}{I_Z} \tag{9–75}$$

The op-amp must supply the base current I_B of the transistor. Let β_{dc} represent the dc current gain of the transistor:

$$\beta_{dc} = \frac{I_C}{I_B} \tag{9–76}$$

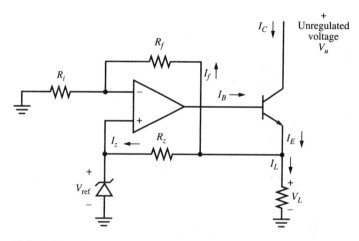

FIGURE 9–28
Addition of Zener voltage reference to basic regulator circuit.

For a given I_C and β_{dc}, the maximum op-amp output current I_B can be determined from (9–76) if I_C is known. Since the emitter current I_E must satisfy $I_E = I_B + I_C$, we have

$$I_E = (1 + \beta_{\text{dc}})I_B \tag{9–77}$$

The actual emitter current I_E consists of the load current I_L plus the Zener current I_Z and the current I_f through the feedback resistor. These last two currents are usually quite small compared with the load current, so that $I_E \approx I_L$. A further approximation with (9–77) results in the following simplified estimate for I_B:

$$I_B \approx \frac{I_L}{\beta_{\text{dc}}} \tag{9–78}$$

The op-amp must be capable of delivering the base current of (9–78) corresponding to the maximum load current and the minimum estimate of β_{dc}. Note that the smaller the value of β_{dc}, the larger the required base current drive from the op-amp.

The unregulated dc voltage is obtained from a conventional rectifier circuit, which typically contains a transformer, rectifier diodes, and one or more filter capacitors, as discussed in Section 9–6. A sufficient amount of filtering of the unregulated supply is required to ensure that the magnitude of the unregulated voltage remains somewhat higher than the regulated voltage at all times. This condition is required to ensure that the pass transistor has the proper bias polarities so that regulation can be maintained.

Power Relationships

The dc power P_i delivered to the regulator circuit by the unregulated power supply is approximately equal to

$$P_i \approx V_u I_L \tag{9–79}$$

where the difference between I_L and I_C and variations in V_u have been ignored in (9–79). The load power P_L is given by

$$P_L = V_L I_L \tag{9–80}$$

The pass transistor must dissipate the power difference between P_i and P_L. Letting P_D represent this dissipation, we have

$$P_D = P_i - P_L = V_u I_L - V_L I_L = (V_u - V_L)I_L \tag{9–81}$$

If V_u is subject to variation, the worst-case dissipation corresponding to the maximum value of V_u and the maximum value of I_L should be used to estimate the maximum heat dissipation. A power transistor capable of dissipating this level of power must be selected, and a suitable heat sink must be provided. Since the value of β_{dc} for a power transistor is typically less than that of small signal transistors, the requirement of (9–78) often necessitates a Darlington compound pass transistor structure where a large load current is required.

Observe from (9–80) that the required pass transistor power dissipation increases as the difference between the unregulated voltage and the regulated voltage increases. The approximate efficiency of the conversion can be expressed as

$$\text{Efficiency} \approx \frac{P_L}{P_i} \times 100\% = \frac{V_L}{V_u} \times 100\% \tag{9–82}$$

Regulator Chips

The preceding development has explained the basic concept of the regulator through a step-by-step evolution. In practice, there is very little incentive to implement a regulator in this fashion. Instead, numerous regulator integrated-circuit chips are available. These chips contain the control op-amp, one or more pass transistors, the voltage reference, and, in some cases, the resistive feedback network. The circuit is designed so that the control op-amp can be powered even from the unregulated power source.

Some of the regulator chips are designed to provide specific dc voltages, and the feedback resistive network is already contained on the chip. Typical of the chips in this category are the LM7800C series regulators, which include the 7805 for 5 V and the 7815 for 15 V. These chips have terminals for the input unregulated voltage, the output regulated voltage, and a common ground connection. The basic wiring diagram is shown in Figure 9–29.

Some regulator chips may be "programmed" for a given load voltage, and means are provided to connect all or a portion of the external feedback network. Typical of such units are the LM117/LM317 regulators whose general features will be discussed in Seciton 9–8.

One feature not considered in the op-amp regulator development was a current-limiting circuit, which ensures that the current rating of the components is not exceeded. However, virtually all regulator chips contain such a feature.

Switching Regulators

A somewhat different approach to voltage control can be achieved with *switching regulators*. The use of switching regulators is increasing rapidly, and many package units are now available. The primary advantage of switching regulators is their high efficiency compared to other types of regulators.

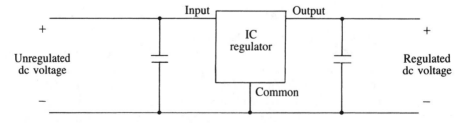

FIGURE 9–29
Basic wiring diagram of fixed-voltage IC regulator (for positive voltage).

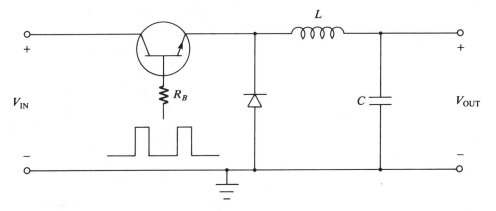

FIGURE 9–30
Concept of changing a dc voltage by modulating with a pulse train.

Buck Switching Circuit

As a prelude to discussion of the switching regulator, consider the circuit of Figure 9–30, referred to as a *buck circuit*. The input voltage V_{in} is assumed to be a dc voltage. The pulse train applied to the base of the transistor alternately switches the transistor between cutoff and saturation. Thus, the transistor is acting like a switch. The diode is reverse biased while the transistor is on, but it becomes forward biased when the transistor is off and thus provides a continuous path for inductive current to flow.

The *LC* filter acts to smooth out the input variations and in effect provide a dc output voltage. Let *D* represent the duty cycle of the pulse train, that is, the ratio of the time that the transistor is saturated to the total period. Neglecting any transistor and circuit losses, the dc output voltage V_{out} is

$$V_{out} = DV_{in} \tag{9–83}$$

This circuit can be quite efficient because a transistor dissipates very little power in either the saturation state (collector-emitter voltage is nearly zero) or the cutoff state (collector current is zero). In effect, the inductance and capacitance store and later release energy to the circuit, but most of the power reaches the load.

A typical switching frequency is chosen to be of the order of several kilohertz. Such a frequency is relatively high for power supply circuits, but it has the significant advantage of requiring only moderate values of inductance and capacitance. This is in sharp contrast to the large values of inductance and capacitance required in traditional power supply filters.

Regulating the Voltage

The buck circuit discussed thus far is capable of providing a dc output voltage less than the input voltage with high efficiency, but it has no regulation. Regulation is provided by feedback, and a block diagram illustrating the concept is shown in Figure 9–31.

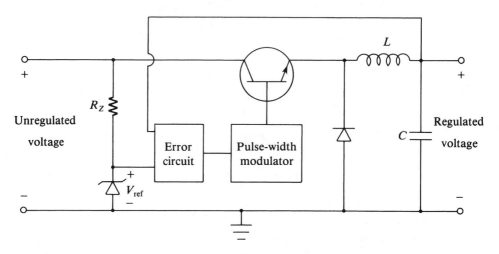

FIGURE 9–31
Basic concept of a switching regulator.

A reference voltage V_{ref} is developed with a Zener diode from the unregulated voltage. This voltage is continually compared to the output regulated voltage through an error circuit. The output of the error circuit is used to control the width of the pulses produced by a pulse-width modulator. (The pulses are being generated at a fixed frequency, but their widths are dependent on a control voltage.)

If the dc output voltage should decrease below the reference voltage, the sense of the error signal causes the widths of the pulses to increase, thus increasing the output voltage. If the output voltage should increase above the reference voltage, however, the error signal causes a reduction in the pulse widths, which acts to reduce the output voltage.

EXAMPLE 9–9

Consider an op-amp regulator circuit of the form shown in Figure 9–28 in which the desired regulated load voltage is to be $V_L = 12$ V. An unregulated voltage whose value is approximately 16 V is available. The Zener diode is a 5.1-V unit, and the desired Zener bias current is 20 mA.

(a) Determine the required value of R_Z. (b) Determine suitable values for R_i and R_f. (c) If the maximum value of load current I_L is 1 A and the β_{dc} range for the pass transistor is between 50 and 100, determine the maximum op-amp output current required for base drive I_B. (d) Determine the load power, the approximate input power, and the approximate efficiency under maximum load conditions. (e) Determine the power dissipation rating P_D for the pass transistor.

Solution

(a) The resistance R_Z must drop the difference between the desired load voltage V_L and the reference voltage V_{ref}. Since the Zener current is to be 20 mA, we have

$$R_Z = \frac{V_L - V_{\text{ref}}}{I_Z} = \frac{12 - 5.1}{20 \times 10^{-3}} = 345 \ \Omega \tag{9–84}$$

(b) The noninverting gain of the amplifier must produce a 12-V output with a 5.1-V input. Thus,

$$12 = \left(1 + \frac{R_f}{R_i}\right)5.1 \qquad (9\text{–}85)$$

Solving for the ratio R_f/R_i, we obtain

$$\frac{R_f}{R_i} = 1.353 \qquad (9\text{–}86)$$

There are an infinite number of combinations of R_f and R_i that have this ratio. Reasonable choices would be

$$R_i = 10 \text{ k}\Omega \qquad (9\text{–}87)$$
$$R_f = 13.53 \text{ k}\Omega \qquad (9\text{–}88)$$

(c) The calculation for maximum op-amp output current will be based on the minimum value of $\beta_{dc} = 50$, since it is this condition that requires maximum base current. First, an "exact" solution will be made. The total emitter current can be represented as

$$I_E = I_L + I_Z + I_f \qquad (9\text{–}89)$$

where I_L is the load current, I_Z is the Zener current, and I_f is the current through the feedback network. This latter current is

$$I_f = \frac{V_L}{R_i + R_f} = \frac{12}{10 \text{ k}\Omega + 13.53 \text{ k}\Omega} = 0.00051 \text{ A} \qquad (9\text{–}90)$$

Substituting $I_Z = 20$ mA, I_f from (9–90), and the maximum load current of 1 A in (9–89), we have

$$I_E = 1 + 0.02 + 0.00051 = 1.0205 \text{ A} \qquad (9\text{–}91)$$

The base current I_B is

$$I_B = \frac{I_E}{1 + \beta_{dc}} = \frac{1.0205}{1 + 50} \simeq 20 \text{ mA} \qquad (9\text{–}92)$$

If the simple estimate of (9–78) is used, the result is

$$I_B \simeq \frac{I_L}{\beta_{dc}} = \frac{1}{50} = 20 \text{ mA} \qquad (9\text{–}93)$$

which produces a virtually identical result in this case. Typically, the two results are different but quite close to each other.

(d) The maximum load power P_L is

$$P_L = V_L I_L = 12 \times 1 = 12 \text{ W} \qquad (9\text{–}94)$$

The approximate input power P_i is

$$P_i \simeq V_u I_L = 16 \times 1 = 16 \text{ W} \qquad (9\text{–}95)$$

The approximate efficiency is

$$\text{Efficiency} \simeq \frac{12 \text{ W}}{16 \text{ W}} \times 100\% = 75\% \qquad \textbf{(9–96)}$$

(e) The pass transistor must have a dissipation rating of

$$P_D = P_i - P_L = 16 - 12 = 4 \text{ W} \qquad \textbf{(9–97)}$$

9–8 LM117/LM317 REGULATORS

As discussed in Section 9–7, a variety of IC regulator chips are available, some of which are preset for specific voltages, while others may be set by the circuit designer to an arbitrary voltage. Among the most popular of the latter type are the LM117/LM317 regulators. There are several variations of these types with different suffixes on the numbers and with different current ratings. Data sheets on the basic types without suffixes are provided in Appendix C. Hereafter this family will be referred to simply as the "117."

The basic 117 is capable of supplying more than 1.5 A in a voltage range adjustable from 1.2 V to 37 V. (Again, some variations have different current ratings.) Only two external resistors are required to set the voltage. The 117 is a three-terminal device containing an *input* terminal, an *output* terminal, and an *adjustment* terminal.

Circuit Connection

The basic circuit connection for the 117 is shown in Figure 9–32. The minimum value of the unregulated voltage should be about 2 V or so higher than the desired regulated voltage. The two capacitors may or may not be required. Capacitor C_1 is recommended by the manufacturer if the regulator is located more than 6 in. from the filter output capacitor. A

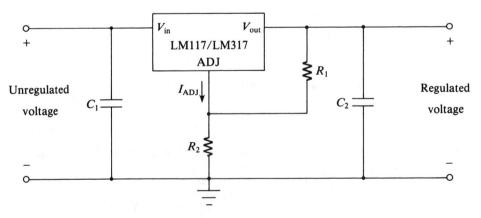

FIGURE 9–32
Circuit connection for LM117/LM317 regulators.

typical value is 0.1 μF. Capacitor C_2 can be used to minimize transients, and a typical value is 1 μF.

Setting the Voltage

The 117 contains an internal reference voltage for which a typical value is $V_{ref} = 1.2$ V. Through internal action within the regulator, this voltage appears across R_1 and results in a current flow I_1 through R_1 of value

$$I_1 = \frac{V_{ref}}{R_1} \tag{9–98}$$

This current also flows through R_2. However, an additional small current I_{ADJ} flows out of the adjustment terminal. The net current I_2 through R_2 is thus

$$I_2 = I_1 + I_{ADJ} \tag{9–99}$$

The voltage V_2 across R_2 is

$$V_2 = R_2 I_2 = R_2(I_1 + I_{ADJ}) \tag{9–100}$$

The net output voltage V_{out} is

$$V_{out} = V_{ref} + V_2 \tag{9–101}$$

Substitution of (9–98) in (9–100) and substitution of the latter in (9–101) yields

$$V_{out} = V_{ref}\left(1 + \frac{R_2}{R_1}\right) + I_{ADJ} R_2 \tag{9–102}$$

The last term represents an error term since I_{ADJ} is subject to some variation and uncertainty. However, a typical value is of the order of 50 μA, and if the resistance level of the circuit is reasonably low, the size of the uncertainty can be kept small. In practice, the output voltage can be set as accurately as necessary by adjustment of either R_1 or R_2 to compensate for uncertainties in V_{ref} and I_{ADJ}.

EXAMPLE 9–10

A 117 regulator of the form shown in Figure 9–32 has $R_1 = 240$ Ω and $R_2 = 1$ kΩ. If $I_{ADJ} = 50$ μA, and $V_{ref} = 1.25$ V, determine the value of V_{out}.

Solution
From (9–102), the voltage V_{out} is

$$V_{out} = V_{ref}\left(1 + \frac{R_2}{R_1}\right) + I_{ADJ} R_2 = 1.25\left(1 + \frac{1000}{240}\right) + 50 \times 10^{-6} \times 10^3$$

$$= 6.458 + 0.05 = 6.508 \text{ V} \tag{9–103}$$

Note that the error term due to I_{ADJ} is only 0.05 V = 50 mV.

EXAMPLE 9–11

A 117 regulator circuit is to be designed to establish a regulated load voltage of 12 V from an unregulated supply of about 15 V. (a) Perform an initial design by determining possible values of R_1 and R_2. (b) If the maximum value of I_{ADJ} is 100 μA, estimate the maximum possible error due to that current.

Solution

(a) The values $V_{out} = 12$ V and $V_{ref} = 1.25$ V are substituted in (9–102), and I_{ADJ} is momentarily ignored. This operation results in

$$12 = 1.25\left(1 + \frac{R_2}{R_1}\right) \tag{9-104}$$

Solving for R_2/R_1, we obtain

$$R_2 = 8.6R_1 \tag{9-105}$$

As somewhat arbitrary choices, the values $R_1 = 200$ Ω and $R_2 = 1720$ Ω will be made. However, since I_{ADJ} has been neglected thus far, it is likely that either R_2 or R_1 must be adjusted to compensate.

(b) Based on $I_{ADJ} = 100$ μA, and $R_2 = 1720$ Ω, the error term would be $100 \times 10^{-6} \times 1720 = 0.172$ V $= 172$ mV.

9–9 MULTISIM EXAMPLES

The examples that follow illustrate how the diode, transistor, and op-amp models in the Multisim library can be used to create realistic simulations of some of the circuits in this chapter.

MULTISIM EXAMPLE 9–1

Develop a Multisim circuit for simulating the simple passive half-wave rectifier circuit of Figure 9–1 with $R = 10$ kΩ, and obtain an input-output characteristic curve.

Solution

The circuit diagram as constructed with Multisim is shown in Figure 9–33. The diode employed is the **1N4148** model, which is typical of general-purpose diode types.

FIGURE 9–33

Circuit of Multisim Example 9–1.

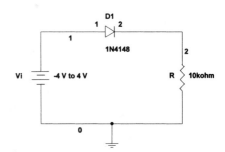

9-1

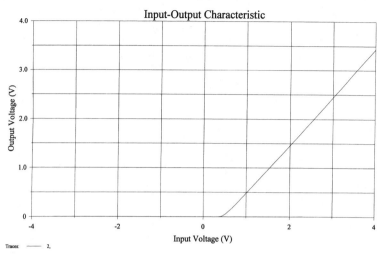

FIGURE 9–34
Input-output characteristic of Multisim Example 9–1.

A **DC Sweep** was employed based on the input voltage varying from −4 V to 4 V, and the results are shown in Figure 9–34. The labeling has been changed somewhat to conform with the text. Note that no output occurs until the input voltage reaches about 0.5 V, which is a little smaller than the our "assumed standard" of 0.7 V. However, at the peak input voltage of 4 V, the output is about 3.4 V, which implies a diode drop of 0.6 V at that level of conduction. This illustrates, of course, the actual nonlinear nature of the diode forward voltage drop, which varies with the current.

MULTISIM EXAMPLE 9–2

Using the 741 model, develop a Multisim circuit for the active nonsaturating precision half-wave rectifier of Example 9–1, and obtain an input-output characteristic curve.

Solution
The form of the original circuit was shown in Figure 9–4(a), and the Multisim schematic is shown in Figure 9–35. The "vertically flipped" version of the op-amp with the inverting input at the top is employed. The two diodes employed are **1N4148** model types.

A **DC Sweep** was employed based on the input voltage varying from −5 V to 5 V, and the results are shown in Figure 9–36. Note the nearly perfect linearity down to 0 V. Since the gain in the active mode is −2, the output voltage is essentially 10 V when the input voltage is −5 V.

MULTISIM EXAMPLE 9–3

Using the 741 model, develop a Multisim simulation circuit for the the op-amp regulator circuit of Example 9–9, and perform a transient analysis at two values of unregulated voltage: (a) 13 V and (b) 20 V.

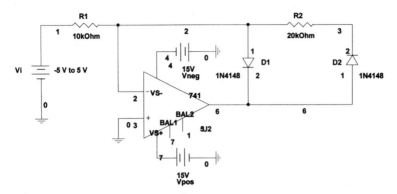

FIGURE 9–35
Circuit of Multisim Example 9–2.

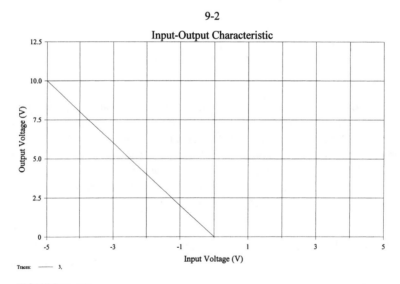

FIGURE 9–36
Input-output characteristic of Multisim Example 9–2.

Solution

The basic form of the circuit was shown in Figure 9–28, and the Multisim schematic is shown in Figure 9–37. A 2N2222A transistor is employed as the pass transistor, and a 1N751 zener diode is used for the reference. The voltage specified in the Multisim library for that diode is 5.125 V, but the resistance values computed in Example 9–9 will be employed without modification.

(a) A **Transient Analysis** was performed with the unregulated voltage set to 13 V and the regulator output voltage is shown in Figure 9–38. The two cursors at different times both show 12.0034 V.

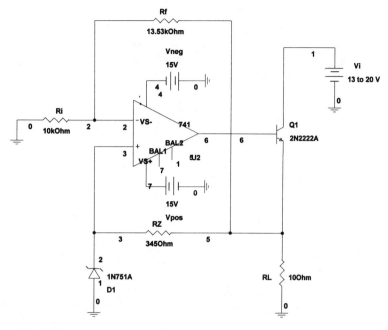

FIGURE 9–37
Circuit of Multisim Example 9–3.

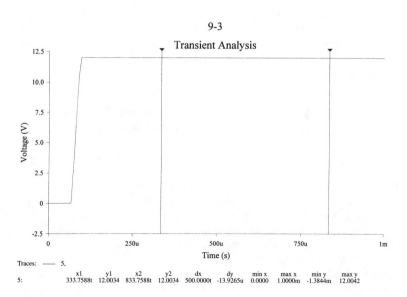

FIGURE 9–38
Transient response of Multisim Example 9–3 with unregulated voltage of 13 V.

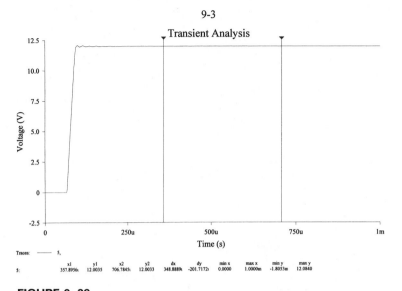

FIGURE 9–39

Transient response of Multisim Example 9–3 with unregulated voltage of 20 V.

(b) A second **Transient Analysis** with the unregulated voltage set to 20 V is shown in Figure 9–39. The two cursors show 12.0035 V and 12.0033 V, respectively. The circuit is obviously accomplishing its purpose. The slightly higher voltage than the design goal of 12 V is probably primarily due to the zener voltage being slightly greater than 5.1 V. Reducing R_f slightly would serve to compensate if desired.

SUMMARY

1. Precision signal-processing rectifiers may be implemented with op-amps and diodes.
2. The primary advantage of precision rectifier circuits is that errors due to the 0.7-V diode drops are virtually eliminated.
3. An ideal peak detector circuit is one that produces an output voltage equal to either the positive or negative peak value of the input voltage.
4. An ideal envelope detector circuit is one that produces an output voltage that follows exactly the envelope of the input while suppressing the high-frequency carrier variations.
5. An ideal sample-and-hold circuit produces an output voltage exactly equal to the value of the input voltage at a sample point and holds that value to the next sample point.
6. Precision peak detectors, envelope detectors, and sample-and-hold circuits may be implemented with op-amps, diodes, and a capacitor.
7. An ideal clamping circuit is one that shifts the signal level either up or down according to the design objectives.
8. Clamping circuits may be implemented with op-amps, diodes, and capacitors.
9. An ideal limiter is a circuit that allows the output voltage to assume only a limited range.

10. Limiters may be implemented with op-amps and diodes.
11. A modern power supply contains a transformer, a power rectifier, a filter, a regulator, and associated input components, that is, switch, fuse, etc.
12. Diode drops in a power supply circuit are typically greater than the 0.7-V level associated with small signal applications.
13. Because of diode drops, transformer losses, and power line variations, it is difficult to predict the exact voltage level at the output of the rectifier in a power supply.
14. With respect to (13), however, as long as the minimum value of the rectified voltage exceeds a certain value, the regulator can establish a fixed load voltage.
15. The basic concept of the regulator is based on comparing a desired voltage with the actual load voltage and providing sufficient power flow control to maintain equality.
16. Many types of IC regulator chips are available, some of which provide fixed voltages, while others can be "programmed."
17. Switching regulators alternately utilize the saturation and cutoff regions of a transistor, thus providing very high efficiency.

PROBLEMS

Drill Problems
Note: Unless indicated otherwise, assume $\pm V_{sat} = \pm 13$ V for all general-purpose op-amps.

9–1. Consider the circuit of Figure P9–1 with $R_i = 10$ kΩ and $R_f = 30$ kΩ. Determine v_o, v^-, and v_o' for each of the following values of input voltage:
    a. $v_i = 4$ V   **b.** $v_i = -4$ V.

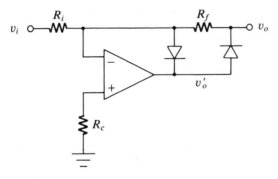

FIGURE P9–1

9–2. Consider the circuit of Figure P9–1 with $R_i = 10$ kΩ and $R_f = 100$ kΩ. Determine v_o, v^-, and v_o' for each of the following values of input voltage:
    a. $v_i = 200$ mV   **b.** $v_i = -200$ mV.
9–3. Consider the circuit of Figure P9–3 with $R_i = 10$ kΩ and $R_f = 20$ kΩ. Determine v_o, v^-, and v_o' for each of the following values of input voltage:
    a. $v_i = 5$ V   **b.** $v_i = -5$ V.

FIGURE P9-3

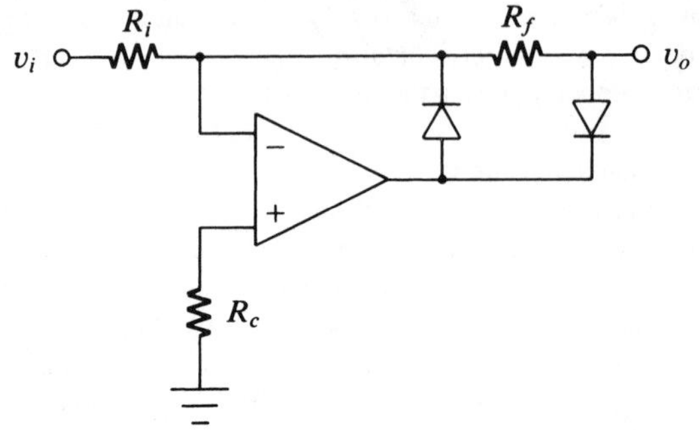

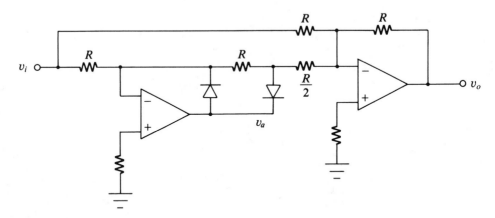

9-4. Consider the circuit of Figure P9-3 with $R_i = 10 \text{ k}\Omega$ and $R_f = 100 \text{ k}\Omega$. Determine v_o, v^-, and v'_o for each of the following values of input voltage:

    a. $v_i = 200 \text{ mV}$ **b.** $v_i = -200 \text{ mV}$.

9-5. Consider the circuit of Figure P9-5 with $R = 10 \text{ k}\Omega$. Determine v_o and v_a for each of the following values of input voltage:

    a. $v_i = 6 \text{ V}$ **b.** $v_i = -6 \text{ V}$.

FIGURE P9-5

9-6. Repeat the analysis of Problem 9-5 if the two diodes are reversed.

9-7. The waveform of Figure P9-7 has $V_p = 4 \text{ V}$, and it is applied as the input to the circuit of Problem 9-1. Sketch the output voltage.

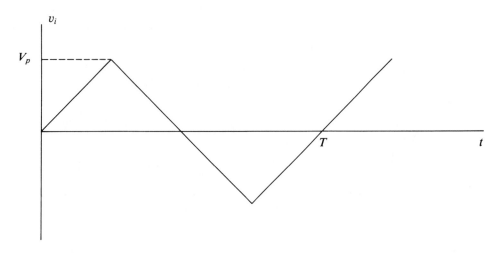

FIGURE P9–7

9–8. The waveform of Figure P9–7 has $V_p = 200$ mV, and it is applied as the input to the circuit of Problem 9–2. Sketch the output voltage.

9–9. The waveform of Figure P9–7 has $V_p = 5$ V, and it is applied as the input to the circuit of Problem 9–3. Sketch the output voltage.

9–10. The waveform of Figure P9–7 has $V_p = 200$ mV, and it is applied as the input to the circuit of Problem 9–4. Sketch the output voltage.

9–11. The waveform of Figure P9–7 has $V_p = 6$ V, and it is applied as the input to the circuit of Problem 9–5. Sketch the output voltage.

9–12. The waveform of Figure P9–7 has $V_p = 6$ V, and it is applied as the input to the circuit of Problem 9–6. Sketch the output voltage.

9–13. Consider the circuit of Figure P9–13. Assume that the input is a signal that varies from -5 V to $+5$ V.

    a. For $v_i = 5$ V, determine v_o when the capacitor is fully charged.

    b. For $v_i = -5$ V, determine v_o after the peak of (a) has occurred if all loading is neglected.

FIGURE P9–13

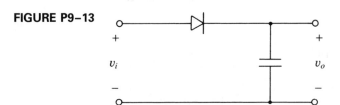

9–14. Consider the circuit of Figure P9–14. Assume that the input is a signal that varies from -5 V to $+5$ V.

    a. For $v_i = -5$ V, determine v_o when the capacitor is fully charged.

    b. For $v_i = 5$ V, determine v_o after the peak of (a) has occurred if all loading is neglected.

FIGURE P9–14

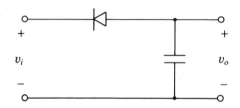

9–15. Consider the circuit of Figure P9–15. Assume that the input is a signal that varies from -5 V to $+5$ V.

a. For $v_i = 5$ V, determine v_o and v_o' when the capacitor is fully charged.

b. For $v_i = -5$ V, determine v_o and v_o' after the peak of (a) has occurred if $R = \infty$.

FIGURE P9–15

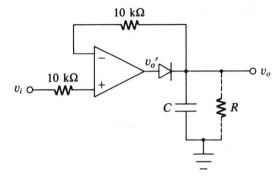

9–16. Consider the circuit of Figure P9–16. Assume that the input is a signal that varies from -5 V to $+5$ V.

a. For $v_i = 5$ V, determine v_o and v_o' when the capacitor is fully charged.

b. For $v_i = -5$ V, determine v_o and v_o' after the peak of (a) has occurred if $R = \infty$.

FIGURE P9–16

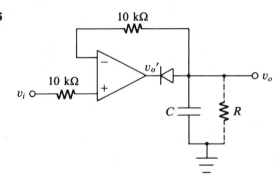

9-17. Assume in the circuit of Problem 9–15 that $C = 1\ \mu F$, $R = 100\ k\Omega$, and the frequency of the input signal is 1 kHz. Determine the approximate value to which the output voltage has discharged when the peak of the next cycle occurs.

9-18. Assume in the circuit of Problem 9–16 that $C = 2\ \mu F$, $R = 200\ k\Omega$, and the frequency of the input signal is 500 Hz. Determine the approximate value to which the output voltage has discharged when the peak of the next cycle occurs.

9-19. Consider the circuit of Figure P9–19. Assume that the input is a signal that varies between -8 V and $+8$ V.

    a. When $v_i = 8$ V, determine v_C, v_o, v^-, v_a, and i_R when the capacitor is fully charged.

    b. For $v_i = -4$ V, determine v_C, v_o, v^-, v_a, and i_R after the peak of (a) has occurred.

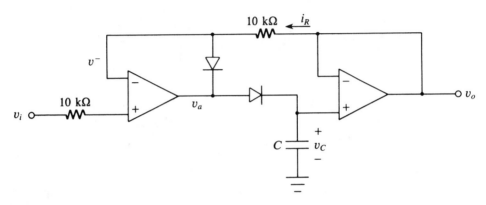

FIGURE P9–19

9-20. Consider the circuit of Figure P9–20. Assume that the input is a signal that varies between -10 V and $+10$ V.

    a. When $v_i = -10$ V, determine v_C, v_o, v^-, v_a, and i_R when the capacitor is fully charged.

    b. For $v_i = 10$ V, determine v_C, v_o, v^-, v_a, and i_R after the peak of (a) has occurred.

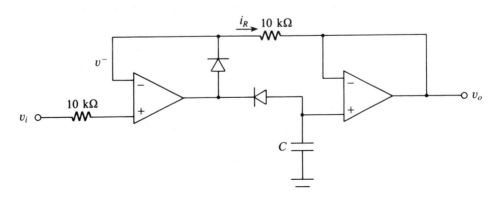

FIGURE P9–20

9–21. Consider the circuit of Figure P9–21. Assume that the input is a random signal that has both positive and negative values. Assume, for analysis purposes, that there is a resistive voltage drop of 0.1 V across the switch *when it is closed*.

    a. At some sampling point, the switch is closed when $v_i = 5$ V. Determine v_C, v_o, v^-, v_a, and i_R when the capacitor is fully charged.

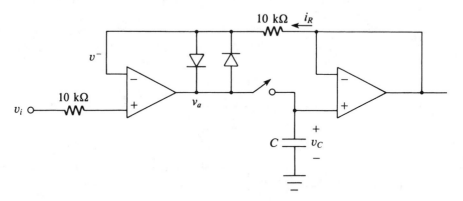

FIGURE P9–21

    b. Following the operation of (a), the switch is opened and the signal voltage drops to a level of 2 V. Determine v_C, v_o, v^-, v_a, and i_R.

    c. If, instead of the input voltage dropping as in (b), suppose that the input voltage had increased to a level of 8 V with the switch now open. Determine the same variable as in (b).

9–22. a. For the circuit of Problem 9–21, assume that the switch is closed when $v_i = -5$ V. Determine v_C, v_o, v^-, v_a, and i_R when the capacitor is fully charged.

    b. Following the operation of (a), the switch is opened and the signal voltage changes to a level of -1 V. Determine v_C, v_o, v^-, v_a, and i_R.

    c. If, instead of the change of (b), suppose that the input voltage had decreased to a level of -9 V with the switch now open. Determine the same variables as in (b).

9–23. Consider the circuit of Figure P9–23 and assume that the input voltage varies from -8 V to $+10$ V. Determine:

    a. V_C

    b. the range of the output voltage.

FIGURE P9–23

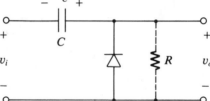

9-24. Consider the circuit of Figure P9–24 and assume that the input voltage varies from -8 V to $+10$ V. Determine:

a. V_C

b. the range of the output voltage.

FIGURE P9–24

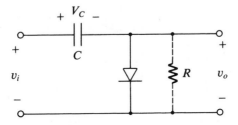

9-25. Consider the circuit of Figure P9–25 and assume that the input voltage varies from -8 V to $+10$ V. Determine:

a. V_C

b. the range of the output voltage.

c. the value of v_o' during the charging interval.

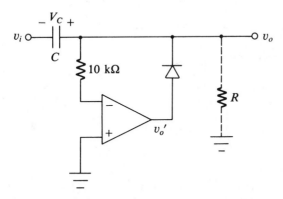

FIGURE P9–25

9-26. Consider the circuit of Figure P9–26 and assume that the input voltage varies from -8 V to $+10$ V. Determine:

a. V_C

b. the range of the output voltage.

c. the value of v_o' during the charging interval.

FIGURE P9–26

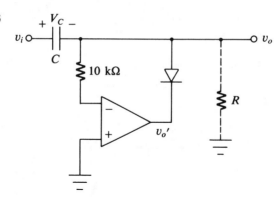

9–27. Consider the circuit of Figure P9–27 with $R_i = 1$ kΩ, $R_f = 20$ kΩ, and $V_Z = 3.3$ V for each diode. Determine:

a. the output limiting voltage
b. the input limiting voltage
c. For $v_i = 0.15$ V, determine v_o, i_i, i_R, i_Z and i_o.
d. For $v_i = 5$ V, repeat the calculations of (a).

FIGURE P9–27

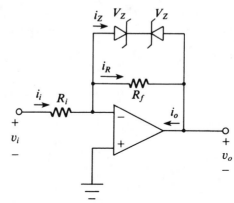

9–28. Repeat the analysis of Problem 9–27 if $R_i = 3$ kΩ, $R_f = 120$ kΩ, and $V_Z = 6.8$ V.

9–29. A certain transformer is specified from primary to secondary as 120 V/12 V with rms units implied. Determine:

a. turns ratio
b. secondary rms current when primary rms current is 0.2 A
c. peak value of secondary voltage.

9–30. A certain transformer is specified from primary to secondary as 120 V/240 V with rms units implied. Determine:

a. turns ratio
b. secondary rms current when primary rms current is 4 A
c. peak value of secondary voltage.

9-31. A certain 60-Hz full-wave rectifier employs a single-capacitor filter. The dc load current to the regulator is 0.8 A and a 2000-μF capacitor is used. Estimate the peak-to-peak ripple voltage.

9-32. A certain 60-Hz full-wave rectifier employs a single-capacitor filter. The dc load current to the regulator is 0.5 A, and the peak-to-peak ripple voltage is measured as 4 V. Determine the value of the capacitance.

9-33. A certain regulator circuit of the form shown in Figure 9-28 has an unregulated voltage of 10 V, a regulated voltage of 6 V, and a load current of 2 A. Determine the following:
   a. input power   **b.** output power   **c.** efficiency.

9-34. A certain regulator of the form shown in Figure 9-28 has a regulated voltage of 5 V. Determine the efficiencies for:
   a. load current = 1 A and unregulated voltage = 7 V
   b. load current = 2 A and unregulated voltage = 9 V

9-35. Consider the circuit shown in Figure P9-35 with $R_1 = 200\ \Omega$ and $R_2 = 3\ k\Omega$. If $I_{ADJ} = 50\ \mu$A and $V_{ref} = 1.25$ V, determine V_{out}.

FIGURE P9-35

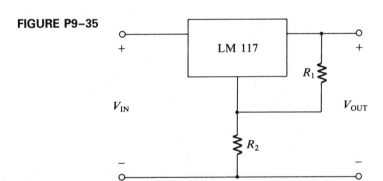

V_{IN} V_{OUT}

9-36. Assume that in a large quantity of LM117s the value of I_{ADJ} varies from 50 μA to 100 μA and that V_{ref} varies from 1.2 V to 1.3 V. Based on the resistance values in Problem 9-35, calculate the range of output voltages that would result.

Derivation Problems

9-37. Consider the passive series half-wave rectifier of Figure 9-1, but assume that the diode is reversed. Sketch the form of the input-output characteristic and write an equation similar to (9-1) to described the output.

9-38. Consider the passive shunt half-wave rectifier of Figure 9-2, but assume that the diode is reversed. Sketch the form of the input-output characteristic.

9-39. Consider the active saturating half-wave rectifier of Figure 9-3, but assume that the diode is reversed.
   a. Write equations to describe the input-output characteristic.
   b. Sketch the form of (a).

9-40. Consider the full-wave precision rectifier circuit of Figure 9-6, but assume that both diodes are reversed.
   a. Write equations to describe the input-output characteristic.
   b. Sketch the form of (a):

9-41. Consider the passive clamping circuit of Figure 9-17, but assume that the diode is reversed. For the input waveform given, sketch the form of the output.

9–42. Consider the active clamping circuit of Figure 9–18, but assume that the diode is reversed. For the input waveform given, sketch the form of the output.

9–43. Consider the active clamping circuit shown in Figure 9–19, but assume that the reference voltage is reversed in polarity. The diode is unchanged. For the input waveform given, sketch the form of the output.

9–44. Consider the active clamping circuit shown in Figure 9–19, but assume that the diode is reversed. The reference voltage is unchanged. For the input waveform given, sketch the form of the output.

9–45. Consider the active clamping circuit shown in Figure 9–19, but assume that the diode is reversed. The reference voltage is also reversed in polarity. For the input waveform given, sketch the form of the output.

9–46. Consider the limiting circuit of Figure 9–21, but assume that the reference voltage is reversed in polarity. The diode is unchanged. Sketch the form of the input-output characteristic.

9–47. Consider the limiting circuit of Figure 9–21, but assume that the diode is reversed. The reference voltage is unchanged. Sketch the form of the input-output characteristic.

9–48. Consider the limiting circuit of Figure 9–21, but assume that both the diode and the reference voltages are reversed. Sketch the form of the input-output characteristic.

Design Problems

9–49. A peak detector circuit of the form shown in Figure 9–10 is to be designed for an application in which a short-term memory of the peak level of an input signal is desired. One design criterion is that the capacitor will drop to 98% of its peak level in a time interval of 12 ms when it is not recharged. The output of the circuit is to be connected to an amplifier that has an input resistive impedance of 600 kΩ. If C is selected as 3 μF, determine the value of R required.

9–50. Assume that the circuit of Figure 9–10 is to be designed for envelope detection of an AM signal used in a test circuit. The carrier frequency is 100 kHz, and the modulating frequency can vary from 100 Hz to 1 kHz. Using the notation established in the text preceding (9–19), we establish the somewhat arbitrary criterion that $\tau = KT_c$ and $T_m = K\tau$, where T_m is the shortest modulation period and K is a constant. (The same constant appears in both equations.)
   a. Determine the required value of τ.
   b. Determine possible values of R and C.

9–51. Consider an op-amp regulator circuit of the form shown in Figure 9–28, in which the desired regulated load voltage is to be $V_L = 10$ V. An unregulated voltage whose value is approximately 13 V is available. The Zener diode is a 3.3-V unit, and the desired Zener bias current is 20 mA.
   a. Determine the required value of R_Z.
   b. If R_i is selected as 10 kΩ, determine the required value of R_f.
   c. If the maximum value of load current I_L is 0.8 A and the β_{dc} range for the pass transistor is between 50 and 120, determine the maximum op-amp output current required for base drive I_B.
   d. Determine the load power, the approximate input power, and the approximate efficiency under maximum load conditions.
   e. Determine the power dissipation rating P_D for the pass transistor.

9–52. Using a PNP pass transistor, draw a schematic diagram similar to Figure 9–28 that could be used to provide a negative regulated voltage from a negative unregulated voltage.

9–53. A 117 regulator circuit is to be designed to establish a regulated load voltage of 24 V from an unregulated supply of about 28 V.
   a. Perform an initial design by determining possible values of R_1 and R_2.
   b. If the maximum value of I_{ADJ} is 100 μA, estimate the maximum possible error due to that current.

9–54. Repeat the design process of Problem 9–51 if a regulated voltage of 5 V is to be established from an unregulated voltage of about 8 V.

9–55. A circuit is desired to determine which of two signals v_1 and v_2 has the largest peak value over a moderate time interval. The desired logic is as follows:

$$v_o = \text{high value if } V_1(\text{peak}) > V_2(\text{peak})$$
$$v_o = \text{low value if } V_1(\text{peak}) < V_2(\text{peak})$$

Sketch the form of a circuit that could be used for this purpose.

9–56. A circuit is desired to determine which of two signals v_1 and v_2 has the largest average value corresponding to the absolute value of the signal; that is, all area is weighted positively. Let $|V_1|_{dc}$ represent the dc or average value corresponding to the absolute value of v_1, and let $|V_2|_{dc}$ represent the corresponding quantity for v_2. The desired logic is as follows:

$$v_o = \text{high value if } |V_1|_{dc} > |V_2|_{dc}$$
$$v_o = \text{low value if } |V_1|_{dc} < |V_2|_{dc}$$

Sketch the form of a circuit that could be used for this purpose. (*Hint:* A simple one-pole RC low-pass circuit can be used to determine the average value of a signal if the RC time constant is sufficiently large.)

9–57. A peak-to-peak voltmeter is a voltmeter that provides a measurement of the peak-to-peak value of the voltage. The voltage is assumed to vary both positively and negatively about a zero level. As a basic signal-processing operation in such a circuit, design a circuit that produces a dc output equal to the peak-to-peak value of the input voltage. Use precision active circuits when possible to avoid errors due to diode drops.

9–58. The properties of two random signals v_1 and v_2 are to be used in a certain system. The signal v_1 is assumed to vary over the range from $-V_1(\text{peak})$ to $+V_1(\text{peak})$, and v_2 is assumed to vary over the range from $-V_2(\text{peak})$ to $+V_2(\text{peak})$. A circuit is desired to provide an output voltage v_0 whose value is

$$v_o = 2V_1(\text{peak}) - V_2(\text{peak}) - 5$$

Design a circuit to accomplish the purpose. Assume that arbitrary dc voltages may be obtained as required.

LABORATORY EXERCISES

Active Saturating Half-Wave Precision Rectifier
LE 9–1.

1. Connect the circuit of Figure 9–3. Some load is required and $R = 10 \text{ k}\Omega$ will suffice, but the value is not critical.

2. Apply a 1-kHz sine wave with a peak value of several volts for v_i. Sketch the forms of v_i, v_o, and v_o' showing their time relationships. Measure all pertinent levels.

Active Nonsaturating Half-Wave Precision Rectifier
LE 9–2.

1. Connect the circuit of Figure 9–4 with $R_i = 10 \text{ k}\Omega$ and $R_f = 20 \text{ k}\Omega$.

2. Apply a 1-kHz sine wave with a peak value of several volts for v_i. Sketch the forms of v_i, v_o, and v_o' showing their time relationships. Measure all pertinent levels.

Active Saturating Peak Detector
LE 9–3.

1. Connect the circuit of Figure 9–10 with $C = 0.1$ μF and with R eliminated. If an electrolytic capacitor is used, the positive terminal should be at the top.
2. Apply a 1-kHz sine wave with a peak value of several volts for v_i. Sketch the forms of v_i, v_o, and v_o' showing their time relationships. Measure all pertinent levels.
3. Investigate the effect on v_o of changing the input peak voltage level both upward and downward from the initial level. (*Note*: Although R is omitted, the scope itself presents some shunt resistance. There may also be additional leakage in the capacitor used.)

Active Isolated Peak Detector
LE 9–4.

1. Connect the circuit of Figure 9–13 with $C = 0.1$ μF. If an electrolytic capacitor is used, the positive terminal should be at the top.
2. Apply a 1-kHz sine wave with a peak value of several volts for v_i. Sketch the forms of v_i, v_o, and v_o' showing their time relationships. Measure all pertinent levels.
3. Investigate the effect on v_o of changing the input peak voltage level both upward and downward from the initial level.
4. Connect a 4.7-Mω resistance in parallel with C and repeat step (3).

Active Clamping Circuit
LE 9–5.

1. Connect the circuit of Figure 9–18 with $C = 0.1$ μF and with R eliminated. If an electrolytic capacitor is used, the positive terminal should be on the right.
2. Apply a 1-kHz sine wave with a peak value of several volts for v_i. Sketch the forms of v_i, v_o, and v_o' showing their time relationships. Measure all pertinent levels.
3. Investigate the effect on v_o of changing the input peak voltage level both upward and downward from the initial level. (See note in LE 9–3.)

117 Voltage Regulator
LE 9–6.

Your objective in this problem is to demonstrate the use of the 117 regulator to obtain a desired dc output voltage from a given higher input voltage. The input voltage will be your available 15-V source, which in this particular case is already likely to be well regulated. The development of Example 9–11 will be employed, so the desired output voltage is 12 V.

1. Connect the circuit of Example 9–11 (Figure 9–32 with R_1 and R_2 determined in the example.) Set $C_1 = 0.1$ μF and $C_2 = 1$ μF.
2. Measure the output voltage and compare with the expected value.

10

ANALOG-TO-DIGITAL AND DIGITAL-TO-ANALOG CONVERSION

OBJECTIVES

After completing this chapter, the reader should be able to:

- Define *analog signal, quantization, quantized variable, discrete-time signal, sampled-data signal,* and *digital signal.*
- State Shannon's sampling theorem and discuss its general significance.
- Define and explain the significance of *aliasing error.*
- Explain the concept and advantages of *multiplexing.*
- Explain the purposes of digital-to-analog (D/A) and analog-to-digital (A/D) converters.
- State the form of a digital word with *n* bits.
- Determine the normalized value of the analog representation of a digital word.
- Define the terms *full-scale voltage* and *one least-significant bit* for a D/A converter.
- Draw the schematic diagrams of several representative D/A converter circuits and discuss their operation.
- Explain the general operation of and compare the relative speeds of *counter, successive approximation,* and *flash* A/D converters.
- Explain the general operation of *single-slope, dual-slope,* and *quad-slope* integrating A/D converters.
- Discuss the significance of the integration time interval for the dual-slope and quad-slope A/D converters.
- Explain the general characteristics of and differences between *unipolar encoding* and *bipolar encoding.*
- Explain the difference between *rounding* and *truncation.*
- Sketch the forms of the quantization characteristics and of the normalized D/A conversion characteristics for both unipolar and bipolar encoding.

- Discuss the general specifications for an A/D converter and determine various step error sizes from the specifications.
- Discuss the concept of an analog switch and its general operation.
- Discuss the operation of a voltage-to-frequency converter and its applications.
- Discuss the operation of a frequency-to-voltage converter and its applications.

10–1 INTRODUCTION

The major objectives of this chapter are to develop the theoretical and operational basis for data conversion requirements and to provide details on some of the actual circuits and systems used in data conversion applications. The major types of data conversion processes we will consider are analog-to-digital and digital-to-analog conversion as reflected in the chapter title.

In Section 10–2, we establish the theoretical basis for representing a signal with a finite number of samples using Shannon's sampling theorem. We also give various definitions in that section.

We present some actual circuit details of representative digital-to-analog converters and then give a general explanation of the operation of and comparison between the different types of analog-to-digital converters. And we explore in detail the specifications required in selecting the various converters to meet prescribed specifications.

As compared to earlier chapters, the reader will find fewer detailed circuit schematic diagrams in this chapter. Instead, we show more of the units in block diagram form. Many of the circuits we have considered thus far are those that can be implemented readily with one or more op-amps and a few components. With the exception of the most elementary types of digital-to-analog converters, however, this procedure is neither desirable nor feasible for most of the circuits in this chapter. Instead, most users will select a complete module off the shelf based on the specified operational performance and acceptable error. It would serve little purpose to try to present complete, detailed schematic diagrams of circuits such as the IC analog-to-digital converters. The block diagram approach is much better and explains the general operation.

10–2 SAMPLING AND DATA CONVERSION

As digital technology has advanced in sophistication and components have decreased in cost, many systems that were previously all analog in form have incorporated digital subsystems in their designs. Digital equipment such as microprocessors permit very complex control and computational functions to be achieved through software with associated flexibility and ease of adjustment.

When digital components are used in systems that are predominantly analog in form, it is often necessary to convert data from one form to the other. The processes of analog-to-digital and digital-to-analog conversion are required for this purpose. Before these specific operations are considered, however, it is essential to understand the basic forms of the data representations and the assumptions required for conversion. The definitions that follow are useful in establishing the various conversion processes.

Definitions

An *analog signal* is a signal that is defined over a continuous range of time and in which the amplitude may assume a continuous range of values. The term *analog* apparently originated in the field of analog computation, but the term has assumed a very widespread connotation in actual usage.

Quantization describes the process of representing a variable by a finite set of distinct values. A *quantized variable* is one that may assume only distinct values.

A *discrete-time signal* is a function that is defined only at a particular set of values of time. Thus, the independent variable time is quantized. If, however, the amplitude of a discrete-time signal is permitted to assume a continuous range of values, the function is referred to as a *sampled-data signal.* A sampled-data signal could arise from sampling an analog signal at discrete values of time.

A *digital signal* is a function in which both time and amplitude are quantized. A digital signal may always be represented by a sequence of *words* in which each word has a finite number of *bits* (binary digits).

Analog-to-Digital Conversion System

A simplified diagram of a possible data conversion system is shown in Figure 10–1. Some of the preceding definitions will be clarified by a discussion of the operation of this system.

The input to the system is assumed to be an analog signal. It is first processed by a low-pass filter, the function of which will be discussed later. The signal is then sampled at a rate f_s as determined by the clock rate. In the context employed, 1 hertz is considered as 1 sample/second. The sampling period $T = 1/f_s$ is the time interval between successive samples.

The sample-and-hold circuit senses the level of the analog signal once per sampling period and "freezes" that value until the corresponding point in the next sampling period.

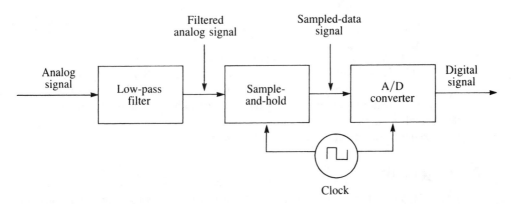

FIGURE 10–1
Block diagram of analog-to-digital conversion system.

(Sample-and-hold circuits were discussed in Chapter 9.) The signal at the output of the sample-and-hold circuit is a sampled-data analog signal since it can assume any analog level, but new samples are obtained only at discrete intervals of time.

The analog-to-digital (A/D) converter converts the constant analog signal to a digital word as represented by a finite number of bits. Assume that each binary word has n bits. The possible number of levels m is given by

$$m = 2^n \tag{10-1}$$

Only a finite number of levels can be represented by the conversion process. Consequently, amplitude quantization will occur in the A/D converter, with each analog sample being assigned to one of the m possible digital levels by some strategy, of which several common forms will be discussed later. The resulting digital values are thus quantized both in time and in amplitude.

The difference between the analog signal amplitude and the nearest standard digital level is called **quantization error** (also called **quantization noise**). In theory, the quantization error can be made arbitrarily small by selecting a sufficient number of bits for each word, but there are practical limits with actual circuits. However, analog systems also contain noise and spurious disturbances, so there are ultimate limits of signal quality that can be attained with both types of systems.

Shannon's Sampling Theorem

A fundamental question of great importance concerns the minimum sampling rate that must be used with the system. In other words, since only a finite number of samples of the analog signal can be taken in a finite time, will information be lost as a result of the sampling operation? The key to this question is **Shannons' sampling theorem,** which establishes the theoretical basis for all discrete sampling operations applied to analog signals.

Assume that the signal has a spectrum containing frequency components extending from near dc to some upper frequency f_u. Shannon's sampling theorem states that the signal can be reproduced by samples taken at intervals no greater than $1/2f_u$ apart. The associated sampling rate f_s must then satisfy

$$f_s \geq 2f_u \tag{10-2}$$

For example, an audio signal extending from dc to 20 kHz could theoretically be reconstructed by taking uniformly spaced samples at a rate of 40,000 samples/second.

In practice, a rate somewhat greater than the theoretical minimum is always used. Operation at the ideal minimum sampling rate would require perfect filters for data recovery. The actual sampling rate is often chosen to be 3 or 4 (or more) times the highest frequency in order to ease the burden of signal recovery.

Aliasing Error

A signal sampled at a rate lower than $2f_s$ will suffer from **aliasing error,** a phenomenon in which frequencies appear to be different from their true values and the signal can never be correctly recovered. A very common example of aliasing is the apparently backward rotation of the wagon wheels in western movies resulting from the inability of the frame rate of the camera to sample the rotation of the wheels at a sufficiently high rate.

Since it is not always possible to define the exact frequency content of random data signals, a common practice is to pass the analog signal through a low-pass filter before sampling, as was indicated in Figure 10–1. The filter characteristics are chosen to reject (or at least to strongly attenuate) all components at frequencies equal to or greater than half the sampling rate. Without such a filter, spurious input components outside the frequency range of the signal could be shifted into the signal frequency range through the sampling and aliasing processes.

The digital words at the output of the conversion system are assumed to be transferred to some type of digital system for which processing is desired. Depending on the application, this system could be a general-purpose computer, a microprocessor, or some special-purpose digital hardware. Computational, communications, or control functions could be performed with the data.

Digital-to-Analog Conversion System

At some point in the overall system, it may be necessary to convert the train of digital data back to analog form. A simplified block diagram of a possible data conversion system for this purpose is shown in Figure 10–2. The sequence of digital words is first converted to analog form by a digital-to-analog converter. Depending on whether serial or parallel forms of the digital words are available, some buffering of the data may be required at the input. In some cases, a holding circuit is used to maintain the restored analog level for the duration of a sampling period. The converted signal has abrupt discontinuities and presents a "staircase" type of pattern. The low-pass filter provides a smoothing of the signal by removing high-frequency content.

Multiplexing

The conversion systems of Figures 10–1 and 10–2 imply one input channel and one output channel in each case. However, one of the most important benefits of data conversion systems is that of *multiplexing.* (Specifically, the focus here will be on *time-division multiplexing,* but because it is the only form discussed, the modifier *time-division* will be omitted in subsequent discussions.)

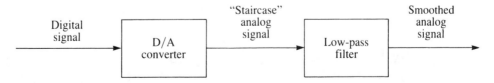

FIGURE 10–2
Block diagram of digital-to-analog conversion system.

Since the sampling theorem permits a signal to be defined in terms of a discrete set of samples, it is possible to accommodate a number of different data signals with the same processing system by the use of multiplexing. For example, one computer can simultaneously monitor and control a large number of different operations in an industrial process. Each signal is sampled in order, and the computer deals only with that particular operation during the time the sample is being processed. As long as each signal is sampled at a sufficiently high rate, one system effectively performs the functions of many separate systems.

Multiplexers are circuits designed to perform the sequential operation, and analog multiplexing units will be discussed in Section 10–9. However, it is possible to employ separate analog-to-digital conversion modules followed by a ***digital multiplexer***. A ***demultiplexer*** is a unit used for separating a group of multiplexed signals into individual data channels.

EXAMPLE 10–1

A certain data processing system of the form indicated in Figure 10–1 is to be used to process an analog signal whose frequency content ranges from near dc to 2.5 kHz. Assume that the actual sampling rate is selected to be 60% greater than the theoretical minimum. (a) Determine the sampling rate and the sampling period. (b) If 8-bit words and real-time processing are employed, determine the maximum bit width if a given word is to be transferred serially during the conversion interval of the next word.

Solution

(a) The minimum sampling rate is twice the highest frequency. However, the actual sampling rate f_s should be 1.6 times the minimum rate (60% greater). Thus,

$$f_s = 1.6 \times 2 \times 2500 = 8000 \text{ Hz} \qquad (10\text{–}3)$$

The sampling period is

$$T = \frac{1}{8000} = 125 \ \mu s \qquad (10\text{–}4)$$

(b) The A/D converter must be capable of converting each analog value to a digital word in a time not exceeding 125 μs if real-time processing is employed. In this time interval, 8 bits from the preceding converted word must be transferred serially from the A/D converter to the remainder of the system. If τ is the bit width, the maximum value is

$$\tau = \frac{125 \ \mu s}{8} = 15.625 \ \mu s \qquad (10\text{–}5)$$

With multiplexed systems, a number of words must be processed during a sampling period if real-time processing is employed. The corresponding conversion time, word lengths, and bit widths must be reduced in inverse proportion to the number of channels.

10–3 DIGITAL-TO-ANALOG CONVERSION

Digital-to-analog conversion is the process of converting a digital word to an analog voltage or current level, with the magnitude and sign of the analog level representing the binary value of the digital word in some sense. A circuit that performs this function is called a **digital-to-analog converter,** which will hereafter be referred to by the common designation **D/A converter.** (Another designation is **DAC.**)

Some of the most common circuit design strategies utilized in D/A converters will be presented in this section. Although many D/A converter chips are readily available, the initial emphasis will be on establishing the basic principles of operation.

Digital Word

For this discussion, assume a digital word $\hat{X}$ consisting of n bits. The quantity $\hat{X}$ will be represented as

$$\hat{X} = b_1 b_2 b_3 \cdots b_n \tag{10–6}$$

where b_1 is the **most significant bit (MSB),** b_2 is the next most significant bit, and so on. The last bit b_n is the **least significant bit (LSB).** A given bit can assume only the two possible states 1 or 0. For example, consider the 4-bit number 1010. We have $b_1 = 1$, $b_2 = 0$, $b_3 = 1$, and $b_4 = 0$. The MSB is 1 and the LSB is 0.

The actual analog value corresponding to a given digital word can be established at any practical level desired. Further, some encoding schemes involve analog levels of both polarities, while others are based on a single polarity. These various encoding schemes will be considered in due course, but for the moment, assume that the digital word is defined as a fractional value with the most significant bit representing the analog weight $0.5 = 2^{-1}$. The next most significant bit then represents a weight of $0.25 = 2^{-2}$, and so on. Finally, the least significant bit represents a weight $(0.5)^n = 2^{-n}$.

Normalized Analog Level

Let X represent the decimal (or analog) representation of the binary word $\hat{X}$. The actual value of X can be expressed as

$$X = b_1 2^{-1} + b_2 2^{-2} + b_3 2^{-3} + \cdots + b_n 2^{-n} \tag{10–7}$$

As the number of bits increases, the numerical value approaches unity. Theoretically, it never quite reaches unity, but it can be made to be arbitrarily close by choosing a sufficiently large number of bits. We will refer to the form of X as given in (10–7) as a **normalized level,** since it is based on a maximum limiting value of unity.

When a digital word such as $\hat{X}$ in (10–6) is assumed to represent a number less than unity, as given by (10–7), a binary point should precede the digital value. However, as long as this normalized or fractional form is understood, the binary point will be omitted in subsequent work.

A D/A converter must provide an analog output voltage that is a weighted combination of the bits in the digital word. A binary 0 for a given bit should produce no contribution to the output, while a binary 1 should produce a contribution, with the weight a function of the relative significance of the bit.

Linear Combination D/A Circuit

The first D/A circuit that will be considered is based on the op-amp linear combination circuit of Figure 10–3. A reference voltage $-V_r$ is simultaneously applied as the input to an array of switches. In practice, electronic switches are used, and they are controlled by the logic levels of the various bits. (Electronic analog switches will be discussed in Section 10–9.) Thus, a logic level of 1 for a given bit connects a given resistance to the source $-V_r$, while a logic level of 0 connects the resistance to ground. In the diagram of Figure 10–3, it is convenient to represent the electronic switching operation by "mechanical" switches.

Observe the binary pattern of the various input resistances. The gain for the b_1 signal path is $-R_f/R$, the gain for the b_2 signal path is $-R_f/2R$, and so on. Finally, the gain for the b_n signal path is $-R_f/2^{n-1}R$. Thus, the weighting of the various bits follows a binary pattern with the MSB receiving the greatest weight, and so on. In view of the inversion of the summer, coupled with the negative values of V_r, the output voltage v_o is either positive or zero and can be expressed as

$$v_o = \frac{R_f}{R} b_1 V_r + \frac{R_f}{2R} b_2 V_r + \cdots + \frac{R_f}{2^{n-1}R} b_n V_r \qquad (10\text{–}8)$$

where each b_k multiplier assumes a value of either 0 or 1.

After some factoring and rearrangement, the expression of (10–8) can be simplified to

$$v_o = \frac{2R_f V_r}{R} (b_1 2^{-1} + b_2 2^{-2} + b_3 2^{-3} + \cdots + b_n 2^{-n}) \qquad (10\text{–}9)$$

Observe that the quantity in parentheses is the analog normalized value of the binary word as given by (10–7), and it approaches a maximum value of unity. The factor $2R_f V_r/R$ is a

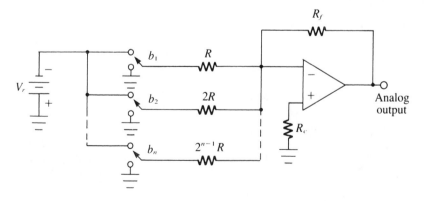

FIGURE 10–3
Digital-to-analog converter implemented with linear combination circuit.

scale setting factor, and it allows the normalized value of the binary word to be converted to any absolute analog level within the operating limits of the op-amp. This quantity can be defined as the full-scale voltage V_{fs}:

$$V_{fs} = \frac{2R_f V_r}{R} \qquad (10\text{–}10)$$

Note that the term *full-scale* is actually a misnomer since the output voltage never quite reaches the value V_{fs}, but it is a convenient reference value.

The output voltage step between successive levels is often referred to as the value of 1 LSB. This value is determined by setting $b_n = 1$ and all other $b_k = 0$ in (10–9). We have

$$1 \text{ LSB} = \frac{2R_f V_r}{R} 2^{-n} = \frac{V_{fs}}{2^n} \qquad (10\text{–}11)$$

While this D/A converter circuit is about the simplest type to understand conceptually, it presents some difficulties in practice. Observe that the largest value of input resistance is 2^{n-1} times the smallest value of input resistance. For 4 bits, this ratio is 8, which does not present any difficulties. However, for 8 bits, this ratio is 128, and for 16 bits, the ratio is 32,768! Finding acceptable resistance values having such large ratios is very difficult, if not impossible. Further, variations in resistance due to temperature changes and other factors are very difficult to control with large resistance spreads. In practice, therefore, this D/A converter is limited to noncritical applications where a small number of bits is used.

R-2R D/A Circuit

One of the most widely employed conversion circuits is the *R-2R D/A converter,* whose basic form is shown in Figure 10–4. The designation "R-2R" is rather obvious in that only

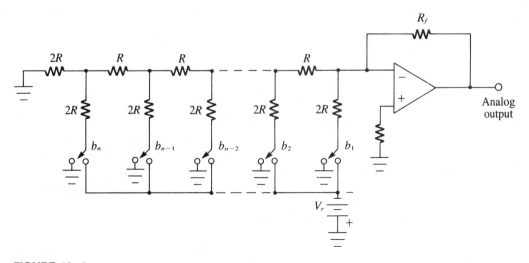

FIGURE 10–4
R-2R digital-to-analog converter.

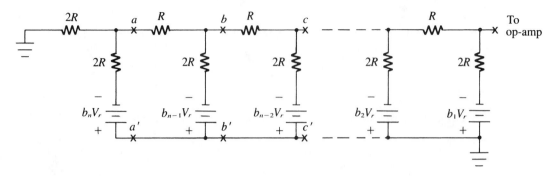

FIGURE 10–5
Equivalent circuit model of R-2R ladder used in analysis, with effect of reference voltage shown in each branch.

two resistance values are employed in the input summing circuit, so the problem with an excessive resistance spread is circumvented with this circuit. Special resistance packages containing the R-2R ladder are commercially available.

Because of the interaction of the different branches in the R-2R ladder, it is not immediately evident that the various bit values combine in the proper ratios. Correct circuit verification is best achieved by successive application of Thevenin's theorem. First, the input ladder circuit is replaced by the equivalent circuit shown in Figure 10–5. Since the single voltage source of value $-V_r$ is connected to all the b_k branches, the circuit is equivalent to one having n separate voltage sources connected to the n branches. The source for the kth branch is represented as $-b_k V_r$, where b_k is the bit value (0 or 1) for that particular branch.

Starting with the points a and a', we determine a Thevenin equivalent circuit looking back to the left. This step is illustrated in Figure 10–6(a). The open-circuit voltage is $(1/2) \times b_n V_r / 2$ in the direction shown, and the equivalent Thevenin resistance is $2R \| 2R = R$, as shown on the right.

Next, the branch corresponding to b_{n-1} is connected as shown in Figure 10–6(b), and the Thevenin equivalent circuit looking back from points b and b' is determined. The best way to determine the open-circuit voltage at this point is by superposition. Note that the b_{n-1} contribution will have twice the value of the b_n contribution, as expected. The resistance looking back again is $2R \| 2R = R$ as shown on the right. This pattern continues as illustrated for points c and c' as shown in Figure 10–6(c).

After the Thevenin transformation is performed n times, the net equivalent circuit looking back from the op-amp inverting terminal is as shown in Figure 10–6(d). This equivalent circuit may then be considered to be applied to the inverting op-amp input, and the basic gain equation for an inverting amplifier may be used. Since the Thevenin voltage is negative with respect to ground and the gain is negative, the output voltage v_o is positive, and it may be expressed as

$$v_o = \frac{R_f}{R}\left(\frac{b_1 V_r}{2} + \frac{b_2 V_r}{4} + \cdots + \frac{b_n V_r}{2^n}\right) \qquad \textbf{(10–12)}$$

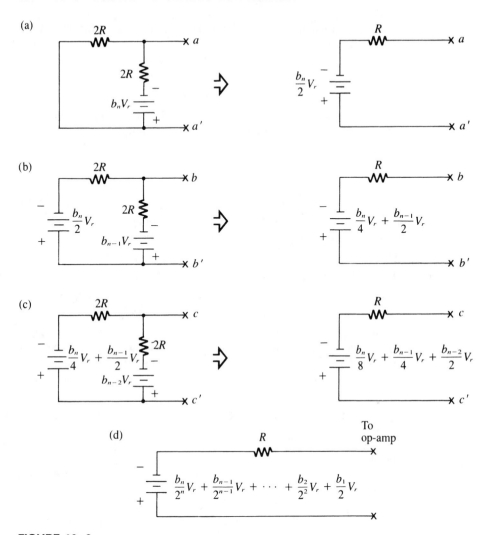

FIGURE 10-6
Successive Thevenin transformations applied to R-2R ladder network.

After some factoring and rearrangement, the expression of (10–12) can be simplified to

$$v_o = \frac{R_f V_r}{R} (b_1 2^{-1} + b_2 2^{-2} + \cdots + b_n 2^{-n}) \qquad (10\text{--}13)$$

Observe that the quantity in parentheses is the normalized value of the binary word as given by (10–7). The factor $R_f V_r/R$ is a scale setting factor, and it allows the normalized value of the binary word to be converted to any absolute analog level within the operating limits of the op-amp. As in the case of the previous circuit, this quantity can be defined as the full-scale voltage V_{fs}:

$$V_{fs} = \frac{R_f V_r}{R} \tag{10-14}$$

The output voltage step between successive levels, which is also denoted as the value of 1 LSB, is determined by setting $b_n = 1$ and all other $b_k = 0$ in (10–13). We have

$$1 \text{ LSB} = \frac{R_f V_r}{R} 2^{-n} = \frac{V_{fs}}{2^n} \tag{10-15}$$

Comparing (10–15) and (10–14) with (10–11) and (10–10), we note that for given values of R_f, R, and V_r, the full-scale and incremental voltage values for the R-2R D/A converter are one-half the values of those for the first D/A converter considered.

Multiplying D/A Converter

Observe in (10–9) and (10–13) that the output voltage for any particular digital word is directly proportional to the value of the reference voltage V_r. While many D/A converters employ a fixed stabilized reference voltage, some converters allow the user to apply an external reference voltage to achieve a multiplication effect. Such a unit is designated as a ***multiplying digital-to-analog converter (MDAC)***. A number of possible applications can be achieved with these converters, including gain control, modulation, and multiplication operations.

Current-Weighted D/A Converter

The R-2R D/A converter previously considered used a voltage source as the basis for the relative weighting. Another widely employed technique uses the relative weighting of constant current sources. This type of unit is referred to as a ***current-weighted digital-to-analog converter (IDAC)***. While there are a number of circuits, the concept is illustrated in Figure 10–7. Note that a variation of the R-2R ladder network is used. The individual current sources are typically implemented with an array of bipolar junction transistors and a precision biasing circuit. For an arbitrary bit b_k, the corresponding transistor represents a constant current source when $b_k = 1$, and the transistor is open when $b_k = 0$. If the value of current is I, the corresponding current source may then be represented as $b_k I$.

Verification of the operation of this circuit can be achieved by a process very similar to the earlier development of the R-2R ladder with a voltage source reference, and the reader is invited to perform such an analysis. The results of successively reducing the circuit and representing the output in terms of a Norton equivalent circuit are shown in Figure 10–8. The value of the Norton current source I_N is

$$I_N = I[b_1 + b_2 2^{-1} + b_3 2^{-2} + \cdots + b_n 2^{-(n-1)}] \tag{10-16}$$

A slight rearrangement of (10–16) yields

$$I_N = 2I[b_1 2^{-1} + b_2 2^{-2} + \cdots + b_n 2^{-n}] \tag{10-17}$$

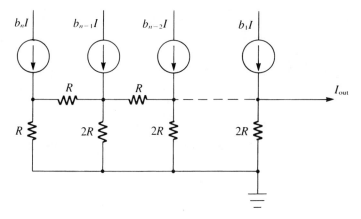

FIGURE 10–7
Basic form of current-weighted D/A converter.

The quantity in brackets is the analog normalized value of the binary word as given by (10–7), and it approaches a maximum value of unity.

The actual output current I_{out} will depend on the equivalent load resistance to which the circuit is attached. If an op-amp inverting current-to-voltage converter is used, the virtual ground permits the entire value of I_N to flow into the current-to-voltage conversion resistance. However, if the resistance is connected directly across the circuit, an appropriate circuit analysis using the model of Figure 10–8 connected to the load must be performed to determine the actual output level. Converters utilizing the current-switching concept will likely be obtained as a package by the user, so the manufacturer's data concerning the output connections should be closely followed.

An advantage of current-weighted D/A converters is their speed. By biasing the transistors in a linear region of operation, speed problems resulting from saturation can be avoided.

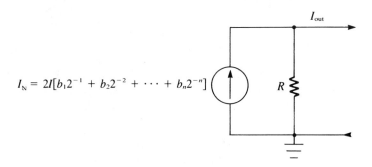

$$I_N = 2I[b_1 2^{-1} + b_2 2^{-2} + \cdots + b_n 2^{-n}]$$

FIGURE 10–8
Norton equivalent circuit of current-weighted D/A converter in Figure 10–7.

**EXAMPLE
10–2**

Consider the R-2R 4-bit converter shown in Figure 10–9 in which the resistances in the R-2R ladder and the dc voltage V_r are fixed, but in which R_f is variable. Determine the separate values of R_f that would be required to achieve **each** of the following conditions: (a) the value of 1 LSB at the output is 0.5 V; (b) a binary input of 1000 results in an analog output of 6 V; (c) full-scale output voltage is 10 V; (d) the **actual** maximum output voltage is 10 V.

Solution

(a) From (10–15), we set $R = 10$ kΩ, $V_r = 10$ V, and $n = 4$, and we determine R_f such that the value of 1 LSB = 0.5 V. We have

$$\frac{R_f \times 10 \times 2^{-4}}{10^4} = 0.5$$

or

$$R_f = \frac{10^4}{10 \times 2^{-4}} \times 0.5 = 8000 \ \Omega \tag{10–18}$$

(b) With a binary value of 1000, the output voltage is determined from (10–13) by setting $b_1 = 1$ and $b_2 = b_3 = b_4 = 0$. Substituting the values of R and V_r and setting the result equal to 6 V, we have

$$\frac{R_f \times 10 \times 2^{-1}}{10^4} = 6$$

or

$$R_f = \frac{10^4}{10 \times 2^{-1}} \times 6 = 12 \text{ k}\Omega \tag{10–19}$$

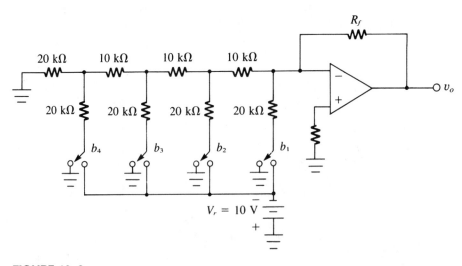

FIGURE 10–9
Four-bit D/A converter.

(c) From (10–14), we substitute the given parameters and require that $V_{fs} = 10$ V; that is,

$$\frac{R_f \times 10}{10^4} = 10$$

or

$$R_f = \frac{10^4}{10} \times 10 = 10 \text{ k}\Omega \qquad (10\text{–}20)$$

(d) Recall that the so-called full-scale voltage is never quite reached, so in (c), the maximum voltage is actually less than 10 V. In this part, however, the actual output voltage maximum is to be 10 V, and this value will be achieved when the input word has a value of 1111. Setting $b_1 = b_2 = b_3 = b_4 = 1$ in (10–13), we require that

$$\frac{R_f \times 10}{10^4}(2^{-1} + 2^{-2} + 2^{-3} + 2^{-4}) = 10$$

or

$$R_f = \frac{10^4}{10 \times 0.9375} \times 10 = 10.667 \text{ k}\Omega \qquad (10\text{–}21)$$

In this case, the "unreachable" full-scale voltage is 10.667 V.

10–4 ANALOG-TO-DIGITAL CONVERSION

Analog-to-digital conversion is the process of converting a sample of an analog signal to a digital word, with the value of the digital word representing the magnitude and sign of the analog level in some sense. A circuit that performs this function is called an ***analog-to-digital converter,*** which will hereafter be referred to by the common designation ***A/D converter.*** (Another designation is ***ADC.***)

Some of the common D/A converter design concepts were considered in the last section. The reason that D/A converters were considered first is that the D/A conversion process is notably simpler in both concept and implementation than the process of A/D conversion. Further, most A/D converters employ a D/A converter as one component within the system, so one should understand the process of D/A conversion before attempting to analyze an A/D converter.

While it is possible for a clever individual to implement an A/D converter using a D/A converter and various logic chips, there is very little incentive for this process with the technology available today. It would be quite difficult for an individual to implement in a short time anything more than the most rudimentary form of an A/D converter, so the use of available A/D converter packages is highly recommended.

In view of their complexity, A/D converters have traditionally been rather expensive. Some of the earliest units, which were implemented using discrete components, cost thousands of dollars. As integrated circuit technology has advanced, however, the price has decreased markedly. Certain A/D converter models are available for less than ten dollars; however, the majority are in the range from tens of dollars to hundreds of dollars.

A/D Converter Types

Some of the most common design strategies used in A/D converters will be discussed in this section, including the following types: (1) *counter* (or *ramp*), (2) *successive approximation,* and (3) *flash.* (A special category referred to as *integrating A/D converters* will be considered in the next section.) Because of the complex interaction of the analog and digital portions of an A/D converter, the discussion will be aimed at a block diagram level rather than a detailed circuit operation level. Since the objective here is to convey understanding of how available circuits work rather than to provide information on building a circuit, this approach is felt to be the optimum one.

In discussions of the different circuits, reference will be made to an analog sample at the input. As suggested in Section 10–2, many data conversion systems employ a sample-and-hold circuit at the input to hold the analog value during the conversion process. It is actually possible in some cases to apply the analog signal directly to an A/D converter, but the sample-and-hold circuit is normally used in most data acquisition systems. The various types of A/D converters will next be considered individually.

Counter A/D Converter

A block diagram of a counter A/D converter is shown in Figure 10–10. Assume that the counter is initially set at zero before conversion is started. A sample of the analog signal appears at one input to the comparator, and the counting process is initiated. Each successive step of the counter causes the digital word at the output to advance one level in the binary sequence. Each of these successive digital words is converted back to analog form

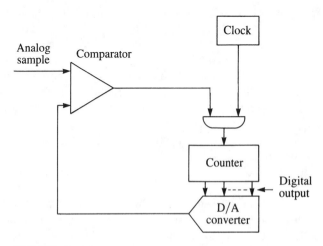

FIGURE 10–10
Block diagram of counter A/D converter.

by the D/A converter, and the output is compared with the analog sample. At the first point at which the D/A converter exceeds the level of the analog sample, the comparator changes states, and by means of the logic gate shown, the counter is inhibited. The value of the digital word at the counter output represents the desired digital representation of the analog signal. Depending on the conversion strategy desired, an additional offset voltage in the feedback circuit may be required to establish the exact point for the comparator at which the counter is inhibited. Further discussion of this concept as it relates to A/D specifications will be given in Section 10–6.

While the counter A/D converter is probably the easiest type to understand conceptually, it is relatively slow for a general-purpose converter. As the counter changes state, settling time must be allowed for the D/A converter, and a comparison must be performed at each step. Sufficient conversion time must be allowed during each conversion cycle for the counter to assume $2^n - 1$ states. As the number of bits increases, the number of states required increases exponentially. For example, if $n = 12$, conversion time must be allowed for $2^{12} - 1 = 4095$ counts. Obviously, only the highest level of the analog signal will necessitate all 4095 counts before the conversion is completed, but with the simplest circuit logic, a total time appropriate for the largest possible value must be allowed.

Certain variations of the counter A/D converter have been devised that make use of the property that a typical analog signal level will not normally change drastically from one sample to the next. With this logic, an up-down counter is used, and the counter follows the change of the analog signal. This type of counter A/D converter is referred to as a *tracking converter.* The tracking converter is faster than the basic counter converter, but the circuit logic is more complex.

Successive Approximation A/D Converter

The most widely employed general-purpose A/D converter is the successive approximation type, for which a block diagram is shown in Figure 10–11. Much of the opera-

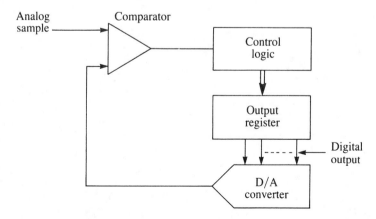

FIGURE 10–11
Block diagram of successive approximation A/D converter.

tion of a successive approximation A/D converter depends on the control logic. Indeed, in some systems employing microprocessors, the control logic is generated by software.

As in the case of the counter A/D converter, a sample of the analog signal appears at one input to the comparator. The digital word being generated is converted to analog form and compared with the analog sample, and the completion of conversion is triggered by a change in the state of the comparator. However, the logic for the successive approximation type is quite different from that of the counter type, as we will see shortly.

The successive approximation logic is illustrated by the flow chart shown in Figure 10–12. Starting with all zeros, a first "trial" digital value is generated by changing the MSB to 1. The D/A converter converts this to an analog value, and this level is compared

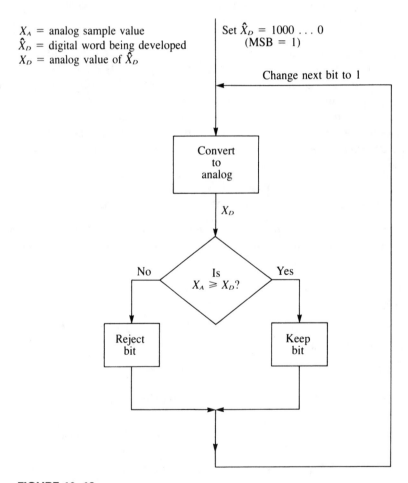

FIGURE 10–12
Flow chart of successive approximation A/D converter.

with the input analog sample. If the analog sample is less than the trial value, the assumed bit of 1 is rejected and replaced by 0. If, however, the input analog sample is greater than or equal to the trial value, the assumed bit of 1 is retained. The MSB remains at the value established in this part as subsequent comparisons are made.

Focus is next directed to the second most significant bit of the trial value, and this value is changed to 1. The trial value is converted to analog form and compared with the input sample. If the analog sample is less than the converted value, the second assumed bit level of 1 is rejected and replaced by 0. However, if the analog sample is greater than or equal to the converted value, the assumed second bit of 1 is retained.

The preceding process is repeated at all bit locations. The digital word appearing in the output register at the end of the process is the converted digital value. As pointed out for the ramp A/D converter, it may be necessary to add a bias level somewhere in the loop to establish transition points appropriate for a given conversion code.

From the preceding discussion, it is seen that for n bit words, only n comparisons are required for the successive approximation A/D converter. Consequently, the successive approximation type can perform a conversion in a much shorter time than the counter type and is thus capable of operating at higher conversion rates.

Flash A/D Converter

The flash (or parallel) A/D converter has the shortest possible conversion time of any available types. As we will see, however, its use is somewhat limited by practical and economic reasons.

A block diagram of the flash converter is shown in Figure 10–13. For $m = 2^n$ possible values, $m - 1 = 2^n - 1$ comparators are required. The analog sample is simultaneously applied as one input to all comparators. The other input to each comparator is a dc voltage whose level is a function of the location on the resistive voltage divider network on the left. The choices of the resistance $R/2$ at the bottom and $3R/2$ at the top are related to transitions midway between quantization levels in accordance with unipolar encoding, and this will be discussed in Section 10–6.

For a given analog sample level, all comparators below a certain point in the ladder will have one particular state, while those above that point will have the opposite state. This pattern of states is applied to a decoder circuit, which produces the required digital word in accordance with the input pattern. A portion of the decoder circuit could be a read-only memory (ROM), in which the input controls the address, and the output represents the digital value stored in the given location.

The short conversion time is evidenced by the fact that all the comparisons are performed in parallel, so the conversion time is based on the time of one complete comparison. The limitation, however, is the large number of comparators required. For example, an 8-bit converter requires $2^8 - 1 = 255$ comparators. However, a 16-bit converter would require 65,535 comparators! Practical implementations of flash A/D converters appear to have been limited to 8 bits, and these command premium prices; however, conversion times less than 20 ns, corresponding to conversion rates exceeding 50 MHz, are apparently possible with flash converters.

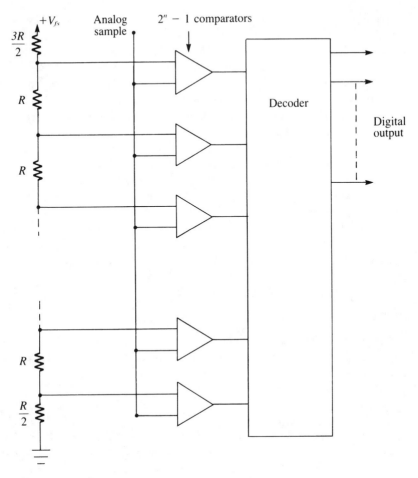

FIGURE 10–13
Block diagram of flash (or parallel) A/D converter.

Comparison

Comparing the three converters considered, we see that the conversion time for each is proportional to the number of comparisons required, and these results are summarized as follows:

Type	Comparisons
Counter (basic form)	$2^n - 1$
Successive approximation	n
Flash	1

The flash type is obviously the fastest, but it is the most expensive (by far), and the number of bits available is limited. The successive approximation type is readily available with a variety of different word lengths and conversion rates, and it is the most widely employed of the types considered here.

EXAMPLE 10–3

Assume that a certain form of integrated-circuit technology allows a comparator operation, amplifier settling, and the associated logic to be performed in about 1 μs. If all other times are assumed to be negligible, compute the approximate conversion times and data conversion rates for the three A/D converters considered in this section for the following word lengths: (a) 4 bits, (b) 8 bits, and (c) 16 bits.

Solution

Recognize that the assumption is based on one type of representative technology and should not be interpreted as any general result. The purpose is to provide a reasonable and meaningful comparison of the relative speeds of the different converter types. Let τ represent the time required for one comparison, and let T_c represent the total conversion time. The following formulas will be assumed:

$$\text{Counter:} \quad T_c = (2^n - 1)\tau \tag{10–22}$$

$$\text{Successive approximation:} \quad T_c = n\tau \tag{10–23}$$

$$\text{Flash:} \quad T_c = \tau \tag{10–24}$$

The conversion rate f_c is determined as

$$f_c = \frac{1}{T_c} \tag{10–25}$$

The results for different values of n are readily calculated and are summarized as follows:

4 bits	T_c	f_c
Counter	15 μs	66.7 kHz
Successive approximation	4 μs	250 kHz
Flash	1 μs	1 MHz
8 bits	T_c	f_c
Counter	255 μs	3.922 kHz
Successive approximation	8 μs	125 kHz
Flash	1 μs	1 MHz
16 bits	T_c	f_c
Counter	65.535 ms	15.259 Hz
Successive approximation	16 μs	62.5 kHz
Flash	1 μs	1 MHz

Observe the extreme range in conversion times and frequencies as the number of bits increases.

10–5 INTEGRATING ANALOG-TO-DIGITAL CONVERTERS

A special class of converters, denoted as ***integrating A/D converters***, will be discussed in this section. In these converters, either a reference voltage or the signal is integrated during the conversion process. When the signal is integrated, an improvement in the signal-to-noise ratio may result with certain types of signals.

It should be emphasized that integrating A/D converters are suitable only for very special types of applications and are usually not suited for the general-purpose functions for which the converters in the last section are intended. The kinds of applications will be clearer after operation is discussed, but as a general rule for starting the development, usage is normally limited to very low-frequency and "dc-like" signals. For example, many digital voltmeters employ an integrating A/D converter in the circuit.

Integrating Converter Types

Integrating A/D converters may be classified as (1) ***single-slope***, (2) ***dual-slope***, and (3) ***quad-slope***. In the single-slope converter, a reference voltage is integrated, while in the dual- and quad-slope converters, the signal voltage is integrated. The possible advantage of integrating the signal will be discussed after the operation of each of the three types has been presented.

Single-Slope Converter

A block diagram of the single-slope integrating A/D converter is shown in Figure 10–14. Assume a reference time $t = 0$ as the beginning of a conversion interval, and assume that

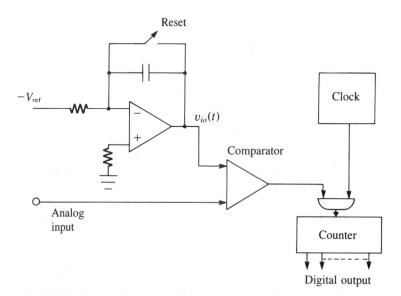

FIGURE 10–14
Block diagram of single-slope integrating A/D converter.

the integrator output voltage $v_{io}(t)$ is initially zero; that is $v_{io}(0) = 0$. Assume also that the counter state is initially zero. A sample of the analog signal is applied as one input to the comparator. Integration is initiated with an input voltage $-V_{ref}$, and the digital counter is started at the same time. For a negative constant voltage input, the output of the integrator at time t can be expressed as

$$v_{io}(t) = -\frac{1}{RC}\int_0^t (-V_{ref})\,dt = \frac{V_{ref}}{RC}t \qquad (10\text{–}26)$$

When the integrator output voltage reaches the level of the analog signal, the comparator changes states. By means of the associated logic, the counter is inhibited. Observe from (10–26) that the integrator output voltage increases linearly with time. Therefore, the time it takes the integrator output to reach the level of the analog signal is proportional to the level of the analog signal. Since the counter has been advancing at a constant rate, the value of the digital word contained in the counter is thus proportional to the level of the analog signal.

The single-slope converter suffers from possible component-value errors as well as clock errors. Referring to (10–26), note that the level of the integrator output voltage is a function of the RC product. Changes of capacitance and/or resistance with temperature will affect the integrator output and introduce error. Further, drifts in the clock frequency will also cause an error.

Dual-Slope Converter

A block diagram of the dual-slope converter is shown in Figure 10–15. Assume that the integrator output voltage $v_{io}(t)$ is initially at zero; that is, $v_{io}(0) = 0$. Assume also that the

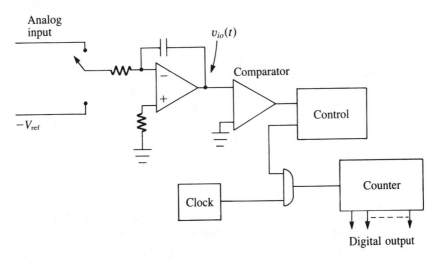

FIGURE 10–15
Block diagram of dual-slope integrating A/D converter.

counter state is initially zero. Finally, assume that the signal voltage is nearly constant and positive during the conversion period with a value V_i. (A sample-and-hold circuit is normally not used with this converter.) The signal voltage is applied to the integrator, and the processes of integration and counting are initiated. Integration is performed for the interval of time required for the counter to advance through all possible states. Let T_1 represent this time, and assume that the analog sample is constant during this interval. The integrator output voltage at $t = T_1$ can be written as

$$v_{io}(T_1) = -\frac{1}{RC}\int_0^{T_1} V_i\,dt = -\frac{V_i T_1}{RC} \tag{10–27}$$

At time $t = T_1$, the input to the integrator is changed from the analog input sample to a fixed negative reference voltage $-V_{ref}$. The counter again starts from zero. The negative integrator input voltage forces the output voltage to move back toward zero, and the voltage during this interval can be expressed as

$$v_{io}(t) = -\frac{1}{RC}\int_{T_1}^{t} (-V_{ref})\,dt - \frac{V_i T_1}{RC} \tag{10–28a}$$

or

$$v_{io}(t) = \frac{V_{ref}}{RC}(t - T_1) - \frac{V_i T_1}{RC} \tag{10–28b}$$

Integration will continue until $v_{io}(t)$ returns to zero, at which time the comparator state will change. Let T_2 represent the time required to return back to zero. Substituting $t = T_1 + T_2$ and $v_{io}(t) = 0$ in (10–28b), we establish the following relationship:

$$\frac{T_2}{T_1} = \frac{V_i}{V_{ref}} \tag{10–29}$$

The significance of this result is explained as follows: The digital value corresponding to the analog sample is proportional to the time T_2. However, since T_1 is always the time required to count through all possible values, long-term drifts in the clock frequency will affect T_2 and T_1 in essentially the same proportion. Therefore, the ratio T_2/T_1 should be relatively insensitive to slow drifts in the clock frequency. Observe also that R and C do not appear in the result of (10–29). Drifts in these components would affect T_1 and T_2 in the same proportion, so the ratio is independent of these parameter values.

Quad-Slope Converter

The quad-slope converter is in reality a dual-slope converter, but with additional logic added so that two dual-slope operations are performed for each data conversion cycle. During the first dual-slope conversion, the data input signal is removed, so the actual A/D input signal represents a combination of noise, drift, and other spurious components. The resulting converted value during this phase is an "error" sample that should be removed. This value is momentarily stored, and the input signal is applied and converted. The error signal is then subtracted digitally from the converted input signal, and the result represents a "purer" and more accurate representation of the analog signal.

Advantages and Disadvantages

Having discussed the different types of integrating A/D converters, the merits of dual-slope versus single-slope, and so on, we now turn to a very basic question concerning possible advantages (and disadvantages) of the integration process as applied to conversion. First, all integrating A/D converters are relatively slow. The precision integrator circuts used in these converters operate best at very slow integration rates.

In general, integration considerably alters the form of a signal. There is, however, a difference between the results of a ***continuous integrator*** and an ***integrate-and-dump*** integrator. The former represents the classical integration that continues over an indefinite interval, for which considerable discussion was given in Chapter 5. However, the form of integration performed in the A/D converters under consideration represents the ***integrate-and-dump*** (I & D) concept. With this approach, integration is performed for a fixed interval, at which time the capacitor is discharged. In the next interval, the capacitor starts with zero voltage again.

The integrate-and-dump operation, when applied to a signal, has the effect of smoothing out random variations of the data being converted. Consequently, if noise or random fluctuations appear with the signal, an enhancement of the signal with respect to the noise can be achieved with the integration process.

One way of describing the smoothing operation is by means of the frequency response. It can be shown that the integrate-and-dump operation is equivalent to a linear filtering operation whose amplitude response is of the form shown in Figure 10–16. The peaks of the amplitude response decrease as the frequency increases, which indicates the general trend of smoothing fluctuations and reducing noise.

A significant property is that there are definite nulls in the amplitude response at periodic intervals. If T_1 is the time over which integration of the signal is performed, the nulls occur at frequencies that are integer multiples of $1/T_1$. With some integrating A/D converters, the choice $T_1 = 16.667$ ms is made. The first null then occurs at $1/16.667 \times 10^{-3} = 60$ Hz. Consequently, extraneous pickup from ac power sources will be suppressed. Thus, dual- and quad-slope integrating A/D converters have the capacity to reduce random fluctuations and even to eliminate certain types of interference.

Although we gain definte noise-suppression advantages with dual- and quad-slope integrating A/D converters, we readily observe from Figure 10–16 that the shape of the frequency response curve could distort the spectrum of a signal. The curve is relatively flat at frequencies well below $1/T_1$, so this is the only frequency range in which the data spectrum is not altered. Thus, ***integrating A/D converters should be used only for dc and slowly varying signals if the integrity of the signal is to be preserved.***

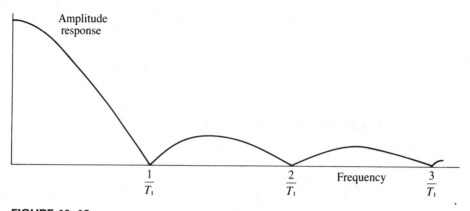

FIGURE 10–16
Amplitude response of dual-slope and quad-slope integrating A/D converters.

10–6 CONVERTER SPECIFICATIONS

The emphasis in this section will be directed toward the analysis and interpretation of A/D and D/A converter specifications. The available package units employ many of the various circuit design techniques that have been discussed in the preceding sections. However, the user is often not able to ascertain exactly how the various components within the unit perform their operations. Instead, he or she is concerned more with relating the given specifications to the requirements of the particular system.

Manufacturers differ somewhat in the manner in which they give specifications. The material given here is not intended to represent any particular manufacturer's approach, but it is intended to provide enough background to deal with a number of possible approaches.

Normalized Value

The discussion will utilize the concept of a ***normalized value,*** which was introduced in Section 10–3. The normalized value can be expressed as

$$\text{Normalized value} = \frac{\text{actual value}}{\text{full-scale value}} \qquad \textbf{(10–30)}$$

Conversely, the actual value can be expressed as

$$\text{Actual value} = \text{normalized value} \times \text{full-scale value} \qquad \textbf{(10–31)}$$

The relationships of (10–30) and (10–31) can be applied to either analog or digital values, depending on whether A/D conversion or D/A conversion is being performed. The "full-scale value" will be either a reference voltage or a current, depending on the converter.

Number of Levels

Let n represent the number of bits in each word. The number of levels m is then given by

$$m = 2^n \qquad\qquad (10\text{–}32)$$

In some of the discussion that follows, the value $n = 4$ will be used for illustration. This value gives $m = 2^4 = 16$ words, which is sufficiently large to illustrate the general trend but small enough to show the results graphically.

Table 10–1 provides information concerning various binary encoding schemes to be considered for 4 bits. The sixteen possible digital words expressed in natural binary and their corresponding decimal integer values are shown in the left-hand columns. Normalization of the values of the binary words to a range less than 1 is achieved by adding a binary point to the left of the values as given in the table. However, to simplify the notation, the binary point will usually be omitted, but it will be understood in all discussions in which the normalized form is assumed.

Two common encoding relationships between the analog and digital values will be considered. These are called (1) *unipolar encoding* and (2) *bipolar offset encoding.* Each process will be discussed separately.

TABLE 10–1

Natural binary numbers, decimal values, and unipolar and bipolar offset values for 4 bits

Natural Binary Number	Decimal Value*	Unipolar Normalized Decimal Value†	Bipolar Offset Normalized Decimal Value†
1111	15	$^{15}/_{16} = 0.9375$	$^{7}/_{8} = 0.875$
1110	14	$^{14}/_{16} = 0.875$	$^{6}/_{8} = 0.75$
1101	13	$^{13}/_{16} = 0.8125$	$^{5}/_{8} = 0.625$
1100	12	$^{12}/_{16} = 0.75$	$^{4}/_{8} = 0.5$
1011	11	$^{11}/_{16} = 0.6875$	$^{3}/_{8} = 0.375$
1010	10	$^{10}/_{16} = 0.625$	$^{2}/_{8} = 0.25$
1001	9	$^{9}/_{16} = 0.5625$	$^{1}/_{8} = 0.125$
1000	8	$^{8}/_{16} = 0.5$	0
0111	7	$^{7}/_{16} = 0.4375$	$-^{1}/_{8} = -0.125$
0110	6	$^{6}/_{16} = 0.375$	$-^{2}/_{8} = -0.25$
0101	5	$^{5}/_{16} = 0.3125$	$-^{3}/_{8} = -0.375$
0100	4	$^{4}/_{16} = 0.25$	$-^{4}/_{8} = -0.5$
0011	3	$^{3}/_{16} = 0.1875$	$-^{5}/_{8} = -0.625$
0010	2	$^{2}/_{16} = 0.125$	$-^{6}/_{8} = -0.75$
0001	1	$^{1}/_{16} = 0.0625$	$-^{7}/_{8} = -0.875$
0000	0	0	$-^{8}/_{8} = -1$

SOURCE: Stanley, *Electronic Communication Systems,* 1982, p. 302. Reprinted with permission of Reston Publishing Company, 11480 Sunset Hills Road, Reston, VA 22090.

* Decimal value for integer representation of binary number.

† Decimal value with binary point understood on left-hand side of binary number.

Unipolar Encoding

The unipolar represenation is based on the assumption that the analog signal is either positive or zero, but never negative. Let X_u represent the normalized unipolar analog signal, and let $\hat{X}_u$ represent the corresponding digital binary representation. In view of the normalized form, the analog level is always bounded by $0 \le X_u < 1$. Note that the analog function is assumed never to equal unity, although it can approach that limit arbitrarily close.

For $n = 4$, the 16 possible values of X_u are shown in the third column of Table 10–1. The binary value 0000 corresponds to true decimal 0, but the binary value 1111 corresponds to the decimal value $15/16 = 0.9375$, which is less than unity, in accordance with the preceding discussion. The binary value 1000 corresponds to the decimal value 0.5.

Let ΔX_u represent the normalized step size, that is, the difference between successive levels. The step size is also the decimal value corresponding to 1 LSB. The normalized step size for unipolar encoding in general is

$$\Delta X_u = 2^{-n} \qquad (10\text{–}33)$$

The largest unipolar normalized decimal value $X_u(\text{max})$ differs from unity by one step size and is

$$X_u(\text{max}) = 1 - 2^{-n} \qquad (10\text{–}34)$$

Observe that $X_u(\text{max})$ approaches 1 arbitrarily close as n increases without limit. The digital word whose MSB is 1 and whose other bits are 0s corresponds to the normalized decimal value 0.5. This point will be conveniently referred to as the **half-scale level.** In addition, the value $X_u = 1$ will be referred to as the **full-scale level** even though it is never quite reached.

Bipolar Offset Encoding

The bipolar offset representation is based on the assumption that the analog signal can assume both positive and negative values with nearly equal peak magnitudes. Let X_b represent the normalized bipolar analog signal, and let $\hat{X}_b$ represent the corresponding digital binary representation. For the bipolar case, the analog function is bounded by $-1 \le X_b < 1$. Note that the analog function can assume the value -1, but it can never quite reach the value $+1$.

For $n = 4$, the 16 possible values of X_b are shown in the fourth column of Table 10–1. The binary value 0000 corresponds to the decimal value -1, but the binary value 1111 corresponds to the decimal value $7/8 = 0.875$, which is less than unity. The binary value 1000 corresponds to a decimal value of 0.

Let ΔX_b represent the normalized step size, which is also the value of 1 LSB. Since the normalized range for bipolar offset encoding is assumed to have twice the range as for unipolar encoding, the step size is twice as big and is

$$\Delta X_b = 2^{-n+1} \qquad (10\text{–}35)$$

The largest bipolar normalized decimal value $X_b(\text{max})$ differs from $+1$ by one step size and is

$$X_b(\text{max}) = 1 - 2^{-n+1} \qquad (10\text{–}36)$$

The digital word whose MSB is 1 and whose other bits are 0s corresponds to the decimal value 0.

From the fourth column of Table 10–1 for 4 bits, it is observed that 8 binary values correspond to negative decimal levels, 7 binary values correspond to positive decimal levels, and 1 binary value corresponds to decimal 0. In general, there are $m/2 = 2^{n-1}$ binary words representing negative decimal values, $2^{n-1} - 1$ binary words corresponding to positive decimal values, and one word corresponding to decimal zero.

One significant property of the bipolar offset representation is that it can be readily converted to the digital twos-complement form. To convert to twos-complement form, the MSB of each bipolar offset word is replaced by its logical complement. In fact, the bipolar form is sometimes denoted as the "modified twos-complement" representation. The close relationship between bipolar offset and twos-complement representations provides advantages in interfacing between analog and digital systems where arithmetic operations are to be performed on the digital data.

Before proceeding further, we will present some discussion about the step size differences between unipolar and bipolar offset representations. We have defined the normalized ranges for unipolar as 0 to 1 and for bipolar as −1 to 1. Consequently, the step size for bipolar has twice the value as that for unipolar. This interpretation seems natural and in accordance with the manner in which such converters would be applied. However, the internal operation of A/D and D/A converters is generally based on a unipolar concept, and the bipolar operation is achieved by adding a fixed bias level. Thus, a given converter designed to operate with an analog signal level from 0 V to 10 V in unipolar would operate from −5 V to +5 V in bipolar if no relative weighting changes were made in the circuitry. To produce this shift, a fixed bias level is added in the circuit.

Because of the actual circuit operational form, some manufacturers prefer to specify error as a function of the peak-to-peak signal range of the converter without regard to whether it is to be used for unipolar or bipolar encoding. This approach is more natural from the standpoint of the circuit operation, but the normalized approach given earlier is more natural from the standpoint of the user who may or may not understand the circuit details.

While confusing, these differences need not cause any difficulty, provided that we are careful. For unipolar encoding, there should be no ambiguity since the two approaches yield identical results. For bipolar offset encoding, however, we will define the full-scale signal range of the converter as one-half the peak-to-peak signal range, and the normalized value will be defined over the range from −1 to +1. The percent error computed using this approach may appear to be twice the value specified by some manufacturers, but it is a result of the difference in the definition and interpretation.

Unipolar Quantization Characteristic

The exact correspondence between the quantized decimal values and the range of input analog levels is represented by a *quantization characteristic curve.* Either *truncation* or *rounding* may be employed. With *truncation,* the encoded value of an analog sample is assigned to the *next lower* quantized level. With *rounding,* the encoded value of an analog sample is assigned to the *nearest* quantized level. For example, assume that two successive digital words correspond to analog levels of 3.1 V and 3.2 V, respectively. With truncation, samples of 3.14 V and 3.16 V would both be assigned the digital word corresponding to 3.1 V. However, with rounding, the value of 3.14 V would be assigned to the digital word corresponding to 3.1 V but the value of 3.16 V would be assigned to the digital word corresponding to 3.2 V. The average error is less with rounding than with truncation, but there are some situations in which truncation is preferred. Rounding will be assumed in subsequent developments unless stated otherwise.

The quantization characteristic of an ideal 4-bit A/D converter employing *rounding* and *unipolar* encoding is shown in Figure 10–17. The actual quantizer relationship is the

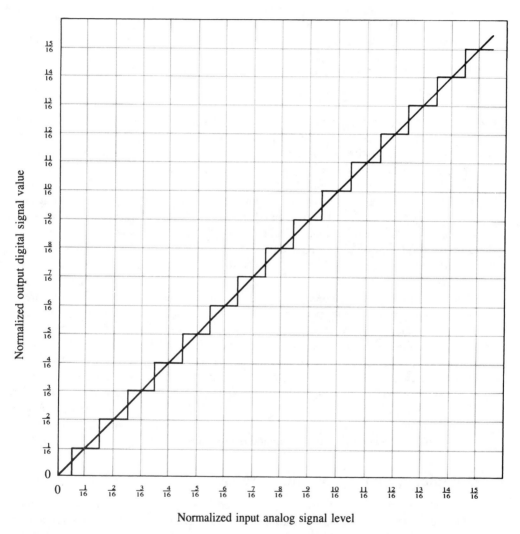

FIGURE 10–17

Unipolar quantization characteristic for 4-bit A/D converter. (*SOURCE*: Stanley, *Electronic Communication Systems*, 1982, p. 306. Reprinted with permission of Reston Publishing Company, 11480 Sunset Hills Road, Reston, VA 22090.)

"staircase" function, and the straight line represents an ideal linear relationship of equality between the output and the input. For a normalized input analog signal x in the range $x < 1/32$, the digital word is 0000. However, for $1/32 < x < 3/32$, the digital word is 0001. The pattern continues to the maximum range $x > 29/32$, in which the output assumes the binary value 1111.

The vertical difference between the quantizer characteristic and the straight line represents the **quantization error.** For specific values where the curves intersect, there is no quantization error. If we assume that the peak normalized input signal is 31/32, the peak quantization error is 1/32, which is one-half the value of 1 LSB.

In general, for unipolar encoding and rounding, the normalized peak quantization error can be expressed as

$$\text{Normalized quantization error} = \pm 2^{-(n+1)} \qquad \text{(10–37)}$$

The actual quantization error in appropriate units is

$$\text{Quantization error} = \pm 2^{-(n+1)} \times \text{full-scale value} \qquad \text{(10–38)}$$

Typical full-scale values for common A/D converters are 2.5 V, 5 V, 10 V, and 20 V. The percentage quantization error is

$$\text{Percentage quantization error} = \pm 2^{-(n+1)} \times 100\% \qquad \text{(10–39)}$$

The value of the percentage error is the same for either normalized or actual values.

Bipolar Quantization Characteristic

The quantization characteristic of an ideal 4-bit A/D converter employing **rounding** and **bipolar** encoding is shown in Figure 10–18. For a normalized input analog signal x in the range $x < -15/16$, the digital word is 0000. For $-15/16 < x < -13/16$, the digital word is 0001. In the mid-range, for $-1/16 < x < 1/16$, the digital word is 1000. Finally, for $x > 13/16$, the digital word is 1111.

The peak error for the bipolar case is now 1/16 on a normalized basis. As explained earlier, this value is a result of the fact that the peak-to-peak range is assumed to be twice as large as for unipolar encoding.

With bipolar offset encoding and rounding, the normalized peak quantization error can be expressed as

$$\text{Normalized quantization error} = \pm 2^{-n} \qquad \text{(10–40)}$$

The actual quantization error is

$$\text{Quantization error} = \pm 2^{-n} \times \text{full-scale value} \qquad \text{(10–41)}$$

Recall that the full-scale value for bipolar offset encoding is defined as one-half the peak-to-peak range in this development. The percentage quantization error is

$$\text{Percentage quantization error} = \pm 2^{-n} \times 100\% \qquad \text{(10–42)}$$

This percentage is referred to as the peak value rather than the peak-to-peak value.

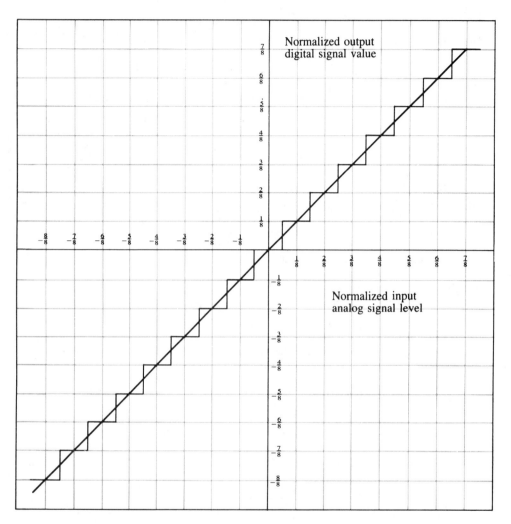

FIGURE 10–18
Bipolar offset quantization characteristic for 4-bit A/D converter. (*SOURCE*: Stanley, *Electronic Communication Systems*, 1982, p. 307. Reprinted with permission of Reston Publishing Company, 11480 Sunset Hills Road, Reston, VA 22090.)

D/A Unipolar Conversion Characteristic

Having observed the quantization characteristics for A/D converters, we will next investigate the corresponding characteristics for D/A converters. Using 4-bit words for illustration again, the normalized characteristics for unipolar encoding are shown in Figure 10–19. The 16 digital word values ranging from 0000 to 1111 are indicated along the horizontal scale, and the corresponding normalized analog values are shown on the vertical

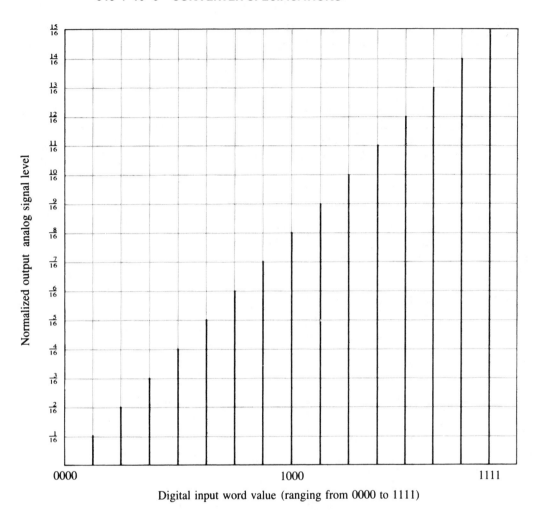

FIGURE 10–19
Normalized unipolar D/A conversion characteristic.

scale. Unlike the A/D conversion, the D/A conversion process is unique; that is, there is a one-to-one correspondence between a given digital word and an analog value once the D/A circuit is established. However, there are gaps between levels, arising from the existing quantized states.

Observe that binary value 0000 corresponds to an analog level of zero, binary 0001 corresponds to a normalized analog level of 1/16, and so on. Finally, the binary value 1111 corresponds to a normalized analog level of 15/16.

In view of the abrupt changes between successive levels, filtering and smoothing of the restored analog signal are normally employed. In many systems, a holding circuit is employed to maintain a given analog level until the next conversion is performed. Further smoothing by a low-pass filter is then used.

D/A Bipolar Conversion Characteristic

The normalized D/A characteristics for bipolar offset encoding with 4 bits are shown in Figure 10–20. Observe that binary 0000 now corresponds to a normalized analog level of -1, binary 0001 corresponds to a normalized analog level of $-7/8$, and so on. At the mid-range, binary 1000 corresponds to an analog level of 0. Finally, binary 1111 corresponds to an analog level of 7/8.

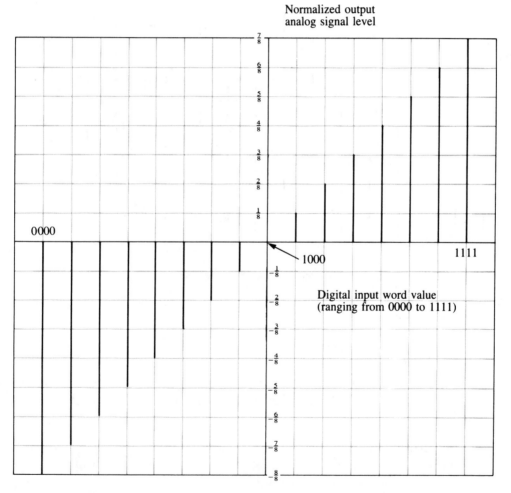

FIGURE 10–20

Normalized bipolar offset D/A conversion characteristic.

EXAMPLE 10–4

A certain 12-bit A/D converter connected for unipolar encoding (with rounding) has a full-scale voltage of 10 V. Determine the following quantities:
(a) normalized step size
(b) actual step size in volts
(c) normalized maximum quantized level
(d) actual maximum quantized level in volts
(e) normalized peak quantization error
(f) actual peak quantization error in volts
(g) percentage quantization error

Solution
The desired quantities may be determined from the unipolar relationships of this section. The full-scale voltage is 10 V. The various results are developed in the steps that follow.
(a) The normalized unipolar step size is

$$\Delta X_u = 2^{-n} = 2^{-12} = 0.00024414 \qquad (10\text{–}43)$$

(b) Actual step size $= 0.00024414 \times 10 = 2.441$ mV $\qquad (10\text{–}44)$
(c) The normalized maximum quantized level is

$$X_u(\text{max}) = 1 - 2^{-n} = 1 - 2^{-12} = 0.99976 \qquad (10\text{–}45)$$

(d) Actual maximum quantized voltage $= 0.99976 \times 10 = 9.9976$ V $\qquad (10\text{–}46)$

(e) Normalized peak quantization error $=$
$$\pm 2^{-(n+1)} = \pm 2^{-13} = \pm 0.00012207 \qquad (10\text{–}47)$$

(f) Actual peak quantization error $= \pm 0.00012207 \times 10 = \pm 1.22$ mV $\qquad (10\text{–}48)$

(g) Percentage quantization error $=$
$$\pm 2^{-(n+1)} \times 100\% = \pm 2^{-13} \times 100\% = 0.0122\% \qquad (10\text{–}49)$$

EXAMPLE 10–5

The 12-bit A/D converter of Example 10–4 is next connected for bipolar offset encoding with the same peak-to-peak voltage range as in Example 10–4. (This range is achieved by adding a positive internal bias of 5 V to the signal before conversion.) The converter is now intended for signals in the range from -5 V to just under $+5$ V. Repeat all the calculations of Example 10–4 for this case.

Solution
For bipolar offset encoding, the normalized range is assumed to be from -1 to $+1$. The full-scale voltage is now assumed to be 5 V. The results are developed in the steps that follow.
(a) The normalized bipolar step size is

$$\Delta X_b = 2^{-n+1} = 2^{-11} = 0.00048828 \qquad (10\text{–}50)$$

(b) Actual step size $= 0.00048828 \times 5 = 2.441\,\text{mV}$ **(10–51)**

Comparing (10–50) with (10–43) and (10–51) with (10–44), we note that the normalized step size for bipolar is twice the value as for unipolar, but the actual step size is the same. The equality of the actual step sizes is a result of the same peak-to-peak voltage range, but the differences in the normalized steps results from the way the normalized range is defined.

(c) The normalized maximum quantized level is

$$X_b(\text{max}) = 1 - 2^{-n+1} = 1 - 2^{-11} = 0.99951 \quad \textbf{(10–52)}$$

(d) Actual maximum quantized voltage $= 0.99951 \times 5 = 4.9976\,\text{V}$ **(10–53)**

(e) Normalized peak quantization error $=$

$$\pm 2^{-n} = \pm 2^{-12} = \pm 0.0002441 \quad \textbf{(10–54)}$$

(f) Actual peak quantization error $= \pm 0.0002441 \times 5 = \pm 1.22\,\text{mV}$ **(10–55)**

(g) Percentage quantization error $= \pm 2^{-n} \times 100\% = 0.0244\%$ **(10–56)**

Comparing (10–54), (10–55), and (10–56) with (10–47), (10–48), and (10–49), respectively, we note that the actual peak quantization error for bipolar is the same value as for unipolar, but the normalized and percentage quantization errors for bipolar are twice the values for unipolar. The percentage comparison is based on the peak (rather than the peak-to-peak) value. Thus, while the *actual* error is the same, the *relative* error is twice as big.

10–7   ADC0801-0805 A/D CONVERTERS

This section introduces one group of popular economical IC A/D converters. The group begins with the ADC0801 and continues through the ADC0805. We will refer to any one of the group or the entire group simply by the number "801." These units are manufactured by National Semiconductor Corporation.

The 801 units are all 8-bit successive approximation A/D converters. The basic operation of all units is similar, but the different identification numbers relate to the accuracy specifications. These A/D converters are compatible with the 8080 microprocessor and its subsequent modifications. Thus, they are ideal for applications in which sampled data are to be processed with a microprocessor.

The 801 converters are designed to operate with a 5-V peak-to-peak analog signal range. Consequently, if the analog signal has a larger range, it is necessary to attenuate the signal before conversion. Likewise, bipolar offset encoding requires the addition of a 2.5-V reference to the analog signal, which would have to be limited from -2.5 V to 2.5 V.

The 801 units have a clock generator on the chip. The speed of the clock can be varied within some limits. Accuracy of the process is assured at clock frequencies up to 640 kHz. It takes a minimum of 66 clock periods to perform a conversion. This means that the minimum conversion time is about $66 \times 1/(640 \times 10^3) = 103\,\mu\text{s}$. This conversion time corresponds to a sampling frequency of about 9697 Hz. On the other hand, the maximum number of clock periods is 73, which corresponds to a conversion time of about $73 \times (1/640 \times 10^3) = 114\,\mu\text{s}$. The sampling frequency in this case is about 8767 Hz.

Free-Running Connection

The 801 converters may be synchronized with an external signal (e.g., a microprocessor), or they may be operated in a free-running mode. A typical free-running connection, based on information provided in data sheets, is shown in Figure 10–21. The choice of V_{CC} = 5.12 V results in convenient values for the various steps, as will be demonstrated in Example 10–6. (The absolute maximum value permitted for V_{CC} is 6.5 V.) The circuit as shown is connected for application with a unipolar analog signal ranging from zero to about 5.12 V.

To ensure that the conversion process starts, the WR and INTR controls are momentarily grounded. This is achieved with the switch on the left. Note that the circuit contains both an A (analog) ground and a D (digital) ground. However, these two separate grounds should be connected together.

The voltage appearing at the $V_{ref}/2$ terminal is half the supply voltage. Thus, for V_{CC} = 5.12 V, a voltage of 2.56 V appears at this terminal.

The 8 bits representing the converted digital words are denoted as DB0 through DB7. The most significant bit is DB7, and the least significant bit is DB0.

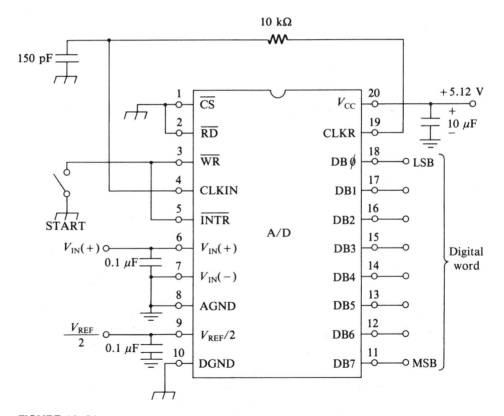

FIGURE 10–21

Typical free-running connection for 801 A/D converters. (Courtesy of National Semiconductor, Inc.)

EXAMPLE 10–6

Consider the 801 A/D converter connection of Figure 10–21. (a) Let $v_1, v_2, v_3, \ldots, v_8$ represent the eight actual analog (or decimal) unipolar voltage levels corresponding to the eight digital words 00000001, 00000010, 00000100, . . . , 10000000. Calculate these voltages. (b) Let $v_{t1}, v_{t2}, v_{t3}, \ldots, v_{t8}$ represent the eight corresponding actual unipolar lower transition voltages at which the previous digital words will be generated. Calculate these voltages.

Solution
(a) For each of the eight voltages being calculated in this part, only one bit has a level of "1," while the other seven have a level of "0." Starting with v_1, this voltage is simply the value corresponding to 1 LSB and is

$$v_1 = 5.12 \times 2^{-8} = 20\,\text{mV} \tag{10–57}$$

The value of v_2 is

$$v_2 = 5.12 \times 2^{-7} = 2v_1 = 40\,\text{mV} \tag{10–58}$$

The value of v_3 is

$$v_3 = 5.12 \times 2^{-6} = 2v_2 = 80\,\text{mV} \tag{10–59}$$

The pattern continues in this fashion until all eight values are calculated. These values are summarized in the first row of the table following part (b).

(b) The transition on the low side for each of the voltages calcualted in (a) occurs at a level of 1/2 LSB below the given voltage level. Thus, in general

$$v_{tk} = v_k - \frac{1}{2}\text{LSB} = v_k - 10\,\text{mV} \tag{10–60}$$

The eight values are readily calculated and are given on the second row of the table that follows.

	$k = 1$	2	3	4	5	6	7	8
v_k	20 mV	40 mV	80 mV	160 mV	320 mV	640 mV	1.28 V	2.56 V
v_{tk}	10 mV	30 mV	70 mV	150 mV	310 mV	630 mV	1.27 V	2.55 V

10–8 DAC0806-0808 D/A CONVERTERS

An introduction to an economical group of IC D/A converters is given in this section. This group consists of the DAC0806, the DAC0807, and the DAC0808 converters. We will refer to any one or the group simply by the number 806. These converters are manufactured by National Semiconductor Corporation, and complete data sheets appear in data manuals.

The 806 units are 8-bit converters and can interface directly with TTL, DTL, or CMOS logic levels. Dual power supply biasing is normally used, and the required ranges are from ± 4.5 V to ± 18 V.

Typical Connection

A typical circuit diagram of a D/A conversion system utilizing an 806 converter is shown in Figure 10–22. The particular scheme shown provides a unipolar positive output voltage. The output current is converted to a voltage through the current-controlled voltage source below the converter.

Assume that $R_{14} = R_{15} = R$. The full-scale voltage V_{fs} for this configuration is given by

$$V_{fs} = \frac{R_f}{R} V_{REF} \tag{10–61}$$

In practice, one of the two resistances can be adjusted to provide a more exact scaling relationship.

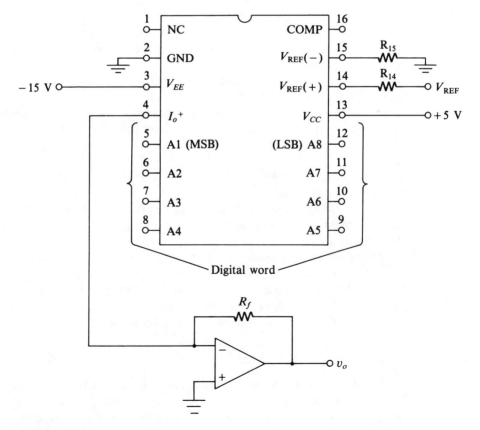

FIGURE 10–22
Typical connection for 806 D/A converter. (Courtesy of National Semiconductor, Inc.)

EXAMPLE 10–7

Consider the D/A connection of Figure 10–22 and assume that $R_{14} = R_{15} = R = R_f = 10$ kΩ and $V_{REF} = 5.12$ V. Determine the full-scale voltage.

Solution
The full-scale voltage is

$$v_{fs} = \frac{R_f}{R} V_{REF} = \frac{10\,\text{k}\Omega}{10\,\text{k}\Omega} \times 5.12\,\text{V} = 5.12\,\text{V} \qquad (10\text{–}62)$$

This particular circuit design matches the A/D design of Example 10–6 in the sense that the output analog levels correspond to the mid-range values of the inputs to the A/D converter.

10–9 ANALOG SWITCHES AND MULTIPLEXERS

Analog switches are components in which an "open" or a "closed" condition between two points can be achieved by means of a control signal without any moving parts. An ideal analog switch would perform electronically the same operation as a mechanical relay. Practical analog switches exhibit a small forward resistance in the "on" state and some leakage in the "off" state. Earlier in the text, references were made to electronic switches in several applications.

Most analog switches utilize field effect transistor (FET) technology. Some employ only junction field effect transistors (JFETs), while others consist of hybrid combinations of JFETs and metal oxide semiconductor field effect transistors (MOSFETs). Still others employ complementary (CMOS) forms. The detailed consideration of the different forms of analog switching is the subject of at least one complete book.

An *analog multiplexer* is a combination of several switches for the purpose of transmitting or processing samples of two or more analog signals in sequential order (serial form) on the same line. While such a unit could be implemented with individual switches, multiplexers are available containing a number of switches with common characteristics.

For a given type of analog switch or multiplexer, there is a certain range of input signal in which the unit will behave properly as a switch. Consequently, a switch or multiplexer must be carefully selected to provide a proper range of operation for a given signal condition. Other parameters of concern are the switching times (both off and on), "on" resistance, "off" leakage current, and the control signal requirements.

P-Channel JFET Switch

The switch concept will be illustrated with one representative type, a *P*-channel JFET. The basic form of such a switch is illustrated in Figure 10–23(a). This simplified form is intended to function only with either positive signals or with a virtual ground form, as will be illustrated shortly.

The *P*-channel type is a useful form because it is compatible with standard transistor-transistor logic (TTL) levels. A TTL 1 applied to the gate turns off (opens) the switch

FIGURE 10–23
JFET *P*-channel analog switch with off and on models.

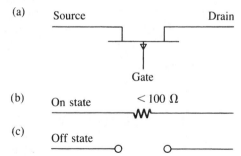

for certain signal levels, while a TTL 0 applied to the gate turns on (closes) the switch. Simplified circuit forms for on and off conditions are illustrated in Figures 10–23(b) and (c). The "on" resistance is typically less than 100 Ω.

4-Channel Multiplexer

A 4-channel multiplexer composed of *P*-channel switches connected in an appropriate multiplexing circuit is illustrated in Figure 10–24. This circuit employs the concept of summing at the virtual ground level, an acceptable condition to ensure proper operation of the JFET switches. Observe the extra JFET connected in the feedback circuit. This unit adds a resistance to the feedback resistance approximately equal to the value con-

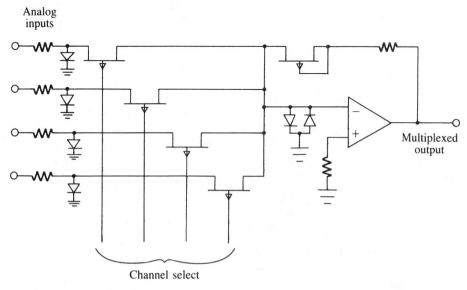

FIGURE 10–24
Circuit utilizing 4-channel JFET *P*-channel multiplexer.

tributed by each FET in the input circuit, and the result is a more accurate gain realization. Channel selection is achieved by a digital circuit in which a TTL 0 is sequentially applied to each of the four gates while 1s are applied to the other three at a given time.

10–10   VOLTAGE AND FREQUENCY CONVERSIONS

Two special types of data conversion circuits are discussed in this section. They are the (1) *voltage-to-frequency (V/F) converter* and (2) the *frequency-to-voltage (F/V) converter.* Some available chips can perform either operation, but the functions are discussed separately in this section. The emphasis here is on the external operating characteristics rather than the detailed circuit considerations.

Voltage-to-Frequency Converter

A V/F converter produces an output signal whose instantaneous frequency is a function of an external control voltage. The output signal may be a sine wave, a square wave, or a pulse train. For the latter case, the instantaneous frequency is often interpreted as the number of pulses generated per unit time.

The resulting frequency variation with signal is a form of frequency modulation (FM), and some of the technology developed in communications may be applied to V/F conversion. The term *voltage-controlled oscillator* (*VCO*) originated with FM, and this term is synonymous with *V/F converter.*

The frequency of many oscillator circuits can be varied by application of a control voltage at certain points. However, any circuit that is to be considered seriously for use a V/F converter must display the proper characteristics; thus it requires some careful design.

The ideal conversion characteristic of a V/F converter is shown in Figure 10–25. The input control voltage is shown on the horizontal scale, and the frequency deviation is shown on the vertical scale. The frequency deviation is the difference between the actual frequency and the frequency with no external control signal applied. The ideal curve is linear. For example, assume that a V/F converter has a frequency of 1 kHz with no signal applied. If an input signal of 1 V causes the frequency to shift to 1050 Hz, a signal of 2 V should cause the frequency to shift to 1100 Hz if the curve is linear. The first case corresponds to a deviation of 50 Hz, and the second case corresponds to a deviation of 100 Hz.

Frequency-to-Voltage Converter

An F/V converter produces an output voltage whose amplitude is a function of the frequency of the input signal. Like the V/F converter, the unit may respond to a sine wave, a square wave, or a pulse train. Such a circuit is essentially an FM discriminator or detector.

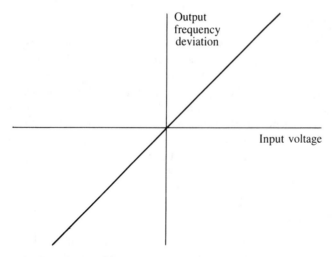

FIGURE 10–25
Ideal V/F converter characteristic.

The ideal conversion characteristic of an F/V converter is shown in Figure 10–26. The input frequency deviation (referred to some reference center frequency) is shown on the horizontal scale. The output voltage is shown on the vertical scale. The ideal curve is again linear. For example, assume that the output voltage is zero with an input frequency of 1 kHz. If an input frequency of 1050 Hz (corresponding to a 50-Hz deviation) causes the output voltage to be 1 V, an input frequency of 1100 Hz (a 100-Hz deviation) should cause the output voltage to be 2 V if the conversion is linear.

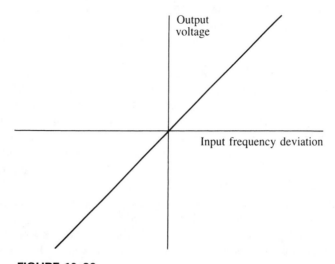

FIGURE 10–26
Ideal F/V converter characteristic.

Applications

In one sense, V/F conversion with a pulse train can be thought of as a form of A/D conversion. This interpretation is based on the fact that pulses assume only two states, and the number of pulses produced per unit of time is a function of the signal level. Voltage-to-frequency conversion is particularly useful for conversion systems in which the data are to be transmitted over some reasonable distance, since the output is in a modulated form. Thus, there is no need to transmit a dc level, and various ac coupling methods, such as transformers and ac amplifiers, may be employed. At the receiving end, an F/V converter matched to the V/F converter at the sending end is used to convert the data back to analog form.

10–11   MULTISIM EXAMPLE

Most of the circuits discussed in this chapter are purchased as off-the-shelf subsystems and do not always lend themselves to the type of circuit modeling that we have employed in earlier chapters. However, Multisim contains some complete subsystems that function in the same manner as certain of the circuits. Both the A/D and D/A converter modules of Multisim will be demonstrated in the example that follows.

MULTISIM EXAMPLE 10–1

Using the Multisim **A/D** and **Voltage D/A** models contained in the library, demonstrate their application by converting a ramp voltage to digital form and converting it back to analog form.

Solution

The A/D and D/A converters are found in the **Mixed** component bin. When the bin opens left-click on the button with **ADC DAC**. The **Component Browser** will then provide several choices. For this example, the basic **ADC** unit was selected for the A/D converter and the **VDAC** (voltage digital-analog-converter) was slected for the D/A converter. The latter was rotated 90° for convenience.

Refer to the Multisim circuit schematic shown in Figure 10–27 for the discussion that follows. The full-scale voltage on the A/D converter is **VREF+ − VREF−.** The first terminal is connected to a 10-V dc source and the latter terminal is grounded, which results in a full-scale voltage of 10 V. Likewise, the + and − terminals on the top of the D/A converter perform the same function and the value set there is also 10 V. Thus, if perfect coding and decoding could be achieved, the final and initial voltage levels would agree.

The input signal to the A/D converter utilizes the **PWL** source and it is connected to the terminal **VIN.** A ramp voltage that increases linearly from 0 to 10 V in a time duration of 1 s was used as the input. This node is labeled as "in". The output of the D/A converter is taken at the node at the bottom, which is labeled "out".

The A/D converter terminal labeled **SOC** represents "start of conversion". A positive pulse initiates this action and it is achieved by the signal source labeled **V clock.** The clock is obtained from the **Sources** bin and is designated as **CLOCK_SOURCE.** The parameters set for this source are the **Frequency, Duty Cycle**, and **Voltage.**

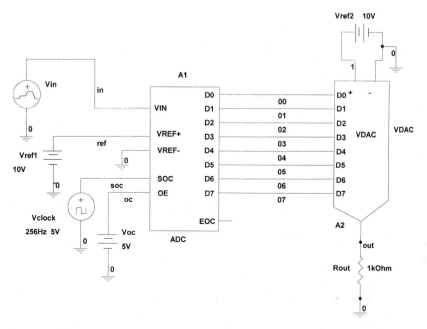

FIGURE 10–27
Circuit of Multisim Example 10–1.

Both the A/D and D/A converters are 8-bit units, so there are 256 possible states. In order to provide a simple format for demonstration, the clock frequency is set at 256 Hz. Thus, all 256 states occur over an interval of 1 second while the input signal ramps from 0 to 10 V. The duty cycle is set to 50% and the clock voltage is set to 5 V.

The value of one significant bit is 10 V/256 = 39.06 mV. The time duration for each change of state should be 1 second/256 = 3.906 ms.

The output of the A/D converter is an open-collector configuration. However, by raising the **OE** terminal high, the output bits are available. This is achieved by connecting a 5-V dc source to this terminal.

The terminals **DO, D1**, . . . , **D7** represent the 8 bits with the highest number representing the most significant bit. These terminals are connected directly to the corresponding terminals of the D/A converter.

The results of two separate transient runs are provided for illustration. In the first case, the transient was run only to 20 ms, and the results are shown in Figure 10–28. The first transition occurs at about 3.9 ms as expected, and the resulting output voltage is about 39 mV. This suggests that the A/D converter uses truncation rather than rounding. Over the interval shown, five steps occur, and there is very close agreement at the transition points. For each transition in which a bit changes from a 1 to a 0, there is a very short negative spike, which has something to do with the programming in the converter software. This could be eliminated by the use of a sample-and-hold circuit at the output of the D/A converter.

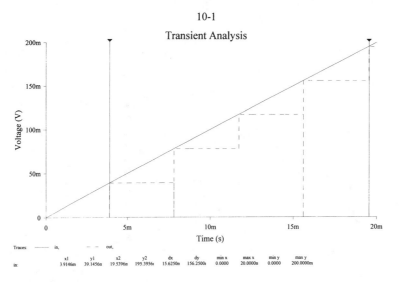

FIGURE 10–28

Response of Multisim Example 10–1 over interval of 20 ms.

The second case involves the process over the complete interval of 1 second, and the results are shown in Figure 10–29. At this "macroscopic" level, the input and output functions appear to be very close except for the spikes previously observed. Of course, even when the spikes are removed, there is always the uncertainty of 1 LSB, which is about 39 mV in this case.

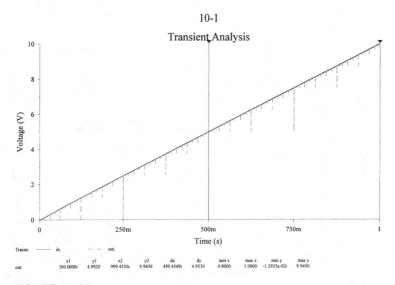

FIGURE 10–29

Response of Multisim Example 10–1 over interval of 1 s.

SUMMARY

1. An analog signal may be converted to a digital signal through the processes of sampling, quantization, and encoding.
2. An A/D converter is used to convert an analog signal into a finite number of digital words.
3. A D/A converter is used to convert a sequence of digital words into an analog signal.
4. The theoretical basis for A/D and D/A conversion is Shannon's sampling theorem, which states that an analog signal may be reconstructed from a sequence of discrete samples if the sampling rate is greater than or equal to twice the highest signal frequency.
5. Multiplexing may be used to process a number of different signals with a single system.
6. A digital word is represented by a number of bits, each of which may assume only two values.
7. It is convenient to represent the analog level corresponding to a given digital word by a normalized variable, whose magnitude is bounded by the range from 0 to 1.
8. A number of circuits utilizing op-amps and related components may be used to perform digital-to-analog conversion.
9. Three types of general-purpose A/D converters are the *ramp, successive approximation,* and *flash* types.
10. Of the three types of A/D converters, the flash is the fastest and the ramp is the slowest.
11. Three types of integrating A/D converters are the *single-slope*, the *dual-slope*, and the *quad-slope*.
12. Integrating A/D converters are primarily used for dc and very low-frequency data.
13. With dual-slope and quad-slope A/D converters, it is possible to completely suppress spurious pickup at integer multiples of a reference frequency by choosing the integration interval to be the reciprocal of the basic frequency.
14. Most A/D converters are obtained as off-the-shelf items by specifying the full-scale voltage and various step and error sizes.
15. The two most common enconding strategies for A/D converters are *unipolar encoding* and *bipolar offset encoding*.
16. Samples may be quantized in an A/D converter according to either *rounding* or *truncation*.
17. An analog switch is a solid-state device capable of achieving nearly an open or a closed condition by means of an electronic control signal.
18. An array of analog switches may be used to implement a multiplexer.
19. A *voltage-to-frequency converter* may be used to generate an ac signal whose output frequency is uniquely determined by an input voltage.
20. A *frequency-to-voltage converter* may be used to generate an output voltage whose level is uniquely determined by an input frequency.

PROBLEMS

Drill Problems

10–1. The basic sampling rate for the *compact disc (CD)* recording industry has been set at 44.1 kHz.

    a. Determine the highest frequency of the analog signal that could theoretically be reconstructed from this sampling rate.

    b. Determine the maximum possible conversion time of the A/D converter if real-time processing is used.

10–2. The highest frequency contained in a certain analog signal is 4 kHz. It is to be sampled and converted to digital form.
    a. Determine the sampling rate if it is chosen to be 50% greater than the theoretical minimum.
    b. Determine the maximum possible conversion time of the A/D converter if real-time processing is used.

10–3. A certain data processing system of the form shown in Figure 10–1 is to be used to process an analog signal whose frequency content ranges from near dc to 800 Hz. Assume that the actual sampling rate is selected to be 25% greater than the theoretical minimum rate.
    a. Determine the sampling rate and the sampling period.
    b. Determine the maximum conversion time of the A/D converter if real-time processing is employed.
    c. If 16-bit words are used, determine the maximum bit width if a given word is to be transferred serially during the conversion interval of the next word.

10–4. Assume in the system of Problem 10–3 that the sampling rate is increased to a rate 50% greater than the theoretical minimum rate and that 8-bit words are used. Repeat the analysis of Problem 10–3.

10–5. Assume that in the system of Problem 10–3, *four* data channels, each with frequencies from near dc to 800 Hz, are to be sampled and multiplexed. Repeat the calculations of parts (b) and (c) if a single A/D converter is used.

10–6. Assume that in the system of Problem 10–4, *eight* data channels, each with frequencies from near dc to 800 Hz, are to be sampled and multiplexed. Repeat the calculations of parts (b) and (c) if a single A/D converter is used.

10–7. The number of bits used for each sample in the CD recording industry is 16 bits. Determine the maximum number of levels that can be encoded.

10–8. In a certain A/D conversion system, it is desired to encode the signal into 2048 possible levels. Determine the number of bits required for each digital sample.

10–9. Consider the 4-bit D/A converter of Example 10–2 (Figure 10–9) with $V_r = 10$ V and $R_f = 10$ kΩ. Determine
    a. full-scale voltage   **b.** value of 1 LSB

10–10. Consider the 4-bit D/A converter of Example 10–2 (Figure 10–9) with $V_r = 5$ V and $R_f = 12$ kΩ. Determine
    a. full-scale voltage   **b.** value of 1 LSB

10–11. For the D/A converter of Problem 10–9, determine the output voltage for each of the following input digital words:
    a. 0000   **b.** 0001   **c.** 1000   **d.** 1010   **e.** 1111

10–12. For the D/A converter of Problem 10–10, determine the output voltage for each of the digital words of Problem 10–11.

10–13. Consider the 5-bit D/A converter of Figure P10–13 with $V_r = 10$ V and $R_f = 10$ kΩ. Determine
    a. number of possible output levels
    b. full-scale voltage
    c. value of 1 LSB

10–14. Consider the 5-bit D/A converter of Figure P10–13 with $V_r = 5$ V and $R_f = 12$ kΩ. Determine
    a. full-scale voltage   **b.** value of 1 LSB

10–15. For the D/A converter of Problem 10–13, determine the output voltage for each of the following input digital words:
    a. 00001   **b.** 00010   **c.** 00100   **d.** 10000   **e.** 11111

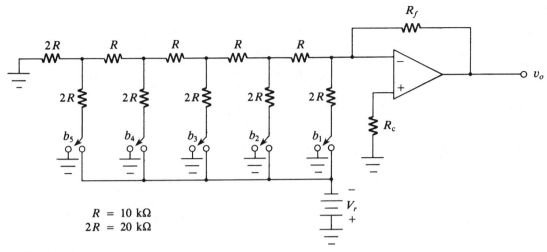

$$R = 10 \text{ k}\Omega$$
$$2R = 20 \text{ k}\Omega$$

FIGURE P10–13

10–16. For the D/A converter of Problem 10–14, determine the output voltage for each of the digital words of Problem 10–15.

10–17. Assume that a certain form of integrated circuit technology allows a comparator operation, amplifier settling, and the associated logic to be performed in about 0.2 μs. If all other times are assumed to be negligible, compute the approximate conversion times and data conversion rates for 6-bit words with the following A/D converter types:
    a. counter   **b.** successive approximation   **c.** flash

10–18. Repeat the analysis of Problem 10–17 for 12-bit words.

10–19. A certain dual-slope A/D converter has a basic integration time interval of 20 ms. Determine the lowest noise frequency that will be completed suppressed by the unit.

10–20. A certain dual-slope A/D converter is found to strongly attenuate signals at 100 Hz, 200 Hz, 300 Hz, etc. Determine the basic integration time interval.

10–21. A certain 16-bit A/D converter connected for unipolar encoding (with rounding) has a full-scale voltage of 10 V. Determine the following quantities:
    a. normalized step size
    b. actual step size in volts
    c. normalized maximum quantized level
    d. actual maximum quantized level in volts
    e. normalized peak quantization error
    f. actual peak quantization error in volts
    g. percentage quantization error

10–22. The 16-bit A/D converter of Problem 10–21 is next connected for bipolar offset encoding with the same peak-to-peak voltage range as in Problem 10–21. The converter is now intended for signals in the range from −5 V to just under +5 V. Repeat all the calculations of Problem 10–21 for this case.

Derivation Problems

10–23. The form of a current-weighted D/A converter for 4 bits is shown in Figure P10–23. Starting with the terminals *a–a'* and continuing to the right, show that the final Norton equivalent circuit looking back from the output is a current source I_N in parallel with R, in which the value of the current source is

$$I_N = 2I(b_1 2^{-1} + b_2 2^{-2} + b_3 2^{-3} + b_4 2^{-4})$$

Hint: Since the current sources are assumed to be ideal, for analysis purposes, the circuit may be redrawn with each current source considered to be in parallel with the corresponding resistance to ground from the junction point.

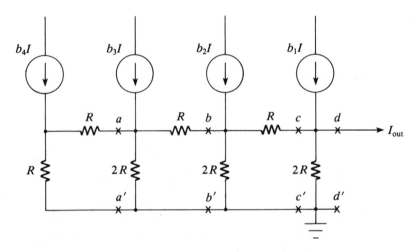

FIGURE P10–23

10–24. An alternate form of a D/A converter utilizing an R-2R network is shown in Figure P10–24 for the case of 4 bits. This form has the advantage that various currents in the network are summed at a virtual ground point. Show that the output voltage is given by

$$v_0 = \frac{R_f V_r}{R}(b_1 2^{-1} + b_2 2^{-2} + b_3 2^{-3} + b_4 2^{-4})$$

10–25. Consider the flash A/D converter of Figure 10–13, and assume 4-bits. Starting with the lowest comparator, let $v_1, v_2, \ldots, v_{15}$ represent the dc voltages at the 15 comparator inputs.
  a. Calculate the 15 voltages.
  b. Let $V_1, V_2, \ldots, V_{15}$ represent the corresponding normalized voltages. Determine these values.
  c. Show that the values of (b) correspond to the transition points for unipolar encoding with rounding.
  d. If a truncation quantization strategy were desired, what changes in the resistive network would be required.
10–26. Assume an A/D converter connected for bipolar encoding with rounding. Assume that the signal is symmetrical and that its peak-to-peak range is exactly equal to the range of the A/D converter. For the purpose of this problem, a quantity SNR will be defined as the ratio of the peak signal voltage to the peak quantization noise voltage.

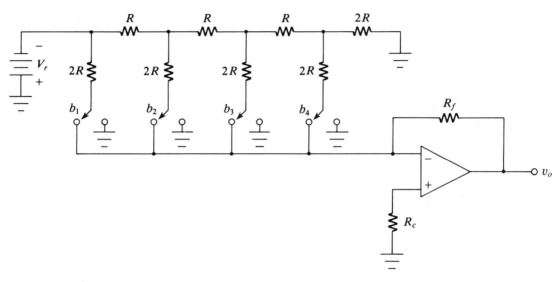

FIGURE P10–24

a. Show the SNR is

$$SNR = 2^n$$

b. Show that the corresponding value in decibels is nearly equal to

$$SNR(dB) = 6n \text{ dB}$$

One significance of this result is that for each additional bit, the signal-to-quantization ratio increases by about 6 dB.

Design Problems

10–27. Consider the 4-bit D/A converter of Example 10–2 (Figure 10–9) with $V_r = 10$ V and R_f variable. Determine the separate values of R_f that would be required to achieve each of the following conditions:

    a. The value of 1 LSB at the output is 0.3 V.

    b. A binary input of 1000 results in an analog output of 5 V.

    c. The full-scale output voltage is 12 V.

    d. The *actual* maximum output voltage is 12 V.

10–28. Consider the 4-bit D/A converter of Example 10–2 (Figure 10–9) with $R_f = 15$ kΩ and V_r variable. Determine the separate values of V_r required to satisfy each of the stated requirements in the four parts of Example 10–27.

10–29. Consider the 5-bit D/A converter of Figure P10–13 with $V_r = 10$ V and R_f variable. Determine the separate values of R_f that would be required to achieve *each* of the following conditions:

    a. The value of 1 LSB at the output is 200 mV.

    b. A binary input of 10000 results in an analog output of 4 V.

    c. The full-scale output voltage is 12 V.

    d. The *actual* maximum output voltage is 12 V.

10–30. Consider the 5-bit D/A converter of Figure P10–13 with $R_f = 15$ kΩ and V_r variable. Determine the separate values of V_r required to satisfy *each* of the stated requirements in Problem 10–29.

10–31. An A/D converter is to be selected in conjunction with the design of a data processing system. The following properties and specifications are given:

1. input analog signal: 0–10 V
2. rounding encoding strategy
3. percent error due to quantization $\leq 0.1\%$ of full scale voltage

For the A/D converter selected, determine

a. full-scale voltage
b. minimum number of bits required.

10–32. An A/D converter is to be selected in conjunction with the design of a data processing system. The following properties and specifications are given:

1. input analog signal range: -10 to 10 V
2. rounding encoding strategy
3. maximum quantization error ≤ 5 mV

For the A/D converter selected, determine the minimum number of bits required.

LABORATORY EXERCISES

ADC0801 A/D Converter
LE 10–1.

1. Using either one of the A/D converters of Section 10–7, connect the circuit of Figure 10–21. You will need to provide some means for monitoring the eight bit values. If you are unable to adjust the power supply voltage to an exact value of 5.12 V, you may use a fixed 5-V supply. This means that the various voltage levels in subsequent parts will need to be rounded downward slightly.
2. For this part, you will need an adjustable dc voltage that can be adjusted from zero to about 5 V, with the means for very fine adjustment at low voltage levels. The dc output mode from many function generators provides this capability on low voltage levels. Start with *zero* volts and observe the digital output. It should be 00000000. *Carefully* adjust the input voltage until the output barely changes to 00000001. This voltage is v_{t1} is defined in Example 10–6. Continue this process until the eight transition voltages defined in Example 10–6 are measured. Compare the measured values with the calculated values.

ADC0806 D/A Converter
LE 10–2.

This exercise is designed to be performed in conjunction with LE 10–1. The eight bits at the output of the A/D converter are to be used as the inputs to the D/A converter.

1. Using either one of the D/A converters discussed in Section 10–8, connect the circuit of Figure 10–22. Use the same voltage employed in LE 10–1 for V_{REF}. Choose $R = 10$ kΩ.
2. Set the dc input voltage of the A/D converter to zero, and connect its output bits as the input bits of the D/A converter. Successively, set the dc voltage to a value within each range between successive transition voltages as determined in step (2) of LE 10–1. Within each range, measure the dc output voltage of the D/A converter, and compare with the voltages of Example 10–6(a).

3. Establish a voltage consisting of a dc level of 2.5 V plus a relatively low frequency ac voltage with a peak value of 2.5 V; that is, the net signal should vary from 0 to 5 V. Apply this signal as the input to the A/D converter. Observe this input and the output of the D/A converter on a CRO for both a sinusoidal waveform and a triangular waveform.

4. Momentarily remove the connections for the four lowest bits and observe the output again.

5. Measure the conversion rate, that is, the number of samples per second.

11

MISCELLANEOUS CIRCUITS

OBJECTIVES

After completing this chapter, the reader should be able to:

- Write an equation describing the input-output relationship for an IC multiplier and discuss its operation.
- Explain the operation and significance of a balanced modulator.
- Discuss the concept of a double-sideband signal, show how it is generated, and sketch its form.
- Explain the operation of a balanced demodulator and discuss its significance.
- Show how a multiplier is connected as a squaring circuit, as a divider circuit, and as a square-root circuit, and discuss the operation of each.
- Discuss the concept of dynamic range compression and how it relates to the application of a logarithmic amplifier.
- State the Shockley equation and discuss its significance.
- Draw the schematic diagram of a basic logarithmic amplifier and of a basic antilogarithmic amplifier and analyze their operation.
- Draw the schematic diagram of a more practical version of a logarithmic amplifier to show how it essentially eliminates the effects of variations in reverse saturation current.
- Define *phase-locked loop* (PLL), discuss its general application, and define *capture range* and *lock range*.
- Show how a PLL is connected for FM detection and how it achieves frequency multiplication and division.
- Discuss the operation and application of the NE/SE 565 PLL IC chip and design a 565 PLL circuit to achieve a specified center frequency and capture range.

11–1 INTRODUCTION

Each chapter so far has had some sort of overall theme to which all circuits and applications in the chapter could be related. This chapter, however, has no single common theme among its topics; in fact, because these topics don't quite fit within the objectives of the earlier chapters, we have gathered them together here.

The first topic is that of IC multipliers. Such circuits provide an output proportional to the product of two input signals. Multipliers may be used as key elements of modulators, demodulators, squaring circuits, dividers, and square-root circuits.

We discuss the concept of dynamic range compression. Then we analyze the operation of a logarithmic amplifier to achieve compression, after which we consider the operation of an antilogarithmic amplifier to achieve range expansion.

We discuss the components of a phase-locked loop (PLL), and we present the applications of a PLL for FM demodulation and for frequency multiplication and division.

Finally, we consider the use of the popular NE/SE 565 PLL IC chip. We close with a discussion of both the analysis and design of circuits containing the 565.

11–2 MULTIPLIERS

Analog multipliers are circuits in which the output voltage at any value of time is proportional to the instantaneous product of two separate input voltages. A discussion of the internal operation of these chips will not be given here, but instead the focus will be on the external operating characteristics.

The dc bias requirements for multiplier chips are typically the same as those for most op-amps, that is, ± 15 V. However, best results for multipliers usually dictate more "back-off" of the signals from the power supply levels than for op-amps, so the typical range for the two input signals is ± 10 V for each. In the illustrations that follow, the bias supply connections will be understood to exist but will not be shown.

Schematic Symbol

A widely used symbol for the multiplier is shown in Figure 11–1. The two quantities v_x and v_y represent the two inputs, and v_o represents the output. The basic input-output relationship is

$$v_o = K v_x v_y \tag{11–1}$$

where K is a constant. A typical value is $K = 1/10$. It can be readily deduced that for $K = 1/10$ and for peak input voltage magnitudes of 10 V, the peak magnitude of the output voltage is $(1/10) \times 10 \times 10 = 10$ V.

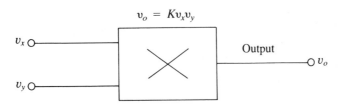

FIGURE 11-1
Schematic symbol of multiplier showing two inputs and output.

Balanced Modulator Circuit

A primary application of the multiplier is to act as a *balanced modulator*. Balanced modulators are widely used in communications systems and in carrier automatic control systems to shift the frequency components of a signal to a higher frequency range to ease the burden of signal transmission and processing.

A multiplier connected for use as a balanced modulator is shown in Figure 11-2(a). In the context of communications applications, some other circuits that are not as "mathematically perfect" as multipliers are also used as balanced modulators, and the form of Figure 11-2(b) is also frequently used on schematic diagrams.

The function v_y on the bottom, for a balanced modulator, is referred to as the *carrier*, and it is assumed to be of the form

$$v_y = A_y \cos \omega_c t \tag{11-2}$$

where $\omega_c = 2\pi f_c$, with f_c the *carrier frequency* in hertz and ω_c the corresponding frequency in radians/second. The function v_x on the left represents the input signal or "intelligence," as it is sometimes called.

(a)   **(b)**

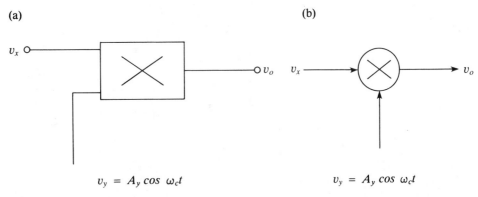

FIGURE 11-2
Connection of multiplier as a balanced modulator and alternate schematic form.

In general, v_x will be a complex signal consisting of a band of frequencies; for example, a voice waveform. However, to illustrate how the balanced modulator works, a single-tone signal will be assumed in the form

$$v_x = A_x \cos \omega_m t \qquad \qquad \textbf{(11–3)}$$

where $\omega_c = 2\pi f_m$, with f_m the *modulating frequency* in hertz and ω_m the corresponding frequency in radians/second. Normally the carrier frequency is much higher in frequency than the modulating frequency: that is, $f_c \gg f_m$.

Based on the assumed inputs, the output of the multiplier is

$$v_o = K v_x v_y = K A_x A_y \cos \omega_m t \cos \omega_c t \qquad \qquad \textbf{(11–4)}$$

This function is a *double-sideband (DSB)* signal. The form of the DSB signal with the single modulating frequency is shown in Figure 11–3. In many applications of the balanced modulator, the carrier frequency will be several orders of magnitude greater than the modulating frequency, but for the sake of illustration in the figure, the difference is not so great.

The envelope follows the lower-frequency modulating signal while the more rapid variations between the envelope levels represent the carrier frequency. There is, however, an important difference between the envelope of the DSB signal and the envelope of a conventional AM signal. Referring momentarily back to the AM signal considered in Figure 9–11(a), we see that the exact modulating signal appears on both the positive and the negative envelopes of the AM signal. However, neither the positive nor the negative envelope of the DSB signal contains the exact modulating signal. In fact, the two envelopes keep

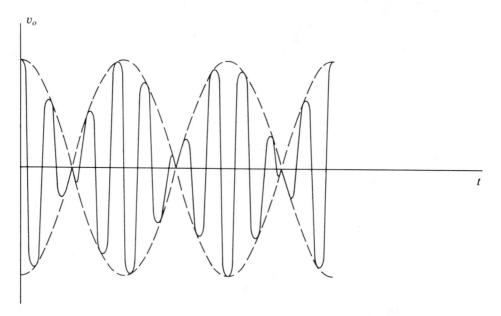

FIGURE 11–3
Illustration of double-sideband signals.

crossing each other, as seen in Figure 11–3. Thus, where exact reproduction of the original signal is required, envelope detection cannot be used with a DSB signal. Later in this section, we discuss the means for detecting a DSB signal.

Sidebands

We now investigate the nature of the frequency components contained in a DSB signal. A basic trigonometric identity is

$$\cos A \, \cos B = \tfrac{1}{2}[\cos(A - B) + \cos(A + B)] \tag{11–5}$$

When this identity is applied to (11–4), the result obtained is

$$v_o = \frac{KA_xA_y}{2} \cos(\omega_c - \omega_m)t + \frac{KA_xA_y}{2} \cos(\omega_c + \omega_m)t \tag{11–6}$$

The two components in (11–6) are referred to as *sidebands*. The first component, whose frequency in hertz is $f_c - f_m$, is called the *lower sideband*, and the second component, whose frequency is $f_c + f_m$, is called the *upper sideband*. Since there is no component with frequency f_c, the carrier is effectively canceled. For that reason, DSB is sometimes referred to as "suppressed carrier-AM." On the other hand, if a carrier component with the correct phase is added to the DSB signal, a conventional AM signal can be generated.

Balanced Demodulator

A multiplier may also be used to implement a *balanced demodulator* or *balanced detector*. The form of such a circuit is shown in Figure 11–4. The function v_{y1} is assumed to be

$$v_{y1} = A_{y1} \cos \omega_c t \tag{11–7}$$

which is a carrier function of the same form as assumed for the modulator.

Assume that the input has the same form as the DSB signal considered earlier, which was obtained as the output of the balanced modulator. Thus,

$$v_{x1} = KA_xA_y \cos \omega_m t \cos \omega_c t \tag{11–8}$$

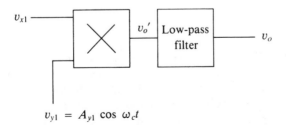

FIGURE 11–4
Connection of multiplier as a balanced modulator or detector.

The output v'_o of the multiplier is

$$v'_o = Kv_{x1}v_{y1} = K^2 A_x A_y A_{y1} \cos \omega_m t \cos^2 \omega_c t \qquad \textbf{(11-9)}$$

where it has been assumed that the K factors for the two units are the same (although this is not necessary). A basic trignometric identity is

$$\cos^2 A = \tfrac{1}{2}(1 + \cos 2A) \qquad \textbf{(11-10)}$$

Applying (11–10) to (11–9), the result obtained is

$$v'_o = \frac{K^2 A_x A_y A_{y1}}{2} \left(\cos \omega_m t + \cos \omega_m t \cos 2\omega_c t \right) \qquad \textbf{(11-11)}$$

The second term in parentheses in (11–11) represents a DSB signal centered at $2f_c$ in hertz. This signal is much higher in frequency than f_m and is easily removed by a low-pass filter as shown in Figure 11–4. Also, for convenience, a quantity α will be defined as

$$\alpha = \frac{K^2 A_y A_{y1}}{2} \qquad \textbf{(11-12)}$$

Assuming an ideal filtering operation and utilizing the definition of (11–12), the filter output v_o is

$$v_o = \alpha A_x \cos \omega_m t \qquad \textbf{(11-13)}$$

This result is a single tone having the same frequency and form as the modulating signal. Its amplitude is proportional to the original amplitude. Thus, the process of demodulation or dectection has been accomplished.

Note that the reference carrier for the demodulator was assumed to have the same frequency and phase as for the modulator. This process is also called *synchronous* or *coherent detection*. Unless the demodulator carrier can be locked in frequency and phase with the modulator carrier, however, this process may not be satisfactory. More details are provided in communications texts.

Squaring Circuit

If both inputs of the multiplier are tied together as shown in Figure 11–5, the circuit reduces to one with a single input. Denoting this common input by v_i, we have $v_x = v_y = v_i$, and the output is

$$v_o = Kv_i^2 \qquad \textbf{(11-14)}$$

This circuit now performs the squaring operation, and thus it can be used for certain power and root-mean-square operations. One communications operation of the squaring process is that of frequency doubling, which is considered in Problem 11–23.

When a particular signal-processing module is appropriately connected in the feedback path of a closed-loop, stable feedback circuit, there is a tendency for the overall circuit to perform the inverse function of the module. We will now demonstrate how this property permits the operations of division and square-root extraction to be achieved with a multiplier.

$$v_o = Kv_i^2$$

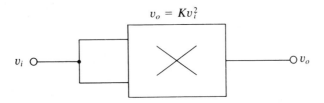

FIGURE 11-5
Multiplier connected as a squaring circuit.

Divider Circuit

Division of one voltage by another voltage can be achieved with the circuit of Figure 11–6. Because the circuit output is now applied as one input to the multiplier, and the multiplier output is fed back to the op-amp input, the multiplier voltage symbols have been modified from those of the basic circuit. The subscripts N and D refer to numerator and denominator, respectively, and their significance will be seen shortly.

The multiplier input-output relationship for this circuit reads

$$v_o' = Kv_o v_D \qquad (11-15)$$

Assuming no signal current flowing in the op-amp input, the two currents i_1 and i_2 must satisfy

$$i_1 + i_2 = 0 \qquad (11-16)$$

Assuming ideal stable linear operation, the inverting terminal will be at ground potential, and

$$i_1 = \frac{v_N}{R} \qquad (11-17)$$

FIGURE 11-6
Divider circuit achieved by connecting multiplier in feedback path of op-amp.

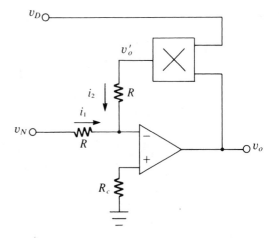

and

$$i_2 = \frac{v_o'}{R} \tag{11-18}$$

Substitution of (11-17) and (11-18) in (11-16) yields

$$\frac{v_N}{R} + \frac{v_o'}{R} = 0 \tag{11-19}$$

or

$$v_o' = -v_N \tag{11-20}$$

The variable v_o' may be readily eliminated in (11-15) by means of (11-20). Subsequent solution for v_o results in

$$v_o = -\frac{v_N}{Kv_D} \tag{11-21}$$

where v_N is the numerator voltage, and v_D is the denominator voltage. The desired operation of dividing one voltage by another is thus achieved.

The range and/or polarities of the voltages for which linear operation is permitted with the divider circuit must be carefully monitored. In general, no combination of voltages that would result in a theoretical value of v_o larger than the saturation level is possible. Since v_D appears in the denominator, its magnitude cannot be too small. The polarity of v_N is arbitrary, but v_D can assume only positive values. This latter restriction is necessary to ensure negative feedback.

Square-Root Circuit

The square-root operation can be achieved with the circuit of Figure 11-7. In Problem 11-24, the reader can show that the output v_o is

$$v_o = \sqrt{-v_i/K} \tag{11-22}$$

FIGURE 11-7
Square-root circuit achieved by connecting squaring circuit in feedback path of op-amp.

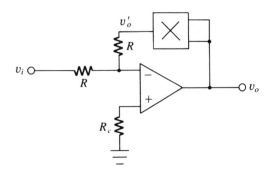

For K positive, the condition of stable linear operation with negative feedback prohibits v_i from being positive. A more meaningful way to express (11–22) is

$$v_o = \sqrt{|v_i|/K} \qquad (11\text{–}23)$$

where only **negative** values of v_i are permitted.

EXAMPLE 11–1

The output of a certain multiplier circuit is given by

$$v_o = 0.1v_x v_y \qquad (11\text{–}24)$$

The ranges of the two input voltages are from -10 V to $+10$ V. Determine the output voltage for each of the following input voltage combinations: (a) $v_x = 3$ V, $v_y = 8$ V; (b) $v_x = -5$ V, $v_y = 7$ V, (c) $v_x = -8$ V, $v_y = -9$ V.

Solution
The various input voltages are substituted in (11–24) and the output voltages are readily determined as follows:

$$\text{(a)} \quad v_o = 0.1 \times 3 \times 8 = 2.4 \text{ V} \qquad (11\text{–}25)$$

$$\text{(b)} \quad v_o = 0.1 \times (-5) \times 7 = -3.5 \text{ V} \qquad (11\text{–}26)$$

$$\text{(c)} \quad v_o = 0.1 \times (-8) \times (-9) = 7.2 \text{ V} \qquad (11\text{–}27)$$

EXAMPLE 11–2

A certain multiplier is used as a balanced modulator to generate a double-sideband signal based on a carrier frequency of 100 kHz. A single-frequency modulating tone of 5 kHz is applied as the input. Determine the output frequencies.

Solution
The form of a DSB output with a single-frequency modulating signal was given by (11–6). The output contains two frequencies that are $f_c - f_m$ and $f_c + f_m$. Thus, the output frequencies are $100 - 5 = 95$ kHz and $100 + 5 = 105$ kHz.

11–3 LOGARITHMIC AMPLIFIERS

Logarithmic amplifiers are a special class of amplifier circuits whose input-output characteristics are carefully controlled to fit a logarithmic characteristic. The resulting signal will be distorted, but by the use of an inverse logarithmic ("antilogarithmic") or exponential characteristic, the original signal can be restored.

Dynamic Range Compression

The rationale behind such a complex procedure needs justification. Many signals of interest have a very wide dynamic range; that is, there is a large difference between the level of the signal during intervals in which it is large compared with the signal during intervals in which it is small. For a number of encoding, recording, and signal transmission applications, this large difference poses a difficulty in maintaining good signal integrity. If an attempt is made to optimize signal quality at high levels, the signal will be degraded during low levels. Likewise, if an attempt is made to optimize at low levels, the high levels may be excessive, and distortion and overloading may occur.

A logarithmic voltage-current characteristic is displayed both with junction diodes and with transistors, and it offers a significant dynamic range compression. With dynamic range compression, small signal levels are amplified by a greater amount than are large signal levels. The resulting signal has less dynamic range between the peaks and the valleys, and it is easier to process in many cases.

Basic Circuit Form

The basic form of a logarithmic amplifier is shown in Figure 11–8. The circuit configuration is that of an inverting amplifier, but a bipolar junction transistor (BJT) is connected as the feedback element. The diode is used to protect the base-emitter junction from possible excessive reverse voltage, so its effect will not be considered. Assuming stable linear operation, the inverting terminal is at ground potential, and the input current i_i is given by

$$i_i = \frac{v_i}{R_i} \qquad (11\text{–}28)$$

Shockley Equation

We must sidetrack briefly to establish the logarithmic characteristic of the transistor. The relationship of the collector current i_C to the base-emitter voltage v_{be} of a BJT is given by the Shockley equation, which is

$$i_C = I_{EO}(e^{qv_{be}/kT} - 1) \qquad (11\text{–}29)$$

where I_{EO} is the reverse saturation current of the base-emitter diode, $q = 1.6 \times 10^{-19}$ C (coulombs) is the magnitude of the charge of an electron, $k = 1.38 \times 10^{-23}$ J/K (joules per kelvin) is Boltzmann's constant, and T is the absolute temperature in kelvins (K). It is convenient at this point to assume the "standard" temperature 25° C (298 K), which is widely used in specifying semiconductor properties. Further, if v_{be} is more than a few

FIGURE 11–8
Basic form of the logarithmic amplifier.

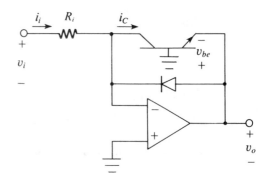

tenths of a volt, the exponential term is much greater than unity, and the latter term may be ignored. With these various assumptions and substitutions, i_C may be expressed as

$$i_C \simeq I_{EO}e^{38.9v_{be}} \tag{11--30}$$

The inverse relationship is

$$v_{be} = \frac{1}{38.9} \ln \frac{i_C}{I_{EO}} = 0.0257 \ln \frac{i_C}{I_{EO}} \tag{11--31}$$

Input-Output Characteristic

Returning to the actual circuit of interest, we note that $i_C = i_i = v_i/R_i$. Further, $v_o = -\dot{v}_{be}$. Making these substitutions in (11–31), we obtain

$$v_o = -0.0257 \ln \frac{v_i}{R_i I_{EO}} \tag{11--32}$$

If desired, the natural logarithmic relationship may be converted to the base-10 logarithm (denoted as "log"). A basic property of logarithms is that $\ln x = 2.302585 \log x$. With this substitution, (11–32) becomes

$$v_o = -0.0592 \log \frac{v_i}{R_i I_{EO}} \tag{11--33}$$

From either (11–32) or (11–33), it is noted that the output voltage is a logarithmic function of the input voltage. For this circuit as given to work properly, v_i must be a positive voltage. Likewise, v_o will be a negative voltage. This circuit produces a change of about 59 mV in the output voltage for each decade of change (10-to-1 ratio) in the input voltage.

Antilogarithmic Amplifier

Once the dynamic range of a signal has been compressed by a logarithmic amplifier, the resulting signal is obviously distorted. The original signal may be restored by means of an antilogarithmic or exponential amplifier. Such a circuit is shown in Figure 11–9. The input v_i of this circuit is shown with the usual positive reference above ground. However, for the circuit to work properly, v_i must be negative in order to forward bias the base-emitter junction of the transistor. Thus, this antilogarithmic circuit would be compatible with the logarithmic circuit previously considered. The diode across the input is used to protect the base-emitter junction from a possible excessive reverse voltage, so its effect will not be considered.

Based on the same assumptions made earlier for the Shockley equation, the collector current i_C is closely represented as

$$i_C = I_{EO}e^{38.9v_{be}} \tag{11--34}$$

Since $v_{be} = -v_i$, the quantity i_C can be expressed as

$$i_C = I_{EO}e^{-38.9v_i} \tag{11--35}$$

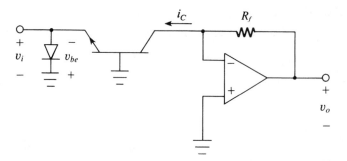

FIGURE 11–9
Basic form of the antilogarithmic amplifier.

Since i_C must flow from the op-amp output through R_f and $v^- \approx 0$, the op-amp output is

$$v_o = R_f I_{EO} e^{-38.9 v_i} \tag{11-36}$$

The argument of the exponential in the proper range of operation is positive since $v_i < 0$. To delineate the fact, $-v_i$ will be replaced by $|v_i|$ in (11–36). It must be understood that the equation with the magnitude form is valid only for $v_i < 0$. With the preceding assumptions, we have

$$v_o = R_f I_{EO} e^{-38.9 |v_i|} \tag{11-37}$$

The positive exponential form of this equation provides the necessary dynamic range expansion to compensate for the logarithmic compression.

More Practical Circuit Form

While the concept of the logarithmic amplifier has been established with the preceding simplifed circuits, there are some serious limitations. The most notable is that the characteristics are dependent on the reverse saturation current I_{EO}, which is extremely temperature dependent and difficult to control.

A circuit that may be used to circumvent this difficulty is shown in Figure 11–10. As in the case of the simpler circuits, the diodes are used to protect the transistors from excessive reverse voltages, so their effects will not be considered in the analysis.

The two transistors shown in the circuit represent a matched pair. Their physical characteristics as well as their temperature are assumed to be the same for all conditions, so a single value of reverse saturation current I_{EO} will be used for both transistors.

Without going through all the details again, note that the analysis of each op-amp circuit on the left follows the same pattern as for the single op-amp logarithmic amplifier. Using the form of (11–32) as a guide, we can express the voltages v_a and v_b as

$$v_a = -0.0257 \ln \frac{v_1}{R_i I_{EO}} \tag{11-38}$$

$$v_b = -0.0257 \ln \frac{v_2}{R_i I_{EO}} \tag{11-39}$$

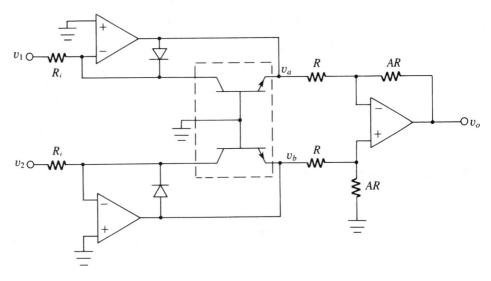

FIGURE 11–10
Practical form of a logarithmic amplifier.

The right-hand side of the circuit is the closed-loop differential amplifier, whose operation was established in Chapter 2. The output voltage v_o is

$$v_o = A(v_b - v_a) \tag{11–40}$$

Substituting (11–38) and (11–39) in (11–40), we have

$$v_o = 0.0257A \left(\ln \frac{v_1}{R_i I_{EO}} - \ln \frac{v_2}{R_i I_{EO}} \right)$$

$$= 0.0257A \ln \left(\frac{v_1}{R_i I_{EO}} \right) \left(\frac{R_i I_{EO}}{v_2} \right) = 0.0257 \ln \frac{v_1}{v_2} \tag{11–41}$$

where the relationship $\ln x - \ln y = \ln(x/y)$ was used. Expressed in terms of base-10, the result of (11–41) may be rewritten as

$$v_o = 0.0592A \log \frac{v_1}{v_2} \tag{11–42}$$

This result shows that v_o is independent of I_{EO}! What is actually happening is that a portion of the collector current for both transistors is dependent on I_{EO}. Since the output is a function of the difference between the two collector currents, however, the components cancel, provided that the I_{EO} values for the transistors are identical.

Note that the output is a logarithmic function of the ratio v_1/v_2. The voltage v_2 may be set to a fixed value, which establishes a reference level for the output. The voltage v_1 then represents the input signal level. The gain constant A can be used to set the dynamic compression ratio.

One limitation of this circuit is that the constant 0.0257 represents a temperature-dependent factor. (Recall that this constant is kT/q and that the value 0.0257 is correct for $T = 298$ K.) It is possible to provide further compensation for this effect with a temperature-dependent, gain-adjustment resistor.

From the development of this section, the reader will likely deduce that one does not simply "throw together" a logarithmic amplifier and expect any meaningful results. Construction and adjustment of a logarithmic amplifier is a delicate process, and it is recommended that the user employ one of the available commercial packages.

EXAMPLE 11-3

Consider a logarithmic amplifier of the form shown in Figure 11–10. For $A = 100$ and $v_2 = 1$ V, determine v_o for the following values of v_1: (a) 0.1 V, (b) 1 V, (c) 10 V.

Solution
Using (11–42) and substituting $A = 100$ and $v_2 = 1$ V, we have

$$v_o = 0.0592A \log \frac{v_1}{v_2} = 5.92 \log \frac{v_1}{1} \qquad (11\text{–}43)$$

The three values of v_1 are then substituted in (11–43), and the results are

$$\text{(a) } v_o = 5.92 \log \frac{0.1}{1} = -5.92 \text{ V} \qquad (11\text{–}44)$$

$$\text{(b) } v_o = 5.92 \log \frac{1}{1} = 0 \text{ V} \qquad (11\text{–}45)$$

$$\text{(c) } v_o = 5.92 \log \frac{10}{1} = 5.92 \text{ V} \qquad (11\text{–}46)$$

Observe that the output increases by 5.92 V for each decade of input change.

11-4 PHASE-LOCKED LOOPS

A phase-locked loop (PLL) is a closed-loop feedback circuit in which a generated signal establishes a synchronization or "lock" with an input signal. Depending on the circuit configuration and the manner in which the input and output are connected, PLLs may be used for FM detection, frequency multiplication and division, tracking, establishing a noise-free reference in the presence of noise, and other applications. The complex operation and applications of the PLL are the subjects of several complete books! The treatment here will emphasize two of the most common applications: (1) **FM detection** and (2) **frequency multiplication and division.** However, we will give two basic definitions as a starting point.

Capture and Lock Ranges

Two important parameters in the operation of a phase-locked loop are the *capture range* and the *lock range*. The *capture range* $\pm f_c$ is the range of frequencies about the center frequency at which the PLL can initially establish synchronization. The *lock range* $\pm f_L$ is the range of frequencies about the center frequency at which the PLL can hold lock, once it is initially established.

Although the lock and capture ranges may be the same in some cases, as a general rule, the lock range is larger than the capture range. In other words, the PLL can hold lock, once established, over a broader frequency range than the range to which it could initially establish lock.

FM Detection

A block diagram of a PLL delineating FM detection is shown in Figure 11–11. The three major components of the system are a voltage-controlled oscillator (VCO), a phase comparator, and a loop filter. In some applications, the loop filter is not required. Most integrated circuit PLLs contain the phase comparator and VCO on a chip, and terminals are provided for connecting an external loop filter. The actual signals in a PLL may either be sine waves or square waves. In the latter case, digital operations are often used to perform the various operations. Operation with sine waves is illustrated in Figure 11–11.

Assume that the input signal on the left contains a modulated angle function $\phi_i(t)$, which represents some form of the "intelligence" signal to be extracted by the PLL. This input signal and the output of the VCO are both applied as inputs to the phase comparator. Note the assumption of a sine function for the input and a cosine function for the VCO output. Most phase comparator circuits utilize a 90° phase difference between the two inputs, and this assumption has been made here.

Assume initially that there is no modulation; that is, $\phi_i(t) = 0$. Assume that the VCO radian frequency with no input voltage applied is ω_c. Under these conditions, the input and the VCO output are both sinusoids of radian frequency ω_c, but with a 90° phase difference. The phase comparator and the loop filter outputs are $v_e(t) = 0$ and $v_o(t) = 0$. The loop is thus locked in an equilibrium state.

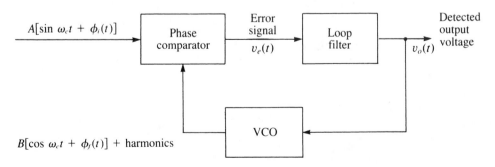

FIGURE 11–11
Block diagram of phase-locked loop (PLL) connected as FM detector.

Assume next that some intelligence signal $\phi_i(t)$ appears at the input. Instantaneously, the two inputs to the phase comparator have different frequencies and phases, and an error voltage $v_e(t)$ appears. This voltage is smoothed by the loop filter to yield $v_o(t)$, and this signal is then applied as the VCO input. This control signal causes the VCO frequency to shift in a direction that causes the VCO frequency to match that of the input, thus reducing the loop "error" toward zero.

One interpretation of the operation of the PLL is that the feedback structure of the loop forces the VCO to generate the same FM signal that was generated back at the transmitter. Thus, the VCO control voltage $v_o(t)$ must assume the same form as at the transmitter, and the intelligence has been extracted. However, the loop coupled with the FM process offers significant reduction in the noise, so a "clean" version of the modulating signal is extracted.

Frequency Multiplication and Division

Operation of the PLL as a frequency multiplier and/or divider is illustrated in Figure 11–12. This circuit contains a "divide-by-N" counter ahead of the feedback loop and a "divide-by-M" counter in the feedback path within the loop. The output of the circuit in this case is the VCO output, and no loop filter is assumed for this illustration. It is also assumed that all or a portion of the loop contains digital signals and components, since the frequency division circuits are most easily implemented with such techniques.

Assume that the input cyclic frequency is f_i. This frequency is divided by N in the first counter, so one input to the phase comparator is a signal with frequency f_i/N. If the VCO output were applied directly as the other input to the phase comparator, a VCO output frequency f_i/N would result in zero frequency error. However, since the divider in the feedback path divides the VCO frequency by M, the VCO must provide an output at M times the input frequency in order to bring the frequency error to zero at the point of comparison. In other words, the division by M in the feedback path forces the oscillator to

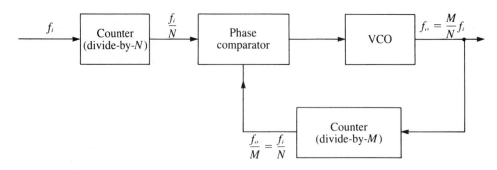

FIGURE 11–12
Block diagram of frequency multiplication and division with PLL.

operate at a frequency M times as large, so the frequency after division will be the same as the input frequency. Thus, the output frequency f_o is

$$f_o = \frac{M}{N} f_i \tag{11-47}$$

Using this concept, it is theoretically possible to generate any output frequency that is expressible as a ratio of integers times an input frequency. This technique serves as the basis for the design of modern frequency synthesizer systems.

EXAMPLE 11-4

In a certain system, it is desired to use a PLL to generate a 2.4-MHz signal from an available reference of 2 MHz. Specify the **smallest** frequency division ratios M and N in Figure 11-12 that can be used to achieve the desired result.

Solution

In order for an **exact** solution to exist, the ratio of the two frequencies must be expressible as a ratio of integers. The desired ratio in this case is 2.4 MHz/2 MHz = 24/20. Thus, one solution would be $M = 24$ and $N = 20$. However, the smallest division ratio is obtained by simplifying 24/20 to its most reduced form; that is, 24/20 = 6/5. The simplest solution is then $M = 6$ and $N = 5$. The input reference frequency is thus divided by 5, and the output of the VCO is divided by 6.

11-5 NE/SE 565 PHASE-LOCKED LOOP

In this section, we consider some of the properties of one of the most popular IC phase-locked loop chips. The unit selected is the NE/SE565, which we will here-after refer to simply as the 565. See Appendix C for detailed specifications and various applications. Here, we give only a simplified presentation of some of the highlights.

Basic Properties

A block diagram of the 565 is given in Figure 11-13. The chip contains a phase detector, an amplifier, a portion of a loop filter, and a VCO. Access to the terminals permits a variety of different circuits to be connected.

The 565 is usable over the frequency range from 0.001 Hz to 500 kHz. It has a highly stable center frequency and is able to achieve a very linear FM detector characteristic. The output of the VCO is capable of producing a TTL compatible square wave. The bandwidth of the PLL is adjustable from less than $\pm 1\%$ to more than $\pm 60\%$.

Power Supply Requirements

The power supply requirements for the 565 are from ± 6 V to ± 12 V. As stated, this would imply two power supplies, and that is the common mode of operation. For some applications,

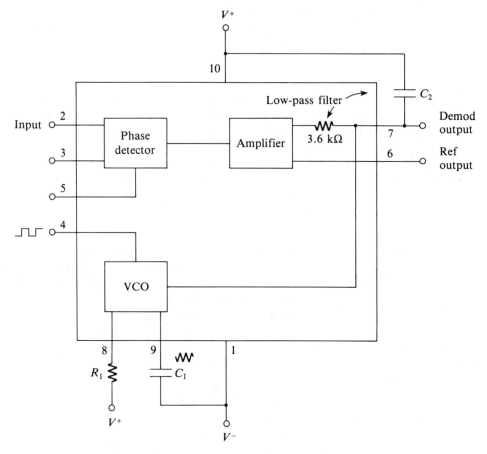

FIGURE 11–13
Block diagram of the 565 PLL.

however, a single power supply can be connected between the V^+ and V^- terminals, and this voltage can be between 12 V and 24 V. Note that for dual power supply operation, the commonly available dual voltages of ± 15 V *could not be used*, since they would exceed the maximum voltage specifications of the 565.

Design Relationships

Referring to Figure 11–13, the VCO free-running center frequency is established by R_1 and C_1. This frequency is given by

$$f_o = \frac{1.2}{4R_1C_1} = \frac{0.3}{R_1C_1} \tag{11–48}$$

Observe that a 3.6-kΩ resistor appears on the chip as a portion of a loop filter. Let τ represent the time constant associated with this resistance (denoted on data sheets as R_2) and the external capacitance C_2. We have

$$\tau = R_2 C_2 = 3.6 \times 10^3 C_2 \tag{11-49}$$

The *lock-range* f_L is given by

$$f_L = \pm \frac{8f_o}{V_{CC}} \tag{11-50}$$

where V_{CC} is the *total* dc voltage between the V^+ and V^- bias terminals.

The *capture-range* f_c is given by

$$f_c = \pm \frac{1}{2\pi} \sqrt{\frac{2\pi f_L}{\tau}} \tag{11-51}$$

FM Detector Circuit

A typical circuit diagram of the 565 connected as an FM detector (or demodulator) is shown in Figure 11–14. The range of R_1 should be selected between about 2 kΩ and 20 kΩ, with an optimum value reasonably close to 4kΩ. Direct coupling can be used at the input if the dc resistances seen from pins 2 and 3 are equal and if there is no dc voltage difference between the pins. In this application, the VCO output (pin 4) is connected to the

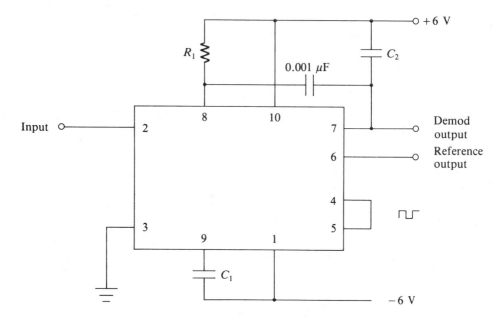

FIGURE 11–14
Typical connection diagram for 565 PLL used as FM detector.

phase detector input (pin 5). A dc reference voltage is provided at pin 6, and this voltage is close to the dc level of the detected output (pin 7).

If a resistance is connected between pins 6 and 7, the gain of the output stage can be reduced with little change in the dc voltage level. The lock range can thus be decreased with little change in the VCO free-running frequency.

The purpose of the 0.001-μF capacitor connected between pins 7 and 8 is to eliminate possible oscillations.

This circuit utilizes a single-pole loop filter, which is achieved with capacitor C_2 and the internal resistance $R_2 = 3.6$ kΩ.

FSK Detector

In the transmission of digital data between computers, terminals, and other data processing equipment, the 1s and 0s are often converted to sinusoidal tones. In this manner, the direct coupling often required for low-frequency pulse transmission is eliminated, and ac coupling may be employed. The process of transferring two or more voltage levels into distinct frequencies is referred to as *frequency-shift keying (FSK)* and is basically a form of FM.

At the receiving end of an FSK link, it is necessary to convert the separate frequencies into distinct voltage levels. One of the most common data transmission rates is 300 *baud.* This terminology refers to 300 bits/second. With this system, and in what is referred to as the *originate mode*, a binary 0 is transmitted as a 1070-Hz tone, and a binary 1 is transmitted as a 1270-Hz tone. (In what is referred to as the *answer mode*, the corresponding frequencies are 2025 Hz and 2225 Hz.)

A 565 connected to receive 300-baud FSK signals encoded at 1070 Hz and 1270 Hz is shown in Figure 11–15. Basically, the PLL locks to the input frequency and provides an output voltage at two separate levels. With the aid of a comparator appropriately connected, the two levels can be changed to desired digital output levels.

In this particular design, the loop filter capacitor C_2 is chosen sufficiently small to eliminate overshoot. A three-stage RC filter is used to remove carrier components at the output. The band edge of the filter is chosen at the approximate midpoint between the maximum signal rate of 300 bits/s and twice the input frequency (approximately 2.2 kHz). Other applications of the 565 are provided in the data sheets of Appendix C.

EXAMPLE 11–5

A 565 PLL circuit is to be designed for an application with the following requirements:

1. VCO center frequency = 100 kHz
2. capture range = ±5 kHz
3. to be powered with ±6-V supplies

Determine (a) appropriate values of R_1 and C_1, (b) the lock range, and (c) the required value of C_2.

Solution
(a) From (11–48), the $R_1 C_1$ product required to achieve a center frequency of 100 kHz is

$$R_1 C_1 = \frac{0.3}{f_o} = \frac{0.3}{10^5} = 3 \times 10^{-6} \qquad \textbf{(11–52)}$$

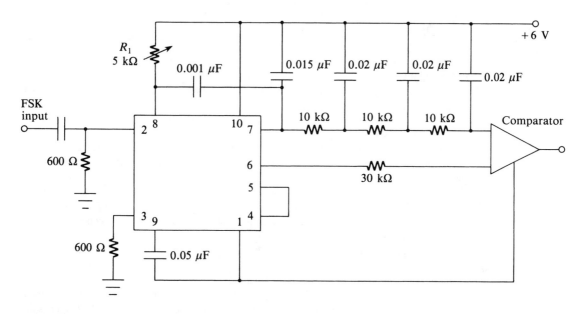

FIGURE 11–15
565 connected for FSK detection at 300 baud.

Based on the recommended range of R_1, it was determined after a little trial-and-error that the following stock values should suffice:

$$R_1 = 3 \text{ k}\Omega \qquad (11\text{–}53)$$

$$C_1 = 0.001 \ \mu\text{F} \qquad (11\text{–}54)$$

(b) Based on power supply voltages of ± 6 V, the voltage quantity in (11–50) is $V_{CC} = 12$ V. The lock range is then

$$f_L = \pm \frac{8f_o}{V_{CC}} = \pm \frac{8 \times 10^5}{12} = \pm 66.7 \text{ kHz} \qquad (11\text{–}55)$$

(c) The capture range expression is given by (11–51). All quantities in that equation are known except τ. Dropping the $\pm$ sings momentarily and solving for τ, we have

$$\tau = \frac{f_L}{2\pi f_c^2} = \frac{66.7 \times 10^3}{2\pi \times (5 \times 10^3)^2} = 0.4244 \text{ ms} \qquad (11\text{–}56)$$

The resistance value is $R_2 = 3.6$ kΩ, so C_2 can be determined as

$$C_2 = \frac{\tau}{R_2} = \frac{0.4244 \times 10^{-3}}{3.6 \times 10^3} = 0.1179 \ \mu\text{F} \qquad (11\text{–}57)$$

A 0.12-μF capacitor would probably be acceptable.

SUMMARY

1. Multipliers are circuits that produce an output voltage proportional to the instantaneous product of two input voltages.
2. A balanced modulator is a circuit used to shift the frequency spectrum of a signal to a different frequency range.
3. A multiplier may be used as a balanced modulator by applying the carrier frequency to one input and the signal to the other input.
4. The output of a multiplier used as a balanced modulator is a double-sideband signal.
5. A multiplier may also be used as a balanced demodulator or detector, which recovers the original modulating signal from the double-sideband signal.
6. A multiplier may be used to create a squaring circuit by connecting the two inputs together.
7. A divider circuit can be created by placing a multiplier in the feedback path of an op-amp.
8. A square-root circuit can be created by placing a multiplier, with both inputs connected together, in the feedback path of an op-amp.
9. Dynamic range compression is the process of reducing the peak-to-peak variation of a signal to a lower range.
10. A logarithmic amplifier is very useful for dynamic range compression.
11. A basic logarithmic amplifier can be created by placing a bipolar junction transistor in the feedback path of an op-amp, with the input established as the collector current and the output established as the base-emitter voltage.
12. An antilogarithmic amplifier can be created by placing a bipolar junction transistor in the input circuit of an op-amp, with the input established as the base-emitter voltage and the output established as the collector current.
13. A more practical logarithmic amplifier circuit utilizes two nearly identical transistors to cancel out the effects of variations in reverse saturation current.
14. A phase-locked loop (PLL) is a circuit in which an internally generated signal can be locked with an external input signal.
15. A complete PLL contains a phase detector, a voltage-controlled oscillator (VCO), a loop filter, and possibly an amplifier and/or a frequency divider.
16. A PLL may be used as an FM detector by using the VCO input voltage as the demodulated output.
17. A PLL may be used to form a combination of frequency multiplication and division by using frequency division ahead of the loop and within the loop.
18. Integrated-circuit PLLs, such as the NE/SE 565, are available as modules and may be tailored to individual circuit applications.

PROBLEMS

Drill Problems

11–1. A certain multiplier has $K = 1/10$. Determine v_o for the following combinations of input voltages:

    a. $v_x = 2$ V, $v_y = 6$ V   **b.** $v_x = 4$ V, $v_y = -8$ V

11–2. For the multiplier of Problem 11–1, determine v_o for the following combinations of input voltages:

    a. $v_x = -9$ V, $v_y = 5$ V   **b.** $v_x = -6$ V, $v_y = -7$ V

11–3. A certain multiplier is used as a balanced modulator to generate a DSB signal based on a carrier frequency of 35 kHz. A single-tone modulating frequency of 2 kHz is applied as the input. Determine the output freqencies.

11–4. The output of a certain balanced modulator has frequencies of 67 kHz and 71 kHz with a single-tone modulating frequency. Determine the carrier and modulating frequencies.

11–5. A certain multiplier used as a balanced modulator has a sinusoidal carrier input of 80 kHz. A two-tone modulating signal has frequencies of 3 kHz and 7 kHz. List the output frequencies.

11–6. The output of a certain balanced modulator with a two-tone modulating signal has frequencies (in kHz) of 122, 125, 135, and 138. Determine the carrier frequency and the two modulating frequencies.

11–7. A multiplier with $K = 0.1$ is connected as a squaring circuit as shown in Figure 11–5. Determine v_o for $v_i = 5$ V.

11–8. Determine the output of the circuit of Problem 11–7 for $v_i = -8$ V.

11–9. A multiplier with $K = 0.1$ is connected in the divider circuit of Figure 11–6. Determine v_o for $v_N = 4$ V and $v_D = 8$ V.

11–10. Determine the output of the divider circuit of Problem 11–9 for $v_N = -2$ V and $v_D = 5$ V.

11–11. A multiplier with $K = 0.1$ is connected in the square-root circuit of Figure 11–7. Determine v_o for $v_i = -9$ V.

11–12. Determine the output of the square-root circuit of Problem 11–11 for $v_i = -4$ V.

11–13. Consider a basic logarithmic amplifier of the form shown in Figure 11–8 with $R_i = 10$ kΩ and $I_{EO} = 10^{-13}$ A. Determine v_o for each of the following values of v_i:
a. 10 mV **b.** 100 mV **c.** 1 V **d.** 10 V

11–14. For the logarithmic amplifier of Problem 11–13, assume that $R_i = 30$ kΩ and $I_{EO} = 5 \times 10^{-13}$ A. Determine v_o for each of the values of v_i in Problem 11–13.

11–15. Consider a basic antilogarithmic amplifier of the form shown in Figure 11–9 with $R_f = 10$ kΩ and $I_{EO} = 10^{-13}$ A. Determine v_o for each of the following values of v_i:
a. −0.4 V **b.** −0.5 V **c.** −0.55 V

11–16. For the antilogarithmic amplifier of Problem 11–14, assume that $R_f = 30$ kΩ and $I_{EO} = 5 \times 10^{-13}$ A. Determine v_o for the following values of v_i:
a. −0.3 V **b.** −0.4 V **c.** −0.5 V

11–17. Consider a logarithmic amplifier of the form shown in Figure 11–10 with A = 5. For $v_2 = 2$ V, determine v_o for each of the following values of v_1:
a. 0.1 V **b.** 1 V **c.** 10 V

11–18. Assume in the logarithmic amplifier of Problem 11–17 that the voltage v_2 is changed to $v_2 = 8$ V. Determine v_o for each of the voltages in Problem 11–17.

11–19. In the PLL circuit of Figure P11–19, the input frequency is $f_i = 160$ kHz. Determine the output frequency f_o.

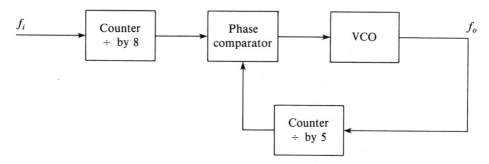

FIGURE 11–19

11–20. In the PLL circuit of Figure P11–19, assume that the counter in the feedback circuit is replaced by one with a different division ratio M. For an input frequency of 96 kHz, the output frequency is 180 kHz. Determine M.

11–21. A 565 PLL demodulator circuit of the form shown in Figure 11–14 has $R_1 = 4.3$ kΩ, $C_1 = 0.001$ μF, and $C_2 = 0.01$ μF. It is powered by ± 6-V voltages. Determine
 a. free-running center frequency **b.** lock range **c.** capture range

11–22. A 565 PLL demodulator circuit of the form shown in Figure 11–14 has $R_1 = 2.2$ kΩ, $C_1 = 560$ pF, and $C_2 = 0.02$ μF. It is powered by ± 12-V voltages. Determine
 a. free-running center frequency **b.** lock range **c.** capture range

Derivation Problems

11–23. Consider the squaring circuit of Figure 11–5, and assume that the input is a single-frequency signal of the form

$$v_i = A \cos \omega_c t$$

The output v_o of the multiplier is connected to a simple RC high-pass filter, which eliminates any dc component from the signal. Using a basic trigonometric identity, show that the output v_o' of the high-pass filter is a signal having twice the frequency of the input. (In communications applications, a band-pass filter is usually employed in order to eliminate other spurious components not evident from this idealized analysis.)

11–24. Consider the circuit of Figure 11–7, and assume stable linear operation. Using an approach similar to that of the divider circuit as given by (11–15) through (11–21), derive the square-root relationship of (11–22).

Design Problems

11–25. A logarithmic amplifier of the form shown in Figure 11–10 is to be used for dynamic compression in a certain data system. The following requirements are given:
 a. output voltage to be zero for $v_i(= v_1) = 6$V
 b. dynamic range compression = 500 mV per decade of input change
 Specify the required values of v_2 and A.

11–26. Repeat the design of Problem 11–2 if the requirements are changed as follows:
 a. output voltage to be zero for $v_i = 2$ V
 b. dynamic range compression = 100 mV per decade of input change

11–27. In a certain system, it is desired to use a PLL to generate a 30-kHz signal from a 27.5-kHz reference. Specify the *smallest* frequency division ratios M and N in Figure 11–12 that can be used to achieve the desired result.

11–28. In a certain system, it is desired to use a PLL to generate a 1.125-MHz signal from an available reference of 2 MHz. Specify the *smallest* frequency division ratios M and N in Figure 11–12 that can be used to achieve the desired result.

11–29. A 565 PLL circuit is to be designed for an application with the following requirements:

 1. VCO center frequency = 30 kHz
 2. capture range = ± 2 kHz
 3. to be powered with ± 12-V supplies

 Determine
 a. appropriate values of R_1 and C_1
 b. the lock range
 c. the required value of C_2

11–30. A 565 PLL circuit is to be designed for an application with the following requirements:

1. VCO center frequency = 250 kHz
2. capture range = ±20 kHz
3. to be powered with ±6-V supplies

Determine
a. appropriate values of R_1 and C_1
b. the lock range
c. the required value of C_2

A

RESISTOR AND CAPACITOR VALUES

A–1 RESISTOR VALUES (5% AND 10%)

The most widely available resistors are the low-power, axial-lead, carbon-composition and film resistors with tolerances of ±5% and ±10%. All possible values are derived from a standard set of two-digit values by adding a sufficient number of zeros or a decimal point. The various two-digit base numbers without decimal points are listed below. Within a given range, all possible values listed are available in 5% tolerances, but only the **boldface** values are available in 10% tolerances. Exact values are obtained by either adding zeros to the right of the numbers or, in the case of very small values, adding a decimal point between the digits.

10	**15**	**22**	**33**	**47**	**68**
11	16	24	36	51	75
12	**18**	**27**	**39**	**56**	**82**
13	20	30	43	62	91

The range of all available values is typically from less than 10 Ω to greater than 10 MΩ.

EXAMPLE A–1

What standard 5% and 10% resistance values are available in the range from 7 kΩ to 14 kΩ?

Solution
In 5% tolerance, the values are 7.5 kΩ, 8.2 kΩ, 9.1 kΩ, 10 kΩ, 11 kΩ, 12 kΩ, and 13 kΩ.
In 10% tolerance, the values are 8.2 kΩ, 10 kΩ, and 12 kΩ.

A–2 RESISTOR VALUES (1%)

Possible values are derived from three-digit values by adding zeros or a decimal point. The three-digit values are as follows:

100	147	215	316	464	681
102	150	221	324	475	698
105	154	226	332	487	715
107	158	232	340	499	732
110	162	237	348	511	750
113	165	243	357	523	768
115	169	249	365	536	787
118	174	255	374	549	806
121	178	261	383	562	825
124	182	267	392	576	845
127	187	274	402	590	866
130	191	280	412	604	887
133	196	287	422	619	909
137	200	294	432	634	931
140	205	301	442	649	953
143	210	309	453	665	976

A–3 CAPACITOR VALUES

Patterns of available capacitance values are too complex to cover adequately in a short space. Available values depend on the types, tolerances, and ranges. However, many capacitors available with 5% tolerances or less have patterns very similar to those of 5% and 10% resistors, as far as the digits are concerned. Some of these standard values were assumed in the text. The reader is referred to various electronic component directories or catalogs for further information.

B

LINEAR INTEGRATED-CIRCUIT MANUFACTURERS

The organizations listed in this appendix represent a few of the major manufacturers of linear integrated circuits and subsystems. The list is not complete, and any omission should not be interpreted in a negative sense.

Most of these organizations publish data and/or applications manuals. Depending on the organization and the status of the interested party, there may or may not be a charge for the manuals. However, the costs involved are usually modest compared with costs of textbooks. Since these manuals are constantly being updated and the titles and formats often change, any detailed list would quickly become outdated. The reader is referred to the individual companies (or their local sales representatives) for further information about possible manuals.

Analog Devices
Two Technology Way
P.O. Box 280
Norwood, Massachusetts 02062

Burr-Brown
P.O. Box 11400
International Airport Industrial Park
Tuscon, Arizona 85734

Data Device Corporation
105 Wilbur Place
Bohemia, New York 11716

Datel-Intersil
11 Cabot Boulevard
Mansfield, Massachusetts 02048

Exar Integrated Systems
P.O. Box 62229
Sunnyvale, California 94088

Fairchild Semiconductor Corporation
313 Fairchild Drive
Mountain View, California 94043

Hybrid Systems
Crosby Drive
Bedford, Massachusetts 01730

569

Intersil, Inc.
10600 Ridgeview Court
Cupertino, California 95014

Micro Networks Corporation
324 Clark Street
Worcester, Massachusetts 01606

Motorola Semiconductor Products, Inc.
P.O. Box 20912
Phoenix, Arizona 85036

National Semiconductor Company
2900 Semiconductor Drive
P.O. Box 58090
Santa Clara, California 95052

Precision Monolithics
1500 Space Park Drive
Santa Clara, California 95050

Siliconix Incorporated
2201 Laurelwood Road
Santa Clara, California 95054

Signetics Corporation
P.O. Box 409
811 East Arques Avenue
Sunnyvale, California 94086

Texas Instruments
Box 3640, M/S58218
Dallas, Texas 75285

C

SPECIFICATION SHEETS

Specification sheets for some of the devices discussed in the text are reproduced in this appendix. Data manuals provided by the organizations listed in Appendix B contain information of this type on hundreds of different integrated circuits.

Appreciation is expressed to the following organizations for permission to reproduce the specific data sheets given here:

1. μA741 Operational Amplifier. *Courtesy of Fairchild Semiconductor Corporation.*
2. LM118/LM218/LM318 Operational Amplifiers. *Courtesy of National Semiconductor Company.*
3. LM311 Voltage Comparator. *Courtesy of National Semiconductor Company.*
4. NE/SE555/SE555C Timer. *Courtesy of Signetics Corporation. Reprinted by permission.*
5. ICL8038 Precision Waveform Generator. *Courtesy of Intersil, Inc.*
6. LM117/LM317 Voltage Regulator. *Courtesy of National Semiconductor Company.*
7. NE/SE565 Phase-Locked Loop. *Courtesy of Signetics Corporation. Reprinted by permission.*

FAIRCHILD

A Schlumberger Company

μA741
Operational Amplifier

Linear Products

Description

The μA741 is a high performance Monolithic Operational Amplifier constructed using the Fairchild Planar epitaxial process. It is intended for a wide range of analog applications. High common mode voltage range and absence of latch-up tendencies make the μA741 ideal for use as a voltage follower. The high gain and wide range of operating voltage provides superior performance in integrator, summing amplifier, and general feedback applications.

- **NO FREQUENCY COMPENSATION REQUIRED**
- **SHORT-CIRCUIT PROTECTION**
- **OFFSET VOLTAGE NULL CAPABILITY**
- **LARGE COMMON MODE AND DIFFERENTIAL VOLTAGE RANGES**
- **LOW POWER CONSUMPTION**
- **NO LATCH-UP**

Connection Diagram
8-Pin Metal Package

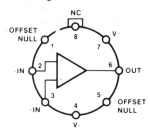

(Top View)

Pin 4 connected to case

Order Information

Type	Package	Code	Part No.
μA741	Metal	5W	μA741HM
μA741A	Metal	5W	μA741AHM
μA741C	Metal	5W	μA741HC
μA741E	Metal	5W	μA741EHC

Connection Diagram
10-Pin Flatpak

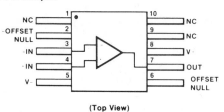

(Top View)

Order Information

Type	Package	Code	Part No.
μA741	Flatpak	3F	μA741FM
μA741A	Flatpak	3F	μA741AFM

Connection Diagram
8-Pin DIP

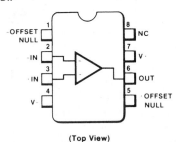

(Top View)

Order Information

Type	Package	Code	Part No.
μA741C	Molded DIP	9T	μA741TC
μA741C	Ceramic DIP	6T	μA741RC

μA741

Absolute Maximum Ratings

Supply Voltage	
μA741A, μA741, μA741E	$\pm$ 22 V
μA741C	$\pm$ 18 V
Internal Power Dissipation	
(Note 1)	
Metal Package	500 MW
DIP	310 mW
Flatpak	570 mW
Differential Input Voltage	$\pm$ 30 V
Input Voltage (Note 2)	$\pm$ 15 V
Storage Temperature Range	
Metal Package and Flatpak	$-65°$C to $+150°$C
DIP	$-55°$C to $+125°$C

Operating Temperature Range	
Military (μA741A, μA741)	$-55°$C to $+125°$C
Commercial (μA741E, μA741C)	$0°$C to $+70°$C
Pin Temperature (Soldering 60 s)	
Metal Package, Flatpak, and	
Ceramic DIP	$300°$C
Molded DIP (10 s)	$260°$C
Output Short Circuit Duration	
(Note 3)	Indefinite

Equivalent Circuit

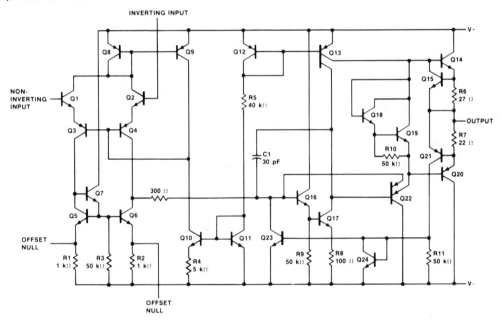

Notes

1. Rating applies to ambient temperatures up to 70°C. Above 70°C ambient derate linearly at 6.3 mW / °C for the metal package, 7.1 mW / °C for the flatpak, and 5.6 mW / °C for the DIP.

2. For supply voltages less than $\pm$ 15 V, the absolute maximum input voltage is equal to the supply voltage.

3. Short circuit may be to ground or either supply. Rating applies to $+125°$C case temperature or $75°$C ambient temperature.

μA741

μA741 and μA741C
Electrical Characteristics $V_S = \pm 15$ V, $T_A = 25°C$ unless otherwise specified

Characteristic	Condition	μA741			μA741C			Unit
		Min	Typ	Max	Min	Typ	Max	
Input Offset Voltage	$R_S \leq 10$ kΩ		1.0	5.0		2.0	6.0	mV
Input Offset Current			20	200		20	200	nA
Input Bias Current			80	500		80	500	nA
Power Supply Rejection Ratio	$V_S = +10, -20$ $V_S = +20, -10$ V, $R_S = 50$ Ω		30	150		30	150	μV / V
Input Resistance		.3	2.0		.3	2.0		MΩ
Input Capacitance			1.4			1.4		pF
Offset Voltage Adjustment Range			± 15			± 15		mV
Input Voltage Range					± 12	± 13		V
Common Mode Rejection Ratio	$R_S \leq 10$ kΩ				70	90		dB
Output Short Circuit Current			25			25		mA
Large Signal Voltage Gain	$R_L \geq 2$ kΩ, $V_{OUT} = \pm 10$ V	50k	200k		20k	200k		
Output Resistance			75			75		Ω
Output Voltage Swing	$R_L \geq 10$ kΩ				± 12	± 14		V
	$R_L \geq 2$ kΩ				± 10	± 13		V
Supply Current			1.7	2.8		1.7	2.8	mA
Power Consumption			50	85		50	85	mW
Transient Response (Unity Gain)	Rise Time	$V_{IN} = 20$ mV, $R_L = 2$ kΩ, $C_L \leq 100$ pF	.3			.3		μs
	Overshoot		5.0			5.0		%
Bandwidth (Note 4)			1.0			1.0		MHz
Slew Rate	$R_L \geq 2$ kΩ		.5			.5		V / μs

Notes
4. Calculated value from $BW(MHz) = \dfrac{0.35}{Rise\ Time\ (\mu s)}$

5. All $V_{CC} = 15$ V for μA741 and μA741C.
6. Maximum supply current for all devices
   25°C = 2.8 mA
   125°C = 2.5 mA
   −55°C = 3.3 mA

μA741

μA741 and μA741C
Electrical Characteristics (Cont.)   The following specifications apply over the range of $-55°C \leq T_A \leq 125°C$ for μA741, $0°C \leq T_A \leq 70°C$ for μA741C.

Characteristic	Condition	μA741			μA741C			Unit
		Min	Typ	Max	Min	Typ	Max	
Input Offset Voltage							7.5	mV
	$R_S \leq 10\ k\Omega$		1.0	6.0				mV
Input Offset Current							300	nA
	$T_A = +125°C$		7.0	200				nA
	$T_A = -55°C$		85	500				nA
Input Bias Current							800	nA
	$T_A = +125°C$		.03	.5				μA
	$T_A = -55°C$		.3	1.5				μA
Input Voltage Range		± 12	± 13					V
Common Mode Rejection Ratio	$R_S \leq 10\ k\Omega$	70	90					dB
Adjustment for Input Offset Voltage			± 15			± 15		mV
Supply Voltage Rejection Ratio	$V_S = +10, -20;$ $V_S = +20, -10$ V, $R_S = 50\ \Omega$		30	150				μV / V
Output Voltage Swing	$R_L \geq 10\ k\Omega$	± 12	± 14					V
	$R_L \geq 2\ k\Omega$	± 10	± 13		± 10	± 13		V
Large Signal Voltage Gain	$R_L = 2\ k\Omega$, $V_{OUT} = \pm 10$ V	25k			15k			
Supply Current	$T_A = +125°C$		1.5	2.5				mA
	$T_A = -55°C$		2.0	3.3				mA
Power Consumption	$T_A = +125°C$		45	75				mW
	$T_A = -55°C$		60	100				mW

Notes
4. Calculated value from BW(MHz) = $\dfrac{0.35}{\text{Rise Time } (\mu s)}$

5. All V_{CC} = 15 V for μA741 and μA741C.
6. Maximum supply current for all devices
   25°C = 2.8 mA
   125°C = 2.5 mA
   −55°C = 3.3 mA

μA741

μA741A and μA741E
Electrical Characteristics $V_S = \pm 15$ V, $T_A = 25°C$ unless otherwise specified.

Characteristic	Condition	μA741A/E Min	Typ	Max	Unit
Input Offset Voltage	$R_S \leq 50\ \Omega$		0.8	3.0	mV
Average Input Offset Voltage Drift				15	$\mu V/°C$
Input Offset Current			3.0	30	nA
Average Input Offset Current Drift				0.5	$nA/°C$
Input Bias Current			30	80	nA
Power Supply Rejection Ratio	$V_S = +10, -20; V_S = +20$ V, -10 V, $R_S = 50\ \Omega$		15	50	$\mu V/V$
Output Short Circuit Current		10	25	40	mA
Power Consumption	$V_S = \pm 20$ V		80	150	mW
Input Impedance	$V_S = \pm 20$ V	1.0	6.0		$M\Omega$
Large Signal Voltage Gain	$V_S = \pm 20$ V, $R_L = 2$ kΩ, $V_{OUT} = \pm 15$ V	50	200		V/mV
Transient Response (Unity Gain)	Rise Time		0.25	0.8	μs
	Overshoot		6.0	20	%
Bandwidth (Note 4)		.437	1.5		MHz
Slew Rate (Unity Gain)	$V_{IN} = \pm 10$ V	0.3	0.7		$V/\mu s$

The following specifications apply over the range of $-55°C \leq T_A \leq 125°C$ for the 741A, and $0°C \leq T_A \leq 70°C$ for the 741E.

				Min	Typ	Max	Unit
Input Offset Voltage						4.0	mV
Input Offset Current						70	nA
Input Bias Current						210	nA
Common Mode Rejection Ratio	$V_S = \pm 20$ V, $V_{IN} = \pm 15$ V, $R_S = 50\ \Omega$			80	95		dB
Adjustment For Input Offset Voltage	$V_S = \pm 20$ V			10			mV
Output Short Circuit Current				10		40	mA
Power Consumption	$V_S = \pm 20$ V	μA741A	$-55°C$			165	mW
			$+125°C$			135	mW
		μA741E				150	mW
Input Impedance	$V_S = \pm 20$ V			0.5			$M\Omega$
Output Voltage Swing	$V_S = \pm 20$ V	$R_L = 10$ kΩ		± 16			V
		$R_L = 2$ kΩ		± 15			V
Large Signal Voltage Gain	$V_S = \pm 20$ V, $R_L = 2$ kΩ, $V_{OUT} = \pm 15$ V			32			V/mV
	$V_S = \pm 5$ V, $R_L = 2$ kΩ, $V_{OUT} = \pm 2$ V			10			V/mV

Notes

4. Calculated value from: BW(MHz) = $\dfrac{0.35}{\text{Rise Time }(\mu s)}$

5. All $V_{CC} = 15$ V for μA741 and μA741C.

6. Maximum supply current for all devices
   25°C = 2.8 mA
   125°C = 2.5 mA
   −55°C = 3.3 mA

μA741

Typical Performance Curves for μA741A and μA741

Open Loop Voltage Gain as a Function of Supply Voltage

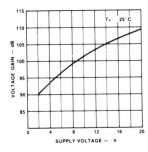

Output Voltage Swing as a Function of Supply Voltage

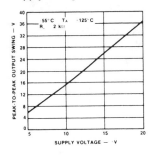

Input Common Mode Voltage as a Function of Supply Voltage

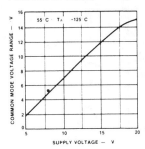

Typical Performance Curves for μA741E and μA741C

Open Loop Voltage Gain as a Function of Supply Voltage

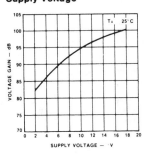

Output Voltage Swing as a Function of Supply Voltage

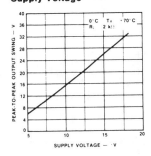

Input Common Mode Voltage Range as a Function of Supply Voltage

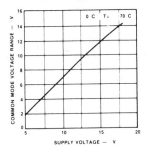

Transient Response

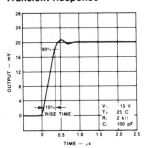

Transient Response Test Circuit

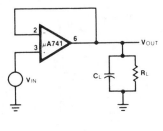

Common Mode Rejection Ratio as a Function of Frequency

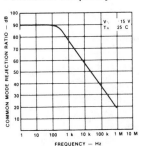

μA741

Typical Performance Curves for μA741E and μA741C (Cont.)

Frequency Characteristics as a Function of Supply Voltage

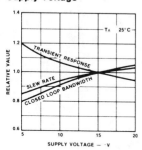

Voltage Offset Null Circuit

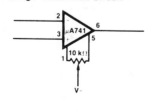

Voltage Follower Large Signal Pulse Response

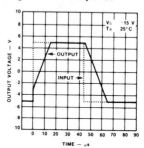

Typical Performance Curves for μA741A, μA741, μA741E and μA741C

Power Consumption as a Function of Supply Voltage

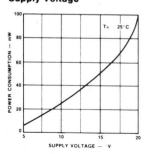

Open Loop Voltage Gain as a Function of Frequency

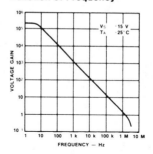

Open Loop Phase Response as a Function of Frequency

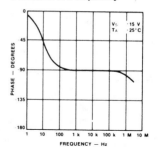

Input Offset Current as a Function of Supply Voltage

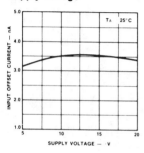

Input Resistance and Input Capacitance as a Function of Frequency

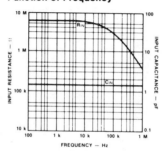

Output Resistance as a Function of Frequency

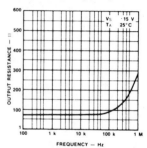

μA741

Typical Performance Curves for μA741A, μA741, μA741E and μA741C (Cont.)

Output Voltage Swing as a Function of Load Resistance

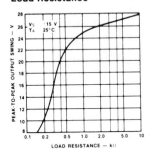

Output Voltage Swing as a Function of Frequency

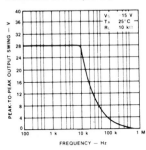

Absolute Maximum Power Dissipation as a Function of Ambient Temperature

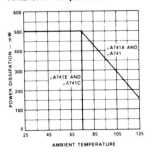

Input Noise Voltage as a Function of Frequency

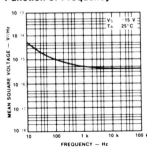

Input Noise Current as a Function of Frequency

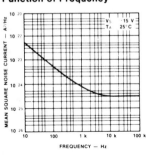

Broadband Noise for Various Bandwidths

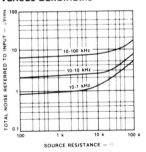

Typical Performance Curves for μA741A and μA741

Input Bias Current as a Function of Ambient Temperature

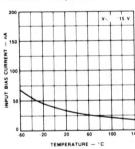

Input Resistance as a Function of Ambient Temperature

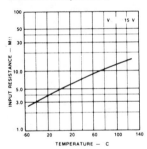

Output Short-Circuit Current as a Function of Ambient Temperature

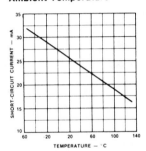

μA741

Typical Performance Curves for μA741A and μA741 (Cont.)

Input Offset Current as a Function of Ambient Temperature

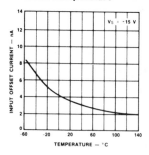

Power Consumption as a Function of Ambient Temperature

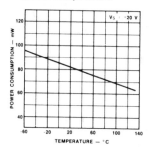

Frequency Characteristics as a Function of Ambient Temperature

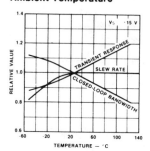

Typical Performance Curves for μA741E and μA741C

Input Bias Current as a Function of Ambient Temperature

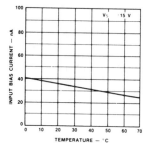

Input Resistance as a Function of Ambient Temperature

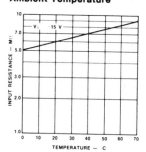

Input Offset Current as a Function of Ambient Temperature

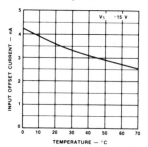

Power Consumption as a Function of Ambient Temperature

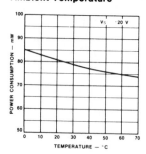

Output Short Circuit Current as a Function of Ambient Temperature

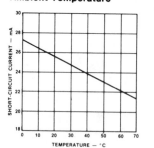

Frequency Characteristics as a Function of Ambient Temperature

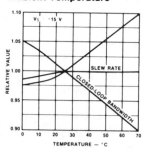

LM118/LM218/LM318

National Semiconductor

Operational Amplifiers/Buffers

LM118/LM218/LM318 Operational Amplifiers

General Description

The LM118 series are precision high speed operational amplifiers designed for applications requiring wide bandwidth and high slew rate. They feature a factor of ten increase in speed over general purpose devices without sacrificing DC performance.

Features

- 15 MHz small signal bandwidth
- Guaranteed 50V/μs slew rate
- Maximum bias current of 250 nA
- Operates from supplies of ±5V to ±20V
- Internal frequency compensation
- Input and output overload protected
- Pin compatible with general purpose op amps

The LM118 series has internal unity gain frequency compensation. This considerably simplifies its application since no external components are necessary for operation. However, unlike most internally compensated amplifiers, external frequency compensation may be added for optimum performance For inverting applications, feedforward compensation will boost the slew rate to over 150V/μs and almost double the bandwidth. Overcompensation can be used with the amplifier for greater stability when maximum bandwidth is not needed. Further, a single capacitor can be added to reduce the 0.1% settling time to under 1 μs.

The high speed and fast settling time of these op amps make them useful in A/D converters, oscillators, active filters, sample and hold circuits, or general purpose amplifiers. These devices are easy to apply and offer an order of magnitude better AC performance than industry standards such as the LM709.

The LM218 is identical to the LM118 except that the LM218 has its performance specified over a −25°C to +85°C temperature range. The LM318 is specified from 0°C to +70°C.

Schematic and Connection Diagrams

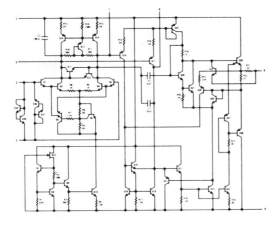

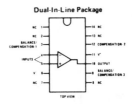

Dual-In-Line Package

Order Number LM118J, LM218J
or LM318J
See NS Package J14A

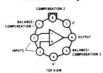

Metal Can Package*

*Pin connections shown on schematic diagram and typical applications are for TO-5 package.

Order Number LM118H, LM218H
or LM318H
See NS Package H08C

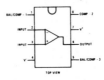

Dual-In-Line Package

Order Number LM118J-8,
LM218J-8 or LM318J-8
See NS Package J08A
Order Number LM318N
See NS Package N08B

3-106

LM118/LM218/LM318

Absolute Maximum Ratings

Supply Voltage	±20V
Power Dissipation (Note 1)	500 mW
Differential Input Current (Note 2)	±10 mA
Input Voltage (Note 3)	±15V
Output Short-Circuit Duration	Indefinite
Operating Temperature Range	
LM118	−55°C to +125°C
LM218	−25°C to +85°C
LM318	0°C to +70°C
Storage Temperature Range	−65°C to +150°C
Lead Temperature (Soldering, 10 seconds)	300°C

Electrical Characteristics (Note 4)

PARAMETER	CONDITIONS	LM118/LM218			LM318			UNITS
		MIN	TYP	MAX	MIN	TYP	MAX	
Input Offset Voltage	$T_A = 25°C$		2	4		4	10	mV
Input Offset Current	$T_A = 25°C$		6	50		30	200	nA
Input Bias Current	$T_A = 25°C$		120	250		150	500	nA
Input Resistance	$T_A = 25°C$	1	3		0.5	3		MΩ
Supply Current	$T_A = 25°C$		5	8		5	10	mA
Large Signal Voltage Gain	$T_A = 25°C$, $V_S = ±15V$ $V_{OUT} = ±10V$, $R_L \geq 2 k\Omega$	50	200		25	200		V/mV
Slew Rate	$T_A = 25°C$, $V_S = ±15V$, $A_V = 1$	50	70		50	70		V/μs
Small Signal Bandwidth	$T_A = 25°C$, $V_S = ±15V$		15			15		MHz
Input Offset Voltage				6			15	mV
Input Offset Current				100			300	nA
Input Bias Current				500			750	nA
Supply Current	$T_A = 125°C$		4.5	7				
Large Signal Voltage Gain	$V_S = ±15V$, $V_{OUT} = ±10V$ $R_L \geq 2 k\Omega$	25			20			V/mV
Output Voltage Swing	$V_S = ±15V$, $R_L = 2 k\Omega$	±12	±13		±12	±13		V
Input Voltage Range	$V_S = ±15V$	±11.5			±11.5			V
Common-Mode Rejection Ratio		80	100		70	100		dB
Supply Voltage Rejection Ratio		70	80		65	80		dB

Note 1: The maximum junction temperature of the LM118 is 150°C, the LM218 is 110°C, and the LM318 is 110°C. For operating at elevated temperatures, devices in the TO-5 package must be derated based on a thermal resistance of 150°C/W, junction to ambient, or 45°C/W, junction to case. The thermal resistance of the dual-in-line package is 100°C/W, junction to ambient.

Note 2: The inputs are shunted with back-to-back diodes for overvoltage protection. Therefore, excessive current will flow if a differential input voltage in excess of 1V is applied between the inputs unless some limiting resistance is used.

Note 3: For supply voltages less than ±15V, the absolute maximum input voltage is equal to the supply voltage.

Note 4: These specifications apply for ±5V $\leq V_S \leq$ ±20V and −55°C $\leq T_A \leq$ +125°C, (LM118), −25°C $\leq T_A \leq$ +85°C (LM218), and 0°C $\leq$ T_A $\leq$ +70°C (LM318). Also, power supplies must be bypassed with 0.1μF disc capacitors.

LM118/LM218/LM318

Typical Performance Characteristics LM118, LM218

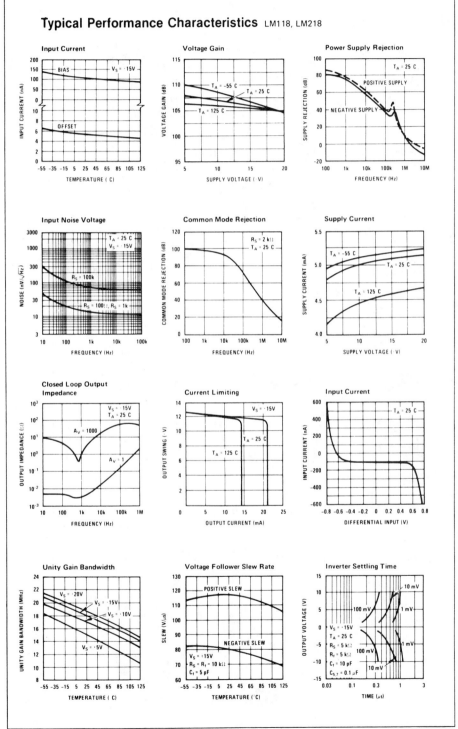

Typical Performance Characteristics LM318 (Continued)

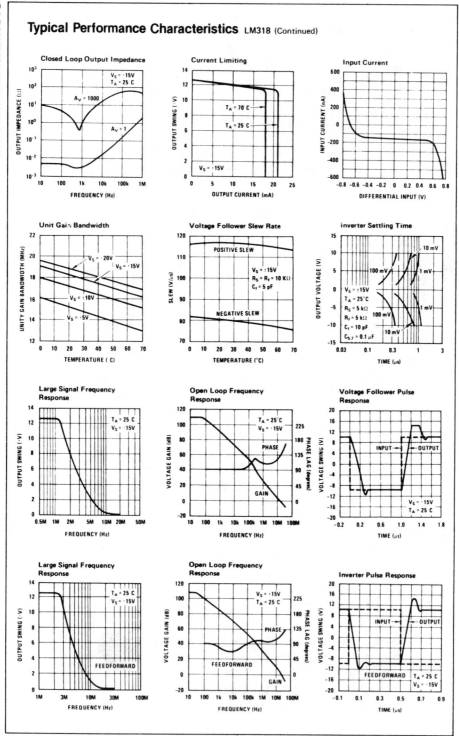

LM118/LM218/LM318

Typical Performance Characteristics LM118, LM218 (Continued)

Typical Performance Characteristics LM318

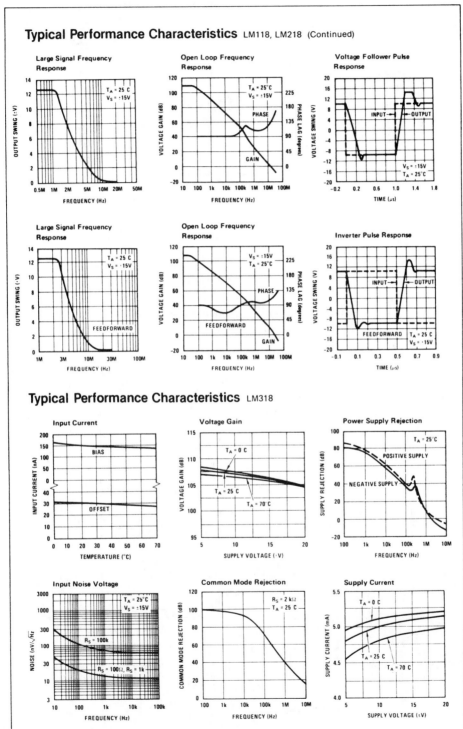

Auxiliary Circuits

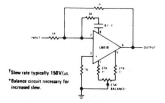

Feedforward Compensation for Greater Inverting Slew Rate†

†Slew rate typically 150V/μs.

*Balance circuit necessary for increased slew.

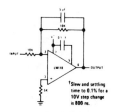

Compensation for Minimum Settling† Time

†Slew and settling time to 0.1% for a 10V step change is 800 ns.

Offset Balancing

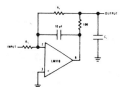

Isolating Large Capacitive Loads

Overcompensation

Typical Applications

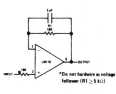

Fast Voltage Follower *

*Do not hardwire as voltage follower (R1 ≥ 5 kΩ)

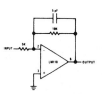

Fast Summing Amplifier

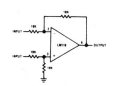

Differential Amplifier

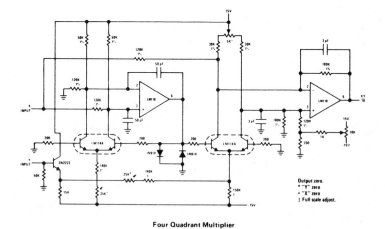

Output zero.
* "Y" zero
+ "X" zero
‡ Full scale adjust.

Four Quadrant Multiplier

LM118/LM218/LM318

Typical Applications (Continued)

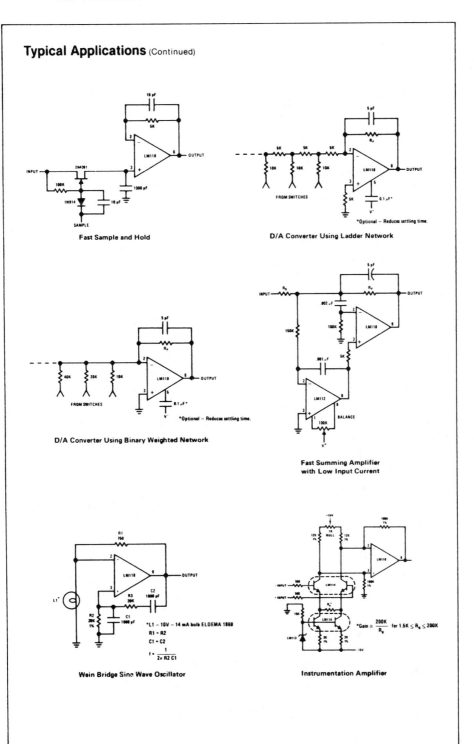

Fast Sample and Hold

D/A Converter Using Ladder Network

*Optional — Reduces settling time.

D/A Converter Using Binary Weighted Network

*Optional — Reduces settling time.

**Fast Summing Amplifier
with Low Input Current**

Wein Bridge Sine Wave Oscillator

*L1 – 10V – 14 mA bulb ELDEMA 1869
R1 = R2
C1 = C2
$f = \dfrac{1}{2\pi\,R2\,C1}$

Instrumentation Amplifier

*Gain $\geq \dfrac{200K}{R_g}$ for 1.5K $\leq R_g \leq$ 200K

LM311

National Semiconductor

Voltage Comparators

LM311 Voltage Comparator

General Description

The LM311 is a voltage comparator that has input currents more than a hundred times lower than devices like the LM306 or LM710C. It is also designed to operate over a wider range of supply voltages: from standard ±15V op amp supplies down to the single 5V supply used for IC logic. Its output is compatible with RTL, DTL and TTL as well as MOS circuits. Further, it can drive lamps or relays, switching voltages up to 40V at currents as high as 50 mA.

Features

- Operates from single 5V supply
- Maximum input current: 250 nA
- Maximum offset current: 50 nA

- Differential input voltage range: ±30V
- Power consumption: 135 mW at ±15V

Both the input and the output of the LM311 can be isolated from system ground, and the output can drive loads referred to ground, the positive supply or the negative supply. Offset balancing and strobe capability are provided and outputs can be wire OR'ed. Although slower than the LM306 and LM710C (200 ns response time vs 40 ns) the device is also much less prone to spurious oscillations. The LM311 has the same pin configuration as the LM306 and LM710C. See the "application hints" of the LM311 for application help.

Auxiliary Circuits**

** Note: Pin connections shown on schematic diagram and typical applications are for TO-5 package.

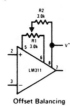

Offset Balancing

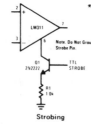

Strobing

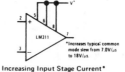

Increasing Input Stage Current*

Typical Applications**

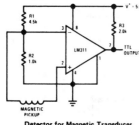

Detector for Magnetic Transducer

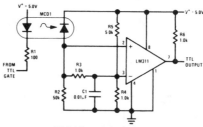

Digital Transmission Isolator

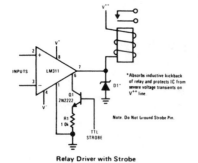

Relay Driver with Strobe

Strobing off Both Input*
and Output Stages

LM311

Absolute Maximum Ratings

Total Supply Voltage (V_{84})	36V
Output to Negative Supply Voltage (V_{74})	40V
Ground to Negative Supply Voltage (V_{14})	30V
Differential Input Voltage	±30V
Input Voltage (Note 1)	±15V
Power Dissipation (Note 2)	500 mW
Output Short Circuit Duration	10 sec
Operating Temperature Range	$0°C$ to $70°C$
Storage Temperature Range	$-65°C$ to $150°C$
Lead Temperature (soldering, 10 sec)	$300°C$
Voltage at Strobe Pin	$V^+ - 5V$

Electrical Characteristics (Note 3)

PARAMETER	CONDITIONS	MIN	TYP	MAX	UNITS
Input Offset Voltage (Note 4)	$T_A = 25°C$, $R_S \leq 50k$		2.0	7.5	mV
Input Offset Current (Note 4)	$T_A = 25°C$		6.0	50	nA
Input Bias Current	$T_A = 25°C$		100	250	nA
Voltage Gain	$T_A = 25°C$	40	200		V/mV
Response Time (Note 5)	$T_A = 25°C$		200		ns
Saturation Voltage	$V_{IN} \leq -10$ mV, $I_{OUT} = 50$ mA $T_A = 25°C$		0.75	1.5	V
Strobe ON Current	$T_A = 25°C$		3.0		mA
Output Leakage Current	$V_{IN} \geq 10$ mV, $V_{OUT} = 35V$ $T_A = 25°C$, $I_{STROBE} = 3$ mA		0.2	50	nA
Input Offset Voltage (Note 4)	$R_S \leq 50k$			10	mV
Input Offset Current (Note 4)				70	nA
Input Bias Current				300	nA
Input Voltage Range		-14.5	$13.8, -14.7$	13.0	V
Saturation Voltage	$V^+ \geq 4.5V$, $V^- = 0$ $V_{IN} \leq -10$ mV, $I_{SINK} \leq 8$ mA		0.23	0.4	V
Positive Supply Current	$T_A = 25°C$		5.1	7.5	mA
Negative Supply Current	$T_A = 25°C$		4.1	5.0	mA

Note 1: This rating applies for ±15V supplies. The positive input voltage limit is 30V above the negative supply. The negative input voltage limit is equal to the negative supply voltage or 30V below the positive supply, whichever is less.

Note 2: The maximum junction temperature of the LM311 is 110°C. For operating at elevated temperatures, devices in the TO-5 package must be derated based on a thermal resistance of 150°C/W, junction to ambient, or 45°C/W, junction to case. The thermal resistance of the dual-in-line package is 100°C/W, junction to ambient.

Note 3: These specifications apply for $V_S = ±15V$ and the Ground pin at ground, and $0°C < T_A < +70°C$, unless otherwise specified. The offset voltage, offset current and bias current specifications apply for any supply voltage from a single 5V supply up to ±15V supplies.

Note 4: The offset voltages and offset currents given are the maximum values required to drive the output within a volt of either supply with 1 mA load. Thus, these parameters define an error band and take into account the worst-case effects of voltage gain and input impedance.

Note 5: The response time specified (see definitions) is for a 100 mV input step with 5 mV overdrive.

Note 6: Do not short the strobe pin to ground; it should be current driven at 3 to 5 mA.

LM311

Typical Performance Characteristics

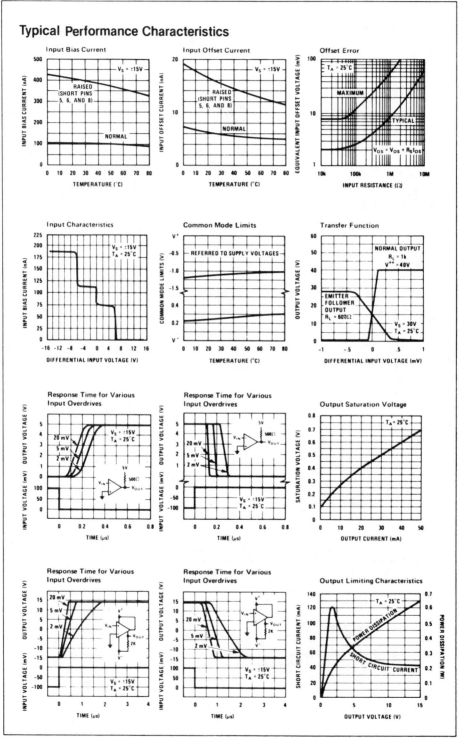

LM311

Typical Performance Characteristics (Continued)

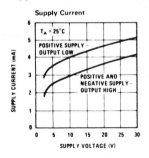

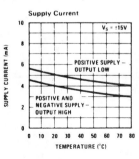

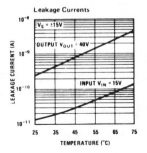

Typical Applications

Zero Crossing Detector
Driving MOS Switch

100 kHz Free Running Multivibrator

*TTL or DTL fanout of two

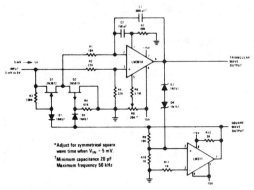

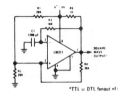

Driving Ground-Referred Load

*Input polarity is reversed
when using pin 1 as output.

10 Hz to 10 kHz Voltage Controlled Oscillator

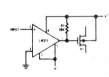

*Adjust for symmetrical square
wave time when $V_{IN} = 5$ mV.
†Minimum capacitance 20 pF
Maximum frequency 50 kHz

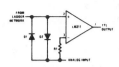

Using Clamp Diodes to Improve Response

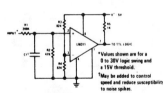

TTL Interface with High Level Logic

*Values shown are for a
0 to 30V logic swing and
a 15V threshold.
†May be added to control
speed and reduce susceptibility
to noise spikes.

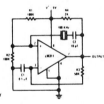

Crystal Oscillator

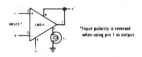

Comparator and Solenoid Driver

LM311

Typical Applications (Continued)

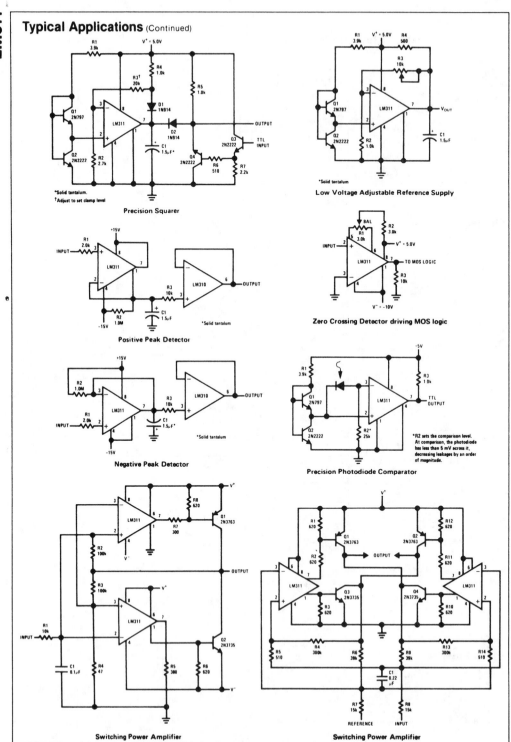

Precision Squarer

Low Voltage Adjustable Reference Supply

Positive Peak Detector

Zero Crossing Detector driving MOS logic

Negative Peak Detector

Precision Photodiode Comparator

Switching Power Amplifier

Switching Power Amplifier

Schematic Diagram

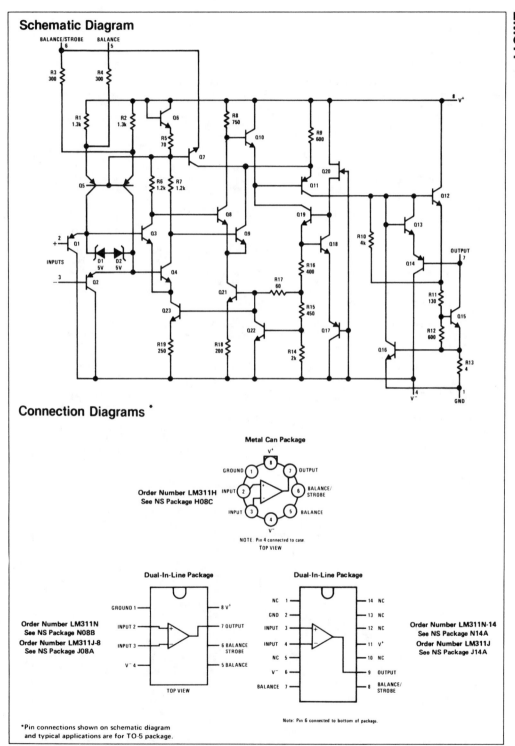

Connection Diagrams *

Metal Can Package

Order Number LM311H
See NS Package H08C

NOTE: Pin 4 connected to case.
TOP VIEW

Dual-In-Line Package

Order Number LM311N
See NS Package N08B
Order Number LM311J-8
See NS Package J08A

TOP VIEW

Dual-In-Line Package

Order Number LM311N-14
See NS Package N14A
Order Number LM311J
See NS Package J14A

Note: Pin 6 connected to bottom of package.

*Pin connections shown on schematic diagram
and typical applications are for TO-5 package.

LM311

LM311

Application Hints

CIRCUIT TECHNIQUES FOR AVOIDING OSCILLATIONS IN COMPARATOR APPLICATIONS

When a high-speed comparator such as the LM111 is used with fast input signals and low source impedances, the output response will normally be fast and stable, assuming that the power supplies have been bypassed (with 0.1 μF disc capacitors), and that the output signal is routed well away from the inputs (pins 2 and 3) and also away from pins 5 and 6.

However, when the input signal is a voltage ramp or a slow sine wave, or if the signal source impedance is high (1 kΩ to 100 kΩ), the comparator may burst into oscillation near the crossing-point. This is due to the high gain and wide bandwidth of comparators like the LM111. To avoid oscillation or instability in such a usage, several precautions are recommended, as shown in *Figure 1* below.

1. The trim pins (pins 5 and 6) act as unwanted auxiliary inputs. If these pins are not connected to a trim-pot, they should be shorted together. If they are connected to a trim-pot, a 0.01 μA capacitor C1 between pins 5 and 6 will minimize the susceptibility to AC coupling. A smaller capacitor is used if pin 5 is used for positive feedback as in *Figure 1*.

2. Certain sources will produce a cleaner comparator output waveform if a 100 pF to 1000 pF capacitor C2 is connected directly across the input pins.

3. When the signal source is applied through a resistive network, R_s, it is usually advantageous to choose an R_s' of substantially the same value, both for DC and for dynamic (AC) considerations. Carbon, tin-oxide, and metal-film resistors have all been used successfully in comparator input circuitry. Inductive wirewound resistors are not suitable.

4. When comparator circuits use input resistors (eg. summing resistors), their value and placement are particularly important. In all cases the body of the resistor should be close to the device or socket. In other words there should be very little lead length or printed-circuit foil run between comparator and resistor to radiate or pick up signals. The same applies to capacitors, pots, etc. For example, if R_s = 10 kΩ, as little as 5 inches of lead between the resistors and the input pins can result in oscillations that are very hard to damp. Twisting these input leads tightly is the only (second best) alternative to placing resistors close to the comparator.

5. Since feedback to almost any pin of a comparator can result in oscillation, the printed-circuit layout should be engineered thoughtfully. Preferably there should be a groundplane under the LM111 circuitry, for example, one side of a double-layer circuit card. Ground foil (or, positive supply or negative supply foil) should extend between the output and the inputs, to act as a guard. The foil connections for the inputs should be as small and compact as possible, and should be essentially surrounded by ground foil on all sides, to guard against capacitive coupling from any high-level signals (such as the output). If pins 5 and 6 are not used, they should be shorted together. If they are connected to a trim-pot, the trim-pot should be located, at most, a few inches away from the LM111, and the 0.01 μF capacitor should be installed. If this capacitor cannot be used, a shielding printed-circuit foil may be advisable between pins 6 and 7. The power supply bypass capacitors should be located within a couple inches of the LM111. (Some other comparators require the power-supply bypass to be located immediately adjacent to the comparator.)

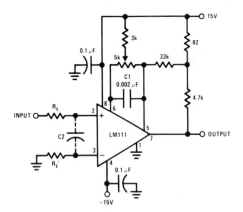

Pin connections shown are for LM111H in 8-lead TO-5 hermetic package

FIGURE 1. Improved Positive Feedback

TIMER   **NE/SE555/SE555C**

SE555F,H,N-14 • SE555C,F,H,N-14 • NE555F,H,N-14

FEATURES

- Turn off time less than 2μs
- Maximum operating frequency greater than 500kHz
- Timing from microseconds to hours
- Operates in both astable and monostable modes
- High output current
- Adjustable duty cycle
- TTL compatible
- Temperature stability of 0.005% per °C
- SE555 Mil std 883A,B,C available M38510 (JAN) approved, M38510 processing available.

PIN CONFIGURATIONS

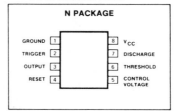

N PACKAGE

GROUND	1	8	V$_{CC}$
TRIGGER	2	7	DISCHARGE
OUTPUT	3	6	THRESHOLD
RESET	4	5	CONTROL VOLTAGE

APPLICATIONS

- Precision timing
- Pulse generation
- Sequential timing
- Time delay generation
- Pulse width modulation
- Pulse position modulation
- Missing pulse detector

ABSOLUTE MAXIMUM RATINGS

PARAMETER	RATING	UNIT
Supply voltage		
SE555	+18	V
NE555, SE555C,	+16	V
Power dissipation	600	mW
Operating temperature range		
NE555	0 to +70	°C
SE555, SE555C	–55 to +125	°C
Storage temperature range	–65 to +150	°C
Load temperature (soldering, 60sec)	300	°C

EQUIVALENT SCHEMATIC

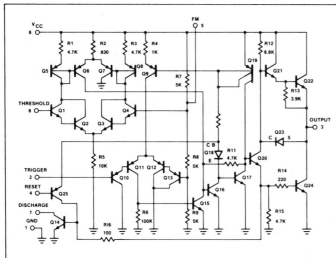

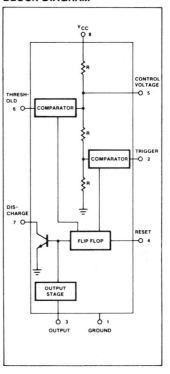

F,N-14 PACKAGE

H PACKAGE

BLOCK DIAGRAM

| TIMER | NE/SE555/SE555C |

SE555F,H,N,N-14 • SE555C,F,H,N,N-14 • NE555F,H,N,N-14

DC ELECTRICAL CHARACTERISTICS $T_A = 25°C$, $V_{CC} = +5V$ to $+15$ unless otherwise specified.

PARAMETER	TEST CONDITIONS	SE555			NE555/SE555C			UNIT
		Min	Typ	Max	Min	Typ	Max	
Supply voltage		4.5		18	4.5		16	V
Supply current (low state)[1]	$V_{CC} = 5V$ $R_L = \infty$		3	5		3	6	mA
	$V_{CC} = 15V$ $R_L = \infty$		10	12		10	15	mA
Timing error (monostable)	$R_A = 2K\Omega$ to $100K\Omega$							
Initial accuracy[2]	$C = 0.1\mu F$		0.5	2.0		1.0	3.0	%
Drift with temperature			30	100		50		ppm/°C
Drift with supply voltage			0.05	0.2		0.1	0.5	%/V
Timing error (astable)	$R_A, R_B = 1k\Omega$ to $100k\Omega$							
Initial accuracy[2]	$C = 0.1\mu F$		1.5			2.25		%
Drift with temperature	$V_{CC} = 15V$		90			150		ppm/°C
Drift with supply voltage			0.15			0.3		%/V
Control voltage level	$V_{CC} = 15V$	9.6	10.0	10.4	9.0	10.0	11.0	V
	$V_{CC} = 5V$	2.9	3.33	3.8	2.6	3.33	4.0	V
Threshold voltage	$V_{CC} = 15V$	9.4	10.0	10.6	8.8	10.0	11.2	V
	$V_{CC} = 5V$	2.7	3.33	4.0	2.4	3.33	4.2	V
Threshold current[3]			0.1	0.25		0.1	0.25	μA
Trigger voltage	$V_{CC} = 15V$	4.8	5.0	5.2	4.5	5.0	5.6	V
	$V_{CC} = 5V$	1.45	1.67	1.9	1.1	1.67	2.2	V
Trigger current	$V_{TRIG} = 0V$		0.5	0.9		0.5	2.0	μA
Reset voltage[4]		0.4	0.7	1.0	0.4	0.7	1.0	V
Reset current			0.1	0.4		0.1	0.4	mA
Reset current	$V_{RESET} = 0V$		0.4	1.0		0.4	1.5	mA
Output voltage (low)	$V_{CC} = 15V$							
	$I_{SINK} = 10mA$		0.1	0.15		0.1	0.25	V
	$I_{SINK} = 50mA$		0.4	0.5		0.4	0.75	V
	$I_{SINK} = 100mA$		2.0	2.2		2.0	2.5	V
	$I_{SINK} = 200mA$		2.5			2.5		V
	$V_{CC} = 5V$							
	$I_{SINK} = 8mA$		0.1	0.25		0.3	0.4	V
	$I_{SINK} = 5mA$		0.05	0.2		0.25	0.35	V
Output voltage (high)	$V_{CC} = 15V$							
	$I_{SOURCE} = 200mA$		12.5			12.5		V
	$I_{SOURCE} = 100mA$	13.0	13.3		12.75	13.3		V
	$V_{CC} = 5V$							
	$I_{SOURCE} = 100mA$	3.0	3.3		2.75	3.3		V
Turn off time[5]	$V_{RESET} = V_{CC}$		0.5	2.0		0.5		μs
Rise time of output			100	200		100	300	ns
Fall time of output			100	200		100	300	ns
Discharge leakage current			20	100		20	100	na

NOTES

1. Supply current when output high typically 1mA less.
2. Tested at $V_{CC} = 5V$ and $V_{CC} = 15V$.
3. This will determine the maximum value of $R_A + R_B$, for 15V operation, the max total R = 10 megohm, and for 5V operation, the max total R = 3.4 megohm.
4. Specified with trigger input high.
5. Time measured from a positive going input pulse from 0 to $0.8 \times V_{CC}$ into the threshold to the drop from high to low of the output. Trigger is tied to threshold.

signetics

TIMER

SE555F,H,N,N-14 • SE555C,F,H,N,N-14 • NE555F,H,N,N-14

TYPICAL PERFORMANCE CHARACTERISTICS

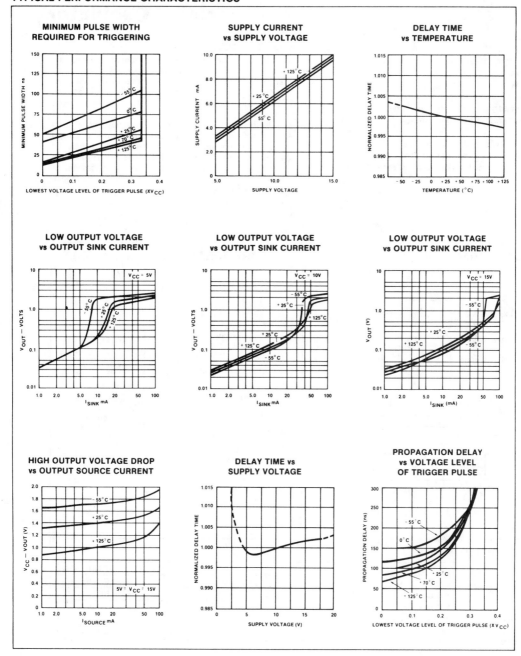

ICL8038
Precision Waveform
Generator/Voltage
Controlled Oscillator

FEATURES

- Low frequency drift with temperature - 50ppm/°C
- Simultaneous sine, square, and triangle wave outputs
- Low distortion - 1% (sine wave output)
- High linearity - 0.1% (triangle wave output)
- Wide operating frequency range - 0.001Hz to 0.3MHz
- Variable duty cycle - 2% to 98%
- High level outputs - TTL to 28V
- Easy to use - just a handful of external components required

GENERAL DESCRIPTION

The ICL8038 Waveform Generator is a monolithic integrated circuit capable of producing high accuracy sine, square, triangular, sawtooth and pulse waveforms with a minimum of external components. The frequency (or repetition rate) can be selected externally from .001Hz to more than 300kHz using either resistors or capacitors, and frequency modulation and sweeping can be accomplished with an external voltage. The ICL8038 is fabricated with advanced monolithic technology, using Schottky-barrier diodes and thin film resistors, and the output is stable over a wide range of temperature and supply variations. These devices may be interfaced with phase locked loop circuitry to reduce temperature drift to less than 50ppm/°C.

BLOCK DIAGRAM

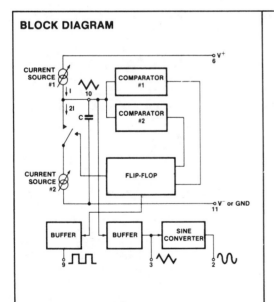

PIN CONFIGURATION (outline dwg JD)

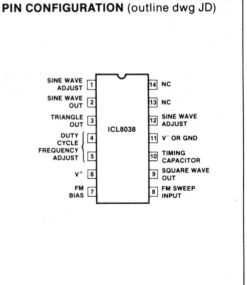

ORDERING INFORMATION

TYPE	TEMPERATURE RANGE	STABILITY	PACKAGE	ORDER PART NUMBER
8038 CC	0°C to +70°C	50ppm/°C typ	CERDIP	ICL8038 CC JD
8038 BC	0°C to +70°C	100 ppm/°C max	CERDIP	ICL8038 BC JD
8038 AC	0°C to +70°C	50ppm/°C max	CERDIP	ICL8038 AC JD
8038 BM	−55°C to +125°C	100ppm/°C max	CERDIP	ICL8038 BM JD
8038 AM	−55°C to +125°C	50ppm/°C max	CERDIP	ICL8038 AM JD

ICL8038

INTERSIL

MAXIMUM RATINGS

Supply Voltage .... ±18V or 36V Total
Power Dissipation[1] 750mW
Input Voltage (any pin) .... Not To Exceed Supply Voltages
Input Current (Pins 4 and 5) 25mA
Output Sink Current (Pins 3 and 9) 25mA
Storage Temperature Range .... –65°C to +125°C
Operating Temperature Range:
   8038AM, 8038BM .... –55°C to +125°C
   8038AC, 8038BC, 8038CC .... 0°C to +70°C
Lead Temperature (Soldering, 10 sec.) .... 300°C

Stresses above those listed under Absolute Maximum Ratings may cause permanent damage to the device. These are stress ratings only, and functional operation of the device at these or any other conditions above those indicated in the operational sections of the specifications is not implied. Exposure to absolute maximum rating conditions for extended periods may affect device reliability:

NOTE 1: Derate ceramic package at 12.5mW/°C for ambient temperatures above 100°C.

ELECTRICAL CHARACTERISTICS

(V$_{SUPP}$ = ±10V or +20V, T$_A$ = 25°C, R$_L$ = ... Test Circuit Unless Otherwise Specified)

SYMBOL	GENERAL CHARACTERISTICS		8038CC		8038BC/BM			8038AC/AM			UNITS
			TYP	MAX	MIN	TYP	MAX	MIN	TYP	MAX	
V$_{SUPP}$	Supply Voltage Operating Range										
V$^+$	Single Supply			+30	+10		30	+10		30	V
V$^+$, V$^-$	Dual Supplies	±5		±15	±5		±15	±5		±15	V
I$_{SUPP}$	Supply Current (V$_{SUPP}$ = ±10V)[2]										
	8038AM, 8038BM					12	15		12	15	mA
	8038AC, 8038BC, 8038CC		12	20		12	20		12	20	mA
FREQUENCY CHARACTERISTICS (all waveforms)											
f$_{max}$	Maximum Frequency of Oscillation	100,000			100,000			100,000			Hz
f$_{sweep}$	Sweep Frequency of FM		10			10			10		kHz
	Sweep FM Range[3]		40:1			40:1			40:1		
	FM Linearity 10:1 Ratio		0.5			0.2			0.2		%
Δf/ΔT	Frequency Drift With Temperature[5]		50			50	100		20	50	ppm/°C
Δf/ΔV	Frequency Drift With Supply Voltage (Over Supply Voltage Range)		0.05			0.05			0.05		%/V$_{SUPP}$
	Recommended Programming Resistors (R$_A$ and R$_B$)	1000		1M	1000		1M	1000		1M	Ω
OUTPUT CHARACTERISTICS											
	Square-Wave										
I$_{OLK}$	Leakage Current (V$_9$ = 30V)			1			1			1	μA
V$_{SAT}$	Saturation Voltage (I$_{SINK}$ = 2mA)		0.2	0.5		0.2	0.4		0.2	0.4	V
t$_r$	Rise Time (R$_L$ = 4.7kΩ)		100			100			100		ns
t$_f$	Fall Time (R$_L$ = 4.7kΩ)		40			40			40		ns
	Duty Cycle Adjust	2		98	2		98	2		98	%
	Triangle/Sawtooth/Ramp										
	Amplitude (R$_{TRI}$ = 100kΩ)	0.30	0.33		0.30	0.33		0.30	0.33		xV$_{SUPP}$
	Linearity		0.1			0.05			0.05		%
Z$_{OUT}$	Output Impedance (I$_{OUT}$ = 5mA)		200			200			200		Ω
	Sine-Wave										
	Amplitude (R$_{SINE}$ = 100kΩ)	0.2	0.22		0.2	0.22		0.2	0.22		xV$_{SUPP}$
	THD (R$_S$ = 1MΩ)[4]		0.8	5		0.7	3		0.7	1.5	%
	THD Adjusted (Use Fig. 8b)		0.5			0.5			0.5		%

NOTE 2: R$_A$ and R$_B$ currents not included.
NOTE 3: V$_{SUPP}$ = 20V; R$_A$ and R$_B$ = 10kΩ, f ≅ 9kHz; Can be extended to 1000.1. See Figures 13 and 14.
NOTE 4: 82kΩ connected between pins 11 and 12, Triangle Duty Cycle set at 50%. (Use R$_A$ and R$_B$.)
NOTE 5: Over operating temperature range, Fig. 2, pins 7 and 8 connected, V$_{SUPP}$ = ±10V. See Fig. 6c for T.C. vs V$_{SUPP}$.

ICL8038

INTERSIL

TEST CONDITIONS

PARAMETER	R_A	R_B	R_L	C_1	SW_1	MEASURE
Supply Current	10kΩ	10kΩ	10kΩ	3.3nF	Closed	Current into Pin 6
Maximum Frequency of Oscillation	1kΩ	1kΩ	4.7kΩ	100pf	Closed	Frequency at Pin 9
Sweep FM Range [1]	10kΩ	10kΩ	10kΩ	3.3nF	Open	Frequency at Pin 9
Frequency Drift with Temperature	10kΩ	10kΩ	10kΩ	3.3nF	Closed	Frequency at Pin 9
Frequency Drift with Supply Voltage [2]	10kΩ	10kΩ	10kΩ	3.3nF	Closed	Frequency at Pin 9
Output Amplitude: Sine	10kΩ	10kΩ	10kΩ	3.3nF	Closed	Pk-Pk output at Pin 2
Triangle	10kΩ	10kΩ	10kΩ	3.3nF	Closed	Pk-Pk output at Pin 3
Leakage Current (off) [3]	10kΩ	10kΩ		3.3nF	Closed	Current into Pin 9
Saturation Voltage (on) [3]	10kΩ	10kΩ	10kΩ	3.3nF	Closed	Output (low) at Pin 9
Rise and Fall Times	10kΩ	10kΩ	4.7kΩ	3.3nF	Closed	Waveform at Pin 9
Duty Cycle Adjust: MAX	50kΩ	~1.6kΩ	10kΩ	3.3nF	Closed	Waveform at Pin 9
MIN	~25kΩ	50kΩ	10kΩ	3.3nF	Closed	Waveform at Pin 9
Triangle Waveform Linearity	10kΩ	10kΩ	10kΩ	3.3nF	Closed	Waveform at Pin 3
Total Harmonic Distortion	10kΩ	10kΩ	10kΩ	3.3nF	Closed	Waveform at Pin 2

NOTE 1: The hi and lo frequencies can be obtained by connecting pin 8 to pin 7 (f_{hi}) and then connecting pin 8 to pin 6 (f_{lo}). Otherwise apply Sweep Voltage at pin 8 (2/3 V_{SUPP} +2V) ≤ V_{SWEEP} ≤ V_{SUPP} where V_{SUPP} is the total supply voltage. In Fig. 2, pin 8 should vary between 5.3V and 10V with respect to ground.

NOTE 2: 10V ≤ V^+ ≤ 30V, or ±5V ≤ V_{SUPP} ≤ ±15V.

NOTE 3: Oscillation can be halted by forcing pin 10 to +5 volts or −5 volts.

DEFINITION OF TERMS:

Supply Voltage (V_{SUPP}). The total supply voltage from V^+ to V^-.

Supply Current. The supply current required from the power supply to operate the device, excluding load currents and the currents through R_A and R_B.

Frequency Range. The frequency range at the square wave output through which circuit operation is guaranteed.

Sweep FM Range. The ratio of maximum frequency to minimum frequency which can be obtained by applying a sweep voltage to pin 8. For correct operation, the sweep voltage should be within the range

$$(2/3\ V_{SUPP} + 2V) < V_{SWEEP} < V_{SUPP}$$

FM Linearity. The percentage deviation from the best-fit straight line on the control voltage versus output frequency curve.

Output Amplitude. The peak-to-peak signal amplitude appearing at the outputs.

Saturation Voltage. The output voltage at the collector of Q_{23} when this transistor is turned on. It is measured for a sink current of 2mA.

Rise and Fall Times. The time required for the square wave output to change from 10% to 90%, or 90% to 10%, of its final value.

Triangle Waveform Linearity. The percentage deviation from the best-fit straight line on the rising and falling triangle waveform.

Total Harmonic Distortion. The total harmonic distortion at the sine-wave output.

TEST CIRCUIT

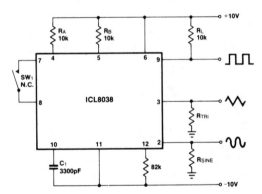

TYPICAL PERFORMANCE CHARACTERISTICS

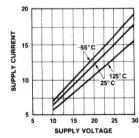

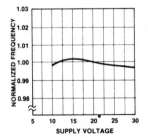

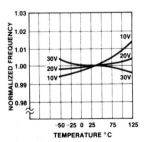

ICL8038

INTERSIL

THEORY OF OPERATION (see block diagram, first page)

An external capacitor C is charged and discharged by two current sources. Current source #2 is switched on and off by a flip-flop, while current source #1 is on continuously. Assuming that the flip-flop is in a state such that current source #2 is off, and the capacitor is charged with a current I, the voltage across the capacitor rises linearly with time. When this voltage reaches the level of comparator #1 (set at 2/3 of the supply voltage), the flip-flop is triggered, changes states, and releases current source #2. This current source normally carries a current 2I, thus the capacitor is discharged with a net-current I and the voltage across it drops linearly with time. When it has reached the level of comparator #2 (set at 1/3 of the supply voltage), the flip-flop is triggered into its original state and the cycle starts again.

Four waveforms are readily obtainable from this basic generator circuit. With the current sources set at I and 2I respectively, the charge and discharge times are equal. Thus a triangle waveform is created across the capacitor and the flip-flop produces a square-wave. Both waveforms are fed to buffer stages and are available at pins 3 and 9.

The levels of the current sources can, however, be selected over a wide range with two external resistors. Therefore, with the two currents set at values different from I and 2I, an asymmetrical sawtooth appears at terminal 3 and pulses with a duty cycle from less than 1% to greater than 99% are available at terminal 9.

The sine-wave is created by feeding the triangle-wave into a non-linear network (sine-converter). This network provides a decreasing shunt-impedance as the potential of the triangle moves toward the two extremes.

Performance of the Square-Wave Output

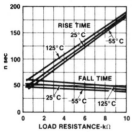
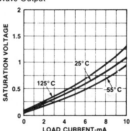

Performance of Triangle-Wave Output

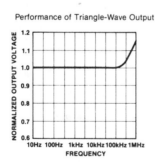
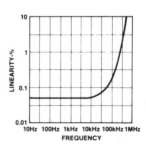

Performance of Sine-Wave Output

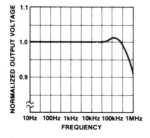
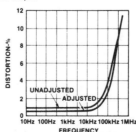

ICL8038

INTERSIL

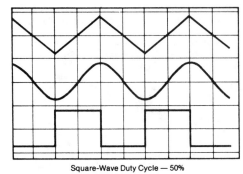

Square-Wave Duty Cycle — 50%

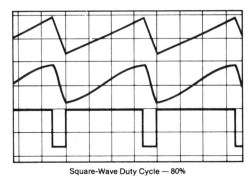

Square-Wave Duty Cycle — 80%

Phase Relationship of Waveforms

WAVEFORM TIMING

The *symmetry* of all waveforms can be adjusted with the external timing resistors. Two possible ways to accomplish this are shown in Figure 1. Best results are obtained by keeping the timing resistors R_A and R_B separate (a). R_A controls the rising portion of the triangle and sine-wave and the 1 state of the square-wave.

The magnitude of the triangle-waveform is set at $1/3\ V_{SUPP}$; therefore the rising portion of the triangle is,

$$t_1 = \frac{C \times V}{I} = \frac{C \times 1/3 \times V^+ \times R_A}{1/5 \times V^+} = \frac{5}{3} R_A \times C$$

The falling portion of the triangle and sine-wave and the 0 state of the square-wave is:

$$t_2 = \frac{C \times V}{I} = \frac{C \times 1/3\ V^+}{\dfrac{2}{5} \times \dfrac{V_{SUPP}}{R_B} - \dfrac{1}{5} \times \dfrac{V_{SUPP}}{R_A}} = \frac{5}{3} \times \frac{R_A R_B C}{2 R_A - R_B}$$

Thus a 50% duty cycle is achieved when $R_A = R_B$.

If the duty-cycle is to be varied over a small range about 50% only, the connection shown in Figure 1b is slightly more convenient. If no adjustment of the duty cycle is desired, terminals 4 and 5 can be shorted together, as shown in Figure 1c. This connection, however, causes an inherently larger variation of the duty-cycle, frequency, etc.

With two separate timing resistors, the frequency is given by

$$f = \frac{1}{t_1 + t_2} = \frac{1}{\dfrac{5}{3} R_A C \left(1 + \dfrac{R_B}{2 R_A - R_B}\right)}$$

or, if $R_A = R_B = R$

$$f = \frac{0.3}{R C} \text{ (for Figure 1a)}$$

If a single timing resistor is used (Figure 1c only), the frequency is

$$f = \frac{0.15}{R C}$$

Neither time nor frequency are dependent on supply voltage, even though none of the voltages are regulated inside the integrated circuit. This is due to the fact that both currents *and* thresholds are direct, linear functions of the supply voltage and thus their effects cancel.

To minimize *sine-wave* distortion the 82kΩ resistor between pins 11 and 12 is best made variable. With this arrangement distortion of less than 1% is achievable. To reduce this even further, two potentiometers can be connected as shown in Figure 2; this configuration allows a typical reduction of sine-wave distortion close to 0.5%.

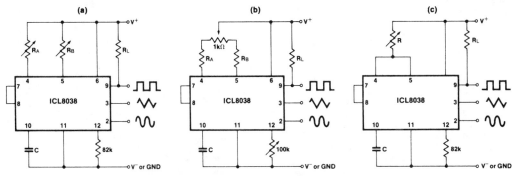

Figure 1: Possible Connections for the External Timing Resistors.

ICL8038

INTERSIL

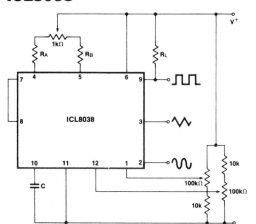

Figure 2: Connection to Achieve Minimum Sine-Wave Distortion.

FREQUENCY MODULATION AND SWEEPING

The frequency of the waveform generator is a direct function of the DC voltage at terminal 8 (measured from V^+). By altering this voltage, frequency modulation is performed.

For small deviations (e.g. ±10%) the modulating signal can be applied directly to pin 8, merely providing DC decoupling with a capacitor as shown in Figure 3a. An external resistor between pins 7 and 8 is not necessary, but it can be used to increase input impedance from about 8kΩ (pins 7 and 8 connected together), to about (R + 8kΩ).

For larger FM deviations or for frequency sweeping, the modulating signal is applied between the positive supply voltage and pin 8 (Figure 3b). In this way the entire bias for the current sources is created by the modulating signal, and a very large (e.g. 1000:1) sweep range is created (f = 0 at V_{sweep} = 0). Care must be taken, however, to regulate the supply voltage; in this configuration the charge current is no longer a function of the supply voltage (yet the trigger thresholds still are) and thus the frequency becomes dependent on the supply voltage. The potential on Pin 8 may be swept down from V^+ by $(1/3\ V_{SUPP} - 2V)$.

SELECTING R_A, R_B AND C

For any given output frequency, there is a wide range of RC combinations that will work, however certain constraints are placed upon the magnitude of the charging current for optimum performance. At the low end, currents of less than 1µA are undesirable because circuit leakages will contribute significant errors at high temperatures. At higher currents (I > 5mA), transistor betas and saturation voltages will contribute increasingly larger errors. Optimum performance will, therefore, be obtained with charging currents of 10µA to 1mA. If pins 7 and 8 are shorted together, the magnitude of the charging current due to R_A can be calculated from:

$$I = \frac{R_1 \times V_{SUPP}}{(R_1 + R_2)} \times \frac{1}{R_A} = \frac{V_{SUPP}}{5R_A}$$

A similar calculation holds for R_B.

The capacitor value should be chosen at the upper end of its possible range.

WAVEFORM OUT LEVEL CONTROL AND POWER SUPPLIES

The waveform generator can be operated either from a single power-supply (10 to 30 Volts) or a dual power-supply (±5 to ±15 Volts). With a single power-supply the average levels of the triangle and sine-wave are at exactly one-half of the supply voltage, while the square-wave alternates between V^+ and ground. A split power supply has the advantage that all waveforms move symmetrically about ground.

The square-wave output is not committed. A load resistor can be connected to a different power-supply, as long as the applied voltage remains within the breakdown capability of the waveform generator (30V). In this way, the square-wave output can be made TTL compatible (load resistor connected to +5 Volts) while the waveform generator itself is powered from a much higher voltage.

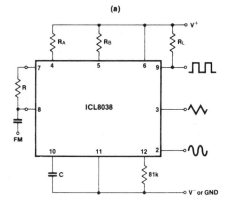

(a)

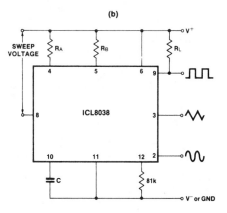

(b)

Figure 3: Connections for Frequency Modulation (a) and Sweep (b)

ICL8038

INTERSIL

APPLICATIONS

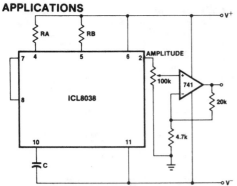

Figure 4: Sine Wave Output Buffer Amplifiers.
The sine wave output has a relatively high output impedance (1kΩ Typ). The circuit of Figure 4 provides buffering, gain and amplitude adjustment. A simple op amp follower could also be used.

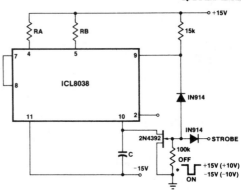

Figure 5: Strobe-Tone Burst Generator.
With a dual supply voltage the external capacitor on Pin 10 can be shorted to ground to halt the 8038 oscillation. Figure 5 shows a FET switch, diode ANDed with an input strobe signal to allow the output to always start on the same slope.

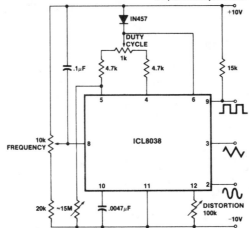

Figure 6: Variable Audio Oscillator, 20Hz to 20kHz.

To obtain a 1000:1 Sweep Range on the 8038 the voltage across external resistors R_A and R_B must decrease to nearly zero. This requires that the highest voltage on control Pin 8 exceed the voltage at the top of R_A and R_B by a few hundred millivolts.

The Circuit of Figure 6 achieves this by using a diode to lower the effective supply voltage on the 8038. The large resistor on pin 5 helps reduce duty cycle variations with sweep.

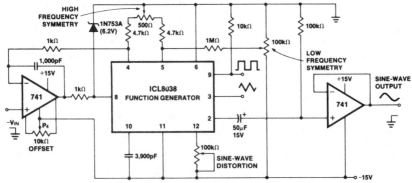

Figure 7: Linear Voltage Controlled Oscillator

The linearity of input sweep voltage versus output frequency can be significantly improved by using an op amp as shown in Figure 7.

ICL8038

INTERSIL

USE IN PHASE-LOCKED LOOPS

Its high frequency stability makes the ICL8038 an ideal building block for a phase-locked loop. In this application the remaining functional blocks, the phase-detector and the amplifier, can be formed by a number of available IC's (e.g. MC 4344, NE 562, HA 2800, HA 2820).

In order to match these building blocks to each other, two steps must be taken. First, two different supply voltages are used and the square wave output is returned to the supply of the phase detector. This assures that the VCO input voltage will not exceed the capabilities of the phase detector. If a smaller VCO signal is required, a simple resistive voltage divider is connected between pin 9 of the waveform generator and the VCO input of the phase-detector.

Second, the DC output level of the amplifier must be made compatible to the DC level required at the FM input of the waveform generator (pin 8, $0.8 \times V^+$). The simplest solution here is to provide a voltage divider to V^+ (R_1, R_2 as shown) if the amplifier has a lower output level, or to ground if its level is higher. The divider can be made part of the low-pass filter.

This application not only provides for a free-running frequency with very low temperature drift, but it also has the unique feature of producing a large reconstituted sinewave signal with a frequency identical to that at the input.

For further information, see Intersil Application Bulletin A013, "Everything You Always Wanted to Know About The 8038."

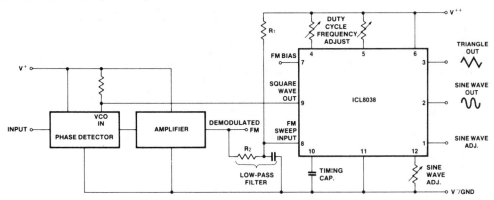

Figure 8: Waveform Generator Used as Stable VCO in a Phase-Locked Loop

LM117/LM317

LM117/LM317 3-Terminal Adjustable Regulator

General Description

The LM117/LM317 are adjustable 3-terminal positive voltage regulators capable of supplying in excess of 1.5A over a 1.2V to 37V output range. They are exceptionally easy to use and require only two external resistors to set the output voltage. Further, both line and load regulation are better than standard fixed regulators. Also, the LM117 is packaged in standard transistor packages which are easily mounted and handled.

In addition to higher performance than fixed regulators, the LM117 series offers full overload protection available only in IC's. Included on the chip are current limit, thermal overload protection and safe area protection. All overload protection circuitry remains fully functional even if the adjustment terminal is disconnected.

Normally, no capacitors are needed unless the device is situated more than 6 inches from the input filter capacitors in which case an input bypass is needed. An optional output capacitor can be added to improve transient response. The adjustment terminal can be bypassed to achieve very high ripple rejections ratios which are difficult to achieve with standard 3-terminal regulators.

Besides replacing fixed regulators, the LM117 is useful in a wide variety of other applications. Since the regulator is "floating" and sees only the input-to-output differential voltage, supplies of several hundred volts can be regulated as long as the maximum input to output differential is not exceeded, i.e., avoid short-circuiting the output.

Also, it makes an especially simple adjustable switching regulator, a programmable output regulator, or by connecting a fixed resistor between the adjustment pin and output, the LM117 can be used as a precision current regulator. Supplies with electronic shutdown can be achieved by clamping the adjustment terminal to ground which programs the output to 1.2V where most loads draw little current.

The LM117K and LM317K are packaged in standard TO-3 transistor packages while the LM117H and LM317H are packaged in a solid Kovar base TO-39 transistor package. The LM117 is rated for operation from $-55°C$ to $+150°C$, and the LM317 from 0°C to $+125°C$. The LM317T and LM317MP, rated for operation over a 0°C to $+125°C$ range, are available in a TO-220 plastic package and a TO-202 package, respectively.

For applications requiring greater output current in excess of 3A and 5A, see LM150 series and LM138 series data sheets, respectively. For the negative complement, see LM137 series data sheet.

LM117 Series Packages and Power Capability

Device	Package	Rated Power Dissipation	Design Load Current
LM117	TO-3	20W	1.5A
LM317	TO-39	2W	0.5A
LM317T	TO-220	15W	1.5A
LM317M	TO-202	7.5W	0.5A

Features

- Adjustable output down to 1.2V
- Guaranteed 1.5A output current
- Line regulation typically 0.01%/V
- Load regulation typically 0.1%
- Current limit constant with temperature
- 100% electrical burn-in
- Eliminates the need to stock many voltages
- Standard 3-lead transistor package
- 80 dB ripple rejection
- Output is short-circuit protected

Typical Applications

1.2V–25V Adjustable Regulator

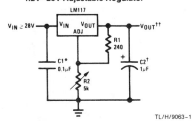

TL/H/9063–1

Full output current not available at high input-output voltages

*Needed if device is more than 6 inches from filter capacitors.

†Optional—improves transient response. Output capacitors in the range of 1 μF to 1000 μF of aluminum or tantalum electrolytic are commonly used to provide improved output impedance and rejection of transients.

$$††V_{OUT} = 1.25V \left(1 + \frac{R2}{R1}\right) + I_{ADJ}(R2)$$

Digitally Selected Outputs

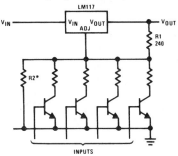

INPUTS

TL/H/9063–2

*Sets maximum V_{OUT}

LM117/LM317

Absolute Maximum Ratings

If Military/Aerospace specified devices are required, contact the National Semiconductor Sales Office/ Distributors for availability and specifications. (Note 3)

Power Dissipation	Internally limited
Input—Output Voltage Differential	$+40V, -0.3V$
Operating Junction Temperature Range	
LM117	$-55°C$ to $+150°C$
LM317	$0°C$ to $+125°C$

Storage Temperature	$-65°C$ to $+150°C$
Lead Temperature (Soldering)	
Metal Package	$300°C$, 10 seconds
Plastic Package	$260°C$, 4 seconds
ESD rating	2k Volts

Preconditioning

Burn-In in Thermal Limit	100% All Devices

Electrical Characteristics (Note 1)

Parameter	Conditions	LM117 Min	LM117 Typ	LM117 Max	LM317 Min	LM317 Typ	LM317 Max	Units
Line Regulation	$T_j = 25°C$, $3V \leq (V_{IN} - V_{OUT}) \leq 40V$ (Note 2), $I_L = 10$ mA		0.01	0.02		0.01	0.04	%/V
Load Regulation	$T_J = 25°C$, 10 mA $\leq I_{OUT} \leq I_{MAX}$		0.1	0.3		0.1	0.5	%
Thermal Regulation	$T_J = 25°C$, 20 ms Pulse		0.03	0.07		0.04	0.07	%/W
Adjustment Pin Current			50	100		50	100	μA
Adjustment Pin Current Change	10 mA $\leq I_L \leq I_{MAX}$ $3V \leq (V_{IN} - V_{OUT}) \leq 40V$		0.2	5		0.2	5	μA
Reference Voltage	$3V \leq (V_{IN} - V_{OUT}) \leq 40V$, (Note 3) 10 mA $\leq I_{OUT} \leq I_{MAX}$, $P \leq P_{MAX}$	1.20	1.25	1.30	1.20	1.25	1.30	V
Line Regulation	$3V \leq (V_{IN} - V_{OUT}) \leq 40V$, (Note 2)		0.02	0.05		0.02	0.07	%/V
Load Regulation	10 mA $\leq I_{OUT} \leq I_{MAX}$ (Note 2) $I_L = 10$ mA		0.3	1		0.3	1.5	%
Temperature Stability	$T_{MIN} \leq T_j \leq T_{MAX}$		1			1		%
Minimum Load Current	$(V_{IN} - V_{OUT}) = 40V$		3.5	5		3.5	10	mA
Current Limit	$(V_{IN} - V_{OUT}) \leq 15V$							
	K and T Package	1.5	2.2	3.4	1.5	2.2	3.4	A
	H and P Package	0.5	0.8	1.8	0.5		1.8	A
	$(V_{IN} - V_{OUT}) = 40V$, $T_j = +25°C$							
	K and T Package	0.30	0.4		0.15	0.4		A
	H and P Package	0.15	0.07		0.075	0.07		A
RMS Output Noise, % of V_{OUT}	$T_J = 25°C$, 10 Hz $\leq f \leq$ 10 kHz		0.003			0.003		%
Ripple Rejection Ratio	$V_{OUT} = 10V$, $f = 120$ Hz		65			65		dB
	$C_{ADJ} = 10$ μF	66	80		66	80		dB
Long-Term Stability	$T_J = 125°C$		0.3	1		0.3	1	%
Thermal Resistance, Junction to Case	H Package		12	15		12	15	°C/W
	K Package		2.3	3		2.3	3	°C/W
	T Package					4		°C/W
	P Package					7		°C/W
Thermal Resistance, Junction to Ambient (No heat sink)	H Package		140			140		°C/W
	K Package		35			35		°C/W
	T Package					50		°C/W
	P Package					80		°C/W

Note 1: Unless otherwise specified, these specifications apply: $-55°C \leq T_j \leq +150°C$ for the LM117, and $0°C \leq T_J \leq +125°C$ for the LM317; $V_{IN} - V_{OUT} = 5V$; and $I_{OUT} = 0.1A$ for the TO-39 and TO-202 packages and $I_{OUT} = 0.5A$ for the TO-3 and TO-220 packages. Although power dissipation is internally limited, these specifications are applicable for power dissipations of 2W for the TO-39 and TO-202, and 20W for the TO-3 and TO-220. I_{MAX} is 1.5A for the TO-3 and TO-220 packages and 0.5A for the TO-39 and TO-202 packages.

Note 2: Regulation is measured at constant junction temperature, using pulse testing with a low duty cycle. Changes in output voltage due to heating effects are covered under the specification for thermal regulation.

Note 3: Refer to RETS117H drawing for LM117H or RETS117K drawing for LM117K military specifications.

LM117/LM317

Typical Performance Characteristics (K and T Packages)
Output Capacitor = 0 unless otherwise noted

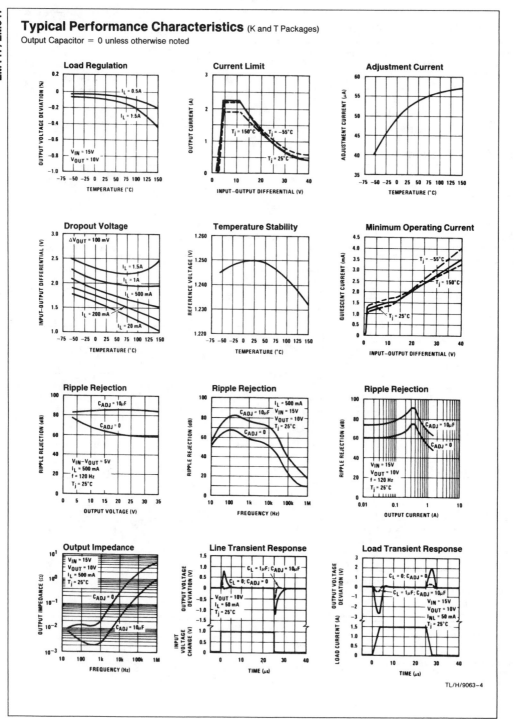

TL/H/9063–4

Application Hints

In operation, the LM117 develops a nominal 1.25V reference voltage, V_{REF}, between the output and adjustment terminal. The reference voltage is impressed across program resistor R1 and, since the voltage is constant, a constant current I_1 then flows through the output set resistor R2, giving an output voltage of

$$V_{OUT} = V_{REF}\left(1 + \frac{R2}{R1}\right) + I_{ADJ}R2$$

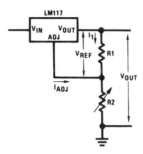

FIGURE 1

TL/H/9063–5

Since the 100 μA current from the adjustment terminal represents an error term, the LM117 was designed to minimize I_{ADJ} and make it very constant with line and load changes. To do this, all quiescent operating current is returned to the output establishing a minimum load current requirement. If there is insufficient load on the output, the output will rise.

External Capacitors

An input bypass capacitor is recommended. A 0.1 μF disc or 1 μF solid tantalum on the input is suitable input bypassing for almost all applications. The device is more sensitive to the absence of input bypassing when adjustment or output capacitors are used but the above values will eliminate the possibility of problems.

The adjustment terminal can be bypassed to ground on the LM117 to improve ripple rejection. This bypass capacitor prevents ripple from being amplified as the output voltage is increased. With a 10 μF bypass capacitor 80 dB ripple rejection is obtainable at any output level. Increases over 10 μF do not appreciably improve the ripple rejection at frequencies above 120 Hz. If the bypass capacitor is used, it is sometimes necessary to include protection diodes to prevent the capacitor from discharging through internal low current paths and damaging the device.

In general, the best type of capacitors to use is solid tantalum. Solid tantalum capacitors have low impedance even at high frequencies. Depending upon capacitor construction, it takes about 25 μF in aluminum electrolytic to equal 1 μF solid tantalum at high frequencies. Ceramic capacitors are also good at high frequencies; but some types have a large decrease in capacitance at frequencies around 0.5 MHz. For this reason, 0.01 μF disc may seem to work better than a 0.1 μF disc as a bypass.

Although the LM117 is stable with no output capacitors, like any feedback circuit, certain values of external capacitance can cause excessive ringing. This occurs with values be-

tween 500 pF and 5000 pF. A 1 μF solid tantalum (or 25 μF aluminum electrolytic) on the output swamps this effect and insures stability. Any increase of the load capacitance larger than 10 μF will merely improve the loop stability and output impedance.

Load Regulation

The LM117 is capable of providing extremely good load regulation but a few precautions are needed to obtain maximum performance. The current set resistor connected between the adjustment terminal and the output terminal (usually 240Ω) should be tied directly to the output (case) of the regulator rather than near the load. This eliminates line drops from appearing effectively in series with the reference and degrading regulation. For example, a 15V regulator with 0.05Ω resistance between the regulator and load will have a load regulation due to line resistance of 0.05Ω $\times$ I_L. If the set resistor is connected near the load the effective line resistance will be 0.05Ω (1 + R2/R1) or in this case, 11.5 times worse.

Figure 2 shows the effect of resistance between the regulator and 240Ω set resistor.

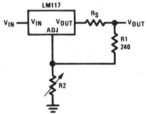

TL/H/9063–6

FIGURE 2. Regulator with Line Resistance in Output Lead

With the TO-3 package, it is easy to minimize the resistance from the case to the set resistor, by using two separate leads to the case. However, with the TO-5 package, care should be taken to minimize the wire length of the output lead. The ground of R2 can be returned near the ground of the load to provide remote ground sensing and improve load regulation.

Protection Diodes

When external capacitors are used with *any* IC regulator it is sometimes necessary to add protection diodes to prevent the capacitors from discharging through low current points into the regulator. Most 10 μF capacitors have low enough internal series resistance to deliver 20A spikes when shorted. Although the surge is short, there is enough energy to damage parts of the IC.

When an output capacitor is connected to a regulator and the input is shorted, the output capacitor will discharge into the output of the regulator. The discharge current depends on the value of the capacitor, the output voltage of the regulator, and the rate of decrease of V_{IN}. In the LM117, this discharge path is through a large junction that is able to sustain 15A surge with no problem. This is not true of other types of positive regulators. For output capacitors of 25 μF or less, there is no need to use diodes.

LM117/LM317

Application Hints (Continued)

The bypass capacitor on the adjustment terminal can discharge through a low current junction. Discharge occurs when *either* the input or output is shorted. Internal to the LM117 is a 50Ω resistor which limits the peak discharge current. No protection is needed for output voltages of 25V or less and 10 μF capacitance. *Figure 3* shows an LM117 with protection diodes included for use with outputs greater than 25V and high values of output capacitance.

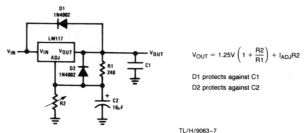

$$V_{OUT} = 1.25V \left(1 + \frac{R2}{R1}\right) + I_{ADJ}R2$$

D1 protects against C1
D2 protects against C2

TL/H/9063–7

FIGURE 3. Regulator with Protection Diodes

Schematic Diagram

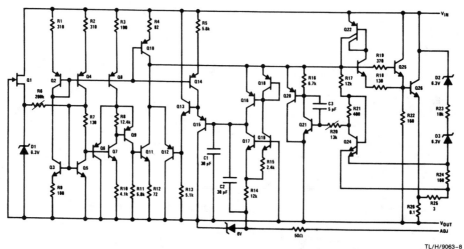

TL/H/9063–8

LM117/LM317

Typical Applications (Continued)

5V Logic Regulator with Electronic Shutdown*

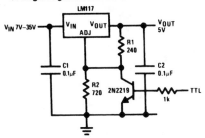

*Min. output ≈ 1.2V

TL/H/9063–3

Slow Turn-On 15V Regulator

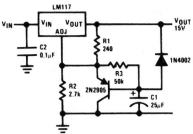

TL/H/9063–9

Adjustable Regulator with Improved Ripple Rejection

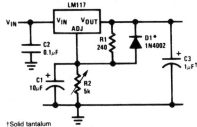

†Solid tantalum

*Discharges C1 if output is shorted to ground

TL/H/9063–10

High Stability 10V Regulator

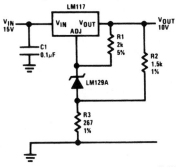

TL/H/9063–11

High Current Adjustable Regulator

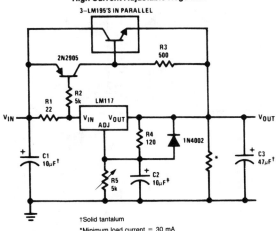

†Solid tantalum

*Minimum load current = 30 mA

‡Optional—improves ripple rejection

TL/H/9063–12

LM117/LM317

Typical Applications (Continued)

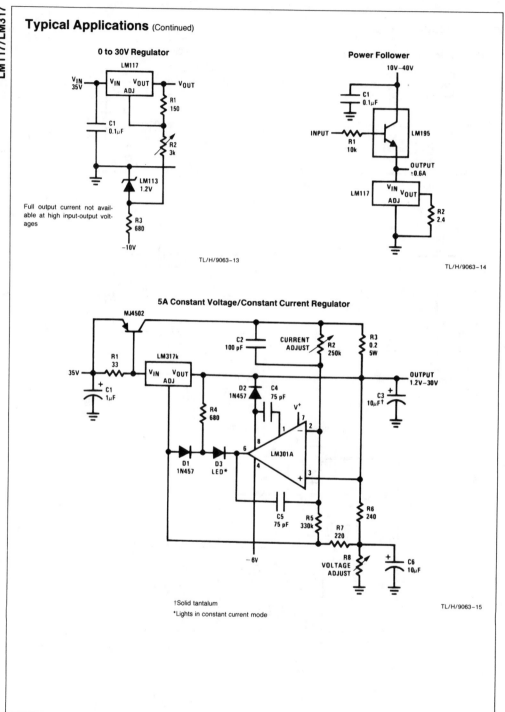

0 to 30V Regulator

Full output current not available at high input-output voltages

TL/H/9063–13

Power Follower

TL/H/9063–14

5A Constant Voltage/Constant Current Regulator

†Solid tantalum

*Lights in constant current mode

TL/H/9063–15

1-56

Typical Applications (Continued)

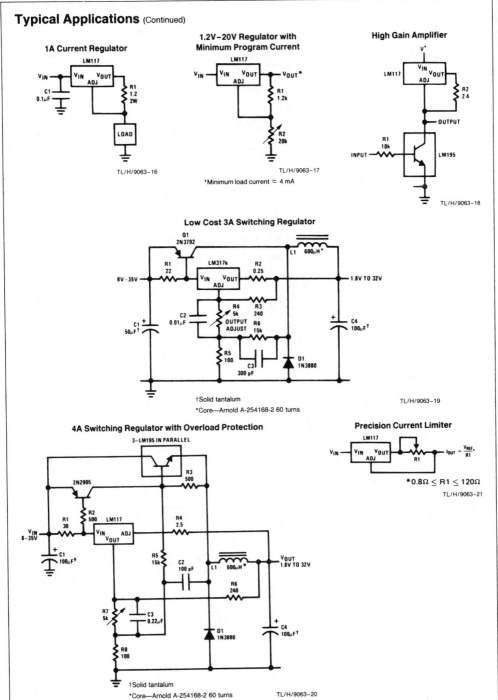

1A Current Regulator

TL/H/9063-16

1.2V–20V Regulator with Minimum Program Current

*Minimum load current ≈ 4 mA

TL/H/9063-17

High Gain Amplifier

TL/H/9063-18

Low Cost 3A Switching Regulator

†Solid tantalum
*Core—Arnold A-254168-2 60 turns

TL/H/9063-19

4A Switching Regulator with Overload Protection

†Solid tantalum
*Core—Arnold A-254168-2 60 turns

TL/H/9063-20

Precision Current Limiter

$I_{OUT} = \frac{V_{REF}}{R1}$

*$0.8\Omega \leq R1 \leq 120\Omega$

TL/H/9063-21

1-57

LM117/LM317

Typical Applications (Continued)

Tracking Preregulator

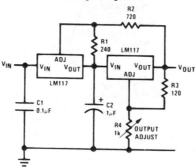

TL/H/9063–22

Current Limited Voltage Regulator

$$V_{OUT} = 1.25V \left(1 + \frac{R_2}{R_1} \right) + I_{ADJ} R_2$$

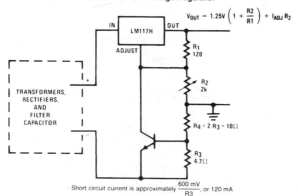

Short circuit current is approximately $\frac{600 \text{ mV}}{R_3}$, or 120 mA

TL/H/9063–23

(Compared to LM117H's 1 Amp Current Limit

At 50 mA output only 3/4 volt of drop occurs in R_3 and R_4

Adjusting Multiple On-Card Regulators with Single Control*

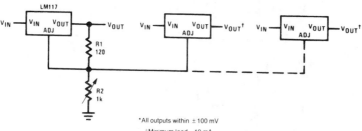

*All outputs within ± 100 mV

†Minimum load—10 mA

TL/H/9063–24

LM117/LM317

Typical Applications (Continued)

AC Voltage Regulator

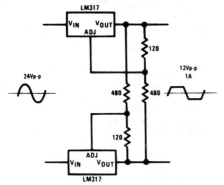

TL/H/9063–25

12V Battery Charger

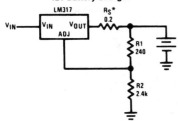

TL/H/9063–26

*R_S—sets output impedance of charger: $Z_{OUT} = R_S \left(1 + \dfrac{R2}{R1}\right)$

Use of R_S allows low charging rates with fully charged battery.

50 mA Constant Current Battery Charger

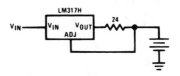

TL/H/9063–27

Adjustable 4A Regulator

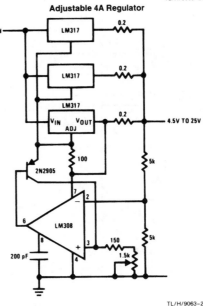

TL/H/9063–28

Current Limited 6V Charger

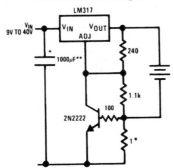

*Sets peak current (0.6A for 1Ω)

**The 1000 μF is recommended to filter out input transients

TL/H/9063–29

LM117/LM317

Connection Diagrams (See Physical Dimension section for further information)

(TO-3 Steel)
Metal Can Package

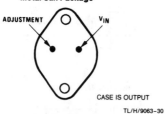

ADJUSTMENT V_{IN}

CASE IS OUTPUT

TL/H/9063–30

Bottom View

Order Number LM117K STEEL, LM317K STEEL
See NS Package Number K02A

(TO-39)
Metal Can Package

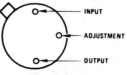

INPUT

ADJUSTMENT

OUTPUT

CASE IS OUTPUT

Bottom View

TL/H/9063–31

Order Number LM117H, LM317H
See NS Package Number H03A

(TO-220)
Plastic Package

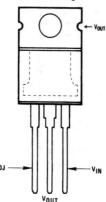

← V_{OUT}

ADJ →   ← V_{IN}

V_{OUT}

TL/H/9063–32

Front View

Order Number LM317T
See NS Package Number T03B

(TO-202)
Plastic Package

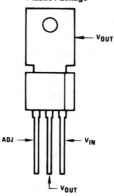

→ V_{OUT}

ADJ →   ← V_{IN}

V_{OUT}

TL/H/9063–33

Front View

Order Number LM317MP
See NS Package Number P03A

PHASE LOCKED LOOP

NE/SE565

NE/SE565-F,K,N

DESCRIPTION

The SE/NE565 Phase-Locked Loop (PLL) is a self-contained, adaptable filter and demodulator for the frequency range from 0.001Hz to 500kHz. The circuit comprises a voltage-controlled oscillator of exceptional stability and linearity, a phase comparator, an amplifier and a low-pass filter as shown in the block diagram. The center frequency of the PLL is determined by the free-running frequency of the VCO; this frequency can be adjusted externally with a resistor or a capacitor. The low-pass filter, which determines the capture characteristics of the loop, is formed by an internal resistor and an external capacitor.

FEATURES

- **Highly stable center frequency (200ppm/°C typ.)**
- **Wide operating voltage range (±6 to ±12 volts)**
- **Highly linear demodulated output (0.2% typ.)**
- **Center frequency programming by means of a resistor or capacitor, voltage or current**
- **TTL and DTL compatible square-wave output; loop can be opened to insert digital frequency divider**
- **Highly linear triangle wave output**
- **Reference output for connection of comparator in frequency discriminator**
- **Bandwidth adjustable from < ±1% to > ±60%**
- **Frequency adjustable over 10 to 1 range with same capacitor**

APPLICATIONS

- **Frequency shift keying**
- **Modems**
- **Telemetry receivers**
- **Tone decoders**
- **SCA receivers**
- **Wideband FM discriminators**
- **Data synchronizers**
- **Tracking filters**
- **Signal restoration**
- **Frequency multiplication & division**

PIN CONFIGURATIONS

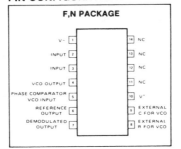

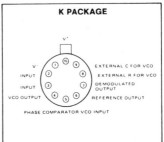

BLOCK DIAGRAM

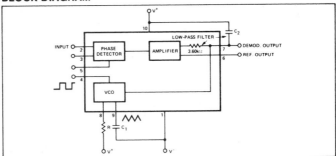

EQUIVALENT SCHEMATIC

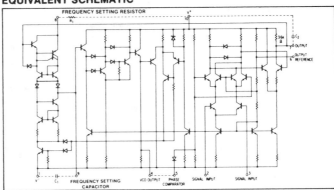

ABSOLUTE MAXIMUM RATINGS $T_A = 25°C$ unless otherwise specified.

PARAMETER	RATING	UNIT
Maximum operating voltage	26	V
Input voltage	3	Vp-p
Storage temperature	–65 to +150	°C
Operating temperature range		
NE565	0 to +70	°C
SE565	–55 to +125	°C
Power dissipation	300	mW

signetics

PHASE LOCKED LOOP — NE/SE565

NE/SE565-F,K,N

ELECTRICAL CHARACTERISTICS $T_A = 25°C$, $V_{CC} = ±6V$ unless otherwise specified.

PARAMETER	TEST CONDITIONS	SE565 Min	SE565 Typ	SE565 Max	NE565 Min	NE565 Typ	NE565 Max	UNIT
SUPPLY REQUIREMENTS								
Supply voltage		12		±12	±6		±12	V
Supply current			8	12.5		8	12.5	mA
INPUT CHARACTERISTICS								
Input impedance[1]	$f_0 = 50kHz$, ±10%	7	10		5	10		kΩ
Input level required for tracking	frequency deviation	10	1		10	1		mVrms
VCO CHARACTERISTICS								
Center frequency Maximum value	$C_1 = 2.7pF$	300	500			500		kHz
Distribution[2]	Distribution taken about $f_0 = 50kHz$, $R_1 = 5.0kΩ$, $C_1 = 1200pF$	-10	0	+10	-30	0	+30	%
Drift with temperature	$f_0 = 50kHz$		200			300		ppm/°C
Drift with supply voltage	$f_0 = 50kHz$, $V_{CC} = ±6$ to ±7 volts		0.1	1.0		0.2	1.5	%/V
Triangle wave								
Output voltage level		1.9	0		1.9	0		V
Amplitude			2.4	3		2.4	3	Vp-p
Linearity			0.2			0.5		%
Square wave								
Logical "1" output voltage	$f_0 = 50kHz$	+4.9	+5.2		+4.9	+5.2		V
Logical "0" output voltage	$f_0 = 50kHz$		-0.2	+0.2		-0.2	+0.2	V
Duty cycle	$f_0 = 50kHz$	45	50	55	40	50	60	%
Rise time			20	100		20		ns
Fall time			50	200		50		ns
Output current (sink)		0.6	1		0.6	1		mA
Output current (source)		5	10		5	10		mA
DEMODULATED OUTPUT CHARACTERISTICS								
Output voltage level	Measured at pin 7	4.25	4.5	4.75	4.0	4.5	5.0	V
Maximum voltage swing[3]			2			2		Vp-p
Output voltage swing	±10% frequency deviation	250	300		200	300		mVp-p
Total harmonic distortion			0.2	0.75		0.4	1.5	%
Output impedance[4]			3.6			3.6		kΩ
Offset voltage (V6-V7)			30	100		50	200	mV
Offset voltage vs temperature (drift)			50			100		μV/°C
AM rejection		30	40			40		dB

NOTES
1. Both input terminals (pins 2 and 3) must receive identical dc bias. This bias may range from 0 volts to -4 volts.
2. The external resistance for frequency adjustment (R1) must have a value between 2kΩ and 20kΩ.
3. Output voltage swings negative as input frequency increases.
4. Output not buffered.

PHASE LOCKED LOOP

NE/SE565

NE/SE565-F,K,N

TYPICAL PERFORMANCE CHARACTERISTICS

POWER SUPPLY CURRENT AS A FUNCTION OF SUPPLY VOLTAGE

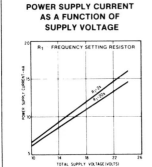

VCO CONVERSION GAIN

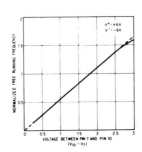

LOCK RANGE AS A FUNCTION OF INPUT VOLTAGE

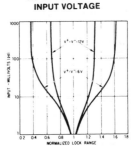

LOCK RANGE AS A FUNCTION OF GAIN SETTING RESISTANCE (PIN 6-7)

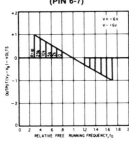

CHANGE IN FREE-RNNING VCO FREQUENCY AS A FUNCTION OF TEMPERATURE

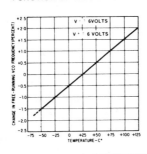

VCO OUTPUT WAVEFORM

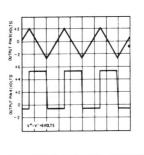

DESIGN FORMULAS
(See Figure 1)

Free-running frequency of VCO: $f_o \approx \dfrac{1.2}{4R_1C_1}$ in Hz

Lock-range: $f_L = \pm \dfrac{8f_o}{V_{CC}}$ in Hz

Capture-range: $f_C \approx \pm \dfrac{1}{2\pi} \sqrt{\dfrac{2\pi f_L}{\tau}}$

where $\tau = (3.6 \times 10^3) \times C_2$

TYPICAL APPLICATIONS
FM Demodulation

The 565 Phase Locked Loop is a general purpose circuit designed for highly linear FM demodulation. During lock, the average dc level of the phase comparator output signal 's directly proportional to the frequency of the input signal. As the input frequency shifts, it is this output signal which causes the VCO to shift its frequency to match that of the input. Consequently, the linearity of the phase comparator output with frequency is determined by the voltage-to-frequency transfer function of the VCO.

Because of its unique and highly linear VCO, the 565 PLL can lock to and track an input signal over a very wide bandwidth (typically ±60%) with very high linearity (typically, within 0.5%).

A typical connection diagram is shown in Figure 1. The VCO free-running frequency is given approximately by

$f_o = \dfrac{1.2}{4R_1C_1}$ and should be adjusted to be at the center of the input signal frequency range. C1 can be any value, but R1 should be within the range of 2000 to 20,000 ohms with an optimum value on the order of 4000 ohms. The source can be direct coupled if the dc resistances seen from pins 2 and 3 are equal and there is no dc voltage difference between the pins. A short between pins 4 and 5 connects the VCO to the phase comparator. Pin 6 provides a dc reference voltage that is close to the dc potential of the demodulated output (pin 7). Thus, if a resistance is connected between pins 6 and 7, the gain of the output stage can be reduced with little change in the dc voltage level at the output. This allows the lock range to be

decreased with little change in the free-running frequency. In this manner the lock range can be decreased from ±60% of f_o to approximately ±20% of f_o (at ±6V).

A small capacitor (typically 0.001 μF) should be connected between pins 7 and 8 to eliminate possible oscillation in the control current source.

A single-pole loop filter is formed by the capacitor C2, connected between pin 7 and the positive supply, and an internal resistance of approximately 3600 ohms.

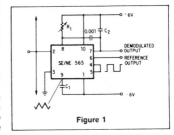

Figure 1

signetics

PHASE LOCKED LOOP
NE/SE565

NE/SE565-F,K,N

Frequency Shift Keying (FSK)

FSK refers to data transmission by means of a carrier which is shifted between two preset frequencies. This frequency shift is usually accomplished by driving a VCO with the binary data signal so that the two resulting frequencies correspond to the "0" and "1" states (commonly called space and mark) of the binary data signal.

A simple scheme using the 565 to receive FSK signals of 1070Hz - and 1270Hz is shown in Figure 2. As the signal appears at the input, the loop locks to the input frequency and tracks it between the two frequencies with a corresponding dc shift at the output.

The loop filter capacitor C2 is chosen smaller than usual to eliminate overshoot on the output pulse, and a three-stage RC ladder filter is used to remove the carrier component from the output. The band edge of the ladder filter is chosen to be approximately half way between the maximum keying rate (in this case 300 baud or 150Hz) and twice the input frequency (approximately 2200Hz). The output signal can now be made logic compatible by connecting a voltage comparator between the output and pin 6 of the loop. The free-running frequency is adjusted with R1 so as to result in a slightly-positive voltage at the output with $f_{IN} = 1070Hz$.

The input connection is typical for cases where a dc voltage is present at the source and therefore a direct connection is not desirable. Both input terminals are returned to ground with identical resistors (in this case, the values are chosen to effect a 600-ohm input impedance).

Frequency Multiplication

There are two methods by which frequency multiplication can be achieved using the 565:

1. Locking to a harmonic of the input signal.
2. Inclusion of a digital frequency divider or counter in the loop between the VCO and phase comparator.

The first method is the simplest, and can be achieved by setting the free-running frequency of the VCO to a multiple of the input frequency. A limitation of this scheme is that the lock range decreases as successively higher and weaker harmonics are used for locking. If the input frequency is to be constant with little tracking required, the loop can generally be locked to any one of the first 5 harmonics. For higher orders of multiplication, or for cases where a large lock range is desired, the second scheme is more desirable. An example of this might be

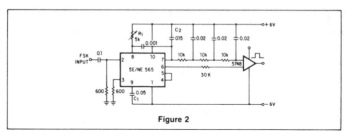

Figure 2

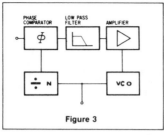

Figure 3

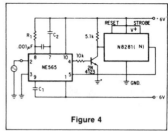

Figure 4

a case where the input signal varies over a wide frequency range and a large multiple of the input frequency is required.

A block diagram of the second scheme is shown in Figure 3. Here the loop is broken between the VCO and the phase comparator, and a frequency divider is inserted. The fundamental of the divided VCO frequency is locked to the input frequency in this case, so that the VCO is actually running at a multiple of the input frequency. The amount of multiplication is determined by the frequency divider. A typical connection scheme is shown in Figure 4. To set up the circuit, the frequency limits of the input signal must be determined. The free-running frequency of the VCO is then adjusted by means of R1 and C1 (as discussed under FM demodulation) so that the output frequency of the divider is midway between the input frequency limits. The filter capacitor, C2, should be large enough to eliminate variations in the demodulated output voltage (at pin 7), in order to stabilize the VCO frequency. The output can now be taken as the VCO squarewave output, and its fundamental will be the desired multiple of the input frequency (f_{IN}) as long as the loop is in lock.

SCA (Background Music) Decoder

Some FM stations are authorized by the FCC to broadcast uninterrupted background music for commerical use. To do this a frequency modulated subcarrier of 67kHz is used. The frequency is chosen so

as not to interfere with the normal stereo or monaural program; in addition, the level of the subcarrier is only 10% of the amplitude of the combined signal.

The SCA signal can be filtered out and demodulated with the NE565 Phase Locked Loop without the use of any resonant circuits. A connection diagram is shown in Figure 5. This circuit also serves as an example of operation from a single power supply.

A resistive voltage divider is used to establish a bias voltage for the input (pins 2 and 3). The demodulated (multiplex) FM signal is fed to the input through a two-stage high-pass filter, both to effect capacitive coupling and to attenuate the strong signal of the regular channel. A total signal amplitude, between 80mV and 300mV, is required at the input. Its source should have an impedance of less than 10,000 ohms.

The Phase Locked Loop is tuned to 67kHz with a 5000 ohm potentiometer; only approximate tuning is required, since the loop will seek the signal.

The demodulated output (pin 7) passes through a three-stage low-pass filter to provide de-emphasis and attenuate the high-frequency noise which often accompanies SCA transmission. Note that no capacitor is provided directly at pin 7; thus, the circuit is operating as a first-order loop. The demodulated output signal is in the order of 50mV and the frequency response extends to 7kHz.

PHASE LOCKED LOOP

NE/SE565

NE/SE565-F,K,N

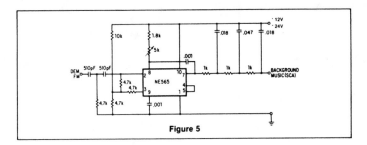

Figure 5

signetics

D

ANALYSIS OF
A DIFFERENTIAL
AMPLIFIER

D–1   OVERVIEW

The detailed analysis of all the internal circuitry of a complete IC op-amp such as the 741 is quite involved, but not really necessary for the objectives of this text. As an aid to overall understanding of an op-amp, however, it is helpful to investigate in some detail the operation of a single-stage discrete component differential amplifier employing two transistors. Many of the properties of a complete IC op-amp can be understood from the properties of this much simpler circuit.

As a prerequisite, we assume familiarity with the basic hybrid-parameter equivalent circuit of a bipolar junction transistor. The properties of that circuit are exploited in the development that follows.

D–2   TWO-TRANSISTOR DIFFERENTIAL AMPLIFIER

One form of a two-transistor differential amplifier circuit is shown in Figure D–1. This particular form has differential inputs, denoted as v^+ and v^-, and a single output, denoted as v_o. It is biased by two dc power supplies, denoted by $+V_{CC}$ and $-V_{EE}$. The circuit thus has terminals that correspond to those of a full IC differential amplifier. With the circuit as given, however, there may be undesirable dc levels at certain points that do not appear in the complete IC op-amp. A small signal analysis will be performed at this point, so the dc levels need not concern us for the objective sought.

The circuit as shown has a *single-ended* output. One variation of this circuit employs resistors in both collectors and provides two possible collector output voltages. That form is said to have a *double-ended* output.

FIGURE D–1

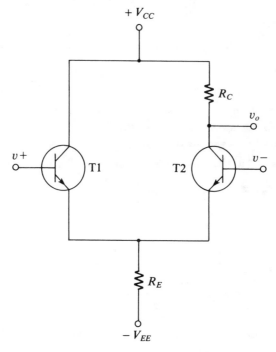

We now give a brief qualitative analysis of the circuit. Assume first that a positive *differential input voltage* is applied between the noninverting and inverting inputs. As we will see later in a quantitative way, the voltage v^+ increases by about half the difference between the two input voltages, and v^- decreases by the same amount. The increased voltage for v^+ causes the collector current of T1 to increase, but the decreased voltage for v^- causes the collector current of T2 to decrease. Since the current through R_E is the sum of the separate emitter currents, this net sum remains essentially constant. The result is that R_E does not produce negative feedback as it would in the case of a single-transistor common emitter circuit. The decrease in collector current for T2 results in an increase of the collector voltage, corresponding to the expected signal amplification.

Consider next the process of applying a *common-mode* signal to the circuit, that is, $v^+ = v^-$. In this case v^+ and v^- change by the same amount, and the corresponding collector currents also change by the same amount. This causes the emitter voltage to essentially follow the two input voltages, and a large negative feedback results. The net gain in this case is very low.

The conclusion of the preceding discussion is that the output is extremely responsive to a differential input voltage, but it is not very responsive to a common-mode input voltage. We now make a detailed analytical analysis of this process.

D-3   SIGNAL DEFINITIONS

In the analysis of the differential amplifier, it is helpful to express the input signals in terms of differential and common-mode components. The *differential input voltage* v_d is defined as

$$v_d = v^+ - v^- \tag{D-1}$$

The *common-mode input voltage* v_{cm} is defined as

$$v_{cm} = \frac{v^+ + v^-}{2} \tag{D-2}$$

Note that v_d is the *difference* between the two inputs, and v_{cm} is the *average* or *mean* of the two inputs.

Equations (D-1) and (D-2) provide definitions of v_d and v_{cm} in terms of v^+ and v^-. It is also meaningful to express v^+ and v^- in terms of v_d and v_{cm}. The reader is invited to show from (D-1) and (D-2) that such expressions are

$$v^+ = v_{cm} + \frac{v_d}{2} \tag{D-3}$$

and

$$v^- = v_{cm} - \frac{v_d}{2} \tag{D-4}$$

D-4   COMPLETE ANALYSIS

A small signal model of the differential amplifier employing the standard hybridparameter equivalent circuit is shown in Figure D-2. The input signals are expressed in terms of differential and common-mode components as given in (D-3) and (D-4).

Application of Kirchhoff's Current Law, at the node for the small signal emitter voltage v_e results in the following equation:

$$\frac{v_e}{R_E} - i_{b1} - h_{fe}i_{b1} - i_{b2} - h_{fe}i_{b2} = 0 \tag{D-5}$$

Solving for v_e, we obtain

$$v_e = R_E(1 + h_{fe})(i_{b1} + i_{b2}) \tag{D-6}$$

The currents i_{b1} and i_{b2} can be expressed as

$$i_{b1} = \frac{v^+ - v_e}{h_{ie}} = \frac{v_{cm} + v_d/2 - v_e}{h_{ie}} \tag{D-7}$$

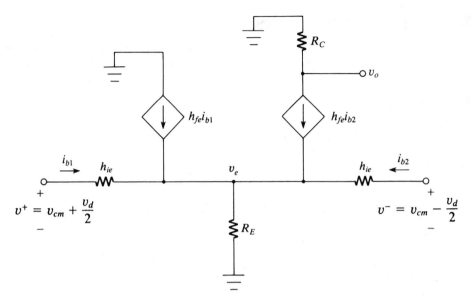

FIGURE D-2

and

$$i_{b2} = \frac{v^- - v_e}{h_{ie}} = \frac{v_{cm} - v_d/2 - v_e}{h_{ie}}$$
(D–8)

Expansion of (D–7) and (D–8) into additive terms and substitution into (D–6) yield

$$v_e = R_E(1 + h_{fe})\left(\frac{v_{cm}}{h_{ie}} + \frac{v_d/2}{h_{ie}} - \frac{v_e}{h_{ie}} + \frac{v_{cm}}{h_{ie}} - \frac{v_d/2}{h_{ie}} - \frac{v_e}{h_{ie}}\right)$$

$$= 2R_E\frac{(1 + h_{fe})}{h_{ie}}(v_{cm} - v_e)$$
(D–9)

where cancellation of the terms involving v_d occurred. The terms involving v_e are next grouped together on the left. After some rearrangement, we obtain

$$v_e[h_{ie} + 2R_E(1 + h_{fe})] = 2R_E(1 + h_{fe})v_{cm}$$
(D–10)

Solution for v_e yields

$$v_e = \frac{2R_E(1 + h_{fe})v_{cm}}{h_{ie} + 2R_E(1 + h_{fe})}$$
(D–11)

Note that *the emitter signal voltage is completely independent of the differential input signal voltage.*

D-5 DIFFERENTIAL OUTPUT VOLTAGE

For the moment, a reasonable approximation will be made with (D–11). In practice, $h_{fe} \gg 1$ and $2R_E h_{fe} \gg h_{ie}$. With these assumptions, (D–11) reduces simply to

$$v_e \simeq v_{cm} \tag{D–12}$$

Thus, to the extent of the validity of the approximation, *the emitter signal voltage is very nearly equal to the input common-mode signal voltage.*

From the circuit of Figure D–2, the voltage v_o can be expressed as

$$v_o = -h_{fe} i_{b2} R_C \tag{D–13}$$

Substitution of (D–8) in (D–13) yields

$$v_o = -h_{fe} R_C \left(\frac{v_{cm} - v_d/2 - v_e}{h_{ie}} \right) \tag{D–14}$$

Next, the approximation of (D–12) is substituted in (D–14), and this results in cancellation of the v_{cm} terms. The output voltage is then given by

$$v_o = \frac{h_{fe} R_C}{2h_{ie}} v_d = A_d v_d \tag{D–15}$$

where

$$A_d = \frac{h_{fe} R_C}{2h_{ie}} \tag{D–16}$$

The quantity A_d is the *differential gain.*

To the accuracy of the earlier approximation, the output voltage is equal to the differential gain factor times the differential input voltage and is independent of the common-mode input voltage; however, this latter deduction is an illusion produced by the approximation. As we will see shortly, the output voltage does contain a common-mode component.

D-6 COMMON-MODE OUTPUT VOLTAGE

To investigate the common-mode output voltage, it will be necessary to use the *exact* form of v_e in (D–11). First, the expression of (D–14) is rearranged in the following manner:

$$v_o = \frac{h_{fe} R_C}{2h_{ie}} v_d - \frac{h_{fe} R_C}{h_{ie}} (v_{cm} - v_e) \tag{D–17}$$

Substitution of (D–11) in (D–17) and some simplification yield

$$v_o = \frac{h_{fe} R_C}{2h_{ie}} v_d - \frac{h_{fe} R_C v_{cm}}{h_{ie} + 2R_E(1 + h_{fe})} \tag{D–18}$$

Assume now that the output voltage is expressed as the sum of a differential component and a common-mode component; that is,

$$v_o = v_{od} + v_{ocm} \tag{D-19a}$$

$$= A_d v_d + A_{cm} v_{cm} \tag{D-19b}$$

where v_{od} is defined as the differential output voltage, v_{ocm} is defined as the common-mode output voltage, A_d is the differential gain, and A_{cm} is the common-mode gain. The individual components are identified as

$$v_{od} = A_d v_d \tag{D-20}$$

where A_d is the factor of v_d in (D–18) and is exactly the same as defined in (D–16) when the earlier approximation was made. Thus the approximation had no effect on the differential output component.

The common-mode output voltage is

$$v_{ocm} = A_{cm} v_{cm} \tag{D-21}$$

where the common-mode gain is identified from (D–18) as

$$A_{cm} = \frac{-h_{fe} R_C}{h_{ie} + 2R_E(1 + h_{fe})} \tag{D-22}$$

Since the common-mode component has been retained, it is appropriate at this point to make some simplifying approximations. In practice, $2R_E(1 \; h_{fe}) \gg h_{ie}$ and $h_{fe} \gg 1$. With these approximations, the common-mode gain reduces to

$$A_{cm} \simeq \frac{-R_C}{2R_E} \tag{D-23}$$

It is interesting to note that when these same approximations were made before substituting the expression for v_e in the expression for v_o, the common-mode component was "lost" in the mathematical process. If the exact expression for v_e is retained during the substitution, but the approximations are made after the resulting simplification, however, a meaningful expression for the common-mode component is obtained. This illustrates one of the many subtle processes relating to the "art of approximating" in electronic circuit analysis.

D–7 COMMON-MODE REJECTION RATIO

The common-mode rejection ratio (CMRR) is defined as

$$\text{CMRR} = \frac{|A_d|}{|A_{cm}|} \tag{D-24}$$

Substitution of (D–16) and (D–23) in (D–24) yields

$$\text{CMRR} = \frac{h_{fe} R_E}{h_{ie}} \tag{D-25}$$

The common-mode rejection ratio is seen to be directly proportional to the emitter resistance R_E, but there is a practical limit to how large R_E can be made, primarily as a result of the excessive dc voltage drop across it as its value increases. In practice, a current source is used, as will be seen shortly.

D–8 CURRENT MIRROR

On integrated-circuit chips, transistors are often used to create diodes. This is achieved by connecting the collector and base terminals together and considering the resulting terminal as the "anode" of the diode. The emitter of the transistor then becomes the "cathode" of the diode. This process is illustrated by transistor T1 in Figure D–3. We refer to this connection as an *active diode*.

We will now consider the process of implementing a current source on an IC chip. Consider transistor T2 in Figure D–3, whose collector current I_C is to be established at a prescribed value. The base of this transistor is connected to the active diode on the left.

The current I_1 is the sum of the base and collector currents of T1 plus the base current of T2. However, for the expected range of the two current gains, the two base currents should be negligible in comparison to the collector current, and it will be assumed that I_1 is very nearly equal to the collector current of T1.

Assuming a base-emitter voltage drop of 0.7 V for T1, the current I_1 can be approximated

$$I_1 \simeq \frac{V_{CC} - 0.7}{R} \tag{D–26}$$

FIGURE D–3

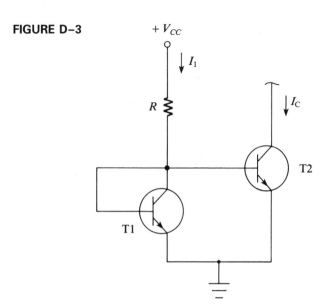

This current is a function primarily of V_{CC} and R, whose values are expected to remain quite stable. Since the two bases are tied together, the base-emitter voltage of T2 is forced to assume the same value as for T1. The transistors are expected to have nearly identical characteristics and operating temperatures. These factors cause the collector current I_C of T2 to be essentially the same as was established for T1, that is,

$$I_C \simeq I_1 \tag{D-27}$$

Thus, I_C is a *mirror* image of I_1.

A current source has a very large dynamic resistance while maintaining a modest dc voltage drop. This property makes it ideal for replacing R_E in the basic differential amplifier circuit of Figure D–1.

D–9 ACTIVE LOAD

Collector resistive loads are usually implemented with transistors on IC chips. The form of an active collector load for a differential amplifier is shown in Figure D–4. This particular active load uses *pnp* transistors, but is intended to be used as the load for a differential amplifier using *npn* transistors.

Transistor T1 is connected as an active diode, and it is used to bias T2 in the active region. In this manner, the effective dynamic resistance of T2 will be very large, thus contributing to a high differential gain for the complete circuit.

D–10 IC DIFFERENTIAL AMPLIFIER STAGE

A complete typical IC differential amplifier stage is shown in Figure D–5. T1 and T2 are the two transistors of the simpler circuit of Figure D–1. This is a single-ended output cir-

FIGURE D–4

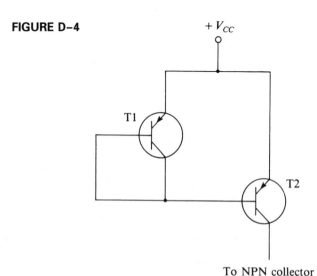

$+ V_{CC}$

T1

T2

To NPN collector

FIGURE D–5

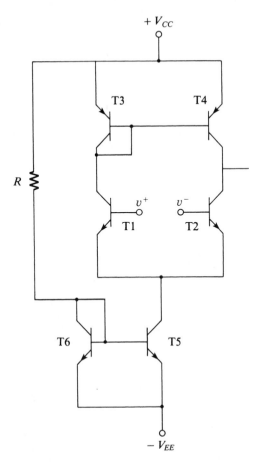

cuit with T3 and T4 serving to establish a dynamic resistance in the collector of T2. Thus, T4 replaces R_C in the simpler circuit, and T5 and T6 constitute a current source, with T5 replacing R_E in the simpler circuit.

D–11 OPERATIONAL AMPLIFIER

A complete IC operational amplifier consists of a combination of several differential amplifier stages of a type similar to the one just discussed, along with a level-shifting circuit. The latter circuit shifts out the dc level that appears at the output of the circuits discussed in detail. A low-impedance output stage (e.g., emitter follower) is also used to provide the desired low output impedance. Finally, in most modern op-amps, a stabilization network is also used to provide closed-loop stability for different operating conditions.

INTRODUCTION
TO MULTISIM

E–1 GENERAL DISCUSSION

Multisim is a comprehensive circuit analysis program that permits the modeling and simulation of a wide variety of electrical and electronic circuits. It offers a very large component database, schematic entry, analog/digital circuit simulation, and many other features, including seamless transfer to printed circuit board (PCB) layout packages. The program is a product of Interactive Image Technologies Ltd.*, and it has evolved from the company's earlier Electronic Workbench (EWB) program. The circuit simulation portion of the program is based on the popular SPICE program (*S*imulation *P*rogram with *I*ntegrated *C*ircuit *E*mphasis). SPICE was developed at the University of California at Berkeley and was a batch-oriented program. However, Multisim is interactive and offers a number of user-friendly features.

A major feature of Multisim is that the schematic diagram is created on the screen using a mouse and various window options. The type of analysis desired is then applied to the circuit, and the results can be viewed in a number of ways.

One of the most valuable features of Multisim is that the source excitation and instrumentation functions closely parallel those of a basic electronics laboratory, and the procedures that are used in obtaining data are very similar to those of the "real world." Hence, it closely approaches the concept of an ideal "virtual laboratory." For example, the test and measurement models contain voltmeters, ammeters, a multimeter, a function generator with several output waveforms, a two-channel oscilloscope, and other instruments. These units must be wired into the circuit in essentially the same fashion as in an actual laboratory. Thus, good laboratory skills can be taught very easily using a computer and software.

---

*Interactive Image Technologies, Ltd. 111 Peter Street, Suite 801 Toronto, Ontario, Canada M5V 2H1

Several points should be made before proceeding on to the actual circuit development steps that follow. First, this treatment is only an introduction to the subject and should not be interpreted as a comprehensive treatment. However, when coupled with the examples provided throughout the chapters, it should suffice to develop proficiency to use Multisim for the types of circuits in the text. Second, like most software, new versions of Multisim will likely appear after this text is printed. All examples in this text employed Version 6.21.

E–2 OVERVIEW

Upon activating Multisim, the initial screen of Version 6.21 assumes the form shown in Figure E–1. The first row of options below the title will be denoted as the **Menu Row.** When the left button of the mouse is clicked on a given word, a window opens to provide a list of options. To select a given option, left-click on the appropriate choice. Some of the menu choices and the options within menus are those encountered in most Windows-based applications, and some are peculiar to Multisim. You do not need to understand all of the options available to get started with Multisim. Experience will lead to familiarity with those features most appropriate to your needs.

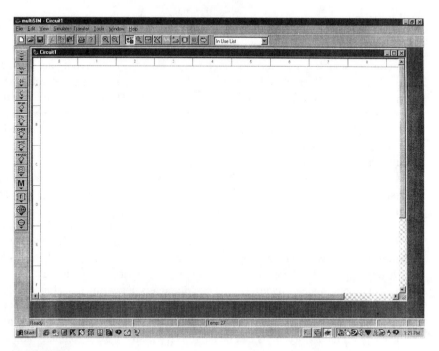

FIGURE E–1
Initial screen of Multisim Version 6.21.

The first eight buttons on the left-hand side of the second row will be denoted as the **System Toolbar.** When the arrow is moved over a particular button, the function of the button appears on the screen after a short time delay. Again, the use of these features will develop over a period of time. Some, such as "New" and "Open," perform some of the same functions as certain operations in the menu row so there is redundancy in some of the features.

The two buttons near the middle with + and − signs are **Zoom** buttons and may be used to expand or contract the schematic on the screen.

The last nine buttons on the right-hand side of the second row form the **Design Bar,** and some of the key operations of Multisim can be activated by left-clicked on the appropriate buttons. The details of these operations will be discussed later.

The various buttons arranged in a column on the left-hand side of the screen are the **Components Toolbars.** Beginning with the battery symbol and continuing down, each button represents a bin of available parts. When the arrow is placed above a given bin, a window opens containing the name and an overview of the types of parts contained within the bin. The form of the **Basic** parts bin (second one from the top) as it initially appears is shown in Figure E–2.

Each of the bars may be moved around on the screen for convenience if desired. To move a bar, left-click on it and drag it to any desired location while holding the left mouse button down. When the button is released, the bar will remain in the new location.

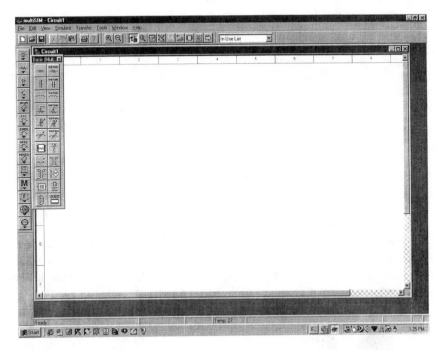

FIGURE E–2
Multisim Version 6.21 screen with Basic parts bin open.

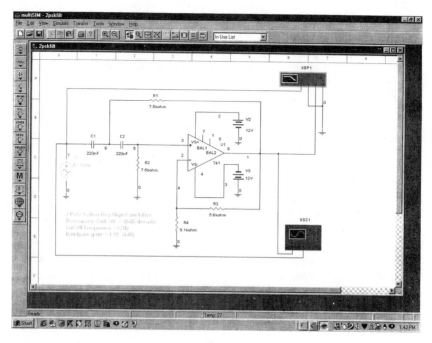

FIGURE E–3

Typical Multisim Version 6.21 schematic diagram.

The large blank area on the screen is referred to as the **Circuit Window.** Using procedures to be discussed in the next section, the circuit schematic will be created in this area. A schematic of a typical circuit containing instrumentation is shown in Figure E–3.

At a very basic level where laboratory skills are being emphasized, the use of the various virtual instruments and sources is very desirable. However, for most of the examples developed in this text, the results are easier to obtain using the **Analyses** menu from the **Design Bar** coupled with graphs of output data.

E–3   CREATING A CIRCUIT DIAGRAM

The process of creating a circuit diagram or schematic consists of *selecting* and *dragging* the components from a parts bin and *connecting* the components together using *wire*. In some cases, it may be desirable to drag all the components out to the circuit window first and then wire them together. However, the procedures may be interchanged at any time, and parts may moved around to make room for other parts as necessary.

Using a Grid for Layout

Although the default option is a blank screen, and it may be used throughout the process if desired, the text author finds the use of a grid to be very helpful in laying out the circuit.

To establish a grid, first left-click on **View** in the **Menu Row.** Next, left-click on **Grid Visible** and a grid pattern appears on the screen. The use of the grid makes it easier to navigate parts and wire around the screen. After the schematic is created, the grid may be deactivated if desired by following the same procedure as for activation.

Parts Bin Toolbars

As previously noted, the **Parts Bin Toolbars** are represented by the column of buttons to the left of the circuit window as shown in Figure E–1. Starting with the battery symbol at the top, each button identifies a particular parts bin based on a common grouping. As you move the arrow across each button, the name is identified. Don't worry if there seems to be an overwhelming number of parts at this point. You will quickly learn to find those that you use often, but it is unlikely that you will ever need more than a small percentage of the available parts.

As you move down the toolbar, a window displaying the components in each bin will appear. In Figure E–2, the **Basic** parts bin containing passive components (*R, L,* and *C*) and a number of other parts appears. Before circuit construction begins, a parts bin may be closed by moving the arrow away from the area. Once you are using a parts bin in building a circuit, a parts bin may be closed by left-clicking on the **X** in the upper right-hand corner.

Selecting and Dragging a Part from a Bin

To obtain a part for the circuit being created, first place the arrow over the parts bin button containing the part. (For certain parts, you may need to use a little trial and error before finding the right bin.) After the part is identified in the appropriate bin window, move the arrow to the correct button and left-click on it. You can then release the mouse button, since that component has now been "locked" to the arrow. With many parts, an additional window will open in which you will select the value or parts number of the component. When the choice is made, click **OK.** The arrow will now appear with a marker attached to it, and it may be moved to the position at which it will be placed. Don't be overly concerned about its exact location at this time, since it can be changed later. When you reach the approximate location at which the component is to be placed, left-click the mouse and the component will be released and left on the screen.

Selecting and Moving Components on the Screen

Previously, the process of obtaining a component from a parts bin has been discussed. However, for various reasons that will be discussed later, it is frequently desirable to "select" and move a component that is already in the schematic workspace. This can be achieved by moving the arrow to the component position and having it come in contact with the component. At the point where contact is established, left-click once and hold the mouse button down. Four small rectangles will now appear as if to create a box around the component. While holding the left mouse button down, you can then drag the component to a new location on the screen. When the proper location is reached, release the left button and the component will be detached. The rectangles may be eliminated by left-clicked outside of the area of the component.

The selection process described in the preceding paragraph may also be used to move the label or value of the component on the screen. This is accomplished by bringing the arrow in contact with the label or value instead of the component itself, and the remainder of the procedure is the same.

Rotating a Component

Components always have a fixed orientation when they come out of the parts bin, and this may or may not be the desired final orientation. All components may be rotated by integer multiples of 90°. First, the component must be selected by left-clicking on it.

Rotation can be achieved most easily by means of options contained in the **Edit** menu near the left-hand side of the **Menu Row.** They are **Flip Horizontal, Flip Vertical, 90 Clockwise,** and **90 CounterCW** (counterclockwise). The titles are somewhat self-explanatory, but the reader may wish to experiment with these operations when an actual circuit is constructed. Two successive steps of either one of the flip operations or four successive steps of either one of the rotate operations will restore the original orientation.

Deleting a Component

To delete a component, first select it. The simplest way to delete it is to press the **Delete** key on the keyboard, although there is a **Delete** option under the **Edit** menu items. If the component is a two-terminal device in which external wires have been connected, the component will be replaced by a short between the terminals.

Ground

All circuits must have at least one ground point. The ground symbol is obtained from the **Source** parts bin, which is the one identified by the battery symbol. The ground establishes the point from which all voltages are measured. To eliminate extra wiring, ground points may be used at various points in the circuit. Again, however, there must be at least one ground symbol, and it is interpreted within the program as node 0.

Wiring the Components Together

At some point, you will be ready to start wiring the components together. You may have all the parts that you need in the circuit window or you may prefer to wire some portion as you go along. For this purpose, you will need wire.

A basic rule to follow is that *all terminals that are to be connected must have a section of wire between them.* There is a temptation in some cases to simply bring component terminals together, and they may appear to be connected. Unless there is a section of wire between them, however, they will not connect. Think of the wire as having solder at each end. Without it, there will not be a connection. This rule applies to the ground symbol as well as all components.

The program automatically establishes node numbers from 1 upward in the order in which you wire them together, although all ground symbols assume node 0. Therefore, if you wish to establish any kind of numbering system, it is recommended that you connect the nodes in the desired numerical order. However, this is not absolutely necessary, since you can perform some renumbering later.

To start a wire from a given terminal, move the arrow as closely as possible to the tip of the terminal. At the point where a connection is possible, the arrow will then change to a set of crosshairs. Left-click at this point, and the process creates a piece of wire that can be moved to some other terminal. The wire will initially appear as a dashed line until the final connection is made. It is recommended that you move in straight lines parallel to either the horizontal or vertical sides of the screen. You can make one 90° turn without any further clicking, but if you make more than one turn, you will need to left-click at the point of the second and subsequent turns. When you reach the desired connection point, left-click at the terminal, and the dashed line will now appear as a solid line, meaning that you have wired the two terminals together. A node number will now appear somewhere along the wire. It will take a little practice to become fully proficient in this process.

To delete a section of wire, first select it by left-clicking on it. Then press the **Delete** key.

Opening a Component Properties Window

In addition to providing initial part numbers and values in the selection process as described earlier, many components will require further parameter values and properties to be established prior to running a circuit simulation. Moreover, you may desire to change some of the parameters without having to go back to the parts bin. This process can be performed at any time after the component is obtained from the parts bin. One of the great virtues of circuit simulation is the ease in which parameter values may be changed throughout a simulation study.

Parameter values and properties are specified in the **Component Properties Window.** To open such a window, proceed as if you were planning to select the given component; that is, place the arrow on the component. However, perform a double left-click on the component and the window should now appear. The name of the window depends on the particular type of component. Depending on the component, a wide variety of different types of properties may be specified. For the simplest types of component, only one or two values need to be specified, while others require a multitude of entries. Within the text examples, many of these windows are discussed in detail.

After you have entered all the data required to specify the component, left-click **OK.** If for any reason, you do not change any values or if you wish to return to the previous state, left-click **Cancel.** The component may now be deselected by moving the arrow away from the component and left-clicking.

Since a section of wire is treated essentially the same as a component, the preceding process may be used to open a properties window for changing its characteristics. This concept will be discussed in the next section as it relates to node numbering.

Node Numbering and Grid Options

The node numbers will appear on the schematic in the same order as they were connected. If you wish to change a node number, first perform a double left-click on the wire section as discussed in the last section. In this case, the window has the title **Node.** Next, type the desired node number and left-click **OK.**

If you are renumbering, the program will not allow you to use a number that is already active. If you have a strong desire to utilize a fixed node numbering order, this would suggest that you connect the nodes in the order planned as previously suggested. You can also label nodes with names such as IN and OUT.

E–4   DESIGN BAR

Although only a few of the operations on the Design Bar are required to support the examples in this text, a brief summary of the various buttons follows:

Components button. This button is automatically set as a default, in which case the **Component Parts** toolbars appear on the screen. When depressed, these toolbars are removed from the screen.

Component Editing button. This button permits modification of various component properties.

Instruments button. This button is used to open the **Instrument** window, in which a number of virtual instruments may be selected.

Simulation button. This button is used to start, stop, or pause a simulation.

Analyses button. This button allows you to select and activate the different forms of analysis that can be performed on the circuit. This operation is the most widely employed procedure for the examples in this text.

Post Processor button. This button lets you perform further operations on the results of your simulation.

VHDL/Verilog Design Bar button. This button allows one to work with VHDL modeling.

Reports button. This button lets one print reports about the circuit, including a bill of materials, list of components, and component details.

Transfer button. This button permits one to communicate with and export to other programs such as Ultiboard and Electronics Workbench. It can also export to programs such as MathCAD and Excel.

GLOSSARY

A/D   Abbreviation for *Analog-to-Digital Converter*. Also abbreviated as **ADC**.

Admittance   Ratio of ac phasor current to phasor voltage. Represented as complex sum of conductance and susceptance. Also the reciprocal of *impedance.*

Aliasing Error   Error resulting from sampling an analog signal at a rate slower than the theoretical minimum based on Shannon's sampling theorem.

All-Pass Circuit   A circuit that has uniform amplitude response but variable phase shift.

Amplification   The process of increasing the voltage, current, or power level of a signal.

Amplifier   An electronic device that performs amplification.

Amplifier, Instrumentation   Description of a composite differential amplifier system possessing high-quality gain and common-mode rejection specifications.

Amplifier, Inverting   An amplifier in which the output voltage has the opposite polarity to the input voltage.

Amplifier, Noninverting   An amplifier in which the output voltage has the same polarity as the input voltage.

Amplitude Response   Ratio of output sinusoidal magnitude to input sinusoidal magnitude.

Antilogarithmic Amplifier   An amplifier whose output voltage is proportional to the antilogarithm ("inverse logarithm") of the input voltage.

Analog Electronics   See *linear electronics.*

Analog-to-Digital Converter   A circuit that converts an analog signal into a sequence of digital words.

Balanced Demodulator   Circuit that recovers modulating signal from a double-sideband signal.

Balanced Modulator   Circuit that generates a double-sideband signal.

Baukhausen Criterion   An analytical criterion for determining the conditions for oscillation in a circuit.

Bipolar Offset Encoding   A form of digital encoding in which the digital words represent both positive and negative levels of the analog signal.

Bode Plot   A special plot of the amplitude or phase response on semilog (or sometimes log-log) scales.

Break Frequency   Frequency on Bode plot at which the breakpoint approximation of the amplitude response changes from a zero slope to either a positive or a negative slope.

Bridge Rectifier   A full-wave rectifier circuit consisting of four diodes connected as the arms of a bridge circuit.

Butterworth Response   A filter response that exhibits the maximum flatness possible in the pass band.

Chebyshev Response   A filter response that exhibits an equiripple amplitude response in the pass band.

Clamping Circuit   A circuit that shifts a signal either up or down about a specified clamping level.

Closed Loop   A term referring to the properties of an op-amp or other amplifier circuit *with* feedback (in contrast to *open loop*).

Common-Mode Gain   Gain parameter for an op-amp providing the ratio of common-mode output voltage to common-mode input voltage.

Common-Mode Voltage   A voltage that appears the same for both magnitude and phase at two separate terminals.

Common-mode Rejection Ratio   Ratio of differential gain to common-mode gain.

Comparator   A circuit that compares the levels of two voltages and provides an output dependent on which of the voltages is larger.

Controlled Source   A source in which the voltage or current is controlled by another voltage or current elsewhere in the circuit. Also called a *dependent source.*

D/A   Abbreviation for *Digital-to-Analog Converter*. Also abbreviated as **DAC**.

Decibel   Logarithmic unit of measurement to compare power, voltage, or current levels.

Dependent Source   See *controlled source*.

Differential Amplifier   An amplifier in which the output voltage is proportional to the difference between two input voltages.

Differential Gain   Gain parameter for an op-amp providing the ratio of differential output voltage to differential input voltage.

Differential Input Resistance   Input resistance of op-amp between the two input terminals.

Differential Voltage   Difference between two voltages.

Digital-to-Analog Converter   A circuit that converts a sequence of digital words into a continuous time-varying analog signal.

Double Sideband   Term used to describe a signal representing the product of a modulating signal and a carrier.

Differentiator   A circuit whose output voltage is proportional to the slope of the input voltage.

Differentiator, Low-Frequency   A circuit that differentiates low-frequency components of the input voltage.

Digital Electronics   The broad division of electronics in which signals are represented in terms of a finite number of digits, including the arithmetic computations and logic operations associated with such signals.

Envelope Detector   A circuit whose output voltage follows the envelope of an amplitude-modulated voltage.

Filter   A circuit that passes certain frequency components while rejecting others.

Filter, Active   A filter circuit that utilizes one or more active devices (e.g., an op-amp) in its implementation.

Filter, Band-Pass   A filter in which the pass band is bounded by stop bands on both sides.

Filter, Band-Rejection   A filter in which the stop band is bounded by pass bands on both sides.

Filter, High-Pass   A filter in which the stop band is at dc and low frequencies with the pass band above a certain frequency.

Filter, Low-Pass   A filter in which the pass band is at dc and low frequencies with the stop band above a certain frequency.

Filter, Passive   A filter circuit that is implemented completely with passive components.

Flash   An A/D conversion technique that establishes the values of all bits in a single comparison interval.

Fourier Analysis   See *spectral analysis*.

Frequency-to-Voltage Converter   A circuit that converts variations in frequency to variations in voltage. Abbreviated as *F/V converter*.

Gain-Bandwith Product   Specification for an op-amp giving product of the dc gain and the 3-dB bandwidth. Equal to the *unity-gain frequency*.

Gain, Current   Ratio of output current to input current. Indicated in text by symbol β.

Gain, Voltage   Ratio of output voltage to input voltage. Indicated in text by symbol A with appropriate subscripts.

Hysteresis   As applied to comparators, a phenomenon in which the transition point is different when changing in one direction than for the other direction.

ICIS   Abbreviation for current-controlled current source.

ICVS   Abbreviation for current-controlled voltage source.

Impedance   Ratio of ac phasor voltage to phasor current. Represented as complex sum of resistance and reactance.

Input Bias Current   Average of the two dc currents at the inputs of an op-amp.

Input Impedance   Impedance seen by a signal source at the input terminals of an amplifier.

Input Offset Current   Difference between the two op-amp input currents.

Input Offset Voltage   Magnitude of dc voltage which, if applied to input terminals, would reduce op-amp output dc voltage to zero.

Input-Output Characteristic   A graph or set of equations describing the relationship between the input and the output of a circuit.

Integrating A/D Converter   An A/D converter in which either the reference voltage or the analog voltage is integrated during the conversion process.

Integrator   A circuit whose output voltage is proportional to the area under the curve of the input voltage.

Integrator, AC   A circuit that integrates the time-varying portion of the input signal.

Inverting Input   Input terminal of op-amp for which the output voltage will change in the opposite direction.

Limiter   A circuit that limits the attainable level of a signal in one or both directions.

Linear Combination Circuit   An electronic circuit whose output voltage is a weighted sum of two or more input voltages.

Linear Electronics   The broad division of electronics in which signals are represented by continuous variables, including all signal-processing operations associated with such signals. Also called *analog electronics*.

Logarithmic Amplifier   An amplifier whose output voltage is proportional to the logarithm of the input voltage.

Loop Gain   Product of all gain factors around a loop in a feedback circuit.

Magnitude Response   See *amplitude response*.

Multiplexer   A circuit that mixes a number of signals together for transmission on a single channel.

Multiplier   A circuit that provides an output voltage proportional to the product of two input voltages.

Multivibrator, Astable   An oscillator circuit that utilizes a timing circuit and that produces a square-wave output.

Multivibrator, Bistable   A circuit having two stable states. Also called a *flip-flop*.

Multivibrator, Monostable   A circuit that has one stable state. Also called a *one-shot*.

Noise   Unwanted disturbance superimposed upon a desired signal.

Noise Gain   A special gain parameter that is useful in predicting the effects of finite bandwidth, dc offset effects, and noise in an op-amp circuit.

Noninverting Input   Input terminal of op-amp for which the output voltage will change in the same direction.

Norton's Theorem   A circuit theorem in which the effect of a complex circuit can be represented in terms of a single current source in parallel with an impedance.

Op-Amp   Abbreviation for *operational amplifier*.

Open Loop   A term referring to the properties of an op-amp or other amplifier circuit *without* feedback (in contrast to *closed loop*).

Operational Amplifier   A high-gain integrated-circuit, direct-coupled amplifier with differential input terminals.

Output Impedance   Impedance seen looking back at the output terminals of an amplifier, that is, the Thevenin equivalent impedance seen at the output.

Pass Band   The frequency range of a filter that passes components with very little or no change in amplitude.

Peak Detector   A circuit whose output voltage is equal to the peak value of a time-varying input voltage.

Phase-Locked Loop   A closed-loop feedback circuit in which a generated signal establishes synchronization with an input signal.

Phasor   Vector quantity used to represent ac sinusoidal voltage or current levels.

Phase Response   Difference in phase angle between output sinusoid and input sinusoid.

Power Supply   A circuit that provides one or more dc output voltages to operate electronic equipment.

Quantization   Process in which the continuous range of values of an analog signal is divided into subranges, with a discrete value of the output assigned for each subrange.

Reactance   Imaginary part of complex impedance.

Rectification   Process of eliminating either the negative or the positive portion of a signal (half-wave), or inverting all portions on one side of zero (full-wave).

Regulator, Switching   A voltage regulator utilizing switching circuits with varying duty cycle to achieve regulation, which results in very high efficiency.

Regulator, Voltage   A circuit that maintains a nearly constant predetermined voltage as operating conditions change.

Ripple   Term describing the ac component appearing at the output of a dc power supply.

Rise Time   The time required for a voltage to rise from 10% to 90% of its final value based on a dc or pulse type input.

Rounding   Process of representing a sample of an analog signal by the closest digital quantized level.

Sample-and-Hold Circuit   A circuit that samples the level of an analog signal and maintains that level for a specified interval of time.

Saturation Voltages   Maximum and minimum voltage levels at which the output of an op-amp can assume based on given power supply voltages.

Schmitt Trigger   A comparator with feedback, which results in hysteresis.

Shannon's Sampling Theorem   A theorem that establishes the minimum sampling rate required in a sampled-data or digital system.

Slew Rate   Maximum rate at which the output of an op-amp can change. Usually stated in volts/microsecond.

Slew Rate Limiting Frequency   The highest frequency that can be processed in a circuit without distortion as a result of the slew rate.

Spectral Analysis   Process of representing a signal in terms of its frequency components.

Spectrum   Frequency components representing a signal.

Stop Band   The frequency range of a filter that strongly rejects or blocks signal transmission.

Successive Approximation   An A/D conversion technique in which the digital word is established one bit at a time.

Summing Circuit   An electronic circuit whose output voltage is proportional to the sum of two or more input voltages.

Susceptance   Imaginary part of complex admittance.

Thevenin's Theorem   A circuit theorem in which the effect of a complex circuit can be represented in terms of a single voltage source in series with an impedance.

Timer   A circuit that permits the control of a specified timing interval.

Transconductance   Ratio of output current to input voltage. Indicated in text by symbol g_m.

Transfer Function   As used in this text, the ratio of system phasor output to phasor input.

Transformer   An electrical device that changes ac voltage and current levels.

Transresistance   Ratio of output voltage to input current. Indicated in text by symbol R_m.

Truncation   Process of representing a sample of an analog signal by the next lower digital quantized level.

Unipolar Encoding   A form of digital encoding in which all digital words represent either positive analog voltages or zero.

Unity-Gain Frequency   Specification for an op-amp indicating the frequency at which the open-loop gain is equal to unity. Equal to the *gain-bandwidth* product.

VCIS   Abbreviation for voltage-controlled current source.

VCVS   Abbreviation for voltage-controlled voltage source.

Voltage Follower   An electronic circuit in which the output voltage is exactly equal to the input voltage, but which provides isolation.

Voltage-to-Frequency Converter   A circuit that converts variations in voltage to variations in frequency. Abbreviated as *V/F converter*.

Wien Bridge Oscillator   A particular type of sinusoidal oscillator widely used in adjustable frequency test oscillators.

ANSWERS TO SELECTED ODD-NUMBERED PROBLEMS

Chapter 1

1–1. 37.5

1–3. 50 mV in series with 10 kΩ or 5 μA in parallel with 10 kΩ.

1–5. 8 V

1–7. 12 V

1–9. 300

1–11. 4 kΩ

1–13. Form of Figure 1–8(b).

1–15. (a) 500 kΩ (b) 50 Ω (c) 80

1–17. (b) 100 (c) 20 mV, 3 V

1–19. (b) 216 (c) 30 mV, 0.48 V, 8.64 V

1–21. (a) 0 dB (b) 20 dB (c) 40 dB (d) $20n$ dB (e) $-20n$ dB

1–23. (a) 31.6 (b) 1000 (c) 6.31 (d) 10^{10} (e) 0.01

1–25. (a) 400, 50 krad/s, 7.96 kHz

1–27. $\dfrac{1000}{1 + jf/50 \times 10^3} = \dfrac{1000}{1 + j\omega/\pi \times 10^5}$

1–29. (a) 59.96 dB, $-5.71°$ (b) 59.03 dB, $-26.57°$ (c) 56.99 dB, $-45°$
(d) 53.01 dB, $-63.43°$ (e) 39.96 dB, $-84.29°$

1–35. 16.88 kΩ, 296.3 Ω

*1–37. $C = 0.01$ μF, $R = 15.92$ kΩ

Chapter 2

2–1. (a) -5 (b) 10 kΩ (c) 0 (d) 2.6 V (e) 0, 5 V, -10 V, 13 V, -13 V

2–3. (a) 6 (b) ∞ (c) 0 (d) 2.167 V (e) 0, -6 V, 12 V, -13 V, 13 V

2–5. 5.2 mA

2–7. (a) 1 mS (b) 10 mA (c) 10 mA (d) 1.3 kΩ

Answers identified with an asterisk (*) have more than one possible correct answer.

2–9. (a) 0.5 mS (b) 3 mA (c) 2.167 kΩ

2–11. (a) 2 kΩ (b) 10 V

2–13. (a) 5 (c) 3 mA

2–15. (a) $v_o = -v_1 - 2v_2 - 4v_3$ (b) 12 V (c) −13 V

2–17. (a) $v_o = 2(v_1 - v_2)$ (b) 0 (c) 12 V (d) 13 V

*2–23. $R_i = 10$ kΩ, $R_f = 200$ kΩ

*2–25. $R_i = 1$ kΩ, $R_f = 59$ kΩ

*2–27. $R_i = 20$ kΩ, $R_f = 499$ kΩ, −24.95

*2–29. $R_i = 10$ kΩ, R_f adjustable from 0 to 200 kΩ

2–31. $R = 10$ kΩ, R_L(max) = 13 kΩ

2–33. $R = 3$ kΩ, R_L(max) = 1625 Ω

2–35. $R = 2$ kΩ, i_i (peak) = 6.5 mA

*2–37. $R_1 = 1$ kΩ, $R_2 = 14$ kΩ

*2–39. $R_f = 80$ kΩ, $R_1 = 20$ kΩ, $R_2 = 10$ kΩ

*2–41. $R = 10$ kΩ, $AR = 30$ kΩ

*2–43. Form $v_A = -2v_1 - 5v_2$ with one linear combination circuit in which $R_f = 50$ kΩ, $R_1 = 25$ kΩ, and $R_2 = 10$ kΩ. Use a second linear combination circuit to form $v_o = -v_A - 10v_3 - 20v_4$, with $R_f = 200$ kΩ, $R_A = 200$ kΩ, $R_3 = 20$ kΩ, and $R_4 = 10$ kΩ.

*2–45. Form $v_A = -5v_1$ with an inverting amplifier in which $R_f = 500$ kΩ and $R_i = 100$ kΩ. Use a second linear combination circuit to form $v_o = -2v_A - 5v_2 - 2v_3$ with $R_f = 500$ kΩ, $R_A = 250$ kΩ, $R_2 = 100$ kΩ, and $R_3 = 250$ kΩ.

*2–47. Two separate inverting amplifiers with $R_i = 10$ kΩ and R_f adjustable from 0 to 100 kΩ are used to amplify the two signals. The two outputs are combined in an inverting summing circuit with $R_1 = R_2 = R_f = 10$ kΩ.

*2–49. $R_1 = 20$ kΩ, $R_2 = 10$ kΩ, $R_3 = 10$ kΩ, $R_4 = 140$ kΩ

Chapter 3

3–1. (a) 0.06 (b) 4.8×10^3 (c) 16.6632 (d) 1.92 GΩ (e) 10.4 mΩ (f) 16.667

3–3. (a) 0.1 (b) 8000 (c) −8.99888 (d) 30,003.4 Ω (e) 6.249 mΩ

3–5. (a) 100 (b) 99.9 (c) 99.01 (d) 90.91 (e) 50

3–7. (a) −51 (b) −50.97 (c) −50.74 (d) −48.48 (f) −33.55

3–9. (a) 35 (b) 34.988

3–11. (a) 16 (b) −9.997, −4.998

3–13. (a) 42 mV (b) 4.08 mV (c) 1943 Ω (d) 0.544 mV

3–15. (a) 19.2 mV (b) 7.2 mV (c) 7.5 kΩ (d) 0.96 mV

3–21. (c) 1.29 mV (d) 1.54%

3–23. (b) 2×10^5

3–25. 2×10^5

3–27. (a) 0.5 mV (b) 416.7 kΩ

Chapter 4

4–1. (a) 4 kHz (b) 40 kHz (c) 400 kHz (d) 2 MHz

4–3. (a) 3.992 kHz (b) 39.22 kHz (c) 333.3 kHz (d) 1 MHz

4–5. (a) 87.5 μs (b) 8.75 μs (c) 0.875 μs (d) 0.175 μs

4–7. (a) 87.68 μs (b) 8.924 μs (c) 1.050 μs (d) 0.35 μs

4–9. 200 kHz

4–11. 300 kHz

4–13. 1.75 μs

4–15. 1.167 μs

4–17. 50 kHz

4–19. 500 kHz

4–21. (a) 0.1 μs (b) 1 μs (c) 10 μs

4–23. (a) 1.2 μs (b) 6 μs (c) 18 μs

4–25. (a) 1.592 MHz (b) 159.2 kHz (c) 15.92 kHz

4–27. (a) 132.6 kHz (b) 26.53 kHz (c) 8.842 kHz

4–29. (a) 2.2 μV (b) 222 μV

4–31. 2.512

4–33. 31.62

4–35. 6.32 V

4–41. (b) 79.6 mV

4–45. (a) 0.983 MHz (b) 1.571 V/μs

4–47. (a) 11.77 MHz (b) 40.21 V/μs (c) Yes

4–49. (a) 17.5 MHz (b) 6.25 V/μs

Chapter 5

5–1. (a) 0.8 V (b) 0.32 mA (c) 3.2 V (d) 8.2 V (e) 1 V (f) 7.2 V

5–3. 689.7 Ω

5–5. (a) 10^{-3} (b) 6 V (c) 1 mV (d) 6000

5–7. Triangular waveform oscillating between +4 V and −4 V.

5–9. A +10-V rectangular pulse from $t = 0$ to $t = 4$ s, followed by −20-V rectangular pulse from $t = 12$ s to $t = 14$ s.

5–11. 1 V

5–13. Triangular waveform oscillating between +0.5 V and −0.5 V.

5–15. −2.35 V

5–17. 12.8 V

5–19. Square wave oscillating between +6.4 V and −6.4 V.

5–21. (a) −64.28° (b) −102.98° (c) −136.61°

5–23. (a) 115.72° (b) 77.02° (c) 43.39°

5–25. (a) $-R/j\omega L$

*5–29. $C = 0.02\ \mu$F, $R_i = 25$ kΩ, $R_f = 250$ kΩ

*5–31. $C = 0.02\ \mu$F, $R_f = 100$ kΩ, $R_i = 1$ kΩ

*5–33. $C = 0.01\ \mu$F, $R = 15.92$ kΩ

*5–35. $C = 0.01\ \mu$F, $R = 15.92$ kΩ

*5–37. $C = 0.01\ \mu$F, $R(\text{min}) = 4265$ Ω, $R(\text{max}) = 59.40$ kΩ

*5–39. $C = 0.01\ \mu$F, $R(\text{min}) = 3296$ Ω, $R(\text{max}) = 76.85$ kΩ

*5–41. **(a)** $R_i = 20$ kΩ, $R_f = 200$ kΩ, $C_i = 3.3$ μF, $C_o = 15$ μF

(b) -13 V $< v_o < 13$ V, -1.3 V $< v_i < 1.3$ V

Chapter 6

6–1. **(a)** Same as Figure 6–3(b).

(b) ± 13-V square wave in phase with input.

6–3. ± 13-V square wave in phase with input.

6–5. **(a)** Same form as Figure 6–5(b), with $V_{ref} = 9$ V.

(b) 13-V positive pulse has width $0.05T$ and is centered at positive peak of input. Remainder of cycle has level of -13 V.

6–7. 13-V positive pulse has width $0.25T$ and is centered at positive peak of input. Remainder of cycle has level of -13 V.

6–9. **(a)** $v_o = -13$ V for $v_i < -9$ V

$\quad\quad = +13$ V for $v_i > -9$ V

(b) -13-V negative pulse has width $0.05T$ and is centered at negative peak of input. Remainder of cycle has level of $+13$ V.

6–11. -13-V negative pulse has width $0.25T$ and is centered at negative peak of input. Remainder of cycle has level of $+13$ V.

6–13. Square wave with levels 0 V and 5 V and with 5-V level aligned with positive half-cycle of input.

6–15. **(a)** 64.7 mV

(b) Same form as Figure 6–11(b) with $\pm V_T = \pm 64.7$ mV.

(c) ± 13-V square wave almost exactly out of phase with input.

6–17. **(a)** 8 V

(b) Same form as Figure 6–11(b) with $\pm V_T = \pm 8$ V.

(c) ± 13-V square wave in which negative half-cycle occurs between the point when v_i first reaches $+8$ V and later reaches -8 V.

6–19. **(a)** 5 V

(b) Same form as Figure 6–12(b) with $\pm V_T = \pm 5$ V.

(c) ± 13-V square wave in which positive half-cycle occurs between the point when v_i first reaches $+5$ V and later reaches -5 V.

6–21. **(a)** $v_o = -V_{sat}$ for $v_i < -V_{ref}$

$\quad\quad = V_{sat}$ for $v_i > -V_{ref}$

6–23. **(a)** $v_o = 12.3$ V for $v_i < -V_{ref1}$ and $v_i > +V_{ref2}$

$\quad\quad = 0$ for $-V_{ref1} < v_i < +V_{ref2}$

(b) $v_o = 0$ for $|v_i| < 6$ V

$\quad\quad = 12.3$ V for $|v_i| > 6$ V

6–25. Circuit of Figure 6–3(a).

6–27. Circuit form of Figure 6–19, but with pin 3 grounded and input applied to pin 2.

*6–29. Obtain 7.5 V with two 10-kΩ resistors connected between $+15$ V and ground as voltage divider. This voltage is connected to inverting input and the signal is applied to noninverting input.

*6–31. Circuit of Figure 6–11(a) with $R_1 = 51$ kΩ and $R_2 = 100$ Ω.

*6–33. Circuit of Figure 6–12(a) with $R_i = 2$ kΩ and $R_f = 13$ kΩ.

*6–35. Employ inverting Schmitt trigger circuit of Figure 6–11(a), but with a 10-kΩ pot employed for R_2. The adjustable arm of the pot is connected to the noninverting input, and $R_1 = 11.67$ kΩ.

Chapter 7

7–1. (a) 3 V (b) 1.151 ms (c) 869 Hz

7–3. (a) 3109 Ω (b) 3.083 V

7–5. (a) 14.55 ms (b) 13.86 ms (c) 28.4 ms (d) 35.2 Hz (e) 51.2%

7–7. 0.44 s

7–9. (a) 0.5333 ms (b) 1875 Hz (c) 13 V (d) 4.33 V

7–11. 20, 180°

7–13. (a) 153.9 Hz (b) 24 kΩ

7–15. (a) 6.5 Hz (b) 2.9 MΩ

7–23. (a) 2.742 ms (b) 274.2 kΩ (c) ±2.55 V

7–25. (a) 0.5 ms (b) 50 kΩ (c) ±12.9 V

*7–27. $R_1 = 10$ kΩ, $R_2 = 43.3$ kΩ, $C = 0.02$ μF, $R = 27.1$ kΩ

7–29. $R_A = 14.43$ kΩ, $R_B = 28.86$ kΩ

7–31. $R_A = 57.71$ kΩ, $R_B = 43.29$ kΩ

*7–33. $C = 0.001$ μF, $R = 9091$ Ω

*7–35. $C = 1$ μF, $R = 363.6$ kΩ

*7–37. $C = 0.01$ μF, $R = 3183$ Ω

7–39. 7958 Ω to 318.3 kΩ

7–41. $R_1 = 1111$ Ω, $C = 0.1432$ μF

7–43. 14 kΩ

7–45. $C = 0.001$ μF, $R = 8122$ Ω, $R_f = 235.5$ kΩ

Chapter 8

8–1. (a) $C = 159.2$ μF (b) 159.2 kΩ (c) 31.83 kΩ to 318.3 kΩ
(d) Same element values

8–3. (b) −21.2 dB, −36.1 dB, −47.8 dB, −57.3 dB, −65.3 dB

8–5. 1912 Hz, 1788 Hz, 1678 Hz

8–7. (b) −24.1 dB, −48.2 dB, −62.3 dB, −70 dB

8–9. (b) 642 Hz, 710 Hz, 780 Hz

8–11. (a) 5 (b) 1357 Hz, 1657 Hz, (d) −33.89 dB

8–17. $C_1 = 0.0225$ μF, $C_2 = 0.01125$ μF

8–19. $R = 22.5$ kΩ, $C_2 = 0.005$ μF

*8–21. $R = 15.92$ kΩ, $R_f = 5858$ Ω, $R_i = 10$ kΩ

8–23. 3 poles

8–25. 3 poles

8–27. $C_1 = 0.0564$ μF, $C_2 = 0.0222$ μF, $C_3 = 0.00322$ μF

8–29. $R = 56.4$ kΩ, $C_2 = 0.00393$ μF, $C_3 = 571$ pF

*8–31. One pole section: $R = 15.92$ kΩ
Two-pole section: $R = 15.92$ kΩ, $R_i = R_f = 10$ kΩ

8–33. $R_1 = 4488$ Ω, $R_2 = 11.43$ kΩ, $R_3 = 78.64$ kΩ

8–35. Feedback $R = 318.3$ kΩ, Input $R = 159.2$ kΩ, Shunt $R = 800$ Ω

8–37. $C_t = 0.00796$ μF,
$R_Q = 11.213$ kΩ with 10-kΩ shunt R

8–39. $C_t = 0.00796 \ \mu\text{F}$,
$R_Q = 2.99 \ \text{M}\Omega$ with 10-kΩ shunt R

Chapter 9

9–1. (a) 0, 0, -0.7 V (b) 12 V, 0, 12.7 V

9–3. (a) -10 V, 0, -10.7 V (b) 0, 0, 0.7 V

9–5. (a) 6 V, -6.7 V (b) 6 V, 0.7 V

9–7. Output is zero during first half-cycle. Output during second half-cycle is a positive triangular pulse with a peak value of 12 V.

9–9. Output during first half-cycle is a negative triangular pulse with a peak value of -10 V. Output is zero during second half-cycle.

9–11. Output during each half-cycle is a positive triangular pulse with a peak value of 6 V.

9–13. (a) 4.3 V (b) 4.3 V

9–15. (a) 5 V, 5.7 V (b) 5 V, -13 V

9–17. 4.95 V

9–19. (a) 8 V, 8 V, 8 V, 8.7 V, 0
(b) 8 V, 8 V, -4 V, -4.7 V, 1.2 mA

9–21. (a) 5 V, 5 V, 5 V, 5.1 V, 0
(b) 5 V, 5 V, 2 V, 1.3 V, 0.3 mA
(c) 5 V, 5 V, 8 V, 8.7 V, -0.3 mA

9–23. (a) 7.3 V (b) -0.7 V to 17.3 V

9–25. (a) 8 V (b) 0 to 18 V (c) 0.7 V

9–27. (a) 4 V (b) 0.2 V
(c) -3 V, 0.15 mA, 0.15 mA, 0, 0.15 mA
(d) -4 V, 5 mA, 0.2 mA, 4.8 mA, 5 mA

9–29. (a) 10 (b) 2 A (c) 16.97 V

9–31. 3.33 V

9–33. (a) 20 W (b) 12 W (c) 60%

9–35. 20.15 V

9–37. $v_o = v_i + 0.7$ for $v_i < -0.7$ V

9–39. (a) $v_o = v_i$ for $v_i < 0$
$= 0$ for $v_i > 0$

9–41. Shifted downward with positive peak at 0.7 V and negative peak at $-V_a - V_b + 0.7$.

9–43. Shifted upward with positive peak at $V_a + V_b - V_{\text{ref}}$ and negative peak at $-V_{\text{ref}}$.

9–45. Shifted downward with negative peak at $-V_a - V_b - V_{\text{ref}}$ and positive peak at $-V_{\text{ref}}$.

9–47. $v_o = v_i$ for $v_i < V_{\text{ref}}$
$= V_{\text{ref}}$ for $v_i > V_{\text{ref}}$

9–49. 300 kΩ

9–51. (a) 335 Ω (b) 20.3 kΩ (c) 16.1 mA (d) 8 W, 10.4 W, 76.9% (e) 2.4 W

*9–53. (a) $R_1 = 240 \ \Omega$, $R_2 = 4368 \ \Omega$ (b) 0.437 V

Chapter 10

10–1. (a) 22.05 kHz (b) 22.68 μs

10–3. (a) 2 kHz, 0.5 ms (b) 0.5 ms (c) 31.25 μs

10–5. **(b)** 0.125 ms **(c)** 7.8125 μs

10–7. 65,536

10–9. **(a)** 10 V **(b)** 0.625 V

10–11. **(a)** 0 **(b)** 0.625 V **(c)** 5 V **(d)** 6.25 V **(e)** 9.375 V

10–13. **(a)** 32 **(b)** 10 V **(c)** 0.3125 V

10–15. **(a)** 0.3125 V **(b)** 0.625 V **(c)** 1.25 V **(d)** 5 V **(e)** 9.6875

10–17. **(a)** 12.6 μs, 79.37 kHz **(b)** 1.2 μs, 833.3 kHz **(c)** 0.2 μs, 5 MHz

10–19. 50 Hz

10–21. **(a)** 15.26 $\times$ 10^{-6} **(b)** 152.6 μV **(c)** 0.999985 **(d)** 9.99985 V **(e)** 7.63 $\times$ 10^{-6}
(f) 76.3 μV **(g)** 7.63 $\times$ 10^{-4}%

10–25. **(b)** 0.03125, 0.09375, 0.15625, . . . , 0.90625
(d) All resistances would have the value R.

10–27. **(a)** 4.8 kΩ **(b)** 10 kΩ **(c)** 12 kΩ **(d)** 12.8 kΩ

10–29. **(a)** 6.4 kΩ **(b)** 8 kΩ **(c)** 12 kΩ **(d)** 12.387 kΩ

10–31. **(a)** 10 V **(b)** 9 bits

Chapter 11

11–1. **(a)** 1.2 V **(b)** −3.2 V

11–3. 33, 37 kHz

11–5. 73, 77, 83, 87 kHz

11–7. 2.5 V

11–9. −5 V

11–11. 9.49 V

11–13. **(a)** −0.4144 V **(b)** −0.4736 V **(c)** −0.5328 V **(d)** −0.5920 V

11–15. **(a)** 5.72 mV **(b)** 0.280 V **(c)** 1.96 V

11–17. **(a)** −0.385 V **(b)** −89.1 mV **(c)** 0.207 V

11–19. 100 kHz

11–21. **(a)** 69.77 kHz **(b)** ±46.51 kHz **(c)** ±14.34 kHz

11–25. $v_2 = 6$ V, $A = 8.446$

11–27. 12, 11

*11–29. **(a)** $R_1 = 10$ kΩ, $C_1 = 0.001$ μF
(b) ±10 kHz
(c) 0.1105 μF

INDEX

DATE DUE

NOV 2 7 2004			